DOMINANCE AND AGGRESSION IN HUMANS AND OTHER ANIMALS

DOMINANCE AND AGGRESSION IN HUMANS AND OTHER ANIMALS

The Great Game of Life

HENRY R. HERMANN, Ph.D.
Florida SouthWestern State College

ELSEVIER

AMSTERDAM • BOSTON • HEIDELBERG • LONDON
NEW YORK • OXFORD • PARIS • SAN DIEGO
SAN FRANCISCO • SINGAPORE • SYDNEY • TOKYO

Academic Press is an imprint of Elsevier

Academic Press is an imprint of Elsevier
125 London Wall, London EC2Y 5AS, United Kingdom
525 B Street, Suite 1800, San Diego, CA 92101-4495, United States
50 Hampshire Street, 5th Floor, Cambridge, MA 02139, United States
The Boulevard, Langford Lane, Kidlington, Oxford OX5 1GB, United Kingdom

Library of Congress Cataloging-in-Publication Data
A catalog record for this book is available from s Library of Congress

British Library Cataloguing-in-Publication Data
A catalogue record for this book is available from the British Library

ISBN: 978-0-12-805372-0

For information on all Academic Press publications
visit our website at https://www.elsevier.com/

Working together
to grow libraries in
developing countries

www.elsevier.com • www.bookaid.org

Publisher and Acquisition Editor: Nikki Levy
Editorial Project Manager: Barbara Makinster
Production Project Manager: Lisa Jones
Designer: Matthew Limbert

Typeset by TNQ Books and Journals

Dedication

This book is dedicated to all heroic people who spend their time attempting to understand and preserve the planet we call Earth and its biota, and have a hand in the very difficult task of maintaining homeostasis among them.

The love for all living creatures is the most noble attribute of man.
Charles Darwin

Contents

1. Defining Dominance and Aggression

2. Traits of Dominant Animals

9. Alternate Human Behavior

10. The Chemical, Physical, and Genetic Nature of Dominance

11. Dominance and Aggression in the Workplace

12. Dominance in Religion

13. Dominance in Politics

14. Human Aggression: Killing and Abuse

15. Killing Humans

16. Are We Our Own Worst Enemy?

17. Attempts to Save the Natural World

18. The Nature of Things

Biography

Henry R. Hermann received his doctoral degree from Louisiana State University in Baton Rouge, LA, in 1967. He spent 30 years as professor of Biology at the University of Georgia, carrying out research and teaching at the under-graduate and graduate levels. He was an advisor for a number of graduate students seeking Master and Doctoral degrees and attracted several postdoctoral scientists from both the United States and foreign countries during his career. He has served in the capacities of associate editor, editor, and consultant on a number of scientific journals. His main research over the years (in both temperate and tropical locations in the world) has been defensive mecha-nisms in social animals. His list of publications, including approximately 20 books, 9 book chapters, and close to 100 journal publications, can be found at www.henryhermannpublications.blogspot.com. Five of his books were instrumental in promoting research in the field of social biology, and his research helped form the cornerstone for an understanding of concepts related to defensive mechanisms in social animals. He has written on diverse subjects. His most recent published book, *Making the Wind Sing, Native American Music and the Connected Breath*, was released by Masterworks Books in 2011. His current book, *The Great Game of Life, Dominance and Aggression in Humans and Other Animals*, represents 10 years of research and preparation. He currently teaches Anatomy and Physiology at Florida SouthWestern State College's Lee County campus and carries out research on wasp defensive mechanisms, focusing on morphology, behavior, and evolution.

Preface

Solomon Naumovich Rabinovich, Yiddish author and playwright (pseudonym Sholem Aleichem), once wrote, "Life is a dream for the wise, a game for the fool, a comedy for the rich [and] a tragedy for the poor" (Waife-Goldberg, 1968). Yet, when we analyze life at the various levels mentioned, we find that there are numerous games we and all other animals play on a daily basis (dominance and aggression games) at each of these levels, and the games generally influence and are influenced by the position we occupy in society throughout life. The differences between people in the world are largely based on where the games are played, the rules they play by, the conditions allotted to them at the games' beginning, the intelligence they have at the onset of the games, the experience they acquire along the route to the games' termination, and how the games are played.

At times, the games are simple and pleasurable, and humans look forward to playing them. Others are more serious, affecting their health and longevity. Some games may happen within minutes, hours, or days, while others take long periods of time to play out. In each, there most often is a winner and a loser. Each player may win at certain games, and others may win at different ones. Winning and losing, however, are often disproportionate because some players are more serious about or talented at playing the games and receiving the rewards they get by being a winner. Some win most of the time, while others lose most of the time. Because of this inequality, winners may dominate their competitors when the game averages are determined.

Sometimes, the rules of the games are fair, made by people who have other's interests and harmony at heart, and sometimes, they are unfair, built on selfishness and greed. In some cases, the winners are ruthless in how they play; they are intolerant of the losers, and the losers suffer from their losses.

Sometimes, the rules are strongly enforced, and sometimes they are not enforced at all. In some areas of the world, the rules may be different, depending on who the rule-makers are, the number of people in the games, and the conditions that exist at the time the games are executed.

Some people follow the rules to the letter, and still others refuse to follow them at all. Sometimes, the rules can be altered, depending on the degree of influence one player has over other players, and sometimes the rules change because of both intrinsic and extrinsic factors. Sometimes, the games can be won, and sometimes, there is no chance of winning. At times, players need help to understand the rules, and sometimes they require assistance in how they should make their moves.

Games of dominance and aggression are present in everything we do and face us every day of our lives, no matter what our position in life. They can be simple, non-threatening, and life-enhancing or they can be life-threatening and difficult to cope with. The games commence at our birth and end at our death. As we enter the world, we, like other animals, inherit behavioral propensities which influence our personalities, and the environment we grow up in influences the development of game enhancers or diminishers we may use and carry to the end. As we express our dominant nature during life's excursion, we often display various forms of aggressive behavior toward one another, which ultimately determine where we will end up in a ranking system we refer to as a dominance hierarchy.

Some of us play the game with a highly competitive spirit, and others lack a competitive nature. Some of us cheat, and others feel cheating is undesirable or unethical. Thus, some of us play fairly, while others take advantage of the weaknesses of other players. Some of us enjoy a high position in the hierarchy and are miserable at the low levels. Others do not care what their position is in the hierarchy.

At times, there are many such hierarchies, and we play endlessly for a position in them in order to maintain our position among others of our population. How we flaunt our position in each of the hierarchies is an expression of the personality we carry through life, and as we meet life's challenges and age, we may forfeit our position in the dominance hierarchy to younger, more able, and smarter competitors.

In the end, no one wins, but we as humans can attempt to set up the playing field and improve the games that will be played by the next generation and hope that they will do the same or better at playing than we did. What may block the success of subsequent generations is the failure of the human population to abide by the rules and the conditions that arise because of this failure.

Throughout this book, an attempt is made to describe these games that humans and other animals play. In the process, we learn something about the players and how similar or different they are. Although humans consider themselves either as the elite of the animal kingdom and/or some form of living creature that occupies a position among living things somewhere between animal status and God, they nevertheless share biological features which link them to many other types of biota and express themselves in both animalistic and humanistic ways.

There are other implications for the powers of dominance. The level of dominance is not merely restricted to individuals who influence other members within their own society. Dominance becomes influential on a much larger scale, determining how we act and work in social groups, families, coalitions, organizations, cities, nations, and as global inhabitants, how we utilize the resources we require for modern living, and how we treat Mother Earth in getting the things we need.

Over long periods of time, humans have changed from an almost totally animalistic existence to one in which they have become more technologically advanced. As such, humans have exploited the Earth to obtain materials for which they can utilize their technological prowess. And with a population that is steadily increasing beyond its environmental carrying capacity, they have overexploited certain of the Earth's resources and in the process, they have altered the *motherboard*, planet Earth, and its animal competitors.

On occasion, generally over long periods of time, the *motherboard* itself changes on its own, and the organisms which occupy it become stressed. This brings on new rules, extinction of old and the diversification of new players, and major changes in the playing fields (ecosystems) to which they belong.

In many ways, the human population functions like a colossal superorganism, taking in materials which it must process in order to obtain energy, grow, and fight infections and other deleterious circumstances and obstacles in order to survive and produce subsequent generations. Its body must recycle materials that are important to metabolic functions and dispose of others that must be eliminated as wastes. In carrying out these functions, it must strive to maintain homeostasis, in spite of abuse, and if homeostasis is unattainable, it will become sick and possibly die. In certain cases, Earth becomes antagonistic to the species which live upon her, and mass extinctions occur.

A look at the Table of Contents will show that the topics covered are quite diverse. Yet, they are all aimed at understanding the complexity of human nature, its origin and evolution, its animalistic and humanistic connections, and how their combined effects have influenced the Earth upon which they have thrived for many thousands of years.

How human nature is expressed varies tremendously as a gradation from dark to light, mild tempered to psychotic, complete subordinance to ultradominant (despotic), altruistic to selfish, cooperative to antagonistic, pleasant to disruptive, and moral to immoral. While we tend to group individuals into well-defined categories, based on their behavioral expressions, the distinction between expressions is actually quite plastic, a gradation or cline from one extreme to another, based on both genetic propensities and learned behavior, and no two people are alike. Thus, the artificial categories we tend to put people in overlap tremendously, sometimes making it difficult to understand a particular behavior and how it is manifested.

Commencing with definitions of dominance and aggression, earlier concepts are presented and redefined, based on contemporary comparative research. When we embark on such an enterprise as attempting to understand human nature, we immediately become overwhelmed with the complexity of the animal, the vastness of pertinent literature, and the number of researchers who work on different phases of the subject. While I will admit that a single volume of work like this one cannot possibly cover the subject of dominance and aggression in a totally comprehensive fashion, it is my hope that the present volume, for which I have spent years in its preparation and writing, will cover it adequately enough to at least stimulate the thoughts of others who feel they must also express themselves on the subject.

When we are born, we are fortunate to begin life on a planet that has supported life for about 4 billion years. Humans have experienced what Earth has to offer for only a fraction of this time. Life over time has not always been pleasant, and life for many humans and other animals continues to be unpleasant today. As organisms evolved on an ever-changing Earth, their populations and ecosystems underwent a variety of gargantuan modifications due to constantly shifting environmental stresses over long periods of time. Many species became extinct in the process. Subsequently, the Earth rebounded and the games and players diversified.

Through natural selection, each species that remained extant developed assorted mechanisms to maximize its advantages for survival and reproduction. Humans, like other successful animals, have developed many of their own mechanisms for this purpose, but at the same time, they have inherited and continue to express numerous animalistic features from a wide assortment of predecessors. We are thus mutants with an amalgamation of animalistic and humanistic qualities.

The human story, including anatomical, physiological, and behavioral aspects which associate and dissociate this animal from others, and a characterization of what this wondrous dominant animal is like in the early 21st century will unfold as we move from one chapter to another. The chapters incorporate such concepts as where they and their personalities arose, their similarities and differences when compared with other animals, their moral or immoral nature, how they express their dominant and aggressive nature, and how they function among other environmental biota.

It elaborates on just how complex, intelligent, and sometimes strange humans are and how they have affected world homeostasis and instability. We find that the social nature of humans, along with human intelligence, dominance, and aggressiveness, has been behind human success in becoming the most dominant animal on the planet. On the dark side, as the human population grows, its dominant and aggressive nature may negatively affect its longevity as a dominant species and its continuing existence on planet Earth.

Acknowledgments

Some people stand out in one's life as exceptional as both friends and colleagues. I would like to express my appreciation for my friend and former colleague, Tobias F. Dirks, recently deceased and former Professor at Dalton Junior College (now, Dalton State College), Dalton, Georgia. He and I have had many stimulating conversations about entomology and biology in general. In the 1980s, when he shared a postdoctoral fellowship with me at the University of Georgia, I was working on the morphology and behavior of ant defensive systems, and he was interested in hymenopteran venoms. Together, we gained an interest in all forms of social behavior and subsequently began work together on social wasps, their dominance hierarchies, and defenses. We spent many hours in the field with graduate students, day and night, marking polistine wasps and observing their behavior during the following days, weeks, and months. It was a memorable time in our careers, and I continue research on both ants and wasps to this day. Dr. Dirks passed away in 2013, but I will always remember our friendship and ventures together.

Likewise, Murray S. Blum, retired from the University of Georgia and passed in 2015, assisted me in building a good foundation in biological research, and our subsequent collective work led to important contributions in the form of scientific papers, chapters, and book volumes on animal defensive mechanisms.

I wish to thank Academic Press for allowing me to edit a four-volume treatment on social insects and Praeger Scientific for allowing me to edit a book on insect defenses during the 1980s. It was at that time that I began to research topics on other animal defenses, which eventually led to the writing of the present volume.

Gary Ross, formerly at Southern College and living in Baton Rouge, Louisiana, is a special friend and colleague whom I have spent much time with in the field. Currently a prolific writer on nature subjects, he is best known for his work on butterflies, and he and I have talked at length about insects especially important in mimicry. I thank him for his friendship and constant input on various subjects, which have sometimes touched on the material in this book.

I have always appreciated my graduate students who received their masters and doctorate degrees at the University of Georgia for their devotion to studies of social and defensive systems. They did not review this book, but many of the concepts that arose from work on social organisms included them and their investigations. Of particular importance are: Donna Willer, who worked on nest establishment and nest switching by gravid females during the nest founding period in colonies of *Polistes exclamans*; Christopher K. Starr, who worked primarily on the behavior of *Polistes annularis*; Jung Tai Chao, who worked on behavior and larval development of *P. annularis*; and Barden Cannamella, who worked on defensive behavior in colonies of *P. annularis* and *P. exclamans*. Together, we investigated many concepts in social biology and discussed them at length.

There are many current associates who have influenced my thinking about a variety of scientific fields. My thanks go to Frank Mraz at Florida Southwestern State College, who has introduced thoughts on a daily basis about the universe and celestial bodies in it that stimulate deep thinking and understanding. Frank and I have had numerous conversations that have stimulated me to learn more about space and human behavior.

Theo Koupelis, former Dean of the School of Pure and Applied Science at Florida SouthWestern State College, Ft. Myers, Florida, Astronomer, Professor, and Author of *In Quest of the Universe*, read a part of the manuscript, which I appreciate. His knowledge of the solar system and other celestial bodies and his book have been important to my understanding of the nature of Earth and the cosmos.

Deborah Misotti, Director of the Talking Monkeys Project in Clewiston, Florida, who specializes in gibbon apes, was especially helpful in including certain behaviors which characterize that primate group. My trips to her facility have always been stimulating, and conversations with her were always extremely informative. I sincerely appreciate her input and review of the chapter on primates.

Bruce Harwood (Cape Coral, Florida) reviewed some of the chapters, especially the introductory ones and chapters on *Human Aggression—Killing, Abuse, and Warfare*. Mr. Harwood is a former Captain/Ranger in the United States Army and decorated war veteran, a recipient of four purple hearts. His first-hand knowledge of modern warfare and the atrocities that result from it have been extremely helpful in preparing the chapter on the human propensity for killing.

Robert Dean Bair, historical novelist and author of The Cloisters of Canterbury, Peace at Lambeth Bridge, Dead Men Talking, The Director, The Overthrow of Dictator Juan Bosch, and The American Held a Gun to His Head, has a special knowledge of war-time history. His writing and conversations have helped me understand the implications of war-time strategies. His review of select chapters was also invaluable in the process of providing clarity for non-science readers.

My appreciation goes to Florida SouthWestern State College for allowing me to complete research projects and this manuscript. Being able to utilize their search engines and library facilities has helped me find and use the manuscripts I needed.

Likewise, my appreciation goes to individuals who generate and maintain Googlescholar.com, Wikipedia, and other search engines. While I have sought scientific papers in scholarly journals and books in preparing this manuscript, they have improved their reports over the years and offer immediate summaries on subjects that otherwise would take eons to investigate on a personal level. They also strive to obtain and display invaluable references which are occasionally missed in other searches.

A Dominant Species on Planet Earth

Beginnings, it's said, are apt to be shadowy. So it is with this story, which starts with the emergence of a new species maybe two hundred thousand years ago. The species does not yet have a name—nothing does—but it has the capacity to name things.

As with any young species, this one's position is precarious. Its numbers are small, and its range restricted to a slice of eastern Africa. Slowly, its population grows, but quite possibly then it contracts again—some would claim nearly fatally—to just a few thousand pairs. . . .

The process continues, in fits and starts, for thousands of years, until the species, no longer so new, has spread to practically every corner of the globe. At this point, several things happen more or less at once that allow *Homo sapiens*, as it has come to call itself, to reproduce at an unprecedented rate. In a single century the population doubles, then it doubles again, and then again. Vast forests are razed . . . , and they shift organisms from one continent to another, reassembling the biosphere.

Having discovered subterranean reserves of energy, humans begin to change the composition of the atmosphere. This, in turn, alters the climate and the chemistry of the oceans. Some plants and animals adjust by moving. They climb mountains and [disperse] toward the poles. But a great many—at first hundreds, then thousands, and finally perhaps millions—find themselves marooned. Extinction rates soar, and the texture of life changes.

No creature has ever altered life on the planet in [the way that humans have] . . . . When it is still too early to say whether it will reach the proportions of the Big Five [extinction periods that Earth has progressed through], it becomes known as the Sixth Extinction.

Elizabeth Kolbert
The Sixth Extinction: An Unnatural History

1

Defining Dominance and Aggression

In an individual-based social hierarchy, individuals might enjoy great power, prestige, or wealth by virtue of their own highly-valued individual characteristics, such as great athletic or leadership ability, high intelligence or artistic, political or scientific talent or achievement. **J. Sidanius and F. Pratlo,** *Social Dominance*

Homo sapiens Linnaeus, 1758, the single extant species of humans occupying planet Earth, arose from predecessor humans in Africa about 200,000 years ago (Henshilwood and Marean, 2003; McBrearty and Brooks, 2000). Since that time, the species has dispersed around the world, living mostly an animalistic existence (Lawlor, 2007; Miller, 1993; Olson, 2008; Fig. 1.1).[1]

Its life as an enlightened organism did not emerge until the Neolithic Revolution and the rise of civilizations in lower Mesopotamia, along the Nile in Egypt, Harappa in the Indus Valley (present-day India and Pakistan), and China (Allchin, 1995, 1997; Ascalone, 2007; Lee, 2002; Rice, 1970; Fig. 1.2). Also referred to as the Neolithic Demographic Transition and Agricultural Revolution, that stage of human existence represents a period of immense social change in the human species, commencing about 10,000 to 12,000 years ago, during which many human cultures began to move from a hunter-gatherer existence to one of agriculture and settlement (Bocquet-Appel, 2011).

Edwin Black (*Banking on Baghdad*, 2004) points to Iraq–Mesopotamia as the Cradle of Civilization, which had a "several-thousand-year advance on the rest of humanity. When the last Ice Age receded, some 10,000 years ago, some peoples [emigrated] from the marshy plain between the Tigris and the Euphrates [Rivers], but it was not the first time that groups of people with cognizant brains had come together. Signs of early human aggregations are seen in cave dwellers in South Africa, 70,000 years ago, at which time they recorded symbolic concepts with geometric designs engraved on ocher stones, revealing organized expression and abstract thinking."

Somewhere between these times and the end of the ice age, people began to gather in specific areas, and "ancient Mesopotamia sprang upon the consciousness of the world." But it was not a simple gathering point for wanderers. "The world's view of the cradle of civilization emerged not from organized communal hunting societies in Siberia that learned to share food and nurture clans, or from the spiritual painters of cave art in France, or from thousands of years of continuous township at Jericho."

While it was necessary for such groups of people to know about the growing and propagation of food plants, Black states that it was more like "the quality of economic life and commerce and its invigoration of all around it that signed the emergence of that most valued social order—civilization." Their new sedentary life style, in turn, initiated such facets of human life as the building of great civilizations, food-crop cultivation, trading economies, political innovations, organized religion, property ownership, and population increase.

Population increase has brought with it such features as an apparent boundless expansion of cities and towns, the laying down of numerous highways and streets, an increase in world tension, fluctuating economies, increasing violence, overuse of resources, destruction of natural habitat, deforestation, environmental pollution, the pasturing and housing of numerous animals, and production of genetically modified plants.

With global and population-related changes, we have entered a period in which most of our grain and water are channeled to agricultural animals, which we consume, and many of the foods we eat may contain toxic or sterility substances (Engdahl, 2007), which we know very little about.[2]

As human populations progressed from an animalistic past, the species eventually became the most intelligent, dominant, aggressive, deceptive, and populous vertebrate animal that has ever existed on the planet. Its expressions of dominance may be found in its very evident social qualities and a wide array of scientific and

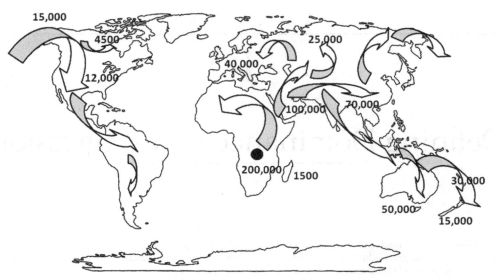

FIGURE 1.1   Map showing dispersal routes taken by humans (*Homo sapiens*) from their initial sites in Africa (*black dot*), along with their respective dates. Most sites of human fossils that have been important in working out the origin of the genus *Homo* have been unearthed on the eastern side of the continent (the currently perceived Cradle of human origin), but recent evidence has indicated that South Africa may be a prime location as well. As they moved, they began to form isolated pockets, each of which developed separate anatomical features. Later, they began to increase their numbers, especially during the Neolithic Revolution and the rise of civilizations, at which time populations initiated the process of settlement. Their existence during their dispersal and subsequent movements was predominantly an animalistic, hunter-gatherer one.

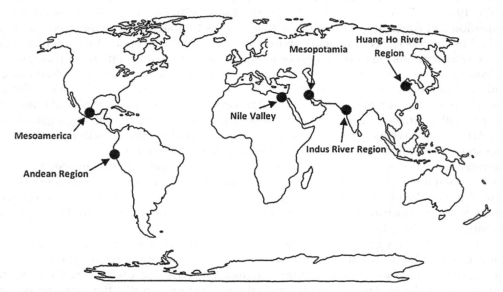

FIGURE 1.2   Map of the world, showing the ancient intensive agricultural civilizations (often known as the Cradle of Civilization). These areas represent a focal point of great social change in the human species, commencing about 10,000 to 12,000 years ago, during which many human cultures began to move from a hunter-gatherer existence to one of agriculture and settlement. *From these* initial points of occupation, humans and settlement erupted throughout the world. Settlement and agricultural development spread through the world during the subsequent thousands of years. Within current times, the need for agricultural land has resulted in land fragmentation, habitat destruction, and deforestation in order to feed a growing human population.

technological accomplishments since the beginning of civilizations, many of which have been quite remarkable (Bunch and Hellemans, 2005; McClellan and Dorn, 2006; Murray, 2004; National Geographic, 2009):

> The 21st century is an age of information, an age of knowledge and certainty. The Earth is mapped, our solar system explored, the universe charted. We've dissected our planet, observed the heavens and put everything in its place. *Are We Alone?* **2015** *National Geographic*

Its aggressiveness is demonstrated in its inclination toward overpopulation, uninhibited destruction of ecosystems, overharvesting of resources, intolerance toward one another, and expressions of warlike behavior. Of these human expressions, overpopulation is undoubtedly the one that is at the source of our major population and environmental tribulations.

Most of our current problems in a modern world stem from our reproductive prowess and inability to recognize the influence overpopulation has on all of Earth's biota and its dwindling resources. Further, most of our dominant political decision-makers appear to be clueless or apathetic about its significance in maintaining environmental homeostasis.[3]

In a statement about the global environment and human population growth, Holdren and Ehrlich (1974) have stated that "three dangerous misconceptions appear to be widespread among dicision-makers and others with responsibilities related to population growth, environmental deterioration, and resource depletion." Their statement was made over 40 years ago, and yet, it has an increasing relevance as the population climbs toward its biotic potential:

1. The first [misconception] is that the absolute size and rate of growth of the human population have little or no relationship to the rapidly escalating ecological problems facing mankind.
2. The second is that environmental deterioration consists primarily of "pollution," which is perceived as a local and reversible phenomenon of concern mainly for its obvious and immediate effects on human health.
3. The third misconception is that science and technology can make possible the long continuation of rapid growth in civilization's consumption of natural resources.

In the process of becoming technologically advanced, the human population has grown to over seven billion individuals and has influenced changes in Earth's biotic and abiotic components wherever it has existed. According to P.M. Vitousek et al. in *Human Domination of Earth's Ecosystems* (1997), "Human alteration of Earth is substantial and growing." Its land surface has been dramatically transformed; its chemistry has been altered; and many species on Earth have been driven to extinction. "By these and other standards, it is clear that we live on a human-dominated planet."

DOMINANCE

Actually, it is not enough to simply say we hold a position of dominance in the world, because the concept of dominance is an enormously complicated affair. In addition, dominance is expressed in a wide variety of ways, and it is influenced by a complex assortment of conditions, most of which involve aspects of aggression (Richards, 1974). Therefore, it would behoove us to begin our examination of dominance and aggression with some sort of explanation of what these concepts are and how they work. Defining dominance and aggression and their miscellaneous ramifications take us through this and the next chapter.

The word "dominance" has many meanings. It has typically been used in biology to describe features such as the way one eye, developing oocyte, hand, or brain hemisphere is dominant over its counterpart, as well as which alleles are the most powerful in determining genetic expressions, and how associations develop and are expressed between organisms. I use it here both in a biological, anthropological, and psychological sense to explain behavioral relationships between certain individuals and groups of animals and in an ecological sense to explain relationships between organisms and their environment (Fig. 2.6).

DOMINANCE IN BIOLOGY, ANTHROPOLOGY, AND PSYCHOLOGY

Richards (1974) states that "definitions of dominance commonly used are numerous and confused." In the fields of biology, anthropology, and psychology and in its most modest form, the term "dominance" is often simply defined as the status or ranking an individual or group has relative to other individuals or groups. While this very brief statement gives us some important clues toward an understanding of what "dominance" is, it is hardly an apt one for a concept as complex and important as that which drives relationships in all groups of animals every day of their lives. Since the meaning of the term "dominance" is the main focal point of this book and of central importance to understanding the behavior of humans and other animals, it would be most fitting to define it more appropriately and analyze its properties.

To do this, we must turn to dictionaries and papers by scientists who delve into the concept of dominance, but when we look at the APA Dictionary of Psychology (2006), which is theoretically based on the work of numerous investigators, we find that dominance is simply defined as an "exercise of major influence or control over others." This purely psychological attempt to define dominance is also too brief, and thus we are left with a feeling that maybe the concept of dominance is not as well understood as we initially thought.

In a comparative investigation of definitions presented by various authors, C. Drews (1993) states that "while the concept of dominance has contributed greatly to an understanding of social structure in animals, there has been a variety of concepts and definitions of dominance which have led to an ongoing debate about the usefulness and meaning of the concept. Criticisms aimed at one definition of dominance do not necessarily apply to other definitions. Existing definitions can be structural or functional, refer to roles or to agonistic behaviour, regard dominance as a property of individuals or as an attribute of dyadic encounters, concentrate on aggression or on the lack of it and be based either on theoretical constructs or on observable behaviour."

One problem investigators have had in defining dominance has to do with the different fields that they most commonly represent, for example, biology, psychology, sociology, anthropology, and ecology. Each field, on occasion, may use separate terms or concepts in describing a phenomenon, or they may ignore the work of others. The chore of properly defining dominance is further complicated by the widely divergent group of animals we have to work with, for example, invertebrates and vertebrates, humans and subhumans. One approach that may simplify our chore is to treat humans as animals, which they most certainly are, with species-specific challenges, which they most certainly have; look at all animals in a comparative fashion; and attempt to utilize and amalgamate concepts of dominance that have been brought out by previous investigators, no matter what their field may be.[4] Chapter 3 complements the first two chapters by delving into why we can learn a lot about ourselves by studying a wide variety of organisms.

By virtue of its highly descriptive value, Drews selected "the original definition of dominance by T. Schjelderupp-Ebbe (1922) as the basis to formulate a structural definition with wide applicability and which reflects the essence of the concept":

> Dominance is an attribute of the pattern of repeated, agonistic interactions between two individuals, characterized by a consistent outcome in favour of the same dyad member and a default-yielding response of its opponent rather than escalation. The status of the consistent winner is dominant and that of the loser subordinate.
>
> Dominance status refers to dyads, while dominance rank, high or low, refers to the position in a hierarchy and, thus, depends on group composition. Dominance is a relative measure and not an absolute property of individuals. The discussion includes reference to the heritability of dominance, application of dominance to groups rather than individuals, and the role of individual recognition and memory during agonistic encounters.

This is an excellent definition with which to begin our understanding of dominance, and yet, it clearly does not expose us to the wide range of mechanisms that lead to dominance. We will consider this definition as a viable expression of what dominance means and later, when we have discussed the various components of dominance and aggression, we will redefine it.

In addition to the power or status a dominant animal possesses, the outcome of dominance interactions establishes: (1) whose genes are most likely going to pass into subsequent generations and (2) what type and level of activity group members will have. In terms of the group to which such individuals belong, a hierarchical ranking of the individuals (3) establishes a homeostasis or allostasis within the group. The status of the most dominant individual (4) may also determine the dominant status of the entire group, with relationship to other groups in the area. At a planetary level, the concept of dominant power and status is more complicated and will be treated in later chapters.

While some investigators may look at a dominance hierarchy as a well-defined and stable relationship between individuals, its features are variously expressed in different groups and societies, and it has the capacity to change. Upon initially examining a dominance hierarchy, the ranking system may appear to have a linear arrangement (Fig. 1.3). However, a closer look at what occurs within the ranking may reveal that the term "linear arrangement" is actually an oversimplified and sometimes incorrect description of a more complex and dynamic system, as it often is found in nature.

$$\boxed{\textbf{Alpha} \implies \textbf{Beta} \implies \textbf{Gamma} \implies \textbf{Epsilon} \implies \ldots \textbf{Omega}}$$

FIGURE 1.3   A linear dominance hierarchy, with an alpha-dominant and progressively less-dominant individuals as we go through the Greek alphabet to Omega, the most subordinate individual in the hierarchy. While a linear hierarchy may express an idealistic view of the ranking systems within a social group, there are variations in behavior that complicate the ranking. Dominance hierarchies change, depending on the degree of dominance and subordination found within a group, as well as the presence of a despotic alpha individual.

Although these definitions are adequate for a blanket explanation of dominance, it does not describe: (1) the precise functional mechanisms for developing and maintaining a hierarchy; and (2) it does not consider the behavioral variability of individuals within their populations and communities.

As indicated by these definitions, behavioral variation (like anatomical variation) is expressed at both the individual and the group level, forming behavioral clines, if you will, and as we discuss the various forms of dominance and aggression, the nature of variability will emerge as a universal phenomenon at all phyletic levels.

DOMINANCE AND SOCIALITY

However, before we get too deeply involved with defining the inner workings of dominance hierarchies, let us discuss where in the animal kingdom they are found. We can return to the definition of dominance and its implications when we understand the group of organisms to which the term is most aptly applied. While the phenomenon of dominance, as defined here, can be applied to any population of organisms whose members come together in a group, it is best demonstrated in social animal groups (Carpenter and Hermann, 1979; Fiske, 1992; Heinrich et al., 2004; Hermann, 1979, 1986; Lin and Michener, 1972; Richerson and Boyd, 1998; Sanderson, 2001; Wilson, 1975, 1999).

When animals aggregate, as in a social group, each animal characteristically relates to one another and enters into agonistic confrontations, which determine who is the most dominant. Confrontations between two individuals generally commence with quasiaggressive displays. If a position in the hierarchy is not determined by these expressions, more physical contact may ensue, sometimes catapulting competing individuals into violent attacks, occasionally even resulting in death. Most confrontations carry on until they result in a winner and a loser. Speaking generally, we say that the winner is dominant and the loser is subordinate, submissive or deferent. Confrontations between all or most members of a social group result in the formation of the ranking system or hierarchy described earlier, from most dominant to most subordinate.

As an example within groups of primates, Allan Mazur (2015) states that "status ranks are allocated among members of a group through face-to-face interaction," and "the allocation process is similar across each primate species, including humans. Every member of a group subsequently signifies its rank through [a] physical or vocal demeanor." We will find that these forms of behavior, in addition to visual and chemical expressions of rank, are exhibited in a wide array of social organisms, invertebrates, and vertebrates alike.

To understand what is meant by "being social," our best approach is to repeat what most behaviorists have done over the years and define its most elaborate form: eusociality (Andersson, 1984; Cameron, 1993; Crespi, 1995; Jarvis, 1981; Jarvis et al., 1994; Nowak et al., 2010; Wilson and Hölldobler, 2005).

Eusociality is defined as a relationship between members of a group of organisms in which there are three outstanding components: (1) cooperative care of the young; (2) reproductive division of labor; and (3) an overlap of generations. These features are found in all eusocial animal groups, no matter what their phyletic position.

Care of Young: The young of eusocial species absolutely require care by adults. Otherwise, they die. They must be fed, cleaned, and protected by parents and/or alloparents from predators, parasites, parasitoids, or other intruders (Hermann et al., 2017) until they reach a point in their life at which they can function on their own. In the case of alloparents, it is often subordinates of the group that perform the chores required.

Reproductive Division of Labor: Having a reproductive division of labor within a social group implies that certain members of the group express a high level of dominant behavior and produce all or most of the offspring. Other members of the eusocial society, which are generally subordinate to the dominant, are often responsible for carrying out other chores within the society. In vertebrates, the dominant individual may join in the processes of obtaining food and protecting the society, whereas in invertebrates (e.g., in eusocial insects), subordinate members of the group generally carry out all activities other than mating. It has been shown that eusocial insect queens are important in defending their nest against parasitoids, but defense against vertebrate intruders is most aptly carried out by workers (Hermann et al., 2017).

While there are cases in social vertebrate groups in which the dominants produce all or most of the offspring, as in mole rats (Clarke and Faulks, 1998; Gillooly et al., 2010; Kent and Zayed, 2013; Susko, 2003), most vertebrate dominants, including humans, choose the mate or mates they wish to breed with and allow breeding between subordinates as well. Subordinates of insect societies, of course, are most often sterile workers and do not breed, although they are sometimes capable of producing males through a haploid–diploid process. In social insects that have fertile cofounding females, cofoundresses that are subordinate to the queen take on the role of workers, unless their dominance hierarchy is other than linear.

Overlap of Generations: Overlap of generations points to the long life of sexual dominants so that they are able to live with (and dominate) their offspring. In a eusocial society, most or all of the offspring are daughters or sons of the queen, the most dominant individual in the society. In many eusocial species, potentially high-ranking sexual individuals are produced which, at some point in their life, pass their dominant genes on to succeeding generations.

While eusocial species must have all of these features, species that express some but not all of them are usually referred to simply as "social."[5] Animals that express very little social behavior may be considered primitively social. Organisms that may border on being social may be referred to as subsocial. There are other terms applied to species that are not quite eusocial that will be brought out later.

To obtain a dominant status in a social group, group members must enter into competitive games (the confrontations mentioned earlier), in which they usually demonstrate agonistic behavior toward one another (Ghosh et al., 1998). Agonistic behavior (a form of aggressive behavior demonstrated by one individual toward another) invariably establishes and maintains a dominance hierarchy (social hierarchy or pecking order).

The Dictionary of Psychology refers to the act of establishing dominance as a "dominance–subordination" process, described as "a form of social relationship within groups in which there is a leader or dominant member who has priority of access to resources over other, subordinate members of the community."

As examples, again among primates, the dictionary mentions dominance–subordination relationships that are highly organized in troops of baboons, in which dominant males have more access to food resources and mates than do subordinate males, and all males often appear to dominate females. However, male domination does not occur in all societies. In hyena groups, for instance, the relationship is reversed, with males subordinate to females. Even closely related primates sometimes show significant differences in whether males or females are dominant. Chimpanzees, for instance, exist in male-dominated societies while their closest relatives, the bonobos, live in female-dominated ones (Furuichi, 1997). In human populations, males are most often dominant, but human societies have changed since women have enjoyed progressively increasing dominant roles since the beginning of the industrial revolution (1760–1890 and through the 19th and 20th centuries; Ashton, 1948; Berlanstein, 1992). Thus, there is variability in human societies with respect to dominance hierarchies, in which most dominant expressions are currently skewed toward the male end.

Based on these definitions and examples, dominance–submission events form "a key dimension of interpersonal behavior, identified through factor analysis, in which behavior is differentiated along a continuum ranging from extreme dominance (active, talkative, extraverted, assertive, controlling, powerful) to extreme subordination (passive, quiet, introverted, submissive, weak)" (Mazur, 2015).

As pointed out by Chase et al. (2003), the influence of confrontations on "the strategies of individuals engaged in conflict has proven integral to establishing social hierarchies reflective of dominant–subordinate interactions. They present three categories of social animal types that engage in such interactions, along with the results of their interactions:

1. *Animals with resource-holding potential*: Animals that have an ability to control resources are better able to win confrontations without having much physical contact. This is a concept that is well defined in human populations in which two of the top resources on the planet are oil and food (Engdahl, 2007).
2. *Animals with resource value*: Animals that are more invested in a resource are more likely to engage themselves in actual fighting, in spite of the potential for incurring higher costs. Stated another way, if the resource is of primary importance to the well-being of a social group, it may be worth fighting for.
3. *Animals that are residents of a resource area*: When participants of confrontations are of apparent equal dominant status, the resident of a territory being competed for is likely to be the victor because they have a high stake in the territory.

Behavior in social species is both innate and learned (Tierney, 1986). Innate expressions (traits) appear to be shared by many vertebrate and invertebrate species as a matter of survival and reproduction. These commonalities will be brought out in the rest of this chapter and in Chapter 2.

While social species lowest on the phylogenetic scale (e.g., ectotherms) seem to rely more on innate responses to their environment, the more cognizant species (humans and other endothermic social species) express both innate and a wide assortment of learned responses.

Examples of invertebrate eusocial species are ants, certain bees, certain wasps, termites, and at least one species of weevil (*Austroplatypus incompertus*). While another beetle, the burying beetle (*Necrophorus* spp.), demonstrates well-developed social behavior, which will be important to us in revealing social traits, it does not live in colonies and thus does not express a reproductive division of labor.

Mole rats (*Heterocephalus glaber* and other species) are eusocial mammals, which, as we shall see, can be shown to share behavioral expressions with those of certain eusocial insects (Bennett and Faulkes, 2000; Faulkes and Abbott, 1997; Gillooly et al., 2010; Jarvis and Bennett, 1993; Kent and Zayed, 2013; Sherman et al., 1991; Susko, 2003). Some of these examples and their extraordinarily similar behavioral expressions are discussed in later chapters. While humans and many other animals demonstrate care of their young and an overlap of generations, their expressions of reproductive division of labor in contemporary societies are more diverse and not as clear as they are in other, less cognizant species.

Social groups of mammals, including humans, express aggregative behavior, including many individuals that are not related. Their groups range from a few individuals to thousands or more, and generally include members that belong to the same social group. Social groups in human populations, for instance, may be in such categories as families, clans, cities, states, countries, political groups, religious groups, patriotic groups, schools, or fans of a particular baseball or football team. Similar groups are found in other social vertebrates. Likewise, aggregative arthropod groups often contain numerous individuals that are not related. Eusocial insects, on the other hand, may have societies of many thousands of individuals, all of which are related.

In addition to the agonistic expressions mentioned previously, displayed in an effort to establish and maintain a position in a dominance hierarchy, social organisms simultaneously express cooperative behavior in many of the society's activities (e.g., care of the young, cooperative hunting, collecting materials for constructing a home, the actual construction process, and defending the group). Together, dominance establishment and cooperation most often function in maintaining societal homeostasis or allostasis (Fisher and Reason, 1988; McEwen and Stellar, 1993).

It is true that certain contemporary aboriginal human societies (as in truly egalitarian groups) express a noncompetitive nature, stressing the value of a humble existence (Boehm, 1999). Such an existence is apparently designed to avert intragroup rivalry and prevent individual embarrassment for a loser, but most humans and other social animals in the contemporary world are decidedly competitive (Case, 2007).

The earliest of human societies were undoubtedly foraging family groups (Johnson and Earle, 2000). Even at that early stage, a dominant member of the society (e.g., an elder) had to direct the society's activities and attempt to keep it homeostatic. As humans evolved, and the human brain brought societies into a progressively more modern world, dominance competitions continued, some guiding dominance hierarchies in life-related situations and others in competitive sporting games.

The local, national, and international games we play are worthy examples of our competitive nature (e.g., boxing, hockey, football, baseball, Olympic games, bowling, archery, cards, board, word, and electronic games). Competition and dominance establishment are also parts of more serious forms of behavior as well when considering such categories as survival, politics, religions, jobs, morality, sexual misconduct, killing, and other violent behavior, as well as war-time games.

A truly egalitarian concept for the development of a moral human society hardly exists anywhere on Earth, except in certain groups that have retained a primitive lifestyle (Erdal and Whiten, 1996). Since human society is predominantly constructed around dominance hierarchies, many individuals who occupy a high position in a hierarchy resort to any methods available to attain and maintain their position within the hierarchy, subsequently experiencing its rewards, even to the extent of using forms of deception, and in extreme cases, killing their adversaries (Ariely, 2012; Harff, 2003; Lane, 1997; Manson et al., 1991; Trivers, 2011).

As we will see, while many members of a society may express themselves in an honest and moral fashion, there are individuals who are clearly immoral. There are legitimate reasons for being honest and there are equally legitimate reasons for lying and cheating in societies, especially in war games, politics, the sales profession, men and women engaged in attracting one another and practicing sexual games, applying for a job, and antisocial behavior, most of which involves expressing ourselves as something we are not. Survival benefits for cheaters, as brought out by Dan Ariely in *The (Honest) Truth about Dishonesty, How We Lie to Everyone—Especially Ourselves* (2012), appear to be on the rise as human societies grow in size and complexity, lose their connection with Earth and its biota, have difficulty in determining where they fit into the global society, ignore their moral obligations to society, and take on a more asocial and sometimes even an antisocial composition.

Determination of dominance status may be influenced by parameters such as size, age, and experience of competitors, as well as the types of displays they provide. Larger individuals often win in physical confrontations, but size may take a second place to strategy or age. Some competitors may hold their dominant positions for long periods of time, while others may win an encounter and enjoy temporary dominance rewards but lose their dominant status later to a younger, more able, competitor or coalition of competitors.

Thus, within all social groups, variations in agonistic expressions (aggression, threats, displays, retreats, placating aggressors, and conciliation) lead to the various dominance levels occupied by competing individuals. At times, positions within a dominance hierarchy are determined after one encounter, and, at other times, agonistic behavior may be repeated numerous times before a position in the hierarchy is established. Repeated confrontations generally occur between individuals who are higher on the dominance scale or equal in their abilities to compete. Subsequent encounters between individuals that have established themselves in a dominance hierarchy are generally of a subtle nature, for example, noticing the posture or other behavioral expressions of one another.

To repeat, a dominance hierarchy is thus an expression of relative dominant power in which certain individuals within a family, colony, community, or population control other individuals, as well as the distribution of rewards available to them. As mentioned earlier, rewards for winning competitions in human and most social groups are generally in the form of status, power, and wealth. With these rewards, dominants can demand the resources they require.

The complete Greek alphabet used by scientists in ranking members of a linear dominance hierarchy (from most dominant to least dominant or most subordinate) are: Alpha, Beta, Gamma, Delta, Epsilon, Zeta, Eta, Theta, Iota, Kappa, Lambda, Mu, Nu, Xi, Omicron, Pi, Rho, Sigma, Tau, Upsilon, Phi, Chi, Psi, and Omega. Such a ranking shows a linear behavioral cline from one extreme to another. The most dominant individual in a hierarchy is an alpha individual. Those individuals who occupy more subordinate positions in the hierarchy are referred to as beta (dominant to all other group members except the alpha) through omega (the most subordinate, submissive, or deferent member of the group; Fig. 1.3).

A set of behaviors, customs, and rituals that have developed as features of establishing and maintaining dominance hierarchies are found well developed in any social group, although their intricate features may vary somewhat from one group to another and even among different individuals within a single group.

In human populations, dominance hierarchies are complex and may develop at many levels, incorporating every aspect of a human's daily life and at any socioeconomic level. Along with variations in human intelligence and the development of widely divergent personality features, the formation of dominance hierarchies has resulted in an increase in the complexity of human customs and rituals while retaining the animalistic features that got them to their current position in animal evolution.

SOCIAL DOMINANCE

Since its introduction by Jim Sidanius and Felicia Pratto (1999), the concept of social dominance in humans has come into use with respect to human social groups that tend to be organized according to group-based social hierarchies in societies that produce economic surplus. Thus, it is rooted in a theory of intergroup relations, which focuses on the maintenance and stability of group-based social status (Grusky and Takata, 1992; Pratto et al., 1997; Sidanius and Pratto, 1999).

While it applies to human populations, it complements other concepts mentioned in the forgoing forms of dominance. According to the authors mentioned earlier, there are two functional types of what they refer to as legitimizing myths, which play a role in intergroup dominance: (1) hierarchy-enhancing and (2) hierarchy-attenuating legitimizing myths. Hierarchy-enhancing myths (e.g., racism or meritocracy) contribute to greater levels of group-based inequality. Hierarchy-attenuating myths (e.g., anarchism and feminism) contribute to greater levels of group-based equality. "People endorse these different forms of ideologies based in part on their psychological orientation toward dominance and their desire for unequal group relations (i.e., their social dominance orientation). People who are higher on social dominance orientation tend to endorse hierarchy-enhancing myths, and people who are lower on social dominance orientation tend to endorse hierarchy-attenuating ideologies. The social dominance theory finally proposes that the relative counterbalancing of hierarchy-enhancing and hierarchy-attenuating social forces stabilizes group-based inequality."

Social dominance attempts to show that group-based inequalities are maintained through three primary intergroup behaviors—specifically: (1) institutional discrimination, (2) aggregated individual discrimination, and (3) behavioral asymmetry. Group-based social hierarchies in human populations (as defined by Sidanius and Pratto, 1999) are based on: (1) age, (2) gender, and (3) group-based relationships:

Age: In its extreme form, adults have more power and higher status than children. However, dominant status of an older individual is often challenged by younger members of a group, especially those in the adolescent years. *Gender*: Generally, men have more power and higher status than women. However, there are human societies in which women have a dominant role. In the United States and many developed countries, dominance relationships have changed over the years, and women now have jobs in which they hold dominant roles. This increase in female dominance in the human population is likely to continue to rise, judging from the number of men and women who are currently attending college class and receiving college degrees.

Group-based relationships: Group-based relationships are culturally defined and do not necessarily exist in all societies. In this third form, hierarchies can be based on ethnicity (e.g., one ethnic group over another), religion, nationality, and so on.

Human social hierarchies consist of a hegemonic group (political, economic, or military dominance or control of one state over others) at the top and negative reference groups at the bottom. More powerful social roles are increasingly likely to be occupied by a hegemonic group member. Since males, in general, have been more dominant than females, they possess more political power. Most high-status positions are held by males. Prejudiced beliefs, such as racism, sexism, nationalism, and classism, are all manifestations of this same system of social hierarchy.

Since the concept of social dominance has been developed for humans, it will be especially relevant to Chapters 8 through 15.

DOMINANCE IN GAME THEORY

Within recent years, a concept called "game theory" has been developed and used in biology and many other fields to describe a wide range of dominance phenomena, currently being considered an umbrella term for the science of logical decision making in humans, animals, and computers. In game theory, the concept of dominance applies to a strategy that is better for one opponent regardless of the opponent's strategy. Game theory has recently been incorporated into certain biological explanations, particularly in a concept referred to as "Prisoners' Dilemma" (William and Dyson, 2012).

While the game theory was first used to explain the evolution (and stability) of sex ratios in populations (Bergstrom and Godfrey-Smith, 1998), it has subsequently been employed as a model to understand many different biological phenomena (Hammerstein, 1981; Kim, 1995; Smith and Parker, 1976; Osborne and Rubinsteun, 1994; Rapoport and Chammah, 1966), leading to games called "chicken," "hawk–dove," and "snowdrift," which are influential models of conflict for two players in game theory (Cressman, 1995). The principle of the game is that while each player prefers not to yield to the other, the worst possible outcome occurs when both players (both with dominant status) do not yield. This, of course, primarily involves individuals who are at the top of the dominance hierarchy and do not have tolerance for their opponent.

DOMINANCE IN ECOLOGY

In addition to the definitions already presented on dominance as a symbol of status and power within a social group, the following definition of dominance and its connection between humans and their environment is of ecological importance. It will point out that certain plants and animals dominate their environment, and they influence the species that exist around them. It will further point out that in addition to the dominance status that particular humans may have due to confrontations with other humans, humans as a group currently represent the most dominant ecological species on the planet, and as dominant individuals or groups, they sometimes are more concerned with losing their position in dominance hierarchies and its rewards than they are about how their dominant status is adversely affecting the health of ecosystems and planet Earth.

Thus, the term "dominance" in an ecological context pertains to the influence of one or a few species in an ecological community over others (Fig. 2.6). Dominant species of plants are considered climax vegetation. Forests that have developed to a stage in which there is climax vegetation are usually relatively stable and homeostatic. Left alone, the ecosystem interactions between organisms in the various trophic levels are more-or-less balanced. The ecosystem loses its homeostatic form when environmental changes come about (Dryzek, 1983; Rolston, 1975).

As in the social groups we have just discussed, ecological dominance may be expressed in an aggressive fashion or in more subtle ways. Aggression in certain animals and under certain conditions may be fierce, resulting in the death of an individual or group of individuals, whereas aggression in most confrontations may be low or poorly defined.

Influence of dominants can be expressed locally, as in discussing climax (the most stable and dominant) vegetation in communities of a particular geographic area (Clements, 1916). For instance, we may speak of an ecosystem called an oak-magnolia forest because those trees are the dominant (and often predominant) trees in a particular area. Knowing that, we may expect to find certain other types of vegetation and particular animals present, which are typically found associated with such a forest.

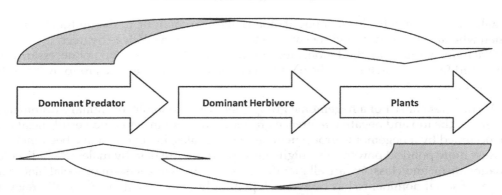

FIGURE 1.4 Trophic levels within a food chain. The straight arrows show typical community relationships, with a dominant predator and the herbivores it feeds upon. In turn, the dominant herbivores feed on a variety of plants. Indirectly, the dominant predator influences the plants that occur in this system by causing changes in the herbivore's population, and the plants available in an area also influence the stability of the ecological dominants by providing varying amounts of food for the herbivores. Thus, each member of the community influences the other. If the food plants suffer a setback from disease or removal, the herbivore population becomes reduced or eliminated. If the herbivores suffer a setback, the predator population suffers from a dwindling prey population, and the plants may begin to increase their numbers. Removing the predators causes a rise in the herbivore population, and the plants suffer from heavier feeding. Generally, these relationships reach a point of stability or homeostasis in the ecosystem to which they belong, although their populations may fluctuate due to daily or seasonal changes. If an exotic species is introduced to this system, it may be invasive and replace one or another of the indigenous trophic levels, and the entire system may collapse. All ecosystems are built on these relationships. Over long periods of time, climatic conditions may change, and the entire ecosystem may respond dramatically by selecting new dominants.

Likewise, we would also expect to find particular forms of biota associated with a pinewoods forest or any other specific ecosystem. Thus, each type of forest defines a certain ecosystem that supports certain key types of organisms and their food chains, including plants, animals, and many other types of organisms.

In all stages of their evolution, humans have been part of this ecological realm. When humans lived in smaller, nontechnological, and poorly technological groups, damage to the environment was not outstanding. Humans, like other animals, did use environmental resources, but whatever damage was done could be readily repaired by nature.

As a dominant contemporary animal species, humans continuously (sometimes dramatically) choose to expand their population and alter environments in the process. With excessive numbers of humans in modern societies, those who see fit to trespass into, alter, and fragment the various habitats occupied by plants and nonhuman animals have been responsible for habitat destruction and species extinction throughout the world (Barbault and Sastrapradja, 1995; Cincotta and Engelman, 2000).

On a larger scale, we may recognize certain biological ecosystems called biomes, large ecological areas on Earth, for example, desert, grasslands, deciduous forests, coniferous forests, tundra, fresh-water lakes, or marine waters, all of which are accompanied by certain species that live and interact with one another, often symbiotically, in various types of habitats within the biomes. Destruction of dominant organisms within these ecological areas destroys not only the forests but also many other organisms that reside within and depend upon them.

Certain of the animals within these biomes, as in the case of smaller communities of organisms, may represent an array of dominant and subordinate forms. These forms, expressed by certain species throughout the world, which have varying degrees of power to alter their environments, generally foster ecological stability. Thus, they have been responsible for a balance in Nature that has brought stabilization to ecosystems over long periods of time (e.g., during thousands or even millions of years of geologic history). Left to themselves, such areas as communities, ecosystems, and biomes are generally well established and relatively homeostatic until components of the ecosystem are changed (Rutledge et al., 1976; Fig. 1.4). Removing predators from an area, for instance, upsets this balance, and the prey species increase in numbers. As another example, bringing an exotic species into an area may replace certain levels of an ecological dominance hierarchy. If the dominant is replaced, the entire ecological hierarchy may change.

Communities and ecosystems undergo natural changes over long periods of time due to physical and chemical changes that take place within the biosphere. As examples of long-term natural changes in Earth which affect ecosystems, tectonic activities result in a modification of the positions of continents (Fig. 1.5); volcanic activity is destructive to communities, adds new land mass to crustal formations, and causes tectonic plates to move away from one another (Keller, 2001; Schumm et al., 2002; Van der Pluijm and Marshak, 2004); tectonic plates sometime collide, forming subduction zones in which the plates ride over and under one another, resulting in mountain formations; polar shifts periodically occur, altering the weather and magnetic field; global warming and cooling fluctuate

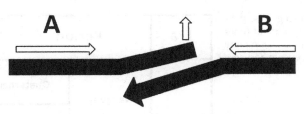

FIGURE 1.5 When two tectonic plates collide, one plate (B) generally moves beneath the other (A), causing one to push up, a common cause of mountain formation. Such tectonic plate upheaval changes ecological relationships between interacting organisms which in turn changes the entire food web. The colliding of plates during the formation of the supercontinent Pangaea resulted in mountain formation, allowing gymnosperms to become the dominant group of plants during the Mesozoic era instead of the mosses and ferns that dominated lowland, swampy environments during the Paleozoic era.

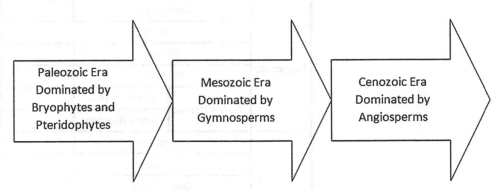

FIGURE 1.6 A progression of the three geologic eras of the Phanerozoic Eon over 542 million years. During these eras, planet Earth underwent dramatic changes, and ecosystems changed with them. Changes may be subtle climatic shifts or catastrophic events, for example, movement, separating and colliding of tectonic plates, global warming, global cooling, or asteroid impaction. Changes may occur over short periods of time as well, often being influenced by human activities, for example, removal of forests, laying down of streets and highways, construction of buildings, building of dams, pasturing food animals, and releasing exotic plants and animals to ecosystems to which they do not belong. All of these latter changes are due to human overpopulation and human activities.

repeatedly (over time), causing not only shifts in temperatures but also the raising and lowering of sea level. Such changes impact Earth, altering climates, causing the modification and destruction of ecosystems, and the rise of different ones.

With Earth changes, populations are constantly faced with adjusting to environmental alterations and attempting to maintain homeostasis. If populations become too large for the ecosystem to which they belong, environmental influences (e.g., hostile conditions, insufficient food, predation, and parasitism) generally function in returning them to their most homeostatic level (Solomon et al., 2007).

Over long periods of time, certain organisms dominate the scene, and eventually, as conditions change, they step aside and other organisms become the dominant forms (Fig. 1.6). As an example, bryophytes (very small nonvascular plants, e.g., true mosses and their relatives) and pteridophytes (larger, vascular plants, e.g., ferns and their relatives) dominated Earth's terrestrial environments, along with amphibians and certain arthropods, during the Paleozoic era (542–251 million years ago; Benchley and Harper, 1998). The climate, topography, evolutionary stages, and general requirements of such plants were suited to them at the time.

In the oceans, as portrayed by Richard Fortey (2000), numerous forms of primitive arthropods known as trilobites arose in the Cambrian period and represented dominant forms of animal life in Paleozoic seas for 300 million years. Early precraniate chordates (e.g., lancelets) were simultaneously taking their place as the roots of chordate evolution, which would yield terrestrial tetrapods (four-legged animals with digits) within that era (Clack, 2009; Hall, 2007; Long et al., 2006).

As Earth changes occurred over vast periods of time, organisms were sometimes unable to adjust, and mass extinctions occurred (Lehrmann et al., 2006; Sahney and Benton, 2008). Many areas on Earth became unsuitable for the continued dominance of bryophytes and pteridophytes. With mountain formation and a cooler climate, many areas of the world became more suitable for gymnosperms (cone-bearing plants, e.g., pine, spruce, and fir), which had arisen during the Paleozoic era. They dominated Earth's terrestrial environment, along with reptiles, during the Mesozoic era (251–65 million years ago; Fig. 1.7).

FIGURE 1.7 Geologic timescale, showing eras, periods, and epochs and their respective position and times as they relate to the species mentioned in the text. A more elaborate timescale, showing important events and species, is found in Chapter 5.

Era	Period		Epoch	Estimated MYA
Cenozoic	Neogene	Quaternary	Recent	.01
			Pleistocene	2
		Tertiary	Pliocene	5
			Miocene	24
	Paleogene		Oligocene	38
			Paleocene	55
Mesozoic	Cretaceous			63
	Jurassic			138
	Triassic			205
	Permian			240
	Carboniferous	Mississippian		290
		Pennsylvanian		330
Paleozoic	Devonian			360
	Silurian			410
	Ordovician			435
	Cambrian			500
	Precambrian			570
				4550

With continuous Earth changes and the rise of flowering plants (angiosperms) and their coevolving pollinators, ecosystems underwent major changes (Doyle and Donoghue, 1986). Flowering plants have dominated Earth's terrestrial environment, along with mammals, birds, and insects, during the Cenozoic era (65 million years ago to the present).

An example of the changing social fauna during these geologic eras can be found in any group of animals and has been clearly expressed in its insects. As Wilson pointed out in *Insect Societies* in 1975 and Frank Carpenter and I repeated in *The Antiquity of Social Insects* in 1981, the geologic history of insects extends from the Devonian and Carboniferous periods, and most of the existing orders of insects had already been developed by the end of the Paleozoic era. The earliest record of the order Hymenoptera (ants, bees, and wasps, including insects relatively high on the invertebrate evolutionary scale) at the time of that writing consisted of three genera from Lower Triassic rocks. Thus, eusociality (the most advanced form of sociality) and the close relationship between flowering plants and pollinators probably arose at some time during the middle of the Mesozoic era.

In *The Social Conquest of Earth*, Wilson (2012) points out that termites (order Isoptera) appear to have arisen about 220 million years ago in the mid-Jurassic, ants (order Hymenoptera) appear to have arisen about 150 million years ago in the Late Jurassic to Early Cretaceous, and honeybees and bumble bees (also order Hymenoptera) appear to have arisen 70–80 million years ago during the Late Cretaceous (Fig. 1.7). The rise of the Hymenoptera and Isoptera (termites) and assorted insects in some of the other orders was accompanied by the development of social behavior.

Human evolution from earlier primates came about much later and are presented in later chapters. As a brief preview, the first prosimian primates diverged from a common ancestor with other mammals about 85 million years ago during the Late Cretaceous period (Fig. 1.7 and Table 5.1). Simians (including monkeys and apes) arose from prosimian stock about 30 million years ago. Apes broke away from Old World monkeys about 20 million years ago (Cartmill and Smith, 2011). The great apes diverged from the gibbon apes 15–20 million years ago; gorilla-like and chimpanzee-like great apes diverged from an orangutan-like ancestor about 14 million years ago; australopithecines, similar bipeds, and chimpanzee-like predecessors separated from the gorilla line about 8 million years ago, and the human line separated from its common ancestor with chimpanzees about 7.5 million years ago (Stringer, 2003).

Local environmental changes also occur within shorter periods of time (over hundreds and thousands of years). Changes in climate or rainfall generally cause ecosystem changes. Conditions that define a desert (e.g., low rainfall and the types of organisms that thrive under such conditions), for instance, may change so that it becomes a semitropical rainforest. Such a change has occurred to Florida within the last 10,000 years. A deciduous forest may be replaced by a coniferous forest or vice versa. Likewise, other plants and animals that live in those areas change as well.

As an example of how changes have taken place in North America over a relatively short period of time (in geologic time), a report by J. Adams in 2005 (*North America During the Last 150,000 Years*) shows that parts of the western half of North America had temperate open woodland 18,000 years ago, while other parts carried vegetation representative of temperate semidesert, indicating that the Southwest received more rainfall and retained a higher humidity than it currently experiences. The eastern half was covered mostly by taiga-like vegetation (moist subarctic coniferous forests), and the coastline had open boreal (northern) woodland and scrub.

These ecological changes generated far-reaching consequences for plant and animal ecosystems, even at the human level. In *Making the Wind Sing* (2011), I discussed how these changes influenced the lives, music, and preservation of artifacts of Native America.

I was first struck with these vast climatological changes and their association with climax vegetation on a trip to Harvard in the 1960s to see the first specimen of *Sphecomyrma freyi* (described by E.O. Wilson, F.M. Carpenter and W.L. Brown in 1967), known at the time as the world's oldest ant, which was collected from amber that oozed out of Sequoia trees in New Jersey during the Triassic period of the Mesozoic era (Wilson et al., 1967). As we know, Sequoia in North America currently occupies a narrow strip of land along the Pacific coast of North America, from Monterey County, California, to extreme southwestern Oregon.

Thirteen thousand years ago, some temperate forests of western North America had been replaced by temperate scrub and woodland, indicating less rainfall. The southern portion of eastern North America was replaced by warm temperate forests, and the Florida desert was invaded by scrub vegetation. Thus, the Southwest was becoming drier, and the Southeast was becoming wetter.

About 11,000 years ago, the southwestern portion of North America had become drier, the upper regions being semidesert, but the eastern regions had become progressively warmer and wetter. By 8000 years ago, following the progression of life's changes since the last ice age (Wisconsin), much of southwestern North America was semidesert, while the central portion became an extensive grassland, and warm temperate forests increased in the east. Vegetation distribution today is much like that found 8000 years ago. Since populations of animals depend on the types of plants that dominate an area, it is likely that their populations changed dramatically along with the plants.

Changing environments and an increasing human population often put an enormous amount of stress on ecosystems. One of many examples is found in the relative amounts of water in western North America. A 2014 article in National Geographic by Michelle Nijhuis, called *When the Snows Fail*, points to current shortages of water in California due to drought and human overuse (Lemberg, 2009; Nijhuis, 2014). Nijhuis stated that, "430,000 acres in California will be left fallow in that year due to drought. Shasta Lake water levels are 65 percent below the historic average. In the driest corners of the valley, aquifers are so overdrawn that fields have sunk by more than 30 feet. The pattern of groundwater use in California practically defines the term unsustainable."

According to Nijhuis, most water in the western United States "arrives in winter storms, which swoop in from the Pacific and dump snow atop the region's mountain ranges. Despite the occasional severe winters, western snowpacks have declined in recent decades. Warmer winters are reducing the amount of snow stored in the mountains, and they're causing snowpacks to melt earlier in the spring." The changing climate in California and most of the West, she says, has not put us in a good position to handle drought. While Los Angeles and other large cities have dramatically improved water efficiency, current California drought conditions and the droughts to come could force the start of a new chapter on the ecological balance between the environment and the human species.

Studies of climax vegetation tell us something about the influence climate changes have on ecosystems. Populations of animals within these ecosystems depend on the presence of certain climax plants that take a place in the hierarchy of food chains. Plants photosynthesize and build tissues that can be fed upon. Herbivores feed upon the plants, predators eat the herbivores, and energy flows through food webs to the rest of the animals, as well as organisms that eventually recycle it back to nature. With changing vegetation, the entire ecosystem changes (Mayhew et al., 2007).

Although humans may have existed as a dominant force within their local territories in early populations, they and their predecessors did not always occupy a position of worldwide dominance. In their rise to global dominance, there were other animals that occasionally preyed upon them, diseases that they faced, and food shortages they had to overcome. Primates and their predecessors, like any developing group of organisms, were initially little fish in a big pond (see Table 5.1), filling roles of predator, prey, and host (Hatcher and Dunn, 2011).

Prior to the building of civilizations, humans had been foraging hunters for a long time. While many writers claim *Homo erectus* was one of the earliest human hunters of animals, a Science Daily report in 2013, called *Giant Prehistoric Elephant Slaughtered by Early Humans*, points out that early humans (possibly *Homo heidelbergensis*) may have had group-hunting instincts and an ability to make hunting tools as early as 420,000 years ago, over 200,000 years before our species arrived (Littlefield and Ozanne, 2011; Spikins et al., 2010).

Francis Wenban-Smith discovered the previously mentioned site in 2003 containing the remains of an extinct straight-tusked elephant (*Palaeoloxodon antiquus*) (2013), revealing not only a deep sequence of deposits containing the elephants, but also numerous flint tools and a range of other species, such as wild aurochs, extinct forms of rhinoceros and lion, Barbary macaque, beaver, rabbit, various forms of vole and shrew, and a diverse assemblage of snails. These deposits date to a warm period of climate during the so-called Hoxnian interglacial period (middle Pleistocene, over 400,000 years ago), when the climate was probably slightly warmer than it is today.

This report actually points to a time that is much later in human evolution than indicated in other studies. That humans were social or were becoming social and had the tools for hunting is seen in much earlier human-like forms. Australopithecines are known to have had crude stone instruments 2.6 million years ago (Briggs and Crother, 2008; Cela-Conde and Ayala, 2003).

Considering the environments in which primitive humans lived and the nature of early humans, it appears that they represented local dominant animal forms that possessed contrasting traits. While much of the literature depicts humans as being mostly aggressive, researchers have suggested that they actually had both a violent and a peaceful (cooperative) nature, much like modern humans, and this diverse nature served them well in their rise to higher forms of dominance.

In *The Human Potential for Peace: An Anthropological Challenge to Assumptions about War and Violence* (2005), Douglas Fry states that along with the capacity for aggression, primitive humans also possessed a strong ability to prevent, limit, and resolve conflicts without violence. Steering away from confrontations is understandably an animalistic survival trait that was possessed by human predecessors and other animals. Exposing one's self and society to violent behavior is maladaptive unless it is important to the preservation of the species or group within the species (e.g., in expressing defensive behavior). Most species avoid conflict whenever possible (McNeely, 2003; Traverse, 1988).

Exploring the philosophy of science issues, Fry draws on data from cultural anthropology, archaeology, sociology, behavioral ecology, and evolutionary biology to show that conflict resolution exists across cultures. By documenting the existence of numerous peaceful societies, Fry's research challenges the concepts brought out by other investigators, for example, Margaret Mead's writings on Samoan warfare (Mead, 2014), Napoleon Chagnon's claims about the Yanomamö (Chagnon, 1967), and ongoing evolutionary debates about whether "hunter-gatherers" have been peaceful or warlike.

The human line, as shown later, evolved through both pleasant and unpleasant environmental changes after its split from its common ancestors with other great apes (Fig. 1.8). As hunters and the hunted, they retained their combined violent and cooperative behavior as part of their survival strategies, traits that are expressed in various degrees even today in contemporary human culture (Koizumi et al., 2012; Stutz, 2012).

- According to the Chimpanzee Genome Project, human (*Ardipithecus*, *Australopithecus* and *Homo*) and chimpanzee/bonobo (*Pan troglodytes* and *Pan paniscus*) lineages diverged from a common ancestor about 5–7 million years ago (De Grouchy, 1987). Our genetic relationships were divulged after 167 different gene sequence sets from humans, chimpanzees, macaques, and mice were examined. Behaviorally, we know that even early prehuman primates possessed both a violent and a cooperative nature. The same is true for humans as they progressed through their early forms in ever-changing environments (Boyd and Silk, 2003).
- *Sahelanthropus tchadensis* is believed by some to be the last common ancestor of humans and chimpanzees, primates that existed primarily in forests (Wolpoff et al., 2006). After the separation, *Orrorin tugensis* is the earliest known human ancestor. As a forest dweller, this species was faced with different challenges than their descendants who ventured into the savanna (Henke et al., 2007).
- *Ardipithecus*, a small-brained, arboreal but bipedal hominine genus, crossed into a savanna habitat almost 4.5 million years ago, subjecting them to new survival challenges for which they required modifications in their defensive system.
- *Australopithecus afarensis*, affectionately known as Lucy, lived between 3.9 and 2.9 million years ago. As savanna-living humans, they were preyed upon by predatory cats, such as *Dinofelis*. *Australopithecus* probably gave rise to *Kenyanthropos platyops*, a possible ancestor of humans in the genus *Homo* (Lieberman, 2001).
- Australopithecines (forest and savanna-dwelling humans in eastern Africa) are known to have had crude stone instruments 2.6 million years ago to use during predatory forays (De Heinzelin et al., 1999); since that time, such

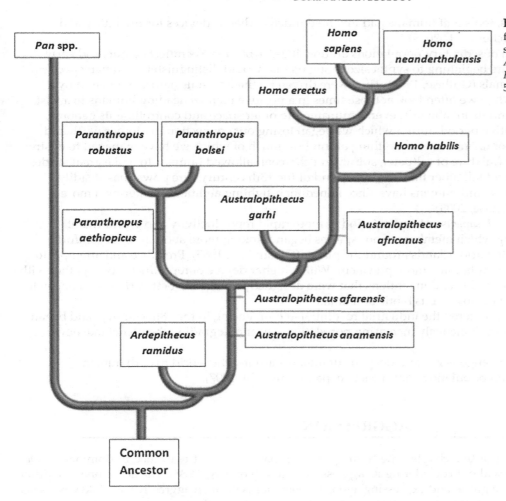

FIGURE 1.8 Evolution of humans from a common ancestor with *Pan* spp., including the genera *Ardipithecus*, *Australopithecus*, *Paranthropus*, and *Homo*, covering a period of about 5–7 million years.

simple tools have progressively become more broadly significant in group defense and in aggressive encounters between societies of humans (both related and unrelated). Humanoid forms were now on the road to achieving greater technological advances and attaining the pan-global dominant status they now enjoy.

- During the Acheulean period (Lower Paleolithic), *Homo erectus* (with its large brain) was beginning to demonstrate extensive tool use while simultaneously dispersing out of Africa (Anton, 2003); once again, their tools were the forerunners of instruments to be later used for a variety of more aggressive purposes.
- *Homo georgious, Homo ergaster, Homo pekinensis,* and *Homo heidelbergensis* belong to the *H. erectus* group (Vekua et al., 2002). As of this writing, *H. georgious* is the oldest fossil human found outside of Africa. The evolution of dark skin and loss of abundant coarse (terminal) body hair had come about by 1.2 million years ago. *H. heidelbergensis* (with a brain 93% the size of *H. sapiens*) may be the common ancestor of modern humans and Neanderthals.
- *Homo sapiens* began leaving Africa about 60,000 years ago and interbred with *H. sapiens neanderthalensis* (Neanderthal) in Europe (D'Errico et al., 1998; Fig. 1.1). After *H. s. neanderthalensis* became extinct some 35,000 years ago, it would be *Homo sapiens sapiens* alone who would arise from the Stone Age to eventually become the thinkers of a very modern world. However, as the primate brain increased in size and complexity, it developed into a more cognizant form that had the potential to dominate everything. Over time (with a continued development of the brain and an increase in population size), human presence and its demonstration of dominance became increasingly more prevalent (Stringer, 2012).
- When groups of humans continued to increase the size of their populations around the world and develop more sedentary habits about 12,000 years ago, they commenced to amplify their territorial stresses and aggressiveness toward one another and subsequently established a well-defined war-time aggression as early as when Mesopotamian and Egyptian civilizations were founded (about 5000–6000 years ago; Fig. 1.2); spears and arrows

were among the first weapons, tools that humans had invented much earlier as devices for predatory and simpler defensive reasons (Archer et al., 2008).

- With the continued expansion of sedentary populations around 400–300 BC, Pre-Socratic, Socratic, and early Post-Socratic philosophers were beginning to synthesize thoughts that would distinguish the human species from the ranks of simpler animals (Guthrie, 1968). They were driven by their curious nature but biased by a theologically based society. While we often look at those times in a positive manner, leading humans to a vast knowledge about the world and its inhabitants, even venturing into outer space and controlling its genome, it also represents a more negative period during which we began losing our connection to Mother Earth and eventually established total dominance over everything around us, much of which we have managed to destroy.

- The invention of gun powder (mixture of saltpeter, sulfur, and charcoal) allowed humans to make great strides in their abilities to dominate and kill other humans by the end of the 13th century. Fire power has steadily increased throughout the planet, and humans have since honed their fighting abilities by entering a more sophisticated electronic age (Chase, 2003).

- In the 15th and 16th centuries, *H. sapiens* entered a period of more rapid reproductivity and world travel (Humanistic Tradition), during which members of our species began thinking more about human comforts and potential and dispersing to various lands around the planet (Giustiniani, 1985). Prejudice and attempts to dominate other groups of humans became more prevalent. With a higher degree of reproductivity, together with the development of modern medicine and inventions that were designed to make life better, *H. sapiens* began to more rapidly increase its numbers toward its biotic potential.

- Modern populations of humans entered the industrial revolution (commencing in the 18th century) and began to improve living standards, simultaneously increasing population size and degrading the planet like never before (Landes,1969).

- In spite of continuous efforts by eugenicists and despotic dominants around the world to curb human population inflation, the human population continues to expand (Engdahl, 2007).

AGGRESSION

As indicated in the earlier part of this chapter, the term aggression has significant relevance to dominance. One cannot really discuss dominance without considering its aggressive nature (Fonberg, 1988). The act of one individual confronting another individual in a group and expressing agonistic behavior is itself an aggressive act. Likewise, since humans are social creatures, it is difficult to discuss dominance without considering their cooperative nature as well.

These are both features that belong to all social animals, but since human and other animal behavior is fluid, they are expressed in varying degrees by each species and even between individuals within a single species. Their cooperative nature has generally been expressed intraspecifically (i.e., between members of their own species), with some degree of an interspecific cooperation (as in tolerating and sometimes even helping other species), while their aggression is expressed both intraspecifically (as in the establishment of dominance hierarchies) and interspecifically (as in interspecific territorial disputes between species).

To show its special characteristics, aggression should be defined separately. The later part of this chapter is devoted to bringing the two concepts together, and the last part of Chapter 2 will allow us to redefine both dominance and aggressiveness.

In *The Psychology of Aggression*, Arnold Buss (1961) states that "literature on aggression is scattered, and researchers in one area may be unaware of what has been happening elsewhere (as in other fields). This isolation prevents researchers from sharing promising methods of investigation, and it tends to make theoretical formulations too narrow." This very same statement may be said about isolation between researchers in fields such as psychology, sociology, and biology. They often exist in their own particular field, each using their own terminology and concepts.

Buss's book is divided into three parts. Part 1 deals with aggression as it is studied in the laboratory, with an emphasis on discrete aggressive responses and reactions to hostile stimuli. Part 2 is concerned with aggressiveness as an enduring response tendency; this section deals with methods of measuring aggressiveness, theories of aggressiveness, and hostility in psychopathology. Part 3 concerns developmental and social aspects of aggression, with the social aspects limited to prejudice.

For our purposes, I chose to introduce the subject of aggression with definitions and features from a variety of dictionaries. While there are many dictionaries and encyclopedias from which to define aggression, most express it in a similar manner. Dorland's Medical Dictionary defines it as:

behavior leading to self-assertion; it may arise from innate drives and/or a response to frustration, and may be manifested by destructive and attacking behavior, by hostility and obstructionism (obstruction of progress), or by a self-expressive drive to mastery.

While each of the definitions presented here share similar concepts, they each sometimes present terms, some of which are important, that are not found in other definitions. Mosby's Medical Dictionary defines aggression as:

a forceful behavior, action, or attitude that is expressed physically, verbally, or symbolically. It may arise from innate drives or occur as a defense mechanism, often resulting [in humans] from a threatened ego. It is manifested by either constructive or destructive acts directed toward oneself or against others.

In this dictionary, kinds of aggression, for example, constructive aggression, destructive aggression, and inward aggression, are listed. Constructive aggression is an act of self-assertiveness, self-protection, and preservation in response to a threatening action. Destructive aggression is an act of hostility that is directed toward an external object or animal and is unnecessary for self-protection or self-preservation. Inward aggression is destructive behavior that is directed toward oneself.

Looking at aggression more from an animalistic viewpoint, *Saunders Comprehensive Veterinary Dictionary* defines aggression (based on canine behavior, which is in some ways similar to other definitions but appropriate when comparing them with humans) as:

behavior that is angry and destructive and intended to be injurious, physically or emotionally, and aimed at domination of one animal by another. It may be manifested by overt attacking and destructive behavior or by covert attitudes of hostility and obstructionism.

This dictionary further lists forms of aggression, most of which may be applied to human expressions of behavior:

- *affective aggression* (based on nervous and chemical stimulation);
- *fear-induced aggression* (which is accompanied by fear, especially when escape is not possible; it may be associated with previous unpleasant experiences);
- *food-related aggression* (which is directed toward people or animals when approached while eating);
- *interfemale aggression* (dominance aggression between females);
- *intermale aggression* (fighting between males; this aggression generally includes elements of competitive, territorial, and sexual aggression);
- *maternal aggression* (a dam's protection of her young; this is a variant of a form of aggression that has been called dominance aggression);
- *nonaffective aggression* (lacking nervous activation);
- *pain-induced aggression* (defensive aggression triggered by pain);
- *play aggression* (biting, nipping, and growling during play);
- *possessive aggression* (a form of dominance aggression; the animal is reacting against someone or another animal trying to remove something, e.g., food);
- *predatory aggression* (directed toward any kind of animal or even inanimate objects; typically, it is elicited by something that is moving quickly);
- *protective aggression* (in which an animal is protecting its territory);
- *redirected aggression* (which occurs when an animal is touched or restrained by a human or another animal, while it is fighting or threatened);
- *territorial aggression* (behavior directed toward the defense of an area by an individual or a group against entry by others, usually members of the same species). This, of course, is similar to behavior listed earlier as protective aggression.

The online *Dictionary of Psychology* (2014) defines aggression in a similar fashion: "a forceful action or procedure (as an unprovoked attack) especially when intended to dominate or master the practice of making attacks or encroachments; unprovoked violation by one country of the territorial integrity of another; or hostile, injurious or destructive behavior or outlook, especially when caused by frustration."

In the APA Dictionary of Psychology, VandenBos defines aggression in a more elaborate form as:

Behavior, motivated by competitiveness, anger, or hostility that results in harm to or destruction or defeat of others or, in some cases, oneself. When the primary goal is intentional injury or destruction of the target, the behavior is called hostile aggression. It is distinguished from instrumental (or operant) aggression, in which the attack is carried out principally to achieve a goal other than the target's injury, such as acquiring a desired resource. Most instances of hostile aggression can also be regarded as affective aggression in that they are emotional reactions to an aversive state of affairs, which tend to be targeted toward the perceived source of the distress but may be displaced onto other people or objects if the disturbing agent cannot be attacked. In the classical psychoanalytic theory of Sigmund Freud, the aggressive impulse is innate and instinctual, but in most non-psychoanalytically oriented psychologists' view, it is a socially-learned reaction to frustration.

Some investigators would not agree that the terms "aggression" and "self-assertion" are synonyms. While self-assertion in humans is defined as an insistence on or an expression of one's own importance, wishes, needs, opinions, or the like (indicating a dominant and narcissistic personality), it, like dominance, does not have to be overtly aggressive. Further, in human society, we tend to accept self-assertion as a good quality in a person, although self-assertion may, at times, be expressed in an aggressive manner. Aggression connotes an impression of possessing some bad or more forceful qualities. Obviously, there is some overlap in these definitions as there are in the acts themselves.

These definitions imply that aggression is sometimes recognized as being defensive in nature. However, as hinted at in the second definition, some forms of aggression in lower animals are either offensive or defensive, and we can be certain that this applies to humans as well.

Inherent/erudite aggression, as expressed by subhuman animals that dominate others and in some human alternative behaviors, is aggression generally from within the aggressor (a genetic propensity; Buss, 1961), although certain environmental stimuli may provide a base that provokes or influences inner aggression. Defensive aggression is aggression that is dependent upon provocation from other organisms or events that threaten a defender (Blanchard and Blanchard, 2010).

For instance, an individual who preys on innocent people through fraudulent behavior may be an inherent aggressor who could have a genetic predisposition for deceptive behavior, but their behavior may also have been influenced by abusive, nonmentored, or poorly mentored behavior in their formative years, during adolescence, and even in adult life.

HUMAN AGGRESSION

In an Annual Review of Psychology article called *Human Aggression*, C.A. Anderson and B.J. Bushman elaborate on the complexity of human expressions of aggression by pointing out five main theories of aggression, which psychologists feel guide most current research (Anderson and Bushman, 2002; Fig. 1.9).

- *Cognitive Neoassociation Theory*: In this theory, "events such as frustrations, provocations, loud noises, uncomfortable temperatures, and unpleasant odors (conditions that are typically found in everyday life) produce what they term negative affect. Negative affect produced by such unpleasant experiences automatically stimulates various thoughts, memories, expressive motor reactions, and physiological responses, which are generally associated with both fight and flight tendencies."

  "Fight associations," they say, "give rise to rudimentary feelings of anger, whereas flight associations give rise to rudimentary feelings of fear." Furthermore, this theory assumes that cues present during an aversive event become associated with the event and with the cognitive and emotional responses triggered by the event.

  Aggressive thoughts, emotions, and behavioral tendencies are linked together in memory. Concepts with similar meanings (e.g., hurt and harm) and concepts that frequently are activated simultaneously (e.g., shoot and gun) develop strong associations. When a concept is primed or activated, this activation spreads to related concepts and increases their activation as well.

FIGURE 1.9 Currently active approaches to understanding aggression, as described by Anderson and Bushman (2002) in *Human Aggression*. This psychological approach is important for all fields of social biology so that individual researchers understand one another and collaborate in their depictions of the human animal.

Five Approaches to Understanding Aggression

Cognitive Neoassociation Theory   Social Learning Theory   Script Theory   Excitation Transfer Theory   Social Interaction Theory

This theory also includes higher-order cognitive processes, such as appraisals and attributions. If people are motivated to do so, they might think about how they feel, make causal attributions for what led them to feel this way, and consider the consequences of acting on their feelings. Such deliberate thought produces more clearly differentiated feelings of anger, fear, or both. It can also suppress or enhance the action tendencies associated with these feelings.

"Cognitive neoassociation theory" provides a causal mechanism for explaining why aversive events increase aggressive inclinations. This model is particularly suited to explain hostile aggression, but the same priming and spreading activation processes are also relevant to other types of aggression.

- *Social Learning Theory*: "According to social learning theories, people acquire aggressive responses the same way they acquire other complex forms of social behavior—either by direct experience or by observing others. Social learning theory explains the acquisition of aggressive behaviors via observational learning processes and provides a useful set of concepts for understanding and describing the beliefs and expectations that guide social behavior. Social learning theory—especially key concepts regarding the development and change of expectations and how one construes the social world—is particularly useful in understanding the acquisition of aggressive behaviors and in explaining instrumental aggression."

- *Script Theory*: According to this theory, "when children observe violence in the mass media, they learn aggressive scripts. Scripts define situations and guide behavior: The person first selects a script to represent the situation and then assumes a role in the script. Once a script has been learned, it may be retrieved at some later time and used as a guide for [further] behavior. This approach can be seen as a more specific and detailed account of social learning processes."

Scripts are sets of particularly well-rehearsed, highly associated concepts in memory, often involving causal links, goals, and action plans. When items are so strongly linked that they form a script, they become a unitary concept in semantic memory. Furthermore, even a few script rehearsals can change a person's expectations and intentions involving important social behaviors.

A frequently rehearsed script gains accessibility strength in two ways. Multiple rehearsals create additional links to other concepts in memory, thus increasing the number of paths by which it can be activated. Multiple rehearsals also increase the strength of the links themselves. Thus, a child who has witnessed several thousand instances of using a gun to settle a dispute on television is likely to have a very accessible script that has generalized across many situations. In other words, the script becomes chronically accessible. This theory is particularly useful in accounting for the generalization of social learning processes and the automatization (and simplification) of complex perception-judgment-decision-behavioral processes.

- *Excitation Transfer Theory*: This theory notes that "physiological arousal dissipates slowly. If two arousing events are separated by a short amount of time, arousal from the first event may be misattributed to the second event. If the second event is related to anger, then the additional arousal should make the person even angrier. The notion of excitation transfer also suggests that anger may be extended over long periods of time if a person has consciously attributed his or her heightened arousal to anger. Thus, even after the arousal has dissipated, the person remains ready to aggress for as long as the self-generated label of anger persists."

- *Social Interaction Theory*: Of particular importance to subjects examined in this book, the social interaction theory "interprets aggressive behavior (or coercive actions) as social influence behavior, that is, an actor uses coercive actions to produce some change in the target's behavior. Coercive actions can be used by an actor to obtain something of value (e.g., information, money, goods, sex, services, and safety), to exact retributive justice for perceived wrongs, or to bring about desired social and self-identities (e.g., toughness and competence). According to this theory, the actor is a decision-maker whose choices are directed by expected rewards, costs, and probabilities of obtaining different outcomes."

Social interaction theory provides an explanation for aggressive acts motivated by higher level (or ultimate) goals. Even hostile aggression might have some rational goal behind it, such as punishing the provocateur in order to reduce the likelihood of future provocations. This theory provides an excellent way to understand recent findings that aggression is often the result of threats to high self-esteem, especially to unwarranted high self-esteem.

This brief look at currently researched aggression models gives us some indication of the complexity of aggression in humans. We will find that they sometimes play a significant role in defining human personality.

What do these approaches have to do with animals in general? Simply, they show that aggression has a purpose and that it may develop for a number of reasons. But we will learn that it develops in certain individuals more than in others, and like dominance, a combination of factors determine its origin, complexity, and expression (Geen, 1990; Schneirla, 1973; Seroczynski et al., 1999; Tremblay, 2000).

IS AGGRESSION INHERITED OR LEARNED?

One of the questions that people have asked ever since humans began to study aggression is: Is it inherited or learned? As indicated in the last couple of paragraphs, it has been determined that it is both. The difficulty in understanding its expressions lies mostly in our misunderstandings of the degrees to which genetic and learned behavior play (Moore, 2003).

General chemical and hormonal components (e.g., the influence of serotonin in aggression) stem, in many cases, from the genetic makeup of an individual, although the chemical nature of an individual certainly may also be influenced by the environment they exist in and even the food they eat. Chemicals that influence the development of aggression are subjects covered in Chapter 9, which elaborates on the importance of chemicals in human behavior (Albert et al., 1993).

Controlled studies of behavior in experimental settings have demonstrated that aggressive behavior is similar to other operant behavior because it is influenced by rewards and punishment. In an article by D.V. De Souza (*Is Aggressive Behaviour Biologically or Environmentally Based?*) (2013), it was pointed out that much aggression in humans is influenced by cultural and social (environmental) factors.

In *The Folly of Fools* (2011), Robert Trivers points to deception as a stimulant for aggression. When animals detect deception that is directed toward them, he says, "they get angry and seek immediate retribution." Trivers points to deception as the cause of many forms of behavioral expression, particularly in self-deception and its role in many human endeavors.

As pointed out in De Souza's article, "Aggressive behavior can [most certainly] be a function of national culture. Residents of some countries show a more pervasive tendency to think of violence as a means of solving problems than persons living in other nations, and when these people move from one country to another, they bring with them the behavior they are accustomed to."

"In some cultures, (for instance, a person's) religious view is expressed so aggressively that the subject may sacrifice his or her life (in some cases risking the lives of others) for the sake of their god" (see Chapter 12). "In other cultures, aggressive behavior may be expressed in sports. American football, wrestling, ice hockey, and boxing, for instance, promote behavior that can physically injure another person. Some sports," as pointed out in that article, "disguise aggressive behavior as part of the art." And yet, game-based aggression is generated within the game and does not necessarily influence the lives of aggressors unless it is carried outside the game (which sometimes happens).

The recognition of regional subcultural differences in human aggression in the United States and other parts of the world is dependent upon different local norms for aggressive behavior (Buss, 1961). Society plays a fundamental role in influencing behavior. Poverty and crime, for instance, have become an intrinsic part of modern society, which unfortunately molds the behavior of subsequent groups of people through imitation and reinforcement, but its expression varies in different cultures.

An example is found in the residents of high crime areas in the world, such as in Laventille, Trinidad (Higman, 1996). Residents form a social order in which their lifestyle reinforces criminal activity as a means for survival. Members of this society are familiar with who the criminals are and choose not to report them. When residents of these communities commit crimes or aggressive acts (e.g., robberies), their criminal activities are reinforced when they escape the law and obtain positive reinforcement, such as material possessions.

Children in these communities naturally learn aggression through social imitation. They also become desensitized toward aggression and view it as a common and acceptable behavior in their community. Should we expect that left alone and without societal rules, humans would naturally express themselves in a similar way? Possibly. Male anhingas and some other birds, for instance, gather (or steal) nesting material from the nests of other birds and present it to the female for their own nest construction (Baughman, 2003).

Another example of intrinsic aggression that affects other cultures is found in the warlike behavior of certain groups of people in the Middle East. According to Sørli et al. (2005), "conflict in the Middle East is a recurring feature in international politics, academic literature, and current news coverage. The fifty-five-year-old Israeli-Palestinian conflict is one of the most enduring conflicts anywhere, but over the past 25 years, the region has also hosted two of the wars with the most international participants (Iraq in 1991 and 2003), as well as the bloodiest interstate war of that period (Iran–Iraq, 1980–1988). The region is also surrounded by other long-term conflict zones: Afghanistan, the Caucasus, the Horn of Africa and Sudan. Internal and regional instabilities have combined with the close ties between Middle Eastern and arms-producing governments to make the Middle East the most militarized region in the world."

These authors point out that the Collier–Hoeffler model of civil war provides the starting point for an analysis about why the Middle East is one of the most conflict-prone regions of the world (Collier and Hoeffler, 2002). In an application to Africa, Collier and Hoeffler, found that "economic development and economic growth, in addition to longer periods of peace, generally decrease the likelihood for conflict." They also found that "ethnic dominance is significant, while social fractionalization is not." Contrary to the findings of Collier and Hoeffler, Gleditsch and Strand state that "regime-type matters." Variables for the Middle East region, Islamic countries, and oil dependence are not significant. "Conflict in the Middle East," they say, "is quite well explained by a general theory of civil war, and there is no need to invoke a pattern of Middle Eastern exceptionalism."

Aggression is not only of cultural importance. It is a worthy subject in studies of individuals and communities in areas that are not typically considered aggressive or prone to violence. Guetzloe (1997) analyzes aggression based on what it is doing to our society and how we can develop a safe environment in the home, school, and community, considering such topics as predisposing factors, early predictors of violent behavior, opportunity factors, precipitating events, family factors associated with youth violence, factors contributing to school violence, and conditions that exacerbate aggression/violence. In a plan for prevention, she states that "the first step toward preventing or reducing violence is to develop a plan for the specific individuals and environment for which the interventions will be designed."

ORIGIN OF AGGRESSION

But where do the different forms of aggression originate? Actually, the roots of much aggression represent a primal emotion like any other fundamental behavior, found in all species that are able to display it (Buss, 1961). While the selection of beneficial features works at the individual level within a species, its preservation is sometimes passed on genetically to other members of its species, and thus, aggressive behavior has evolved to varying degrees in each species as a means of increasing its survival and reproductive fitness (Grigorenko and Sternberg, 2003). Survival and aggression, in turn, depend on specific environmental, social, reproductive, and historical circumstances.

Based on our behavior (as a species acting toward the Earth and its inhabitants), we humans most certainly rank among the most violently aggressive species (Anderson and Bushman, 2002). At the same time, it has been said that we (as a group) also rank among the most altruistic and empathetic. Could this be true? We will find that at the individual level, these opposing features in human traits are expressed in various ways in different individuals and in different cultures.

Evolution did not haphazardly shape us to be violent or peaceful. As with all animals, it shaped us through natural selection, allowing our populations to respond flexibly and adaptively to different conditions and circumstances. Most individuals will not risk expressing violence unless the risks are low or the benefits are high. Thus, the species to which aggressive potential belongs also will not generally express an aggressive nature unless it makes adaptive sense to do so. Even social wasps often warn an intruder before they commence a violent defensive display. Being violent without adaptive benefits is generally maladaptive. Of course, there are exceptions to this rule at the individual and group level.

Through many of the conflicts that humans have had, the distinction between peace and violence is separated by a finer line than we may realize. With the population of humans growing around the world, we as a species may be in mass denial about the magnitude of the problem that violence represents for the future (Dunne et al., 2006; Travis, 2012). Many of us may think of ourselves as peace-loving and that the violence and transgressions of the past will not return, but recent history and current events demonstrate how easy it is for humans to respond to certain circumstances with interpersonal and intergroup violence.

Of course, we may want to analyze and determine who the aggressors of the world are. If people within a population are relatively peaceful and would rather negotiate a problem, we must look to our leaders, the dominants who send us to war in spite of our personal feelings, relying on our patriotism to carry out their wishes. It is true that we elect our dominants to office, but we must realize that dominants are often unresponsive to carrying out the wishes of their subordinates. They would prefer to obtain whatever resources are at stake at any cost (Meiertöns, 2010).

According to behaviorists, future suspected fertile ground for intragroup violence (violence between members of a conspecific group, e.g., within the human species or subgroups within it) will be especially important in places where key natural resources are becoming scarce (Hipp et al., 2009). As our population continues to grow beyond its environmental carrying capacity, basic resources (e.g., food, clean water, and fuel) will become more limiting, and the environmental and social drivers of violence may become more difficult to control.

DOMINANCE AND AGGRESSION

Dominance and aggression, like nature and nurture, go hand in hand to mold a species. Certain individuals may be born with a strong or weak predisposition for peace or violence, but the culmination of developing a personality of one sort or another also depends on environmental input, for example, how an individual is cared for and treated in the formative through the adolescent years. Traumatic experiences and an assortment of life-learning ventures after adolescence also may play a role in personality development (Caspi and Roberts, 2001).

There are many other ways of defining dominance and aggression. However, the definitions and examples employed here are especially applicable to an understanding of the diversity of dominance and aggression in humans and other organisms within a biological context. It will be modifications of these definitions that will apply to most of the concepts we peruse throughout the chapters of this book. As we progress through the chapters, we add other circumstances that influence modifications in dominance hierarchies and how they affect our life as dominant animals. Before we redefine dominance, we must investigate the ramifications of behavioral variation that are covered in Chapter 2.

As extraordinary as it may seem, many of the things humans do in everyday life, whether in the process of making a way in the world, interacting with fellow humans in social gatherings or playing games, dominance interactions play a significant role in all of us. Wherever humans compete, a victor occupies the alpha position, and many people, depending upon their personalities, strive to be dominant in one thing or another. If for no other reason, being dominant in something makes an individual feel good about themselves, no matter what their socioeconomic or phyletic position, and being dominant also brings rewards.

In behavioral contests, no matter what form of dominance we are talking about, an individual or group of individuals (coalition) overpowers another. In the process, it ranks individuals or groups with relationship to other individuals or groups.

As pointed out in the beginning of this chapter, ranking can and often is considered to be linear, that is, positioning individuals or groups from alpha (the most dominant) to omega (the least dominant or most subordinate; Fig. 1.3). Other members of the group generally fit somewhere in between alpha and omega in a progressively dwindling (linear) dominant status. Such dominance ranking involves agonistic (aggressive) behavior of one individual or group toward the other.

Dominance interactions also may be expressed as a wide assortment of behaviors from complete docility to despotic. Despotic dominance is expressed in individuals or groups that dominate others with absolute power or authority, and in the process, linearity is lost. This type of dominance may exhibit more violent forms of aggression than those expressed in a homeostatic linear dominance hierarchy, and through deception, a despotic dominant may choose to hide his or her aggressive nature.

In both linear and despotic forms of dominance, varying degrees of tolerance toward other members of the group are expressed by the most dominant individuals (Dunn et al., 2009; Hammersley-Fletcher and Brundrett, 2005). As we will see, degrees of tolerance, whether in human society or otherwise, often determine not only the outcome of dominance interactions but also the survivability of the group to which the alpha individual belongs.

Some forms of dominance are interspecific (between members of different species) in which one species overpowers another. Exotic species (species brought into an area and compete with indigenous species), for instance, often outbreed indigenous forms and dominate the acquisition of resources that are also important to an indigenous species, and it may be a destructive (invasive) force in the survival and reproduction of indigenous forms (Corn, 2003).

There are numerous examples of exotic animals that put undue pressure on indigenous populations (e.g., numerous plants that are brought into an area, which overpower native vegetation, tropical and other exotic fish that are dumped into the environment when pet owners grow tired of them and the introduction of giant toads to Australia and other areas of the world, which plays havoc with both marsupial and eutherian carnivores). We could stretch this concept somewhat and use the movements of humans around the world, for example, the movement of Europeans into North America in the late 1400s and early 1500s, as an example of an exotic species introduced to new lands, a species that has consistently proven to demonstrate invasive features.

From examples of ecological dominance presented at the beginning of this chapter, we know that dominant vegetation influences the types of other vegetation, as well as the types of animals that occur in a particular environment. When such vegetation enters a new area, it often becomes invasive and overtakes indigenous forms and changes the ecology of an area.

An example of how exotic vegetation can overtake indigenous plants, a hurricane in southwest Florida during the early 2000s devastated Sanibel and Captiva Islands. During the clean-up of destroyed vegetation, the Sanibel/Captiva Conservation Foundation seized the opportunity to remove invasive exotic plants, and when it was all over, large portions of land were denuded of vegetation because exotic plants had replaced many of the indigenous species.

As with plants, humans fragment and otherwise dominate space and resources within an environment in which other organisms initially lived, often negatively affecting their survivability (Kolar and Lodge, 2001).

Intraspecific dominance hierarchies (dominance hierarchies between members of the same species) can be expressed between unrelated members of the same species, or in the case of social species, between closer members of a social group (many of which may be genetically related).

BENEFITS AND DETRIMENTS OF BEING DOMINANT

If dominance hierarchies are characteristically established in groups of social species, we would assume that they have some particular function that is beneficial to the species and the individuals that make up the hierarchies (Fig. 1.10). The main function of dominance hierarchies, other than distinguishing between dominant and submissive forms, is in the maintenance of stability within the group.

Based on computer simulations, L.A. Dugatkin and R.L. Earley (2003) point out that agonistic behavior and dominance hierarchy establishment are often dependent upon both intrinsic and extrinsic factors found associated with each species. Intrinsic factors refer to traits, such as body size, which often correlate with an animal's fighting ability in terms of physical prowess. However, these authors claim that intrinsic factors alone do not seem to produce the linear hierarchies that are often in nature. When extrinsic factors, such as winner and loser effects, were added to their computer models, hierarchies were much more similar to those found in nature. We must remember, though, that hierarchies that appear linear in nature may be more complex upon close examination.

Does it make any difference where in the dominance hierarchy an individual is? To the individual or group, the position in the dominance hierarchy means a great deal. There are many benefits to being an alpha individual or group. As J. Sidanius and F. Pratto say in *Social Dominance* (2001), "In an individual-based social hierarchy, individuals might enjoy great power (Fig. 1.10), prestige or wealth by virtue of their own highly-valued individual characteristics, e.g., great athletic or leadership ability, high intelligence or artistic, political or scientific talent or achievement."

Being in a group-based social hierarchy, on the other hand, "refers to that social power, prestige, and privilege that an individual possesses by virtue of his or her ascribed membership in a particular socially constructed group or social class." Examples of socially constructed groups in human society are seen in families, clans, ethnic groups, estates, nations, races, castes, social classes, religious sects, regional groupings, teams, or any other socially relevant groups.

There are also social detriments associated with occupying a dominant position. Responsibilities for maintaining societal homeostasis fall on the shoulders of an alpha individual, and the alpha may be called upon to defend the society. In human and insect societies, the alpha and others in a high position of dominance choose to delegate subordinates to the chore of defense. In addition, the alpha position may constantly be in jeopardy from younger members of the group or individuals who refuse to be subordinated.

Another motivation for group living is mutual defense (Krause and Singer, 2001). Even though subordinates may receive less food or have fewer opportunities to mate, they may have greatly increased chances of escaping predation or other forms of aggression if they work together. Group defense also often but not always involves relatedness between individuals in the group and the sharing of genes with offspring. In human society, national pride and patriotism replaces close relationships (Blatberg, 2000).

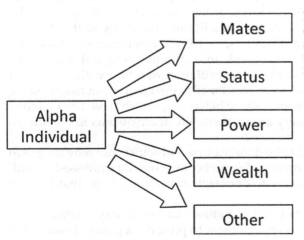

FIGURE 1.10 Alpha individual and the rewards it receives because of its degree of dominance. The quality and degree of dominance expressed by the alpha individual varies among social animals from despotic to poorly defined. In natural populations of nonhuman animals, the most productive degree of dominance appears to be somewhere between these two extremes.

What about other benefits in occupying a less-than-dominant position in a hierarchy? Actually, the position of dominance by an individual, in terms of its society, is often insignificant. Since one of the main beneficial features of establishing a hierarchy is to develop and maintain harmony within the society, social hierarchies provide a means by which animals can live in groups and exploit resources in an orderly manner. Thus, all members of a society are important to the society's survival and reproductivity. This thought should also be considered for human society. However, in despotic dominance (Chapter 2), intragroup harmony and the distribution of resources is not always achieved.

In many animal species, the position an individual occupies within the dominance hierarchy tends to be relatively fixed throughout life, although there are circumstances which influence a change in position, for example, age and an accompanying change in an individual's ability to defend, or in human societies, societal stipulations about how long a person can be dominant.

In most cases within vertebrates, males dominate females (Keller, 1985). However, in certain societies, females are dominant. Agonistic interactions among females are often not as overtly aggressive as those among males. A fixedness of status position is particularly dramatic with respect to the gender system.

We will find that the act of dominance in most social species is relatively straightforward. In humans and certain other animals, on the other hand, the concept of dominance is considerably more complex.

DOMINANCE IN HUMANS

Hominid evolution (Family Hominidae, including humans and other members of the family referred to as great apes) has progressed through significant changes since they separated from a common ancestor: A common ancestor of chimpanzees and bonobos, for instance, split from the line about 5–7 million years ago (Alexander,1990). Chimpanzees have a male-dominated society that has well-defined aggression-related dominance hierarchies. Bonobos, the closest relatives of chimpanzees, on the other hand, are female-dominated and live a relatively nonaggressive existence.

That these two species are so closely related but have a significantly different dominance system shows that their dominance systems have developed independently and differently in spite of their genetic closeness. Thus, their dominant systems developed with relationship to their individual customs.

Hominids leading to contemporary humans evolved through many predecessors. Before our species arose, all humans and human-like forms lived an animalistic life and thus responded to their environment like nonhuman animals would. With the rise of our species, the development of a bigger, more complicated brain and more innovative skills led to increasingly more complex human thought processes and activities.

Our species has changed considerably over time. Our cognitive capabilities have advanced far beyond those of our primate relatives (Von Eckardt, 1996; Blomberg, 2011). Our ability to discern 92 natural elements in the periodic chart and to combine them to not only understand life but also to change it to our will, to create new organisms and to manipulate the ones that got to where they are by natural selection, is taking our dominant status to an extreme. We are powerful, very powerful, but even Spiderman knew that great power comes with great responsibility.

Our influence upon world ecosystems, national and international politics, religions, local communities, and individuals within single populations, the workplace, and other social groups will be subjects examined in this book. A comparative sociobiological approach is taken, especially looking at humans and other social animal species by examining the interactions that determine their and our ranking systems. We will find that many of the different forms of behavior we display as humans have analogs in other animal species, vertebrates, and invertebrates alike.

Within our species, the variation of interactions that affect dominant rank among individuals and groups, the norms, and aberrant nature of individuals who often make the rules, which others of us must follow will be looked at critically. Our societal demands, rules and morals, and our willingness or reluctance to deal with them help to make up a widely divergent population of individuals who utilize both animalistic behaviors, which have been handed down from a long line of nonhuman predecessors, and contemporary innovations, which are distinct in the human population.

In the process, we have sometimes developed an increased need to be domineering and selfish, not only for group survival but for very personal reasons as well. In fact, our selfish traits appear to be most overtly expressed in individuals who are our leaders, and consequences of selfish traits appear to outweigh any altruistic acts that infer an interest in group survival.

Our wide-ranging personalities express ourselves in social, asocial, and sometimes antisocial ways to gain access to particular resources. We sometimes utilize ideas that threaten our environment to preserve a national and world economy, and at the same time hold back technological breakthroughs that would alleviate some of the pressures we put on our dwindling resources.

While we become more technologically advanced, we, our children, and their children are faced with an increasingly stressful world, which we and they must learn to exist in, providing a behavioral environment that is sometimes unfit for a healthy psychological existence (Oehlberg, 2014). Through our technological expertise, our world has become extremely complex, open to social misfits who steal, lie, deceive, and commit murder and rape toward individuals in their own and other societies.

Is this a harsh, overly negative description of our species and its influence on our surroundings? Of course it is, but is it a true and honest one of who we are? The chapters of this book present a picture of the human species that is all of this and more.

Some of us would point out that in spite of overwhelming evidence that we may be on the verge of experiencing colossal personal, societal, and environmental changes, our innovative skills have provided many of us with surroundings that are better than ever before. Many of us in contemporary society have homes that provide heat in the winter and cooling in the summer, although many people have lost most or all of their possessions in recent years for economic and war-associated reasons. Most of us have jobs to go to, refrigerators to keep culinary treats in, television, stereos, waste disposal systems, toilets, dish washers, showers, coffee makers, ovens, electronic devices that provide unlimited information and enjoyment, and a multitude of other societal luxuries.

We have become an electronic society, with computers that link us to cyberspace and unlimited sources of information. We can call up information on virtually any subject within minutes on a wide array of electronic devices. And we can spend much of our idle time texting one another or communicating through Facebook or other social channels.

In many ways, we have it too good, so good that we have come to ignore what life-related features we have lost in years gone by and our logical thinking by specializing in fact-demanding fields and, in the process, losing our connection with Earth and its biota (Crane, 2014). Many of us have no particular feelings for the environment, unless it directly concerns us.

In our demanding lives, we often do not consider the following: recycling everything that can possibly be reused, putting emphasis on the negative aspects of an overuse of pesticides, putting too much emphasis on making as much money as possible (often at an expense to other individuals and the environment), not respecting life by killing every animal we see or continuing to wage war on our fellow humans (often for selfish and deceptive reasons), practicing alternate and deceptive behaviors (e.g., putting learning on the back-burner), removing writing and speaking skills from our curricula, texting every waking minute, smoking, shooting, sniffing, and imbibing toxic substances, committing physical and emotional abuse on our offspring and other people and driving with a phone to our ears.

One cannot argue that for many of us, we never had it so good, but there are conditions that have been developing for years in the contemporary world, which will undoubtedly threaten our existence in the future, the signs steadily increasing over the years. We all have been part of it, and we all are responsible for whatever happens in the future. Thus, this presentation of information is less of a criticism about what we have not done and more about what we have done and the circumstances which have arisen that may eventually cause us or our descendants to lose the pleasures we now experience.

What will we get out of it when it is all over? Hopefully, we will realize that being dominant carries with it certain very important responsibilities that dominant and aggressive animals with human attributes should recognize. Being dominant and aggressive and not recognizing them can result in serious consequences.

Note

Superscript numerals appearing in this chapter refer to additional text/explanation given in the appendix.

CHAPTER

2

Traits of Dominant Animals

Human beings are by nature social animals and have always formed themselves into groups. Affiliations are organized on the basis of common rules and networks of relationships—be they family, power, religion or trade. Such a community is known as a society. *R.Winston and D.E. Wilson*, Human, 2004

GREGARIOUSNESS

Sociality originates with aggregative behavior, an attraction between members of a species. While certain animals aggregate in groups of related individuals, many animals aggregate in groups of individuals that are unrelated (Hermann, 1984b; Stamps, 1988; Wegner and Erber, 1993). During periods of sexual reproduction, for example, males of certain bird species sometimes exhibit elaborate aggregative courtship behaviors and compete with other conspecific males for mates and territories (Klopfer, 1969; Stokes, 1974). Once mated, at least one of the adult pair remains at the nest and cares for its offspring for a period of time. In some species of birds and mammals, both sexes and/or alloparents share postnatal responsibilities (Clutton-Brock et al., 2001; Reeve et al., 1998; Sherman et al., 1995).

Certain other (often widely divergent) animals, for example, reptiles, snails, and lady-bird beetles, are attracted to other members of their species or to a particular site in large numbers in order to hibernate or aestivate together (Allee, 1926, 1927; De Ruiter et al., 2005). Offspring of some organisms (e.g., certain spiders, scorpions, certain reptiles, mammals, and birds) may remain together as a group with other members of their species for a while until they reach a certain stage of development or become adults, at which time they may go off on their own to live a more solitary life or develop societies of their own. Some birds and mammals are altricial and require continued care in their early postnatal development while others are more precocial.

Many species of bees and wasps aggregate on certain plant stems to simply spend the night (Eickwort, 1981). Gregarious roosting occurs among many tropical butterflies and birds. Monarch butterflies, for instance, have enormous overwintering aggregations of diapausing and reproductive individuals (Ross, 2001, 2010) (Fig. 2.1).

Many necrophagus flies aggregate at oviposition sites (Yeates and Wiegmann, 2005). Conspicuous feeding aggregations are formed by a wide variety of plant-feeding insects (Sword, 2008). Bedbugs and soft ticks aggregate during the day near their host and disperse to feed at night. Cockroaches are typically gregarious when not foraging. Some carabid and tenebrionid beetles also aggregate under shelter during the day and forage at night (Erwin et al., 2005).

Springer et al. (1996) have reported on multispecies aggregations in a highly productive habitat, or Green Belt, along the edge of the continental shelf in the Bering Sea. "Physical processes at the shelf edge, such as intensive tidal mixing and transverse circulation and eddies in the Bering Slope Current, bring nutrients into the euphoric zone and contribute to enhanced primary and secondary production and elevated biomass of phytoplankton and zooplankton. Fishes and squids concentrate in this narrow corridor because of favourable feeding conditions and because of a thermal refuge from cold shelf-bottom temperatures that can be found at the shelf edge from fall to spring. The abundance of zooplankton, fishes, and squids, in turn, attracts large numbers of marine birds and mammals. In aggregate, the observations suggest that sustained primary productivity, intense food web exchange, and high transfer efficiency at the shelf edge are important to biomass yield at numerous trophic levels and to ecosystem production of the Bering Sea." Moore and DeMaster (1997) point out that "marine mammals can be used as indicators of environmental productivity because they must feed efficiently and therefore aggregate where prey is plentiful."

Dominance and Aggression in Humans and Other Animals
http://dx.doi.org/10.1016/B978-0-12-805372-0.00002-X

FIGURE 2.1 Huge aggregations of monarch butterflies migrate to locations such as El Rosario, near the town of Anganguero, state of Michoacan, Mexico. The site is about 100 miles west–northwest of Mexico City, at elevations of 9000 to 12,000 feet in the Transverse Neovolcanic Belt. According to G. N. Ross (pers. comm.) and Wells et al. (1990), aggregations play an indispensable role in the reproductive dynamics of overwintering *Danaus plexippus*. Unlike the migrations of birds, matings take place during the long trips, during which caterpillars (A) feed upon milkweeds, pupate (B), and emerge as new adults (C), which complete the trip. The butterflies that fly south are not the same as ones that fly back into the United States on the return trip.

P. Møller and T.A. Mousseau (2013) have found that predatory mammals aggregate in areas with abundant prey, especially when prey species are exposed to high levels of radiation. Matthiopoulos et al. report that "many far-ranging bird and mammal species aggregate in colonies to breed, and most individuals remain faithful to one colony." They suggest that, under local density dependence, site fidelity slows down the colonization process and can temporarily trap the entire population in a subset of the available potential colonies. When site fidelity is strong, the metapopulation follows a step-like trajectory. Population growth occurs only rarely because individuals must overcome their site-fidelity to found new colonies. Even though this effect is temporary, it renders the entire metapopulation vulnerable to rare catastrophic collapses.

"Under global density dependence," they say, "site fidelity imposes competition between colonies for the limiting resource. Stochastic events lead to the dominance of certain colonies and the temporary extinction of others. If site fidelity is strong, it can permanently prevent the metapopulation from occupying all available potential colonies." They conclude that, "irrespective of the mechanism of population regulation, colonially breeding species that show strong site fidelity are likely to occupy only a portion of the breeding habitat available to them."

It will be later shown that a wide variety of vertebrates aggregate, many of them living a social existence. Understanding the degree of sociality in vertebrates has been variously approached within recent years, primarily relating to the size, chemistry, and complexity of the brain (Barton and Dunbar, 1997; Dunbar, 2002; Dunbar and Shultz, 2007a,b; Reader and Laland, 2002; Roth and Dicke, 2005; Shultz and Dunbar, 2007).

Aggregations become social when they take on some or all of the features designated under the definition of eusociality (care of the young, reproductive division of labor, and overlap of generations, Chapter 1 and this chapter), as well as colony defense. A social existence and aggregating in large numbers have survival benefits in the acquisition of resources and defense against intruders.

Even humans developed as aggregative animals in their trek toward a more modern civilization (Lieberman and Shea, 1994; Winterhalder, 1981). In *Communal Hunts, Human Aggregations, Social Variation and Climatic Change* (1987), William B. Fawcett explains that many communal hunts by Plains people in North America served to mediate social and political tensions by providing food and exchangeable items for human aggregations. "Feasting, ceremonies, and exchanges could instill a sense of solidarity among participants and contribute to the mediation of tensions created by differences in power, wealth, gender, and age."

Human aggregations commenced with small family units and progressed over the years to increasingly larger assemblies of hunting and gathering people with common inheritance and interests (Ash and Robinson, 2010). Thus, they were social groups that had not yet established a sedentary lifestyle (Barnard, 1998; Hamilton et al., 2001, 2007).

The habits of such groups on the verge of becoming more settled can be found in antiquity in the early meanderings of North American Indians and represent the probable process followed by many wandering and trading groups that eventually became sedentary (Hermann, 2011). The region we now refer to as the United States, for instance, "served as a transition zone between the gregarious and foraging nations of Canada and the more settled empires of Mesoamerica." As an example, a network of trading between the ancient Pueblo and the Aztecs of Mexico is well known. It is suspected that many Mesoamerican religious concepts reached the Southwest by way of such trading, and it is believed that Mesoamerican traders roamed as far north as the San Juan Basin in the 11th and 12th centuries.

Within this range, there were variations in the way that people existed. For instance, once a knowledge of agriculture was established in the area now referred to as the United States (about 3000–3500 BC), early people in the north-central and western regions of the United States continued to hunt and gather wild foods, and people in the south-central and eastern regions usually became [more gregarious, developing into] agriculturists. The dividing line ran roughly in a diagonal fashion from the Northeast to the Southwest (Fig. 2.2). They were both composed

FIGURE 2.2 Location of early nomadic and sedentary indigenous populations in North America. While there is a degree of artificiality in making divisions between northern and southern tribal traditions, certain divisions appear to hold true. This map, showing a division between northwestern and southeastern areas, points to nomadic and agricultural customs, respectively. *Redrawn from Hermann (2011).*

of social individuals because they demonstrated cooperative and agonistic behavior, and they lived in multifamily groups.

Those people who commenced practicing agriculture and their descendants included the Iroquois and Huron of the Northeast, the Algonquian tribes of the Great Lakes and Atlantic Coast, and the five eastern tribes of Creek, Cherokee, Choctaw, Chickasaw, and Seminole. The hunting and foraging side of the line included the Plains nations of the Dakota, Lakota, Assiniboin, Crow, Kiowa, Arapaho, and Cheyenne, as well as the southern nations of Comanche, Apache, and Navajo.

Even though the majority of Native Americans had been living in villages, towns, and cities long before Columbus's landfall in the Americas, many early tribes migrated and emigrated considerably before and after the appearance of Europeans, often resulting in a mixing of cultures. Indeed, one of the significant features of the Archaic Period in North America (5000–1000 BC) was the beginning of long distance travel to find a suitable location to settle and for trade.

As an example, descendants of very early Native Americans who settled in the Southwest claim that their earliest ancestors roamed the Earth restlessly until they learned to cultivate the soil and glean wealth and wisdom from the land. The Mogollon, Hohokam, and Anasazi cultures became established in the Southwest between 300 BC and AD 1500.

Hopi stories point to a long period of movements by various clans, extending in the cardinal directions as far as people could go. There was a distinct manner, as depicted on four sacred tablets, in which they were to carry out their emigrations, how they were to recognize the place they were to settle permanently, and the way they were to live when they got there. They claim that certain landmarks well outside of their current location of settlement are associated with these emigrations.

From Mesoamerica, knowledge of seeds and planting (established between 7000 and 2000 BC) moved northward over a period of time and eventually took hold among the hunter-gatherers of the Southwest, eventually altering the character of life in that region:

> The introduction of agriculture changed the lives of people wherever it appeared, subsequently altering their ceremonial procedures. Agriculture spread north from Mexico to the Pueblos, then toward the Southeast, and finally to the Great Lakes. The Cherokees began practicing corn farming around the time that the Roman Empire fell. At some time prior to the year AD 1000, corn, beans, and squash (plants often referred to as the three sisters) were planted in the deep alluvial soil near Lake Ontario.

According to Roland Manakaja, Natural Resource and Cultural Research Director of the Havasupai Tribe in Supai, Arizona (pers. comm.), "there was much movement of tribes between Mesoamerica and Arizona, resulting in trading and an exchange of ideas." Such an exchange was common between many societies. In addition to their gregarious behavior, tribes sometimes made permanent moves over great distances, taking along with them their culture and ideas.

Siouan-speaking groups, for instance, are believed to have undertaken an epic journey from what we know today as the Carolinas to Minnesota, where they established themselves as the Eastern Sioux. Bands of Athabascan-speaking people from northwestern Canada entered present-day New Mexico (about AD 1025) and became the Navajo. A branch of Missouri River Hidatsa traveled westward to become the Montana Crow in the 17th century. The Kiowa subsequently moved from the northern to the southern Great Plains. The long and eventful migrations and emigrations of the Hopi have already been mentioned. Other emigrations and migrations took place as well.

In addition to moving groups of people, a loose trade network spanned the entire North American continent, and there were important trading hubs that carried Native Americans and products over wide areas. Some such hubs were Cahokia in Illinois, Poverty Point in Louisiana, and Chaco Canyon in Arizona.

Trading hubs often centered around waterways. The Mississippi River, for instance, has always been the main artery for north–south travel through the populous central section of what is now the United States.

Large rivers were not the only routes taken by indigenous people, however. Both river systems and mountain ranges formed Native American transportation routes, and many such routes were later used by engineers to plot routes for railroads and highways.

Nevertheless, streams and rivers were the major indigenous highways in the interior of North America, and portages were well marked to detour travelers around rapids, waterfalls, and other obstacles to connect watercourses.

Chaco Canyon lies at the center of a network of roads." Anasazi roads are straight paths 13–40 feet wide and cleared of stones, in a fashion similar to roads in Peru called Nazca lines and those in Mexico referred to as sacbe. While these roads were used by people, they also may have been constructed for the movement of spirits.

Cahokia, one of the most important areas representing the Mississippian Temple Mound Building cultures from AD 700–1700, was a well-known hub. It was located at what is now Ohio on the Mississippi River (its urban center was at the presently located town of Collinsville, IL) and thus was at the center of a water network stretching from the Caribbean to Hudson Bay. Based on materials found at Cahokia burial sites, there were items from the Great Lakes, Oklahoma, Arkansas, North Carolina, the Gulf of Mexico, Illinois, and Yellowstone National Park in Wyoming.

"To get to this and other trading areas, Native Americans sometimes covered great distances. Tecumseh (Shawnee Chief in the late 1700s) and his fellow travelers covered an area from the Seminoles of Florida to the headwaters of the Missouri River. The Iroquois of the current central New York area were familiar with the country as far west as the Black Hills of the Dakotas and as far south as Florida. Objects from the vicinity of the Pacific and Atlantic Oceans, Gulf of Mexico, Rocky Mountains, Minnesota, Wisconsin, and Mexico have been found in mounds in Ohio, Tennessee, and elsewhere. These are but a few indications of Native American movements" and their aggregative behaviors.

ISSUES CONCERNING SOCIAL ANIMALS

It is obvious that gregariousness is not always associated with a social state but that it strengthens social traits (traits that bring and keep individuals together, resulting in their interaction). However, the term sociality (the act of being social) is generally reserved for multicellular animals that come together into a more-or-less colonial style of life (Jackson and Coates, 1986). In addition, the group or colony (especially in invertebrates) is often composed of related individuals (Hermann, 1984a; Wilson, 1975). We find that numerous widely divergent species have independently developed a social state as a strategy for survival and reproduction, and we can detect commonalities in them in spite of their distant phyletic relationships. Let us look at an unusual invertebrate example.

BURYING BEETLES

Beetles are generally solitary in their lifestyle, and thus having social traits is quite unusual and developed independently from social species in other animal groups. A group of necrophagous insects called burying beetles, for instance, are unusually social for members of their order (Coleoptera) (Easton, 1979). Mention was made in Chapter 1 about another social beetle, a weevil (*Austroplatypus incompertus*). Both beetles developed their social traits independently.

Although burying beetle larvae are able to feed themselves, both parents generally engage in feeding them as well. Adults feed upon and digest flesh from dead animals and subsequently regurgitate liquid food to their larvae. This type of feeding is a form of progressive provisioning, which is typically found in one form or another in animal societies, and facilitates immature development. Adults may even produce secretions from head glands (another feature typical of many social animals) that have an antimicrobial quality, inhibiting the growth of bacteria and fungi on the prey's corpse.

At an early stage of development, burying beetle parents may cull their young, a form of infanticide that apparently functions to match the number of larvae to the size of the carcass upon which they feed so that there is enough food to support the brood's size. Infanticide in other animals, including humans, is discussed further in Chapter 15.

Without parental assistance, offspring may be underfed, hindering development and reducing their chances of surviving to adulthood. Because of their degree of social behavior (including brood care), the most successful beetle parents establish a good balance between the size of offspring and the amount of brood.

Adult burying beetle parents continue to protect their larvae throughout their immature life, which take several days to mature. Many types of competitors that are generally attracted to their food source make this task difficult, for example, calliphorid (blow) flies, ants, or burying beetles of either the same or another species. Last-stage larvae that survive move into the soil and pupate, subsequently transforming into adult beetles.

Aside from eusocial (truly social) and other species with social qualities (described in Chapter 1), parental care is uncommon among insects. However, exceptional cases of brood care are found in certain social members of the order Hymenoptera (ants, certain bees, and certain wasps) and Isoptera (termites). A good example is found in a honeybee colony in which there is a single queen and thousands of workers who are all daughters of the queen.

When together, colony members exhibit some forms of collaborative behavior (e.g., a division of labor in which the queen deposits eggs and workers engage in such activities as collecting food, feeding the young, constructing the nest, and defending the colony). Fertile females also exhibit behavior that is outstandingly competitive (agonistic behavior), predominantly during the early stages of nest construction.

SOCIAL ISSUES

Some issues that are focused upon by scientists interested in social behavior are:

- the kind and degree of sociality that is expressed;
- the group of social animals that are being observed;
- the size of the group;
- the reason the group was formed;
- the kin relationship between individuals in the group;
- the manner of interaction between members of the group;
- the resources individuals in the group are competing for;
- the degree of territorial expression (defending a particular area);
- the form of dominance establishment expressed between competing individuals.

DOMINANCE STRUGGLES, CHARACTERISTIC OF ALL SOCIAL ANIMALS

Besides eusocial species, there are many animals, invertebrates, and vertebrates alike that have been recognized with a varying number of social traits (Runciman, 2012; Sahlins, 1960; Smelser and Baltes, 2001). Chapter 4 presents examples of some of these species. The species we talk about most in this chapter are eusocial invertebrates, but strangely, lessons we learn from them can be applied to other animal populations, including those of humans.

In reviewing material already presented on dominance behavior in social species, members of a society at some time in their social life demonstrate agonistic behavior toward one another. No matter what the phylogenetic position of social animals, the outcome of establishing the hierarchy to which they will eventually belong is the same: (1) to bring status, power, and/or resources to the alpha individual; (2) to establish a position within the hierarchy for each of the competing individuals; and (3) to establish homeostasis or allostasis within the society to which the competitors belong. The behavior expressed by competing organisms in the process of establishing a dominance hierarchy determines where in the hierarchy they will be.

In many cases, what appears to be a linear hierarchy is established, and group homeostasis is hastened. However, we describe in this chapter that the hierarchy is not always linear because of the degrees of dominance expressed by various members of the society (especially the alpha) and the tolerance competitors demonstrate toward one another.

THE NATURE OF SOCIAL SPECIES

It was on a warm spring morning during April of the late 1960s, in the foothills of the Smokey Mountains, when I first witnessed what I felt was a revelation in social behavior. It was not with humans, other primates, or even another vertebrate, and yet, it had the earmarks of a concept that would relate to all social animals. It was with a group of social wasps.

My research at the time involved the behavior of a group of organisms referred to as paper wasps, a family of eusocial insects (Family Vespidae) (Carpenter, 1996) that live in a female-dominated society, such as ants and honeybees (East and Hofer, 2001, 2002; Schino, 2001). They got their name from the behavior of scraping wood from trees, buildings, and other sources of cellulose-rich materials, mixing it with oral secretions and constructing their nest with the masticated pulp.

There are several types of these paper-making creatures (hornets, yellow jackets, and what we may initially refer to as open-nesting paper wasps), all living a colonial life. Most of them aggressively impact the lives of humans while defending their nests, and most of us have learned directly or indirectly about their demonstrative defensive behaviors (Hermann et al., 2017).

All paper wasps live in colonies or eusocial groups composed of numerous closely related individuals, and there are various forms of interaction among the colony members (Jeanne, 1980). In some ways, different species share similar behavior, and in other ways, there are behavioral differences in the methods they use to express some of their social traits.

All paper wasp species exhibit a well-defined reproductive division of labor. Hornets and yellow jackets practice a form of dominance in which there is a single fertile female who is the one and only dominant individual, the queen (a monogynic system). Temperate zone forms of other paper wasps usually live a monogynic existence as well, but there are variations of dominance status among different species and between nests, and this alters the relationship between colony members (Willer, 1988; Willer and Hermann, 1989).[2]

In the United States and other temperate climatic zones of the world, a single queen generally is the nest founder. She typically deposits all of the colony's eggs. Workers that arise from these eggs, all of which are infertile females and daughters of the queen during the active stage of a functioning colony, care for the young, as well as continue to build, maintain, and defend the nest.

They do these chores during a period of spring, which follows nest founding (initiation of nest building). This period of time, when the nest undergoes its maximum growth, is referred to as the ergonomic period. Males (generally all sons of the queen) are produced toward the fall and generally do not work in the colony, but they provide spermatozoa during a fall mating ritual for the insemination of females that function as queens or potential queens during the following spring. Male paper wasps, like the males of other wasps, bees, and ants, are produced by an asexual process called haploidy in which they develop from unfertilized eggs.

Our indigenous hornet (bald-faced hornet, *Dolichovespula maculata*, which is more closely related to yellow jackets than to the true hornets) (Carpenter, 1996), constructs a large nest of paper and protects its brood cells by covering them in a paper bag–like envelope (*referred to as a calyptodomous nest*). Within the envelope, two or more horizontal tiers of numerous paper cells lie above and below one another, the opening of the numerous cells within each tier facing downward. It is within these cells that immature wasps (the larvae) receive the care they need to grow and become adults.

Their nest characteristically hangs from the branch of a tree, although it occasionally is found to be constructed in such places as the hollows of trees, under the eaves of houses, upon the sides of buildings, on windows, and in other protective places. There is a single (monogynic) queen that produces the season's offspring.

The exotic European hornet (*Vespa crabro*) is a much larger species, often using holes in trees and attics of buildings to construct its nest (Butts et al., 1991). Their social life also centers around a single queen and a multitiered (calyptodomous) nest.

Wasps referred to as yellow jackets, represented by several species of yellow and black paper wasps in North America that are closely related to hornets (both in the subfamily Vespinae), construct a similar bag nest but most often build it within the confines of a hole in the ground. Some are known to utilize other places for their nests. As with hornets, there is a single queen that produces the season's offspring.

One of the only variations in single-queen nests for these insects seems to be in some of the gigantic multi-queen nests of a yellow jacket species in Florida (*Vespula squamosa*) which are constructed over long periods of time (Butts et al., 1991; Pickett et al., 2001). Yet, the nests are so large (up to about 3 m or so) that individual queens are isolated by unused sections of the nest, and they each maintain a monogynic (single-queen) existence.

Both hornets and yellow jackets are highly defensive, a behavior expressed to protect their colonies. Constructing a bag around the tiers of their nest represents another measure of protection for the colony's queen and brood (Jeanne, 1980; Parrish, 1984). However, because of the bag, much of the behavior within the colony is out of view.

The paper wasps I primarily worked with belong to a group that is generally referred to by biologists as polistine wasps (subfamily Polistinae) (Carpenter, 1996; Espelie and Hermann, 1990; Rau, 1940; Strassmann, 1981, 1989; West, 1967). In temperate zones, they are not so elaborate in their nest building as hornets and yellow jackets. In other parts of the world, especially throughout the tropics, more elaborate (bag-like) nests are sometimes constructed by other polistine species.

Nests constructed by this group in the United States consist of a single tier of cells that look much like one of the tiers within the nest of a hornet or yellow jacket, but the nest lacks the bag-like cover (they are referred to as open or *gymnodomous* nests), and adults and brood cells are quite visible (Fig. 2.3). Nests are attached to a substrate by one or more stalks called pedicels or petioles (a condition referred to as *stellocyttarous*). They often nest under the eaves of buildings, but many nest in other locations, including on plants and in holes.

Colleagues, graduate students, and I studied the nests day and night throughout their active season (spring, summer, and fall) to determine the features of their social behavior. We marked them at night when they were all on the nest and least defensive and watched them during the day in an attempt to understand the function of each individual and how they interacted with one another within the colony.

Unlike bag-making (calyptodomous) paper wasps, species that have open nests demonstrate an incredible array of defensive displays, in addition to stinging. We recorded such displays and published what we found (Hermann, 1984a,b; Hermann and Dirks, 1975), but during the early spring of every year in a temperate location, workers, which are the nest defenders, are not yet available (the ergonomic, nest enlargement phase has not commenced). Other researchers have reported similar behaviors (Strassmann, 1981, 1989; West, 1967).

Queens are poorly defensive toward vertebrate nest intruders, and it was not only defensive displays that caught our eye. It was a time of the year (during the nest-founding phase in early spring) when potential queens, all fertile from mating the previous fall, left their hibernacula, returned to the previous year's nest and entered a series of agonistic confrontations with one another, which would eventually determine who was to obtain the most desirable nest

(A)   (B)

FIGURE 2.3   Two types of female-dominated social wasps. (A) *Polistes annularis*, is most typically a pleometrotic social wasp species. It demonstrates agonistic behavior in the early spring prior to constructing a nest. The dominant female chooses the spot for the nest and deposits all of the eggs during the nesting season. During the ergonomic period, when the nest is in full force, there would be at least one dominant queen (the alpha) with several fertile cofoundresses (subordinates to the queen) and sterile workers. Cofoundresses that are subordinate to the queen have stored spermatozoa from their fall mating and reproductive systems, which atrophy when they lose agonistic encounters with the dominant female. They remain that way unless the dominant queen is lost. Later in the season, sterile workers can deposit eggs that produce males. (B) A hornet nest. Hornet queens, like honeybees and most other social wasp species in the United States, are despotic and thus do not tolerate other fertile females on their nest.

site (usually nearest the previous season's nest) and become the dominant (alpha) egg layer (the queen) of a newly constructed nest (Hermann and Dirks, 1975).

These wasps are all eusocial. They demonstrate behavioral characteristics that deem them so (explained earlier and further elaborated upon later), and like other social creatures, they closely interact with one another and establish a dominance system in which individuals that belong to the social group can subsequently be ranked according to their influence upon one another and the society as a whole.

RELATIONSHIP TO OTHER SOCIAL SPECIES

The species I was particularly interested in was *Polistes annularis*, commonly referred to as the red paper wasp (Fig. 2.3). It is found across the eastern United States from New York to Florida and west to North Dakota and Texas. I was attracted to it because it often commenced nesting with more than one fertile female, a behavior referred to as pleometrosis. Most temperate species of polistine wasps commence nest building with a single queen, a condition referred to as haplometrosis. Pleometrotic species are especially interesting with reference to their early season agonistic behavioral expressions.

If we would have ignored the fact that they were wasps, it could have been any social animal, vertebrate, and invertebrate alike. Their act of establishing some sort of dominant or subordinate status toward one another is a form of behavior that is omnipresent and well established in all social forms of life, including humans.

When we determined the organization of dominance hierarchies in paper wasps, we simply numbered their position in the hierarchy, #1 being the alpha female. Through repeated trials, we found that the earliest conflicts between cofoundresses (all fertile females) resulted in the establishment of a dominance order that appeared at first to be a simple linear ranking (Fig. 1.3). Their ranking subsequently became more interesting as we obtained data on their intricate lives.

The first and most important resource acquired by the dominant female was the right to select her particular nest site. Other rights, such as being the principal or sole egg layer and thus an individual that could provide her genes to all offspring, naturally followed. Once established, the dominant position of the alpha female (queen) generally continued throughout the season.

Similar competitive behavior and the establishment of a dominance hierarchy can readily be seen in numerous other social animals as well. In *Barnyard Democracy in the Workplace*, Céleste M. Brotheridge and Linda Keup (2005) state that both farm animals and humans "have a tendency to form fairly stable social structures that are characterized by a dominance hierarchy" in which there is:

• an established pecking order, offering differential access to resources;
• hazing (scrutinization) of new members;

- penalties for nonconformance;
- a loss or gain of personal space.

Dominance is not always expressed as fighting. Often, it is cues (some of which are subtle) that are recognized by both the dominant and subordinate members of the group, and actual fighting is most often avoided. Dominance establishment for many social species may even come about without evident aggression.

While we humans, as with all living creatures, may not always consciously recognize the cues of dominance establishment, it is expressed by us in every walk of life (Cummins, 1996; Rowell, 1974; Sidanius and Pratto, 2001). It is part of our daily behavior and etched in our genetic makeup.

If strong aggression is expressed between gravid female wasps, it is expressed in the early stages of developing a dominance hierarchy. After initial encounters, recognition of dominance status is generally dependent upon more subtle behavioral cues. Dominant females lift their body and look down upon their subordinated adversary. They open their mandibles (the largest, most threatening of the mouthparts), lunge forward, and sometimes engage in battles to assert their dominance. Battles between closely ranked females may result in their falling from the nest, and it may even end in occasional death or near-death.

The degree of aggressiveness expressed during agonistic confrontations depends on how closely ranked the competing females are and how close they are to the alpha position. If they share a high position in the dominance hierarchy and neither is willing to take a subordinate position (i.e., they are intolerant of one another), the level of aggression will most likely be high. Extreme altercations usually result in death of one female, or one will switch to a nest where the competition is not as threatening. In the latter stages of establishing dominance, when all the battling has been completed, postural cues and a behavior called abdominal wagging, mostly by the queen, are generally enough to maintain her position as the alpha female.

BENEFITS AND DETRIMENTS OF DOMINANCE AND SUBORDINANCE

In terms of an individual's position in the social dominance hierarchy, we need not always worry where an individual fits. Sometimes, the position an individual occupies supplies some sort of benefits whether they occupy a high or low position (Drews, 1993; Ellis, 1995). As E.O. Wilson has pointed out in *Consilience*, studies have shown that membership in dominance orders pays off for both dominants and subordinants alike. "Membership in either class gives animals better protection against enemies or societal stress and better access to food, shelter, and mates than does a solitary existence." It is through such membership that team work is expressed, as long as it is a well-balanced hierarchy that is also built on tolerance by the alpha.

It is my endeavor in this and later chapters to approach the concept of dominance from various points of view and to comment on each of the ways it affects our lives, as well as the lives of other living members of our ecosystems. It is by no means a simple concept, and in humans, it affects much of what we do within our conscious moments.

DOMINANCE IN *POLISTES ANNULARIS* COLONIES

In colonies of *P. annularis*, all fertile females returning to a particular nest site and vying for dominance during the springtime founding period were related and thus members of a familial group, being sisters that were raised from a single nest the year before. Through hundreds of dissections of sperm-storing chambers (spermathecae) and ovaries, we had determined that all such females had been inseminated during the previous fall and initially appeared equally capable of becoming an egg-laying female (a queen) (Hermann and Dirks, 1975). Yet, over a period of a few days to a week or two, cofounding red wasps engaged in dominance struggles until one female generally became the alpha (most dominant) individual on the nest.

The Alpha Female (The Queen)

Under "normal" circumstances, it would subsequently be the alpha female that chose her nest site and dominated the new nest as it was being constructed by several subordinate fertile females (a concept very similar to group founding, most prevalent in tropical species, in which a group of females fly to a new area and commence nest construction together). In addition, it would also be this dominant female that oversaw events that involved running the nest and depositing all of the eggs. It would be her progeny that would become the colony's workers (all of which are sterile) who cared for the young (their younger sisters), continued with nest construction and defended the nest against intruders (Hermann et al., 2017; Hermann and Dirks, 1975).

It would also be her later progeny that toward the end of the nesting season developed into males and potential queens, who mated in the fall, stored sperm in the spermatheca, hibernated, and entered dominance struggles during the following year's nesting season. In essence, it would be her genes and her daughters' genes that most likely would be directly represented in the colony and future colonies. Whatever she does during her reign cements the passage of her behavioral features for generations to come.

Subordinate Gravid Females (The Cofoundresses)

More subordinate females, all of which were initially potential queens (they had mated and stored sperm in their spermathecae as well), became the first worker force for a colony with a well-defined queen. While it appears that the alpha female is the only winner in such struggles, her sisters share a similar genetic makeup and thus are also closely related to their sister's daughters (their nieces). Their energy in supplying the colony's needs, the queen, and their nieces is important in preserving the entire group's genetics.

Depending on the dominance status of individual females, there are ways for subordinate females to introduce their genes as well. If subordinated cofoundresses chose to remain on the nest during the nest-founding period, they helped the queen, but if they chose to depart, they often established their own nest on which they sometimes became the dominant queen (depending on dominance struggles). There were cases in which a very dominant fertile female traveled from one nest to another (referred to as nest switching), attempting to exert her status over other females, but their internest visitations were discontinued after eggs were deposited and the offspring had to be cared for. It appeared that these wandering fertile females failed to dominate the alphas on the nests that they visited and eventually found a nest upon which they were able to become the alpha.

The fertile worker period overlapped with the ergonomic period in which newly emerging infertile workers (their nieces) would take over their chores. The fertile females (sisters of the queen) not only were subordinated by the queen and functioned as regular infertile workers (behavioral subordination), their reproductive systems atrophied to a point at which it was impossible for them to deposit eggs (physiological subordination) as long as they were in the presence of the alpha female.

Mating by subordinate females the previous fall was not always a total loss in terms of directly spreading their genes, however. If the nest's queen (the alpha female) was somehow lost, the female next highest on the dominance scale (the beta female and sister to the previous queen) would step into the dominant position, her reproductive system would redevelop, and she would become the egg layer. As indicated earlier, beta females also sometimes became the queens (alphas) of other nests rather than occupying a secondary position on the nest they initially belonged to.

Ovary suppression and redevelopment in insects may be under the influence of insect growth hormones, for example, ecdysone and juvenile hormone (JH). Röseler et al. (1985) have shown that modulation of hormone levels in *Polistes dominulus* may be associated with the establishment of dominance hierarchies. When foundresses were injected with JH, a hormone responsible for regulating growth and development in many insect species, foundresses exhibited an increase in dominant behavior when compared to those who had not been injected. They point out that foundresses with larger corpora allata, structures in the brain in females that are responsible for the synthesis and secretion of JH, were naturally more dominant than those with smaller corpora allata.

These authors also found that injection of 20-hydroxyecdysone, a form of the hormone ecdysone known to enhance maturation and size of oocytes, showed an increased dominance when compared to the dominance level in females treated with JH, indicating that 20-hydroxyecdysone may be more important in establishing dominance.

Gesquiere et al. (2011) point out that hormones are also important in determining dominance in vertebrates. In the dominance hierarchy of naked mole rats, luteinizing hormone (from the pituitary gland) and testosterone in males (produced in the testes) and the entire ovarian cycle in females are suppressed in subordinates. In savanna baboons, high levels of testosterone were found in dominant males. Since high-ranking males are under high stress, they suggested that it is the beta males that are most fit, avoiding the stress found in the alpha while still maintaining some of the reproductive and nutritional benefits of moderate rank.

Also, there have been studies to show that hormone production may somehow be influenced by the degree of dominance a female displays. There also is sufficient evidence to suggest that the degree of dominance displayed by queens variously affects ovarian development of workers and subordinates in insects, such as bumble bees and halictine bees.

DEGREES OF DOMINANCE

Through years of research, we found that other events sometimes complicated what initially appeared to be a flawless picture of the establishment of a linear dominance hierarchy within colonies of gymnodomous paper wasps, particularly in the red wasp (*P. annularis*). There were degrees of dominance in alpha females, that is, the dominance hierarchy showed variation, just as you would expect in any group of social organisms that enter into competition for dominance because the individuals themselves show variation. Within this variation, there were females that appeared to be overly dominant (despotic), females who entered a well-defined linear hierarchy and others who appeared to be overly subordinate.

Females that remained on the nest with a dominant queen and represented the most subordinate levels always assumed worker status. Therefore, their personal genes could never enter the gene pool unless all females higher on the dominance scale were somehow lost, and they assumed the dominant role. Yet, they assisted their sisters (who they shared their genetics with) to indirectly pass on their genes.

Females that failed to establish a well-defined dominance scale (in which there was no clear-cut alpha female) and continued vying for dominance within their nest environment (the triangular arrangement mentioned by investigators who have studied numerous social species) lacked the leadership qualities required to dominate their peers (although they repeatedly entered low-level dominance struggles with one another) (Fig. 2.4). Lacking a true dominant, such females faced a problem of establishing colony harmony (Ehrlich and Ehrlich, 2008; Hayaki, 1983; Houpt et al., 1978; MacLachlan et al., 2010; Scheel et al., 1977).

Failing to establish themselves in the hierarchy, when one female deposited an egg, another female of equal or almost equal status often ingested the egg (a process referred to as oophagy, which, like dominance itself, is a selfish act) and deposited her own (Fig. 2.5). Oophagy repeatedly occurred on such nests with a poorly-defined dominance hierarchy. Thus, nests with abundant females at this dominance level (all pseudodominant and lacking a clearly-defined alpha female) had less organization, they wasted energy that could have been used in establishing a better-defined division of labor, and nest productivity suffered.

Other researchers have gotten similar results when working with social wasps. D.E. Willer reported on a similar system of dominance establishment for *Polistes exclamans*, a species that is more typically haplometrotic (despotic, founding their nest with single queens) (Willer, 1988). In that species, there was a clear difference in pleometrotic/haplometrotic behavior, depending on whether nests were constructed on buildings or plants. Females that were

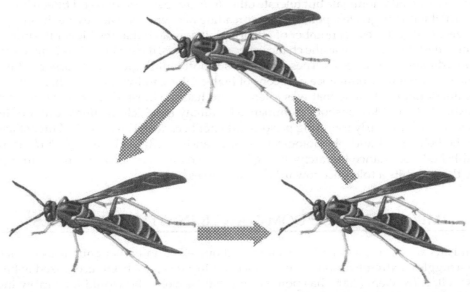

FIGURE 2.4 Unestablished dominance in a group of social wasps in which each individual makes an attempt to be dominant but is overpowered by other members of the group which are also attempting to be dominant. Such a dominance predicament is found on nests which lack an alpha individual. Such pseudodominants on the nest are all poor leaders, demonstrating oophagy of the eggs deposited by other females and depositing their own eggs, which will, in turn, be eaten by yet other females. Thus, a true dominant individual that makes the rules is not present to establish a homeostatic condition in the group. This condition, often referred to a triangular dominance association, has been found to be present in other social animals, including humans (Ehrlich and Ehrlich, 2008; Hayaki, 1983; Houpt et al., 1978; MacLachlan et al., 2010; Scheel et al., 1977).

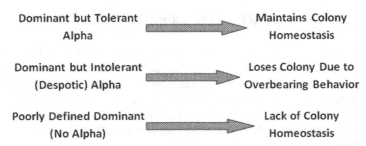

FIGURE 2.5 Consequences of different dominant states in colonies of *Polistes annularis*, a eusocial species of wasps that (in northeast Georgia) most often founds a colony with multiple foundresses. Tolerant queens keep their subordinate cofoundresses and develop a homeostatic colony that is productive in producing a large number of cells, many colony workers, and reproductives. This leads to colony cofoundresses during the next season and the spreading of the dominant female's genes. Intolerant queens lose their cofoundresses (and a homeostatic colony) and thus have a reduced chance of spreading their genes to subsequent generations. Poorly-defined dominants lack the ability to maintain colony homeostasis, often resulting in considerable oophagy and poorly-defined roles of subordinate cofoundresses. As pointed out in the text, teamwork among the subordinate individuals is behind the most successful colonies.

overly dominant (despotic) were intolerant of their more subordinate sisters. The end result was that they often engaged in fierce battles with their sisters, chasing their subordinates away or killing them, and the victor remained on the nest by herself, without a cofounding worker force. The scenario for alpha polistine females becoming queens has also been reported by M.J. West for *P. annularis* (West, 1967).

Intolerant and Tolerant Females

Females that chased all others from the nest exhibited what can be referred to as a strong despotic dominance (Fig. 2.5). Such solitary queens had to collect wood, build the nest, deposit the eggs, care for the young, and defend the nest without assistance.[2] Consequently, their nest was relatively small throughout the season, if it survived that long. Many such nests became defunct within days to a month from when the initial building commenced because of poor defense.

Some biologists argue that despotic females are most productive because their intolerant behavior leads to more colonies. I agree that they lead to more colonies, but I disagree that they are more productive. Yes, there were more nests initially, but (as mentioned earlier) most became defunct within a short while.

It was a female wasp that could dominate but tolerate other females (between two and fifteen but most ideally five to seven) that was rewarded with the greatest potential for spreading her genes and achieving the greatest individual nest productivity (the largest nest and greatest number of offspring). It appeared that the higher the number of subordinate females there were on a nest, the greater was the chance of oophagy. Thus, the nest was less homeostatic than one with a lower number of subordinate cofoundresses. These findings support Huntingford and Turner's statement that "group size and composition can offset the dominance decisions of high-ranking individuals and hierarchy type (1987)."

Nests with a dominant (alpha) female that was tolerant and had the most ideal number of subordinates to carry out colony chores exhibited a weakly despotic dominance hierarchy in which the linear order of females (a pecking order) was initially visible (in the early founding period) but later became hidden (in the latter stages of the founding period, because of the behavioral and physiological subordinance mentioned earlier). Thus, important characteristics of a well-established dominance hierarchy that generates colony harmony and maximum productivity are a well-defined leader that exhibits a tolerance toward cofoundresses.[3]

IMPORTANCE OF DOMINANCE IN OTHER ANIMALS

While these organisms are wasps, we can learn a lot about ourselves and other animal societies from the outcome of their dominance struggles. Extreme despotic dominance, for instance, has been recognized to be disadvantageous in other animals as well. F. De Waal (2005) has pointed out that "a wolf who would let narrow individual interests prevail would soon find himself alone chasing mice," a lesson just as valid in human society. He also points out that societies have generally worked out a system of tolerance and that "even the harshest societies, such as those of baboons and macaques, [make an attempt to] limit internal strife [through tolerance]."

An extraordinarily similar system to that in colonies of red wasps is exhibited in colonies of African naked mole rats (Burland et al., 2002) in which they are dominated by a despotic queen. They live in colonies with as many as 100 to 300 members, and only the queen (along with a male referred to as a king) reproduces. Her form of dominance suppresses fertility in both females and males, but she tolerates other females in the colony, and that system appears to work well for them. It is comparable to group founding in paper wasps.

Baker et al. (1998) point out the risks of going it alone. They state that dispersal is often associated with increased mortality, possibly because of subordination, which may decrease the benefits of leaving the group. "Given the opportunity to desert, subordinate red foxes often do not leave due both to the risk of death and to the low possibility that they would establish themselves as dominant members in a new group."

As in subordinate fertile female wasps, subordinate female mole rat reproductive structures are underdeveloped. Queen dominance over males may even affect breeding males as well, keeping testosterone levels down until she is ready to mate. When naked mole rat queens die, high-ranking females become reproductively active. I would imagine that there are different degrees of tolerance expressed among alpha females in these societies and different degrees in the success rate of colony life.

In studies of yellow-bellied marmots, G.E. Svendsen (1974) used particular terms for similar types of dominance, although they appear to have demonstrated behavior for each in a slightly different way. "Aggressive females," he said, "tend to live as solitary individuals or monogamous pairs, and have the highest fitness when they are monogamous. Social females tend to live in social groups and have their highest fitness in a larger group." Females subordinated by the alpha female tend to live by themselves in peripheral burrows.[4]

Suppression of reproduction has been reported to also be similar in marmoset monkeys (Barrett et al., 1993). It has been suspected that ovary suppression in these and other vertebrates are probably under the influence of certain hormones, but they could also be affected by stress put upon them by the behavioral actions of an alpha individual.[5] Thus, suppression may be the result of nervous or hormonal control or both.

Suppression of maturation has been reported in ravens that were subordinated by females who were in a dominant position (Heinrich, 2006). Thus, breeding privileges are bestowed on the dominant bird by acting upon the reproductive potential of subordinates.[6]

It is shown in later chapters that varying degrees of dominance are evident in all groups of social animals and that a range of tolerance and intolerance, as found in these paper wasps, as well as in the other species discussed here, is extremely important in human and other animal societies.

Selection of What Works Best

So how do we apply this information to societies of other animals? If we consider the degree of dominance and tolerance as actions important to survival and reproductivity, what works best (makes the species most successful or fittest) generally is the behavior selected if, indeed, such an adaptation is available to them through natural selection. Haplometrosis works well in societies of hornets, yellow jackets, and most polistine wasps in temperate zones, but pleometrosis and group founding appear to work best in certain temperate polistine colonies and colonies of most tropical social wasps where predation is heaviest. Thus, human and other animal societies may also respond to population pressure in a similar way.[7]

We all know people who are overly dominant toward other people. Such despotic dominants show up in all walks of life. A few become political, industrial, and corporate leaders, some of whom (dictators) are intolerant of anyone who does not follow their lead. Dictators of the various countries in the world throughout history have filled this position (Chapter 13), and it is unfortunate that many individuals at this level within a business environment become the managers of workers who must attempt to follow them or suffer the consequences (see Chapter 11).

To individuals who strive for a high position in a dominance hierarchy, the alpha position is the most desired. It is in the process of striving for the alpha position that some individuals willingly deceive other individuals who are striving for the same position. Dishonesty, cheating, and lying are examples.

As in the wasps discussed, there are people who are extremely subordinate and must be the followers of others who occupy a higher position in the dominance hierarchy. The segment of the human population that represents the workforce in the world may be found at all levels in the dominance hierarchy, but most of them are still followers nonetheless. Some are wannabe dominants, others have some degree of dominant (leadership) potential, and still others are distinctly subordinate.

Individuals who cannot occupy the alpha position in their jobs may choose to be alphas in other situations, for example, woodworking, athletics, art, culinary projects, and so on. Others may seek none of these and enjoy status at a lower position in the hierarchy.

Within the jobs that they hold, individuals who demonstrate some degree of dominance behavior may take a back seat to one or more individuals who dominate them, and they may seek dominant status within their job structure in other ways, such as various managerial positions.[8]

What works best in human society is difficult to say, but it may depend on the segment of society and its population size. In a moral approach, human leaders who dominate and yet are tolerant may be the most desired form of leadership by their companies and followers. Many dominants may disagree with this assessment, but tolerant leaders may be empathetic to the needs of their workers or followers, or they may at least understand that it is through toleration that they can get the most out of them. Either way, dominants who are tolerant, understanding and cordial provide an atmosphere in the workplace in which a subordinate may find a feeling of belonging.

On the other hand, what works best in a subhuman organism may not work well for humans. While animals show degrees of toleration in dealing with their subordinates, their degree of variation may not be as great as that within a human population. Since humans are very cognizant animals, they may be more skilled at cheating and benefiting from leniency (tolerance). We see this in the teaching profession. The more tolerant a professor is, the greater is the chance for a student to take advantage of the professor's lenient attitude. Thus, while toleration may be a good thing, the most favored degree of tolerance may vary under different circumstances.

Despotic Humans

In human society, we may associate despotic dominance with a lack of caring or an asocial personality (avoiding or averse to the societal importance of others; not sociable) (Brosnan, 2006; Matsumura, 1999; Morris, 2009). Of course, there are various levels of asociality as well, in which there may be some degree of or a total lack of caring. The most intolerant individual may border on an antisocial personality, leading to what sociologists and psychologists refer to as sociopathic and psychopathic personalities (discussed in Chapter 9). Since there is tremendous variation in human behavior, the lines demarcating one degree of dominance and tolerance with another are not always clear, and thus, the gradation between asociality and antisociality at one point may be difficult to recognize.[9]

Do their bosses ever make an attempt to understand the people who work for them? Some, of course, do, but people in high positions usually have a dominant personality which may be expressed as asocial or antisocial (Gudykunst and Matsumoto, 1996) (Chapter 9). This is what we have learned to expect of such a person. Of course, the potential for acquiring an alpha position between the recognized alpha and a subdominant who wants to be alpha most likely will result in a clash of personalities, just as it is in social wasps.

Perhaps we are not being fair in stereotyping dominants. As we may suspect, it is not always the dominant at fault when the personalities of dominants and subordinates collide, but problems often arise because of dominance confrontations, nevertheless. Detrimental dominance situations in the workforce may be due either to the presence of an intolerant, and thus inept, alpha supervisor, or to poor following skills by individuals who are subordinates or want-to-be alpha. Varying dominance levels can help tolerant individuals work together as a team, or they can shatter a teamwork effect through being intolerant. The secret to homeostasis is having the right combination of individuals and personality types working together (Berger and Riojas-Cortez, 2004; Reagans et al., 2005).

There are other variables that should be considered. For example, certain positions in human society may require a more intolerant alpha while others may require an alpha with more tolerance (Fig. 2.5). In red wasp societies, once a tolerant alpha individual has determined her position in the colony, she directs subordinates with simple cues. Teamwork among the subordinate individuals is behind the most successful colonies.[10]

An examination of surgical and ICU units in a hospital that provides excellent care is an example of a social unit that works best with tolerance and teamwork between people in the medical profession and their patients (O'Leary et al., 2012; Sexton et al., 2006). A close look at the various employees in such a business shows that there is a tremendous system of teamwork, with each individual member of the team knowing precisely what they have to do. Without it, they would not be the same caregivers. If we move to a less intensive section or floor in the hospital, caregivers are not always tolerant of one another or their patients (Barker et al., 2002; Devlin and Arneill, 2003).

It appears that the antithesis of a system with tolerance is a workshop in which workers work under intolerable conditions and receive abuse from an intolerant despot (Fig. 2.5), the main emphasis being to get products out. However, in such a condition, we are left with the question: Is it a moral environment?

Dominance and Antisocial Behavior

Variations in aggression expressed by members of our society are also quite commonly related to dominance establishment. Violent crimes, including murder, serial killing, and rape, as well as spousal and child abuse, are examples (Chapters 9, 14, and 15). Such aberrant behavior often combines the dominant or wannabe dominant personality with characteristics of antisocial behavior (Figs. 8.9 and 9.1). In the fast-moving societies that we belong to, there is a tendency for many individuals to lose their individualism and express behavioral tendencies toward psychopathy (antisocial behavior, selfishness, cunningness, deceit, and intolerance) (Blair et al., 2005).[11]

Let us be bold and go out further on the limb we have crawled upon and say that in addition to nourishing the development of antisocial personalities, our dominant status and stress in the world has also led to an increase of ecosystem ignorance (Ferson and Ginzburg, 1996). Selfish attitudes of individuals in dominant positions (such as certain governmental leaders, Chapter 13) and the stresses associated with maintaining a position in our various dominance hierarchies often take their attention away from thoughts of cohabitation and skew them toward domination and personal gain, power, and profit.

Therefore, it goes through life, the constant battle in dealing with a tremendous variety of dominance types and levels. It begins in infancy, plays a role in our formative and adolescent years, and subsequently enters situations involving most of what we do as adults. Eventually, we all assume some position in the various hierarchies that are available to us in life.

As we have stated, if we cannot occupy a dominant position in government, religion, or our workplace, we may seek fulfillment somewhere else. The topics that are available to us so that we may occupy a dominant position in something are endless. With such a wide selection of dominance games to choose from, almost everyone who seeks dominant status (inadvertently or through cognition, low or high socioeconomic level) can experience it in one way or another.

However, just as in colonies of social wasps, everyone does not seek a dominant position. While some strive to be dominant, others may choose to assume a more subordinate role, a position they are most comfortable with or fits their particular personality. At times, some members of our population may tend to think that such a person stops short of fulfilling their inherent abilities. Yet, the position a person occupies in a dominance hierarchy is usually determined by choice, and human society other than in despotic rule is most ideally built upon tolerance and teamwork (Sidanius and Pratto, 2001). Everybody has their place.

This appears to be a point passed over by worldwide eugenicists who wish to build a supreme species, breeding the top echelons of human society and eliminating subordinates (Engdahl, 2007; Black, 2012). Many of us have a position in life, and we are a part of one team or another. Because of conflicts in modern society, based on the determination many of us have to occupy a high position in a dominance hierarchy, what many of us may lack most is tolerance.

Is it proper that we should all strive for dominance, sometimes at the loss of other qualities that may be beneficial to us, our loved ones, or our society, or do we have a choice in the matter? Could it be that the degree of dominance and tolerance a person, group, or country has is more important to a society and the world than becoming involved in activities that are more altruistic? I would think that the answer to this depends on what we feel our ultimate goals are as a society.

As mentioned earlier, some believe that the purpose of a human should be to improve the species, breeding the dominants and eliminating the subordinates (Engdahl, 2007). Such an approach is a process of artificial selection, kin to bringing out certain features in dogs that we feel are desired. This, of course, is a humanist view and not an animalistic one, based on the cognizant abilities of dominant humans. An animalistic approach would not consider the betterment of the species. It would be more involved with simpler levels of survival and reproduction. Whether it becomes better is a process of natural selection.

Idealistically, the process of establishing a dominance hierarchy in most groups terminates in some level of tolerance and cooperation among society members, leading to a homeostatic condition. However, we show that aberrant behavior (an expression of personality extreme) may interfere with the hierarchy's function of establishing order and creating havoc in the society (Chapters 9, 14, and 15).[12]

While acting altruistically or showing empathy or compassion toward another individual or group may at first appear to be the idealistic approach to life, we as humans (as with all other social animals) are faced with one eye-opening fact. There are others in our ranks who are dominants and ultradominants (despots) who do not care about these things, and they thus can sometimes be a major threat to all members of society who are less than alpha.

This concept may also be carried over to group dominance. Certain groups or countries in the world may appear altruistic and tolerant while others choose to dominate other groups or countries by any means necessary, and they proceed to do this with a lack of moral stability.

OTHER SOCIAL TRAITS

There are numerous other traits that are associated with sociality and influenced by dominance and aggression, which should be brought out here. Some have more animalistic roots while others are clearly expressed primarily or exclusively by humans.

SELFISHNESS (PSYCHOLOGICAL EGOISM)

Dominant behavior in social species, including humans, is selfish (Baier, 1990; Broad, 1971). Being selfish benefits the individual or group demonstrating it but at an expense to others, especially those that are considered competitors. In many instances, this is considered natural and desirable in a wild population because it is a means for individuals to increase their chance of surviving the challenges of natural selection. This is true for humans as well. Certain behavior in humans, for instance, is largely infiltrated with selfish and deceptive expressions.[13]

Even in the absence of interactive behavior, the activities of one organism may adversely affect other individuals. In feral populations, selfishness in one form or another is often an absolute necessity for survival because it benefits the individual that demonstrates it by contributing to personal fitness. Such behavior that maximizes an individual's reproductive success will be favored by natural selection.

In dominance interactions, both individuals demonstrate some degree of aggression toward one another. The winner becomes alpha at the detriment of the other, and whatever the dominant individual does to remain dominant is a selfish act, designed to control its surroundings, have first choice for available resources, and spread its genes. Likewise, territoriality is selfish. So is any attempt to survive and reproduce at the expense of others. Based on these features, we would expect selfish behavior to be more characteristically expressed in the most dominant organisms of each society.

DECEPTION

Deception, another act of selfishness, which generally benefits the animal demonstrating it, is common at many levels in the animal kingdom (Trivers, 1971). Deception is defined as the act of deceiving or misleading another individual. However, there is self-deception as well.

Types of Deception

Deception between different individuals is a specialized aspect of communication, typically involving a sender of a deceptive signal and a receiver who interprets the message. Deception, like any other behavior, may have a morphological, phyletic, and behavioral base. R.W. Mitchell and N.S. Thompson (1986) organized various types of deception into four behavioral levels:

- *Sender-Dominant Deception*: In level one, involving mimicry, the actions of the receiver have no influence on the sender. Examples are found in the cryptic patterns and colors of butterflies, certain other insects, and other animals that blend in with their surroundings (crypsis). A similar effect is seen in the camouflaged attire worn by military personnel in war zones.
- *Sender-Deceiver Deception*: In level two, the receiver's actions influence the behavior of the sender in ways other than form. Broken wing behavior, demonstrated by certain ground-nesting birds, is an example, where an intrusion by one animal elicits the broken wing demonstration in another and in doing so, the demonstrator typically leads the intruder away from its nest. Pretending to be dead, a condition generally referred to as akinesis, is another example. Akinesis is a common response to intrusion by many social and nonsocial animals, which may deceive a potential predator to the point of saving the demonstrator's life.
- *Trial-and-Error Deception*: In level three, trial-and-error behavior is expressed, for example, with dogs that demonstrate behavior to get the attention of their owners, as well as similar behavior in birds, human babies who attempt to draw the attention of their parents, or reconciliation behavior in primates in general.
- *Modified Sender-Deceiver Deception*: In level four, the sender's behavior may be modified, depending on the actions of the receiver. Feigning interest in something that has nothing to do with the sender's interests or hiding while engaging in an illicit activity, all expressed by humans and captive chimpanzees, are examples.

Signals that represent honest communication between animals result when both the sender and receiver have the same interest in the result. On the other hand, deceit is beneficial to the animal that desires to be dominant, and thus it may exploit another animal in order to improve its fitness.

Since natural selection strongly favors any feature that increases an animal's survival and reproductivity, deceitful or dishonest acts that enhance these features can also be selected to the advantage of a species under the proper conditions. Two features that are commonly involved in deceit are mate choice (in which individuals seeking mates sometimes produce deceitful signals) and parental care (in which young may exaggerate their needs to receive a higher degree of parental interaction).

In the following paragraphs, I have briefly described some other forms of deception found in the animal kingdom. There are many more examples that are elaborated on by Trivers in *The Folly of Fools: The Logic of Deceit and Self-Deception in Human Life*. My purpose is not to provide all the known forms of deception but to present a few examples to emphasize that they are found in a diverse array of animal groups, including humans. When reading about these examples, the reader should mentally compare them to what they know about human behavior. The reader is also directed to the book, *Deception: Perspectives on Human and Nonhuman Deceit*, by R.W. Mitchell and N.S. Thompson (1986).

Promiscuous Males and Females

Males in many species (including humans) have demonstrated that they are the most promiscuous sex and thus are the champions of sexual deceit (Coltman et al., 1999; McPeek and Gavrilets, 2006). Behaviorists also have reported females in several feral species, which were in apparent monogamous mating systems (one male and one female) but sought copulations by other males (they were promiscuous) (Keil and Sachser, 1998; Muller et al., 2007). To accomplish this, they sought outside males at times and locations that made detection by their usual mates unlikely (behavioral deception). In human society, males have been considered the all-time promiscuous sex, but new liberties have been acquired by females of most societies, and they currently are exhibiting more promiscuous behavior as well.

Behavioral Deception

In addition to the types of behavioral deception mentioned earlier, female fireflies are known to produce various species-specific lighting patterns to attract mates (Lloyd, 1986; El-Hani et al., 2010). Carnivorous fireflies are known to mimic the bioluminescent mate-attraction signals of other species and use it to attract and prey upon males. Humans lie and cheat, for instance, in their dating games, inventing stories about being someone they are not in their pickup lines.

Other Forms of Deception

Various arthropods, which are guests in ant colonies (generally referred to as myrmecophiles), appear anatomically similar to the ants with which they share the nest or produce chemicals that, in effect, tell the ants that they are one of them (anatomical and chemical deception) (Kronauer and Pierce, 2011). In a similar way, bolas spiders utilize chemicals similar to moth sex attractants in order to get prey (Haynes et al., 2002). In amphibians, small male green frogs change the frequency of their croaks to resemble signals that are produced by larger frogs (auditory deception) (Wollerman and Wiley, 2002). In humans, males and female sport artificial sex pheromones (perfumes, colognes, and so on) to attract a mate.

Deception in Birds

As indicated earlier, deceptive behavior is commonly found in birds (Munn, 1986). Birds may establish a territory and deprive others of its resources or a place to breed. In the example brought out in a previous paragraph, birds that build their nest on the ground, such as stilts, killdeer, and nighthawks, often display a broken-wing behavior to attract an intruder and draw its attention away from the nest (Deane, 1944).

Certain deceptions are constantly expressed. For instance, birds in general do not fly directly to their nest when carrying nest materials or food for their young. They spend time following an indirect path, scanning the environment for potential intruders in the process.

Female marsh harriers court males in order to obtain access to food which they subsequently take to feed chicks that are fathered by another male. Cuckoo birds practice an extreme form of deceit when they deposit their eggs in the nests of other species, resulting in the rearing of their young by foster parents. Pied flycatcher males deceive females into thinking they are unmated and available.

While Trivers' book focuses on self-deception in humans, he offers a number of examples of deception in nonhumans (Trivers, 2011): frequency-deception selection in butterflies, epic coevolution struggles, intelligence and deception, female mimics, false-claim calls, camouflage, death and near-death acts, randomness as a strategy, deception that induces anger, the consciousness of deception, deception in the game of evolution, and the depth of deception.

He says that, "It is often easier to see patterns of importance [in deception] if we cast our net of evidence widely—in this case to include all species, not just our own." "Deception," he says, "fares well when rare and poorly when frequent, and detection of deception fares well when deception is frequent but not when it is rare."

Deception in Butterflies

Deception in butterflies can be quite extraordinary. Batesian mimicry, in which the model is unpalatable and the mimic is palatable, works best if mimics are less abundant than their models (Emlen, 1968). Trivers points out that, "the more frequent the deceivers are, the more they begin to diversify in order to avoid detection." He presents an example of extreme diversification in which females of certain mimetic species deposit five kinds of eggs, each of which grows up to resemble one of five different unpalatable species.

Some other examples of deception in nature discussed by Trivers are:

- Parasitic bird species that lay eggs in nests of multiple species, with individual species specialized to lay eggs that match egg coloration produced by the host.
- Similar mouth coloration in the young of parasitic and host bird species.
- Butterfly larvae that enter ant nests and use sounds and pheromones to be accepted by their hosts.
- Calls used by birds to gain entrance into host nests where they gobble up the young.
- Firefly females that use courtship flashes of another species to prey on unsuspecting males.
- Orchids that look and smell like female wasps to entice males for copulation, which facilitates pollination.
- Birds that use warning calls to make other birds drop their prey so they can eat it.
- Plants that develop structures that resemble fake butterfly eggs to keep butterflies from depositing eggs on them.
- Construction of fake caches by squirrels.

Much deception in nonhuman primates involves withholding or providing incorrect information. Falsification is found in such behaviors as chimpanzees hiding a weapon behind their back as they approach another, and monkeys holding water in their mouth and waiting to spit it at an unwary human. As in human society, it has been demonstrated that chimpanzees show empathy for each other in a wide variety of contexts, and they possess an ability to engage in a variety of forms of deception, including a form of behavior that has been referred to as asocial politics. Various explanations have been suggested for the evolution of more complex moral behavior in humans, for example, an increasing need to avoid disputes and injuries when moving to open savanna, developing stone weapons and its dependence upon increasing group and brain size.

HUMAN DECEPTION

Deception is a common mammalian game, and with their intellectual capabilities, humans are likely the all-time masters of deception. Based on comments by R.W. Mitchell and N.S. Thompson (1986), "human deception develops out of previous expectations which are built on communication and knowledge. The deceptive person designs a pretension tailored to the beliefs of the victim, thereby succeeding in extending a lie or changing existing convictions."

According to these authors, human communication is based on conventions that are regularities that are conformed to by a group. As they state it, "we presuppose that a speaker is telling the truth . . . and is acting in accordance with norms."

We do not expect conventions to be insincere and false, and our desire to believe people are telling the truth helps a deceiver accomplish his or her goal. There are many examples.

Fishermen engage in short-term deceptions by claiming they are not doing well, whereas in reality they may be experiencing excellent results that make them return the following day. Later, when telling stories about their catch, their description of the size of the fish they caught or the one that got away may be greatly exaggerated.

Hollywood filmmakers create stories to deceive moviegoers about many subjects that appear believable but are untrue. Writers do the same. Politicians often deceive the public (their voters) in speeches they make, even before they are voted into office, and they often continue their deceptions while in office. Some bankers and CEOs of investment corporations often deceive their clients while attempting to increase their fitness in the working world. As we see in Chapters 9 and 15, psychopathic killers often appear on the surface to be regular, nonthreatening citizens.

Wartime and other forms of deception may often employ the use of stories. "The first fictional novels were deceptions." Daniel Defoe, regarded by many as the first true novelist, passed off many of his novels as historic fact. He also fabricated facts in his journalistic ventures (West, 1998).

"To appear truthful," Mitchell and Thompson say (1986), "a deceiver should simulate innocence." An example of using cues of innocence to avoid detection of deception is seen in Clifford Irving's false claims to have interviewed Howard Hughes and received authorization to publish the interviews as Hughes's autobiography. In another case, homosexual lawmakers validated their image as heterosexuals by supporting antigay legislation.

These authors conclude that "because signs of deception and truth-telling can themselves be hidden or used deceptively, the victim has no absolute basis for inferring deception or honesty, and deceivers and victims can believe stories which have no evidential basis." As a result, deception, as well as honesty, may go unnoticed.

Humans exhibit a form of self-deception in which they have beliefs contrary to the evidence at hand. Wishful thinking and bluffs in sporting and war games are such forms of deception.

There are many levels of deceptive behavior that result in differing costs and benefits for the sender and receiver. While one important aspect of successful deception is making it plausible, especially assisting it to fall in line with the receiver's desires, deception does not always work. There are other features that may be important in order to carry out a deception. Although the receiver may watch the sender's eyes, lips, head, and body, skilled deceivers have control over these features. They are often in less control of their lower legs and feet because they are frequently ignored. Foot movement and body orientation, as well as minor pupil dilation, are indicators of deceit.

Additional comments on deception in humans are covered in Chapter 8, *Human Nature*; Chapter 9, *Alternate Human Behavior*, where aberrant behaviors lead to antisocial tendencies; Chapter 11, *Dominance and Aggression in the Workplace*; Chapter 12, *Dominance in Religion*; Chapter 13, *Dominance in Politics*; Chapter 14, *Human Aggression— Killing and Abuse*, and Chapter 15, *Killing Humans*.

COOPERATION AND AGONISTIC BEHAVIOR

These apparently opposing forms of behavior have already been discussed as important features of social behavior. All social species have both cooperation and agonistic behavior built into their daily lives, one solidifying the unit to work as a team and the other to determine who the leader is.[14]

ALTRUISM

The phenomenon of altruism, in which organisms behave in ways that seem to reduce their individual fitness but increase the fitness of another organism, is sometimes difficult to understand, and it and the concept of fitness have recently come under scrutiny from theorists. Altruism is a phenomenon in which an individual performs what appears to be a selfless act for another individual or its colony at great potential threat to itself (Nagel, 1978). It is based on what has come to be referred to as "Hamilton's rule," which states that the benefit of an altruistic act toward a relative must be greater than the cost suffered by the altruist in order for selection to favor the altruist.

Examples are given by N.A. Campbell and J.B. Reece in *Biology* (2002) for belding's ground squirrels, honeybees, and mole rats. Belding's ground squirrels in mountainous regions of the western United States are vulnerable to predators, such as hawks and coyotes. When one of these predators approaches, one of the squirrels gives off a high-pitched alarm call that alerts other members of its species who retreat to their burrows. By alarming, it puts its own life at risk.

Honeybees (*Apis mellifera*) have a worker force of infertile females that give up their life during the process of stinging while defending their colony (Ratnieks and Helantera, 2009). Flying from the colony to defend it from an intruder may possibly be construed as altruistic. However, it is difficult to see the loss of the sting and subsequent death as truly altruistic components of colony defense since honeybee workers do not naturally know they are going to die when stinging. Sting autotomy (self-amputation of the sting) is not a learned act. It occurs because of enlarged barbs on the sting lancets, and each bee dies in the process without passing on the message to other members of its colony (Hermann, 1971).

Any stinging species of insects puts its life on the line when defending its colony. Although they may not lose their sting in colony defense (like honeybees, certain wasp species, and certain ant species) (Hermann, 1971), an encounter with an intruder represents a personal threat. Thus, the act of leaving the colony to defend may be seen as an altruistic act in itself, just as in demonstrations when a dog defends its master.

Colonies of eusocial mole rats each have only one reproducing female, referred to as the queen. Nonreproductive females may sacrifice their own lives in trying to protect the queen or males from invading predators (Jarvis, 1981). While it may, at first, appear difficult to understand that protecting one's offspring or the offspring of another individual at one's expense maximizes one's genetic representation in the population, this, according to many researchers today, is precisely what may happen. Based on the work of William Hamilton, altruistic behavior is best expressed in individuals that are closely related to the recipient of an altruistic act (Hamilton, 1963). Hamilton stressed that since the altruist closely shares a genetic makeup with the recipient, it is, in effect, defending its own genes in the process.

According to this theory, selection of such a trait can result in an animal increasing its genetic representation in the next generation by helping close relatives. This phenomenon in which the total effect an individual has on proliferating its genes through its own offspring and enabling close relatives to increase the production of their offspring is called inclusive fitness.

The rule of kin selection states that the closer the relationship between individuals, the more natural selection would favor altruistic behavior. Thus, kin selection weakens with hereditary distance.

As pointed out earlier, the concept of inclusive fitness has recently received criticism. In *The Social Conquest of Earth* (2012), E.O. Wilson points out that "the foundations of the general theory of inclusive fitness, based on the assumptions of kin selection have crumbled, while evidence for it has grown equivocal at best." As he states, "the beautiful theory never worked well anyway, and now it has collapsed."

A new theory points to a process that is "perceived as neither kin selection nor group selection, but individual-level selection, from queen (in the case of ants and other hymenopteran insects) to queen, with the worker caste being an extension of the queen phenotype." He concludes, "In this approach, it is possible to reduce the entity of the selective process to its effect on the genome of each colony member and its direct descendants," the result being "achieved without reference to the degree of relatedness of each colony member to members, other than between parent and offspring."

Wilson further suggests that the genetic code prescribing social behavior in contemporary humans is a chimera. "One part prescribes traits that favor [the] success of individuals within the group. The other part prescribes traits that favor group success in competition with other groups."

Thus, conflicting ideas about the importance of fitness have arisen, indicating that much attention will be focused on them in the coming years to resolve the apparent argument. Could it be that both approaches have a degree of validity?

RECIPROCITY

Initially, it appears more difficult to understand why individuals that are not directly related to others would exhibit altruistic behavior. However, social organisms may perform certain behaviors to help another individual while expecting something in return at some later time. Such an act has been referred to as reciprocal altruism in the past but more recently as reciprocity. Reciprocity, for instance, occurs between organisms that are not closely related. One animal helping an unrelated one in a fight or offering food to another who is not kin may be adaptive if the individual that is aided returns the favor in the future. This phenomenon was originally offered by Robert Trivers (1971), but the concept is found in such children's stories as the lion and mouse mutualism in Aesop's Fables.

The phenomenon of reciprocity in nature typically functions to ensure a reliable supply of essential resources, especially for animals living in a habitat where food quantity or quality fluctuates unpredictably. As an example, some vampire bats occasionally fail to feed on prey while others manage to consume a surplus of blood. Bats that eat well may regurgitate part of their blood meal to save a conspecific from starvation. Since these animals live in close-knit groups over many years, an individual can count on other group members to reciprocate the favor on nights when it goes hungry.

In natural populations of animals other than humans, it is difficult to determine whether an act is truly altruistic, a function of reciprocity, or otherwise. In humans, it may be less difficult since we understand the neurological processes of humans best. The concept of altruism in humans, for instance, is supported in situations such as the Sandy Hook Elementary School Shooting, in which teachers hid students in closets and bathrooms, and even threw themselves in the line of fire. Some paid with their lives. These are altruistic acts that are not associated with relatedness or reciprocity. They were on-the-spur-of-the-moment decisions to help students who had been entrusted to the care of teachers, showing that behavior displayed by individuals do not always have a selfish origin or benefit the altruist. The same could be said for soldiers who throw themselves in harm's way to protect other members of their group.

TERRITORIALITY

Territoriality is a form of behavior in which one or more individuals actively defend a home range against other members of their own species. Others have listed the causes of territoriality as an expression of site attachment, aggression, and sexual behavior (Alcock, 2001; Ardrey, 1966; Beebe et al., 2008; Malmberg, 1980). Territories are generally used for mating, rearing young, and feeding. Familiarity with a territory may assist individuals in avoiding threatening situations, such as the presence of predators and trespassers. Territories are established through agonistic behavior of different conspecific groups and offer benefits that increase personal or group fitness.

Expressions of territoriality may be produced as aggressive behavior in one individual or a group of individuals toward other individuals, usually while inside their territory but not so outside. Such agonistic interactions generally show an inverse relationship between behavioral intensity and distance from the territorial focal point.

MORALITY

Morality is a suite of interrelated thoughts and behaviors that cultivate and regulate complex interactions within social groups. It generally includes such features as empathy, reciprocity, altruism, cooperation, and a sense of fairness. In *The Righteous Mind* (2012), Jonathan Hardt states that there is "more to morality than harm and fairness." Humans, he says, compete within every group and are "descendents of primates who excelled at that competition," presenting what he refers to as "the ugly side of our nature." On the other hand, he says that "human nature was also shaped as groups competed with other groups." While we are sometimes selfish hypocrites, we "also have the ability, under special circumstances, to shut down our petty selves and become like cells in a larger body, or like bees in a hive, working [cooperatively] for the good of the group." In *Social Worker*, John Parrington (2015) states that "working together is part of what makes us human." A more fitting statement would be working together is part of what makes certain species social.

A moral code is a system of morality that is established with reference to a particular philosophy, religion, or culture. Descriptively, "morality" refers to personal or cultural values, codes of conduct, or social mores (norms that are widely observed and have great moral significance), referring to actions that are considered right or wrong. Ideally, human behavior in its most homeostatic social form is moral. However, moral codes are often abused in all societies and by many types of individuals.

Morality appears to function at both the individual and group levels to encourage cooperation in order to provide possible survival and/or reproductive benefits. Thus, human morality, though sophisticated and complex relative to the moral character of other animals, is essentially a natural phenomenon that evolved to restrict an individual's ability to undermine a group's cohesion.

IMMORALITY AND AMORALITY

Immorality is the active opposite of morality, while amorality is variously defined as an unawareness of, indifference toward, or disbelief in any set of moral standards or principles. We find that both immorality and amorality are expressed in antisocial personality disorders, a subject covered in Chapter 9.

RELIGIOUS MORALS

There are many forms of religious morals in humans. Modern monotheistic religions, such as Christianity, Islam, Judaism, and certain others define right and wrong by the laws and rules set forth by their respective gods and prophets, as interpreted by religious leaders (dominants) within each respective faith (Chapter 12). Polytheistic religious traditions tend to be less absolute. For example, within Buddhism, the intention of the individual and the circumstances should be accounted for to determine if an action is right or wrong (Powers, 2007).

In certain religions, for example, Hinduism, right and wrong are decided according to the levels of social rank, kinship, and stages of life (Flood, 1996). Also, there is no absolute prohibition on killing in this religion. It recognizes that it may be necessary in certain circumstances.

It has been found that religion is not always positively associated with morality.[15] Major crimes have been found, in many instances, to be compatible with a superstitious piety and devotion. Thus, it is justly regarded as unsafe to draw any inference in favor of a person's morals from the fervor or strictness of his or her religious exercises.

It is difficult to understand that if morality is part of religious belief systems, a religion would condone, or to use a more forceful word, recommend, killing someone simply for believing in another form of religion. It remains genocide, whatever its form.

DIVERGENCE FROM MORAL BEHAVIORS

Morals associated with religions sometimes diverge from commonly held contemporary moral positions, for example, those on murder, rape, other mass atrocities, and slavery. As examples, followers of Hinduism defend its treatment of the caste system, and followers of Islam defend its harsh penal code or attitude toward women and what they feel are infidels (Fernea, 1985). In Christianity, the Bible may be interpreted as giving its followers a carte blanche for harsh attitudes toward children, ill treatment of the mentally handicapped, and abuse toward animals and the environment, the divorced, unbelievers, people with deviant sexual behavior, and elderly women.

In Exodus 22:18, for instance, statements such as "Thou shalt not suffer a witch to live" have led to the burning alive of numerous women in Europe and America. Statements in the Old Testament indicate that God apparently condones a slave-owning society, considers birth control a crime punishable by death, and condones child abuse. Researchers have noticed that there are morally suspect themes in the Bible's New Testament as well.

In the first place, with the dying of the Wicked Witch of the West and Voldamort, we no longer have to worry about bad witches or wizzards; morality rests on our shoulders. Secondly, slavery and abuse to children, women, and any other human should be considered immoral in any society. Thirdly, not considering birth control as a viable option for slowing the birth rate in a world that is facing innumerable threats for survival is unquestionably an immoral act of immense proportions.[16]

The overall relationship between religious faith and crime is unclear. Some researchers argue for a positive correlation between the degree of public religiosity in a society and certain measures of societal dysfunction. Others have concluded that a complex relationship exists between religiosity and homicide, with some dimensions of religiosity encouraging homicide and other dimensions discouraging it (Cline, 2015).

Still other studies seem to show positive links in the relationship between religiosity and moral behavior, even altruism. Such studies are clouded by antisocial individuals who portray themselves as religious followers and pillars of the community.

Criminological research also acknowledges an inverse relationship between religion and crime. Analysts on religion and crime have pointed out that religious behaviors and beliefs generally exert a moderate deterrent effect on an individual's criminal behavior (Cline, 2015). The challenge for a behaviorist is deciphering who is truly religious and practices moral behavior and who is pseudoreligious and while seemingly moral in actuality practices hidden immoral behaviors and thoughts.

MORALITY AND NONRELIGIOUS BELIEFS

In *Ethics & Morality: Philosophy of Behavior, Choice, and Character*, Austin Cline (2015) states that, "Atheists and theists frequently debate morality on several levels: What is the origin of morality? What are proper moral behaviors? How should morality be taught? What is the nature of morality?"

The terms ethics and morality are often used interchangeably and can have the same meaning in casual conversation, but morality refers to moral standards or conduct while ethics refers to the formal study of such standards and conduct. For theists, morality typically comes from some sort of a god or gods, and ethics is a function of theology; for atheists, morality is a natural feature of reality in human society, and ethics is a part of that philosophy. Between these two extremes lies the majority of the human population, showing a gradation from one end to the other:

> "An important feature of morality is that it serves as a guide for people's actions, whether associated with religion or nonreligion. Consequently, it is necessary to point out that moral judgments are made about those actions that involve choice. It is only when people have possible alternatives to their actions that we conclude those actions are either morally good or morally bad. This has important implications in debates between atheists and theists." Cline states, "if the existence of a god is incompatible with the existence of free will, then none of us have any real choice in what we do and, therefore, cannot be held morally accountable for our actions."

REDEFINING DOMINANCE

The two forms of dominance we have spent time describing in Chapter 1 and this chapter have been through biological and ecological approaches (Fig. 2.6). The ecological definition becomes important later in the book when we address how humans have related to the environment and its ecosystems. Dominance establishment as a biological

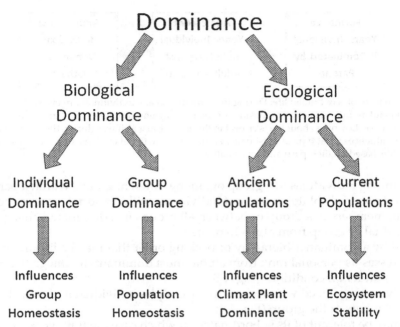

FIGURE 2.6 Two forms of dominance brought out in Chapter 1 and this chapter, biological and ecological dominance. Both relate to ecosystem homeostasis. Under biological dominance, dominance among social group members influences group stability. The dominant attributes of a group influence population homeostasis. Under ecological dominance, dominant plant types have been recognized as relatively stable dominant forms in the Paleozoic, Mesozoic, and Cenozoic eras, while current populations of climax vegetation change with local environmental changes. All such changes affect populations of other organisms.

process is redefined as follows, based on the work of Drews (1993) and Schjelderup-Ebbe (1922) and on the descriptions brought out in these chapters.

Biological dominance is an attribute or characteristic established by a competitor or group of competitors after having expressed *agonistic behavior* toward another competitor or group of competitors. The winning individual or group is referred to as the *dominant or alpha individual or group*, and the losing competitor is referred to as the *subordinate, deferent, or beta individual or group*. To fully understand the concept of dominance, this definition must be accompanied by a number of collateral statements:

- Dominant status may refer to both individuals within a group and to the group itself (Fig. 2.6).
- The most dominant (alpha) individual has the highest status and control over other societal individuals and receives the most rewards.
- Status and control enable the dominant (alpha) individual to select the highest ranking mates, as well as obtain the right to shelter, space, food, and other resources.
- Although dominance hierarchies are often thought to be a system of stable linear variations in prestige (Fig. 1.3), status, and authority among group members, rank in a hierarchy may be determined or influenced by a wide array of intrinsic and extrinsic factors, including personal traits (genetically determined characteristics, for example, size and strength, degree of aggressiveness or cognizant abilities of the species, and relationships between group members) or differences in age. Rank also may be dependent on kin relationships, conditions in which some subordinate individuals are able to outrank more dominant ones by acting together (as a coalition), experience of the competitor, or the degree of interest or investment an individual has in the area in which they exist.
- Dominance status has different meanings, depending on the degree of sociality a species has. A dominant individual in a eusocial species may demonstrate more control of its group, even influencing the reproductive output of its subordinates, than a dominant individual in a group that has less of a social nature (Fig. 1.6).
- Dominance rank of an individual depends on the result of agonistic encounters within a group. The status, power, and other rewards an individual has depends on the rank it has with relationship to other members of the group.
- Dominance rank of a group depends on the result of agonistic encounters between groups. The status, power, and other rewards a group has depends on the rank it has with relationship to other groups within a particular area.

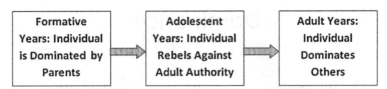

FIGURE 2.7   Changing dominance roles in human life. During the formative years, individuals are under the domination of parents. How the individual turns out may be dependent upon genetic propensities or the learning environment at this time. Dominance roles begin to change as an individual approaches adolescence. Learning about life is critical in this period, just as it was during the formative years. Psychological trauma during the learning years often influences what type of personality an individual will develop. As an adult, an individual dominates others and will influence their outlook on life, based on their particular personality.

- Because of variability among individuals in a group or among different groups within a particular area, dominance is a relative measure and not an absolute property of individuals within a social group or between groups.
- Encounters between all members of a group or between all groups in a designated area generally results in a ranking for each individual or group from alpha to omega.
- Ranking is represented by a dominance hierarchy or pecking order that may be linear, meaning that each individual clearly expresses its personal rank from alpha (most dominant) to omega (least dominant, submissive, or deferent) under all conditions (Fig. 1.3).
- Ranking may be decisive or organized within a group if the alpha individual is clearly dominant and controls the behavior of other members of the group (Fig. 2.3).
- A decisive dominant may be tolerant of its subordinates, in which case the unit works as a team. Teamwork generally leads to greater productivity by the unit or group.
- A decisive dominant may be intolerant of its subordinates, becoming despotic at its extreme, in which case the unit loses its team effect. In such a case, productivity is generally low and may lead to intolerable conditions and the loss of individuals within the group.
- A dominant attribute may be determined after one encounter or a number of encounters between the same dyadic competitors. Failure to establish dominance, competitors may escalate their agonistic aggressiveness and terminate their encounters with or without a winner. Such encounters may occasionally be life threatening. Fierce competition may result in the beta individual leaving the group. What happens to it depends on whether it enters another group and how it ranks in that group.
- Ranking may be indecisive or poorly organized or fluid if the alpha individual is poorly dominant (poor leadership). If dominance rank is not established after a display of agonistic behavior or repeated encounters, that is, dominance rank is poorly demonstrated, the entire society may have difficulty determining their rank in the hierarchy, and societal life may lack homeostasis.
- A dominant individual within a eusocial species produces most or all of the offspring of the society.
- In certain cases, a dominant may influence the reproductive output of subordinates, sometimes even causing atrophication of subordinate reproductive systems.
- Although dominance may initially be dependent on assorted expressions of aggression, stable dominance relationships, once developed, are in fact generally maintained in a more subtle fashion, often through the use of vocalizations or visual communication, with minimal aggression.
- Personality develops under the influence of dominance status (Fig. 2.7). In an individual's formative years, they are dominated by parents and other adults. During adolescence, many individuals rebel toward dominant influence from adults. During adult life, an individual's personality influences how their form of dominance will be expressed toward their children and others.

This definition of biological dominance points to the complex nature of dominance establishment and maintenance in social animals. It should be of value in understanding the variable and complex nature of animal societies, including that of humans. It is hoped that investigators will evaluate these conditions and possibly add their comments to determine the complete nature of dominance in all animals, including humans (Fig. 2.8).

REDEFINING AGGRESSION

We now have an opportunity to redefine aggression in a biological context as it relates to dominance behavior:

Aggression is an attack by an individual or group (the *aggressors*) toward another individual or group (the target). It may be *instigative* or *offensive*, as demonstrated toward a target group or individual without former

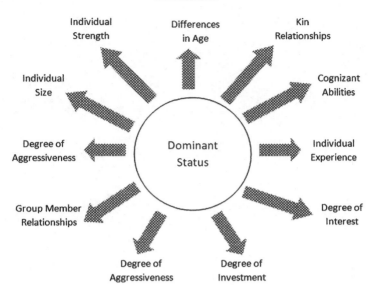

FIGURE 2.8 Dominant status is influenced by a smorgasbord of factors, as shown in the earlier figure. Some may relate to traits handed down through generations while others may be learned (experience) or related to age. Social species display a wide assortment of dominance interactions that they face every day during their lifetimes.

provocation; or it may be *defensive*, as demonstrated by members of a targeted group or individual that has been attacked and/or provoked.

Instigative or offensive aggression may be an *expression of dominance behavior* in which a member or members of a group express *agonistic behavior* toward another individual or group. In other cases, it may be a *predatory or parasitic expression* in which *antagonistic behavior* or *hostility* is demonstrated toward another member or members (those that have been targeted). Such aggression generally terminates with the destruction or defeat of either the target or offensive aggressor.

When the primary goal of instigative aggression is intentional injury or destruction of the target, the behavior is called *hostile aggression*. Hostile aggression is distinguished from *instrumental* (or *operant*) *aggression*, in which the attack is carried out principally to achieve a goal other than the target's injury, such as acquiring a desired resource. Most instances of hostile aggression can also be regarded as *affective aggression* when it involves emotional reactions to an aversive state of affairs.

Defensive aggression is a reaction by a targeted individual or group to instigative or offensive aggressive behavior to protect a territory, social unit, resource, or investment (e.g., offspring or property). Defensive aggression may be preceded by *warning behaviors* if instigative aggression is poorly expressed toward the target. In such cases, certain postures and body movements may be demonstrated to prevent or bring an instigative attack to termination. If an instigative attack is strongly expressed, defensive aggression may be explosive, in which an all-out defensive attack may be demonstrated by the attacked toward the attacker.

PERSONALITY

Having defined dominance and aggression and having listed other attributes of animal behavior in the first two chapters, it is appropriate to use the attributes to briefly describe human personality because of its significance to dominance and aggression. In its simplest form, personality is defined as the visible aspect or description of one's character as it appears to others. Personality is affected by temperament (innate), as well as learned behavior throughout life, but particularly in the formative years, during adolescence, and all traumatic experiences. Both personality and self-schema (the way we look at ourselves) are expressions of our dominant and aggressive nature and are discussed in detail in Chapter 8 (Beck, 1975; Cialdini, 2003; Reiss, 2009; Sherif, 1936; Valentino et al., 2008).

Note

Superscript numerals appearing in this chapter refer to additional text/explanation given in the appendix.

3

The Significance of Comparative Studies

A comparative approach to studying biology uses natural variation and disparity in populations of organisms to understand patterns of phylogenetic history and life-related mechanisms at all levels. For our purpose, comparative analyses are important in showing homologies and analogies among social biota.

ATTRIBUTES SHARED WITH OTHER ORGANISMS

While this book is mostly about humans, reference has been made to a wide assortment of vertebrate and invertebrate animals that share a vast array of biological features. In fact, some of our discussion about the significance of dominance in animals to this point have related to what many people would consider lowly insect forms. Why is this?

Why would we devote any time and energy at all to even reflect on the likes of those or any other subhuman creatures? As practiced by many contemporary biologists and other scientists, a comparative approach is taken to point out that many biological concepts associated with most species are the same or very similar, no matter which animals we are speaking about. But is this really an acceptable approach to understanding who we are? With respect to this question, let us ask another: Why not?

In *The Human Animal, Mystery of Man's Behavior* (1971), Hans Hass brings up the question of whether we can learn much about our own biology from studying lower animals. This should be obvious when we read the works of many excellent and well-informed contemporary writers who contemplate the workings of biological systems at every level. As Hass points out, many of us consider human actions as "the products of conscious mental acts," which make them essentially different from other organisms that we generally refer to as animals. But are we really that different?

It may be humiliating, against religious beliefs and otherwise seemingly out of line for some of us to consider the idea of comparing humans with other animals. Many humans are still avoiding the concept of evolution and denying that it has anything to do with humans, the children of God.

With a large and extremely complex brain, we think of humans as being special, uniquely the only beings to possess an awareness of self, the only ones endowed with a complex form of reason. With recent investigations, however, for example, those reported by Christopher Boehm in *Moral Origins* (2012), recognition of self may be found in certain other animals as well. It seems that the more science investigates and compares humans and other animals, the greater our similarities become apparent.

We are multicellular animals, with cells, tissues, organs, and organ systems that are very similar to those of other animals. Comparing our cardiovascular, respiratory, skeletal, muscular, endocrine, urogenital, reproductive, and nervous systems with those of other animals, for instance, demonstrates how similar we are to other animals. In addition, similarities in DNA and embryological development offer overwhelming support for our animalistic nature and relationship to other organisms.

There is no doubt that we have some traits that separate us from other animals. However, we could say this about any group of organisms. Each species has its own characteristic features. We will see that our separateness among the many species on Earth is a matter of degree. While many species have their own particular attributes, they also possess features that link them to other life forms.

Dominance and Aggression in Humans and Other Animals
http://dx.doi.org/10.1016/B978-0-12-805372-0.00003-1

Many of our most human attributes seem to be a direct result of the development of our superior brain and social organization. We have an ability to invent and understand abstract thoughts and art. We can consciously direct our actions with an eye to the future, although our lack of attention to how we may progress into the future without destroying ourselves is worth contemplating. We have developed immensely complex codes of religious, moral, and aesthetic behavior and use complex forms of language, although these attributes have sometimes led to major conflicts within our society. By using our ability to think and converse, we have created vast empires and complex political and economic systems, which seem to have no counterparts in other species of the animal kingdom:

> To you is given a body more graceful than other animals, to you powers of apt and various movements, to you most sharp and delicate senses, to you wit, reason, memory like an immortal god. **Leon Battista Alberti**, *On Painting*, *1956*

We have many truly grandiose cultural and technological qualities. We know enough about biology these days to determine the genomes (entirety of hereditary information) of organisms, manipulate genes, develop clones, even create new organisms.[1] Most of us live by moral rules which Boehm (2012) suggests developed in egalitarian hunter–gatherer groups during the rise and change of our species. What can lowly invertebrates teach us that might add to our self-knowledge?

To begin with, we are an arrogant species indeed to think that everything we do or think about should have an anthropocentric connection. But is this an improper attitude? Not really. Most organisms that belong to the animal kingdom think of themselves first. Self-interest is an animalistic trait that is beneficial in the process of survival. It is animalistic to ignore other species unless it represents potential food, a potential mate, our offspring, or is threatening. It is also animalistic to be devoted not only to reproducing one's kind but to overproducing (Fig. 8.5). Each individual and each population is most important to itself, its survival, and reproductivity.

Reproductivity has developed as asexual and sexual forms, its nature becoming more complex as we progress through the phylogenetic ranks. Behavior required to obtain a mate and perform sex, as well as to survive among other organisms in the environment and under changing conditions, on the other hand, requires more imagination and depends on the competition an organism is faced with and the stress put on populations. Dominance and aggression help them achieve a status that helps them accomplish these things.

I repeatedly get questions from students and others who ask, "What good is a mosquito?" when they actually should mean, "What good is a mosquito to me?" or "What good is a mosquito to itself?" To the mosquito, it is very important. It is a form of animal that must survive and reproduce just like individuals of other species do. They (the females) must drink blood, which has particulate matter in it in order to reproduce. Thus, they live a life as an ecto-parasite, and cause us and other animals some discomfort in the process. Other organisms sometimes live on or in nonhuman animals as parasites, and occasionally, the organisms that live in them are transmitted to us as pathogens. They do not have a choice. That is what they do.

If they could, mosquitoes and other organisms that harm us, or any other animal for that matter, might look at us and ask the same question. What good are humans to them? When we think of answering this very complex question, we should realize that we are not good at all to them, except as an occasional meal, which is important to them for their survival and reproductivity. Otherwise, we trap, poison, and electrocute them. We spray the adults with pesticides and use chemicals or bacteria to kill their infants. In the process, we pollute the environment, resulting in the death of untold numbers of organisms, including ourselves.

And, as animals, we do not think twice about these things, as long as we retain the comforts we feel we deserve in life. It also means that our existence on Earth is still being run mostly by animalistic rather than cognizant forces. As Morgan Freeman suggested in the movie *Lucy*, our societies are run for power and profit, and thus humans appear to be more interested with having than being.

What good is a house fly, a horse fly, a tick, or a snake? The same answer applies to them as well. Biologically, they are important to themselves, and they are often important to the homeostasis of the ecosystem to which they belong. They are species like we are, having progressed through millions of years of evolution, encompassing innumerable trials and tribulations of natural selection, to arrive in their present condition. And, like us, they will continue to evolve unless they lose their ability to cope with a changing world. As a species, each has the responsibility of surviving and reproducing in order to avoid extinction. Humans are no different.

Let us gather our thoughts from the previous paragraphs and examine ourselves from a biological perspective. Contemporary humans represent but one species among many animals, a species that, like all other species, has arisen from a long line of ancestors, the most recent of which were predecessor humans. Table 5.1 and sections of text in other chapters provide information on how many human and human-like forms from the past have been discovered. Before that, it was prehumans, lower primates, nonprimate mammals, and so on.

Comparing humans with other mammals clearly shows that their cells, tissues, organs, and bodily systems look and work the same, sometimes with slight modifications. By comparing mammals (including humans) with other types of organisms, we have to agree that we share many features. As elite as we would like to think we are, we remain a mammalian member of a highly evolved group called primates, which includes numerous species with varying degrees of animalistic features.

As Richard Fortey states in *Horseshoe Crabs and Velvet Worms* (2012), "we carry our vertebrate pedigree back 525 million years. We have pursued our own trajectory for as long as mollusks or brachiopods or arthropods." To not accept, disregard, or forget our biological roots is somewhat comparable to an affluent person who forgets his or her more impoverished beginnings.

In view of evidence assembled during the last 100 years, as Hass and other biologists have pointed out (1971), there can be no serious doubt that all higher forms of life (including humans) are initially descended from unicellular organisms.

We can now state, with a confidence verging on certainty,

- that the Cambrian Explosion (an event characterized by mass diversification during the beginning of the Paleozoic era, commencing some 542 million years ago) resulted in the rise of most of the major groups of organisms that are on Earth today, including the line that has led to humans (Table 5.1; Fig. 1.7; Fig.3.1);
- that creatures such as insects and spiders are more directly descended from marine arthropods;
- that existing terrestrial vertebrates evolved through a long series of natural-selective changes from tiny lancet-like invertebrate creatures (prochordates), which possessed basic chordate features but lacked jaws, a well-defined brain, and an elaborate endoskeleton;
- that the early chordates gave rise to jawless fish (some of which were the first vertebrates), which in turn gave rise to fish with jaws and a cartilaginous endoskeleton (Fig. 3.1);
- that certain cartilaginous fish arose from jawless fish and gave rise to bony fish, with certain members giving rise to organisms with a lung, tetrapod (four-legged) body style, and chiridian limbs (limbs with hands and feet that possess digits; Fig. 5.4);

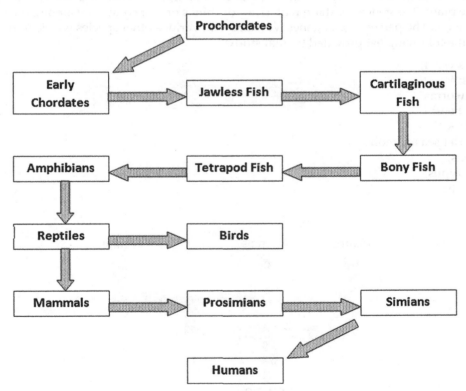

FIGURE 3.1 Phylogenetic pathways from early prochordate to modern humans, a sequence of life forms that took over 500 million years to play out. All members of the chordate group have features, such as a notochord, dorsal hollow nerve cord, and a postanal tail. While postnatal humans have a well-defined, complex nervous system, endoskeleton, and short os coccyx, the features that define a chordate are present mostly in their prenatal form. Each one of the groups shown earlier has occupied a dominant position at one time or another during geological history. Ecological dominance changes when environmental changes occur.

- that certain tetrapod fish gave rise to amphibious forms (Fig. 5.4), leading to such animal types as salamanders, newts, frogs, toads, and caecilians;
- that reptiles are descended from certain types of tetrapod amphibians, with modifications for a terrestrial life due mostly to the development of an amniotic egg (an egg that possesses a protective covering that keeps the embryo in an aquatic environment during its developmental stages; Fig. 3.6);
- that from certain reptiles there evolved birds on the one hand and mammals on the other, both of which are endothermic (producing their own body heat) and share a development within an amniotic chamber;
- that humans are descended from more primitive social mammals and belong to a rather interesting, brainy group called hominines (also spelled hominins), their most immediate predecessor being a common ancestor that links us to chimps and gorillas (Fig. 1.8).

We are thus related not only to other apes but also indirectly to a wide array of organisms—even to plants, if we go far enough back in time. Actually, although plant-like and animal-like organisms split from one another early in the evolution of life on Earth, they remain quite similar in many ways (Fig. 3.2).

Cells of plants, animals and all eukaryotic organisms are quite similar in structure, although plant cells typically have cell walls while those of animals do not, and plant cells have chloroplasts, with which they can make their own food. Their DNA is similar, being composed of a series of chemicals called nucleotides, which have the same chemical makeup, involving the same nitrogenous bases, pentose sugars, and phosphate groups. Cell reproduction is much the same as well.

According to Daniel Chamovitz (*What a Plant Knows, A Field Guide to the Senses*; 2012), plants are like animals in many ways, including being aware of complex light environments, intricate aromas, different physical stimulations, having preferences, and utilizing a memory. In addition to having these attributes, the mechanisms and chemical functions that make them work are often strikingly similar. As he states it, "we share biology not only with chimps and dogs but also with begonias and sequoias."

In a National Geographic article called *Genes Are Us, And Them*, Carl Zimmer (2013) makes the following statement: "A human and a grain of rice may not, at first glance, look like cousins. And yet, we share a quarter of our genes with that fine plant. The genes we share with rice—or rhinos or reef coral—are among the most striking signs of our common heritage. The percentage of genes we share with various other species is a clear indicator of our common origin," as in the following list provided by that study:

18% with baker's yeast;
24% with grapes;
38% with roundworms;
44% with honey bees;
47% with fruit flies;
54% with the starlet sea anemone;
65% with chickens;
69% with the platypus;
73% with zebrafish;
84% with dogs;

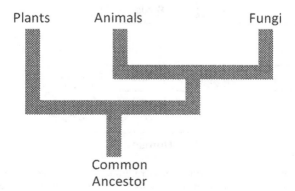

FIGURE 3.2 Simple cladogram of plants, animals, and fungi. All organisms on Earth are linked by similar DNA. As organisms evolve, they develop more personal features that characterize each species. Anatomical and behavioral attributes shared by different species give us some idea about how closely related they are. However, since the discovery of the structure of DNA in 1953, comparative studies of DNA have been most important in determining true relationships between different species and higher taxonomic groups.

85% with cows;
88% with mice;
90% with chimpanzees.

"All animals, plants, and fungi," he reminds us, "share an ancestor that lived about 1.6 billion years ago" (Fig. 3.2). Every lineage that descended from that progenitor retains parts of its original genome. Since evolution has conserved so many genes, exploring the genomes of other species can shed light on genes involved in human biology and disease. Even yeast has something to tell us about ourselves.

"Our catalog of genes themselves has changed. Genes can disappear, and new ones can arise from mutations in DNA." Others that previously served some other function or no function at all may suddenly become active.

"Other novel genes have been delivered into our genome by invading viruses. It's hardly surprising that we share many more genes with chimpanzees than with yeast because we've shared most of our evolutionary journey with those apes. And in the small portion of our genes with no counterpart in chimpanzees, we may be able to find additional clues to what makes us uniquely human."

These data point out that there is substantial evidence for these and other relationships between all forms of organisms. Consider the comparative data:[1]

1. As living beings, we all have bodies composed of cells, over 37 trillion of them making up the human body. Our cells, like the cells of other multicellular organisms, consist of a variety of types, most of them differentiated to do a specific job, but they all have similar intracellular structures and arose from a single undifferentiated stem cell called a zygote.
2. Our cells include a variety of subunits referred to as organelles, which look and function like those in the cells of other organisms. Organelles and their chemicals accomplish all the functions required of a cell, including the acquisition of food through forms of diffusion and the process of endocytosis (taking in food substances by wrapping them in plasma membrane, somewhat like the feeding behavior of amoebae), the processing of food, digestion to obtain anatomical building blocks and ATP, getting rid of body wastes (exocytosis), replicating DNA, and taking part in the process of protein synthesis.
3. The DNA (deoxyribonucleic acid) of our cells and those of all other living organisms provide an incredibly important genetic blueprint, which is responsible for genetic traits and the synthesis of proteins. Different forms of RNA (ribonucleic acid, another nucleic acid) and proteins are the workhorses of such reactions.
4. Our DNA, RNA, and the nucleic acids of all other organisms have the same basic structure, consisting of long chains of chemicals called nucleotides which, in turn, are composed of the same four nitrogenous bases (with the exception of uracil, which replaces thymine in RNA), pentose sugars, and phosphate groups.[2] The differences among organisms is not in changes in the basic chemical nature of DNA but in the different sequences of repeating nucleotides, the banking of noncoding DNA around protein histones, and the consequent active proteins that are constructed from coding segments referred to as genes. Proteins are the main chemicals that run our bodies.
5. As indicated earlier, the DNA of all organisms includes genetic segments (often referred to as exons) and nongenetic segments (sometimes referred to as introns). Introns consist of DNA that has variably been referred to as selfish DNA, junk DNA, genetic detritus, noncoding DNA, and pseudogenes. DNA introns may carry what appears to be defunct genetic information that belonged to organisms millions of years ago, and it may sometimes function in the epigenetic reconstitution of "old genes."
6. The basic chemicals that make up our bodies and are consumed on a daily basis in our foods are much the same, whether we are bacteria, archaeans, protistans, fungi, plants, or animals, although there are individual differences that are important in characterizing each type of organism.
7. We all are members of food chains and thus depend on one another for our daily needs. Commencing with the sun, CO_2, and water, plants produce glucose ($C_6H_{12}O_6$; Fig. 3.3), and glucose undergoes a dehydration synthesis (hooking them together), the resulting molecule being stored in their tissues as starch. Glucose molecules are also linked to make structural macromolecules, such as cellulose.

$$Energy + CO_2 + H_2O \rightleftharpoons C_6H_{12}O_6 + O_2$$

FIGURE 3.3 Photosynthesis (from left to right) and respiration (from right to left). In photosynthesis, the sun's energy is used, along with CO_2 and H_2O, to produce glucose ($C_6H_{12}O_6$) and oxygen (O_2). Plants further hook glucose molecules together in a process called dehydration synthesis to make chemicals, such as starch (a storage polysaccharide) and cellulose (a structural polysaccharide). Herbivores feed on the plants and break these products down in the process of hydrolysis, taking and using energy from the chemical bonds during the process of digestion and cellular respiration (the reaction mentioned earlier, but in reverse order). The energy derived in cellular respiration is in the form of ATP (adenosine triphosphate).

Plants also pick up and store an assortment of nutrients to build their bodies and help them establish homeostasis. Herbivores feed upon plants, break the starch down, reconstruct carbohydrate macromolecules, store them as glycogen, and eventually synthesize ATP (the most immediate source of energy for all organisms) to be used by their body; many of the nutrients pass into the body of the herbivore, helping them to create a homeostatic system; subsequently, a sequence of other consumer organisms feed upon one another and pass the energy and nutrients along from one trophic (feeding) level to another in food chains.

By eating plants directly, humans (especially those who restrict their diet to vegetation) function as herbivores. Yet, humans in general eat meat and thus function as carnivores as well. As a group, we can best define ourselves as omnivores, animals that eat a variety of food types:

8. To get ATP, we must all process food materials through a digestive and respiratory sequence of events, which terminates in newly constructed cells, tissues, organs, and organ systems, or in the cell's mitochondria. Wastes that accumulate after digestion and assimilation must be gotten rid of, no matter what the organism is.

9. As biological species, we must all struggle to survive in a complex, changing world. To accomplish this, we must compete with one another for different types of resources, often entering complex relationships with one another, especially when we exist in social groups. To maintain homeostasis in social groups, our relationships involve dominance and aggression, levels of which vary among its members.

10. As a group, we must undergo a form of reproduction in order to carry on our species. Forms of reproduction vary considerably, generally being expressed collectively in asexual and sexual ways.[3] Human somatic cells, of course, reproduce by an asexual method we call mitosis. Humans themselves produce gametes by a reduction division called meiosis, but individuals, of course, are restricted to a sexual means.

11. Our differences and degree of complexity in anatomy, physiology, chemistry, and behavior as multicellular animals are the result of approximately 600 million to a billion years of evolutionary change that has arisen due to mutations, epigenetic changes, variation, and natural selection. In spite of their apparent differences, there are a great number of similarities in the anatomy, physiology, and chemistry of different organisms, especially between members of closely related groups.

12. As we will see, the survivability of all organisms and their populations is built upon and functions around a vast array of truths and deceptions and passing on genes to succeeding generations.

Evidence of common descent among all living things has been repeatedly investigated and confirmed by scientists working in many biological fields over the years since Charles Darwin published *On the Origin of Species by Means of Natural Selection*, or *the Preservation of Favoured Races in the Struggle for Life* in 1859 and *The Descent of Man, and Selection in Relation to Sex* in 1871. While new hypotheses arise on occasion with respect to organismic evolution and lengthy, sometimes heated, debates follow, most deal with intricate details of the larger picture, periodically testing Darwin's original concept of how evolution works. Evidence obtained through scientific investigation has supported the concept of modern evolutionary synthesis, the rationalization that explains how and why life has always changed and will continue to change over time as long as the sun shines and Earth remains alive.

Comparing genetic sequences in organisms has revealed that those which are phylogenetically (evolutionarily) close generally have a higher degree of similarities than those which are phylogenetically distant (Zimmer, 2013). Further evidence for common descent (epigenetic evidence) comes from what scientists sometimes refer to as genetic detritus, for example, the nongenetic DNA mentioned earlier. Such DNA sequences may have been active as genes in the distant past but subsequently became clustered around protein histones, causing them to become ineffective as genes. While such DNA appears to be undergoing some form of degeneration over long periods of time, much of it appears to still remain as part of DNA in contemporary species, including humans.

Under environmental duress, these DNA sequences may return to an active state and be passed down to succeeding generations as active genes. They sometimes appear to temporarily influence the development of certain features that are expressed in our embryological stages (as in totipotent and pluripotent genes) but which are subsequently turned off so that they are no longer apparent in our postnatal form.

Ontological (embryological) studies point out that as embryos, humans and other animals go through developmental stages that began to appear before and in the Paleozoic era over 500–600 million years ago, including the following:[4]

- *Unicellular* (zygote). Our start in life, like the beginning of any unicellular, colonial, and multicellular organism, begins as a single cell, an undifferentiated (stem) cell which contains the ingredients and ability to feed and replicate itself (Fig. 3.4). Many organisms on Earth have remained unicellular, for example, bacteria, archaeans, and most protistans. Others have evolved to more complex levels in the process to handle the constant appearance of environmental threats.

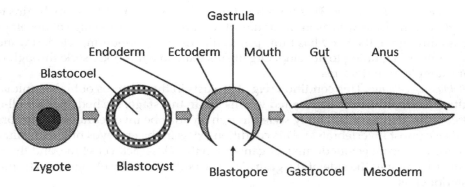

FIGURE 3.4 Early embryonic development as seen in all animals. The zygote, a single cell that has resulted from the fusion of a spermatozoan and an oocyte, begins to divide. Once the cells in the embryo reach about 32 in number, the embryo is called a morula. Further cell division brings the embryo to a point at which its cells move to the periphery to form a blastocyst. The blastocyst subsequently develops an invagination that pushes into the blastocoel to form a gastrula. Jellies (Phylum Cnidaria), for instance, reach the gastrula stage but never develop a third primordial tissue. Their body form is called diploblastic. Later stages of development occur with the development of a third primordial tissue, the mesoderm, and the developmental stage is referred to as triploblastic. While all multicellular organisms undergo the development shown in this figure, linking them together developmentally, differences are found in whether the blastopore becomes the mouth or anus. Humans and all animals down to echinoderms develop as deuterostomes, in which the blastopore becomes the anus. Phyletically lower organisms are protostomes, in which the blastopore becomes the mouth.

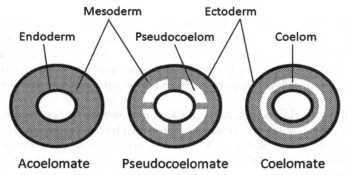

FIGURE 3.5 Changes in the third primordial tissue, the mesoderm, which develops between the ectoderm and endoderm. The most primitive condition is represented by a solid mesoderm, as found in flatworms (acoelomic condition) (Phylum Platyhelminthes). Scattered pouches develop in the mesoderm, as seen in the roundworms (pseudocoelomic condition; Phylum Nematoda). A well-defined coelom develops when the pseudocoelomic pouches fuse and form the body cavity or coelom (coelomic condition), as seen in segmented worms (Phylum Annelida) and all organisms above them, including humans (Phylum Chordata). Humans pass through the stages pictured in their early embryonic development, showing similar development in all multicellular organisms, including humans.

- *Colonial* (morula, blastula, or blastocyst; Fig. 3.4). This condition (a round ball of cells, either solid or with a hollow space inside) exists in embryos of multicellular animals, as well as in adults of certain colonial animals. Colonial organisms consist of cells that remain with one another, although each cell may function in the same or very similar manner. Such organisms do not develop true tissues, organs, or organ systems, and some cells have an ability to change from one form to another. While there are such colonial organisms in the world, many organisms have evolved even more complicated forms.
- *Multicellular/diploblastic* (gastrula; Fig. 3.4). This condition (a ball of cells with a hollow interior, which develops an invagination and two primordial tissues) exists in the embryos of higher animals, as well as in adult jellies. The cells of multicellular organisms which have reached the gastrula stage in development begin to undergo differentiation to form tissues, organs, and organ systems. Yet, only two primordial tissues (ectoderm and endoderm) make up the body of such an organism. It is here, though, that a primitive mouth, anus, and digestive system emerges. Organisms that have this body style through life are invertebrates.
- *Multicellular/triploblastic/acoelomic.* This condition (worm-like and with the three primordial tissues, ectoderm, endoderm and mesoderm) exists in the embryos of higher animals, as well as in adult flatworms (Fig. 3.5). It is in this stage that the third primordial tissue (mesoderm) arises so that the body is no longer diploblastic (with two primordial tissues), but the body lacks a body cavity (a coelom). Organisms which have this body style through life are invertebrates, e.g, flatworms (e.g., tapeworms, flukes and planarians).

- *Multicellular/triploblastic/pseudocoelomic.* This condition (Fig. 3.5; in which a body cavity begins to form) exists in the embryos of higher animals, as well as in the adults of roundworms. In this stage of development and in the organisms that pass through it, the body has three primordial tissues (endoderm, ectoderm, and mesoderm), and the coelomic cavity is beginning to develop. Organisms that have this body style through life are invertebrates, for example, roundworms.
- *Multicellular/triploblastic/coelomic.* This condition (Fig. 3.5) exists in the embryos of higher animals, as well as in the adults of multicellular forms from segmented worms up. In this stage, the body has a well-defined coelom (body cavity). Organisms that have this body style through life may be invertebrates or vertebrates.
- *Invertebrate/protostomic* (most invertebrates). This condition exists in the embryos of lower animals, including most invertebrate forms (except echinoderms). Organisms in this class lack vertebrae, and their mouth develops from the blastopore (opening in the side of the gastrula). The body generally elongates to form well-defined anterior and posterior ends.
- *Invertebrate/deuterostomic/nonchordate.* This condition exists in the embryos and adults of higher animals from echinoderms up (including chordates). Animals that have these features develop an anus from the gastrula's blastopore.
- *Invertebrate/deuterostomic/prochordate.* This condition exists in the embryos of higher animals, as well as in the adult stage of multicellular animals, for example, tunicate larvae and lancelets, two rather small, extant marine life forms. These organisms have developed chordate features: an elongate body with mouth and anus, pharyngeal slits, a dorsal/hollow nerve cord, a notochord and postanal tail. These and the following groups are listed in Table 5.1, along with important human species and events that have occurred throughout geologic history.
- *Invertebrate/deuterostomic/vertebrate/jawless fish,* for example, including the derived hagfish and lamprey. This condition exists in the embryos of higher animals, as well as in jawless fish. It is within the jawless fish group that vertebrae first develop in a cartilaginous endoskeleton. Many early jawless fish, like the extinct ostracoderms and contemporary hagfish, for instance, lack vertebrae, but lampreys have them. Food is sucked into the mouth, and some have developed a parasitic role and keratinous tooth-like structures within their mouth.
- *Cartilaginous fish,* for example, sharks, skates, rays, ratfish, and sawfish. This condition exists in the embryos of higher animals, as well as in cartilaginous fish adults. These fish have developed jaws and bony teeth, which can be used to bite with. While most of the skeleton of these fish is composed of cartilage, different members of the group have varying amounts of bone development. A lateral line develops in this group, which is sensitive to water pressure changes. Fins on fish in this group are referred to as ray fins, which are used in swimming.
- *Bony fish,* including most of the fish we are familiar with. This condition exists in embryos of higher animals, as well as in adults of bony fish. Skeletons in this group are mostly bony in adult forms, although some degree of cartilage exists between bones and in joints. This group develops an air sac (swim bladder), a device that functions like a primitive lung, although alveoli are not present in most. Fertilization of ova remains external in most cases. While most members of this group possess ray fins, certain members developed lobe fins that helped them move through their aquatic environment. Certain members had undergone further development of the distal lobes of their appendages, a group often referred to as tetrapodomorphs. Certain of these organisms increased the use of their modified appendages, developing them into walking appendages with an assortment of bony digits (the chiridian limb, used to maneuver through their aquatic environments). The group in which lobe fins and walking digital appendages developed is best referred to as tetrapods.
- *Amphibians,* including frogs, toads, salamanders, newts, and caecilians (very secretive, legless amphibians). This condition exists in the embryos of higher animals, as well as in adult amphibians. These organisms arose from certain of the tetrapod fish and represent the first group that began to live in a partially terrestrial environment. However, their ova are fertilized externally (in most cases), as in most fish, their eggs are not covered by a protective shell, and their immature stages must undergo development in water, breathing through internal or external gills. Most of the adults lose their gills and breathe air through lungs.
- *Reptiles, birds, and lower mammalian stages* (which deposit eggs or produce live young), including snakes, lizards, tuatara, crocodilians, turtles, birds, and monotremes. This condition exists in the embryos of higher animals, as well as in adults of the previously mentioned groups. Reptiles represent the first group that truly occupied a terrestrial landscape, having developed an amniotic egg (Fig. 3.6). The amniotic egg represents the form of development followed by all reptiles, birds, and mammals, although the precise mechanism of fetal development and nourishment varies considerably.[5] Humans remain a member of the group that develops in an amniotic egg.

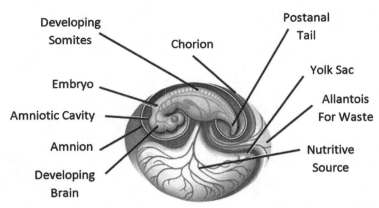

FIGURE 3.6 Amniotic egg. Reptiles, birds, and mammals all develop in such an egg. The only mammal group that retains the shell around the egg is the Monotremata, which includes the platypus and echidna, both indigenous to the Australian region. The amniotic egg developed in reptiles and gave them and future generations a way to deposit their eggs in a terrestrial environment. The embryo within the egg remains in an aquatic environment consisting of amniotic fluid. In humans and most other mammals, the yolk sac (which delivers food to the embryo) and allantois (which stores waste materials from the embryo) are replaced by the placenta. Hematopoietic tissues, which produce blood cells, are also found in the yolk sac of early human embryos, but blood cell formation is later taken over by other tissues in the fetus and finally in red bone marrow prior to birth. These developmental features tie reptiles, birds, and mammals together.

There is no longer a question about our ties to other organisms, and thus experimentation with organisms lower on the phylogenetic scale provides vast amounts of information that help us to understand them and ourselves.

INVESTIGATIONS ON OTHER ANIMALS

Over the years, significant amounts of information have been provided through research on lower organisms that can be applied to humans, including investigations on viruses, prokaryotes, protists, fungi, plants, invertebrates, and vertebrates. What would we know about human cell biology and genetics without research on a fungus we call baker's yeast (*Saccharomyces cerevisiae*) (Botstein et al., 1997), certain bacteria (e.g., *Escherichia coli*) (Lawrence and Ochman, 1998), fruit flies (*Drosophila melanogaster*; Adams et al., 2000), and certain roundworms (e.g., *Caenorhabditis elegans*; Kosinski and Zaremba, 2007)? What would we know about molecular genetics and genetically modified organisms without the use of *E. coli*, fruit flies, mosquitoes, bollworms, fish, and lower mammals? How would we know about transposing DNA without studies on Indian corn (Schnable et al., 2009)?

How would we teach each other the fundamentals of comparative anatomy and physiology without the use of amphibians, reptiles, rats, and other mammals (e.g., mice and nonhuman primates)? How could we have experimented with and cured human ailments without investigations on numerous types of subhuman organisms? How would we know about relatedness, evolution, natural selection, and fitness without observations and experimentation on a wide variety of lower organisms? And how would we know about the concept of sociality without comparative studies on insects and a variety of other animal forms?

Note

Superscript numerals appearing in this chapter refer to additional text/explanation given in the appendix.

Social Nonprimate Animals

If we are to understand the nature of our aggressive urges, we should look at them in terms of our animal origins. **Desmond Morris, 1967,** *The Naked Ape*

This chapter includes examples of invertebrate and nonprimate vertebrate species that share a social existence with us. The reader will find that sociality exists at many levels (as hinted to in Chapter 1), from barely social to eusocial (truly social). What this means is that sociality has arisen independently in many organisms, and the degree of sociality possessed by a species depends on where in the natural selection process the species exists.

In *The Social Conquest of Earth* (2012), Wilson points out that eusociality "arose in ants once, three times independently in wasps, and at least four times … in bees." While rare outside of insects, it has arisen in gall-dwelling aphids, thrips, marine sponges, sponge-dwelling shrimp, and naked mole rats. Within the class Insecta, it has arisen in the termites, certain insects in the order Hymenoptera (8–11 times, including ants and certain bees and wasps) and ambrosia beetles (Andersson, 1984; Honeycutt, 1992; Nowak et al., 2010). Many other invertebrates and vertebrates have evolved variable degrees of sociality, the most fundamental characteristics being parental investment, cooperativeness, and competition among colony members and group defense.[1]

Excerpts from Sally Boysen's book, *The Smartest Animals on the Planet* (2009), have been added to information on each type of animal whenever possible, along with occasional comments which compare prehuman animals with humans. Boysen's comments point out features that may be used to indicate intelligence in nonhuman animals, such as tool use, communication, imitation and social learning, mirror self-recognition, numerical abilities, animal language, cooperation, and altruism.

INVERTEBRATES

Arachnids: Spiders and Their Relatives

In *Sociality in the Arachnida* (1981), Ruth Buskirk points out that "although most arachnids (chiefly spiders, scorpions, ticks, and mites) are solitary animals, a number of species show extended parental care or relatively permanent social groupings." As an example, "a number of mite species called tetranychids occur in colonies of variable size. Some overwinter as diapausing adult females, and dispersal patterns result in mite aggregations on the host plant. Aggression between males [in these groups] is common."

Mites called dermanyssids that inhabit the ears of certain noctuid moths sometimes live in what Buskirk refers to as very large colonies. However, if there are no social interactions occurring in these aggregations, the term "colony" may not be appropriate.

A mygalomorph spider (Orthognatha) in Gabon, West Africa, lives in groups. Many spiders in the Labidognatha (the group to which most spiders belong) express some degree of social behavior as well. In summary of the arachnid literature, Buskirk points out that there are numerous accounts of spider species in various parts of the world that have aggregative behavior. There are also spiders that have additional social qualities, such as a mutual attraction between members of the same species. A significant element that plays a role in such attraction is the web. Social spiders apparently show a preference for congeneric silk and an avoidance of the silk of other species.

Social spiders have built-in mechanisms for tolerating members of their group (Agnarsson et al., 2006). In some spider groups, increased prey capture actually results from cooperation between several individuals in subduing prey. They sometimes cooperate in web building and in certain other behavioral events. Some demonstrate cooperative brood care. Symbiotic relationships develop in which young spiders feed on food that has been captured in the web of another spider.

Buskirk suggests that sociality in spiders may have arisen from parental care. As in some social insects, social spiders engage in group foraging, communal food storage, and other forms of cooperative behavior. She further states that "there is no evidence for reproductive division of labor in social arachnids and thus no documentation of eusociality."

The Rise of Sociality

In *Antiquity of Sociality in Insects* (1981), F.M. Carpenter and I pointed out that "most of the existing orders of insects had already been developed by the end of the Paleozoic era. It was during the most recent part of the Jurassic period of the Mesozoic era (about 175 million years ago) that the first termites are known to have existed, arising from social, wood-eating cockroaches. Ants, members of the order Hymenoptera, appear to have arisen about 25 million years later."

Termites

Termites live their lives underground or in nests that generally maintain a connection with the ground. Being closely related to cockroaches, some share similar gut symbionts; they feed upon wood but rely upon their gut symbionts to digest the cellulose that they ingest.

In *Territoriality in Social Insects* (1979), Cesare Baroni Urbani points out that "both cockroaches and termites show a tendency toward aggregation. They both appear to be attracted by a fecal aggregation substance, and there are also aggregation pheromones secreted by the mandibular glands of cockroaches which are similar to those in termites."

While there are many behavioral differences between the various groups of termites, colony activities are built around a single queen in which her abdomen becomes distended (physogastric) as a result of an enlarged reproductive system where she produces many thousands of eggs. There is a male (generally referred to as a king) that periodically mates with the queen.

Mating swarms generally follow a drenching rain. Males and females terminate their swarming behavior by landing on the ground and immediately breaking off their wings. Females attract males with a sex pheromone. When a male locates a female, he follows her in tandem to a site in which they collaborate in digging into the Earth and constructing a chamber to initiate their nesting behavior.

Care of the young is subsequently taken over by worker termites (infertile females that emerge in the colony) who feed them by a regurgitatory process called trophallaxis. There is no hierarchical ranking, dominance being despotic where the queen is the sole egg layer, and workers do the chores. Agonistic behavior is not required or is subtle within the colony because workers (infertile females) never compete with the queen. There are sometimes anatomical and behavioral distinctions between sterile workers and soldiers.

Paper Wasps

Hornet, yellow jacket, and many temperate-zone open-nesting paper wasp colonies are dominated by despotic queens (Hermann et al., 2017; Hermann and Dirks, 1975). Certain temperate open-nesting paper wasp species (such as *Polistes annularis*, the red paper wasp) demonstrate dominance behavior, which ideally terminates in a weakly despotic hierarchy between co-founding females, the alpha female dominating but tolerating other cofoundresses.

Dominance varies in this species from poorly defined to despotic, resulting in a selection process that determines the survivability of the colony. Many tropical species have a similar cofounding behavior. Subordinate cofoundresses assume all worker duties during the nest-founding stage until adult workers (infertile daughters of the alpha female, the queen) are produced, commencing the ergonomic phase of nest life.[2]

Cofounding behavior is extremely important in species that occupy areas with an especially high threat from predation (as in tropical areas of the world). For instance, *Parachartergus azteca*, a wasp species I studied in acacia trees in Mexico, would not have been able to establish their nests in trees occupied by pseudomyrmecine ants if they did not commence nesting with a number of cofoundresses (Hermann, unpublished). A single female may be able to land on the tree, temporarily keep pseudomyrmecine worker ants away and commence nest-building, but once she deposits an egg and leaves the nest for provisioning, foraging ants would move in and prey upon the eggs and young. With cofounding females present, certain of the females that remain on the nest at all times protect it from ants while others forage for food and building materials. There is no doubt about why tropical areas have a high number of cofounding paper wasp species. Predation in tropical countries (especially with regards to invertebrate predation) is greater than in temperate regions, and cofounding females are needed on the nest while others collect nest-making materials and food.

Cofounding among paper wasps is in certain respects similar to groups of early human settlers in North America working together to build a village. Safety was in numbers. The group size and degree of cooperation were important to their survival.

Bumblebees

Bumblebees are considered to be primitively eusocial because there is very little anatomical diversity between a queen and colony workers, and considerable reproductive competition exists between them (Cardinal and Danforth, 2011; Dornhaus and Chittka, 2005; Huggins et al., 2012). The colony generally has a small physical size, and many colonies are occupied by fewer than 50 individuals.

Nests generally are formed in depressions in the ground, in tunnels, or beneath clumps of grass. Some species construct a wax canopy, called an involucrum, which functions as insulation and protection.

After emerging from hibernation in temperate zones, despotic bumblebee queens begin to search for a nest site. Once found, she prepares wax pots, many of which are used to store pollen and nectar. Some of the pots are used to deposit eggs. Pollen and nectar reserves are fed to the growing larvae, which pass through four stadia (larval stages). Last-instar larvae spin a silken cocoon and subsequently pupate, emerging as adults by chewing their way out of the cocoon.

Unlike honeybees, bumblebee workers are known to deposit sterile eggs that develop into haploid males. Only the queen is mated and capable of depositing female (diploid) eggs.

Queen bumblebees suppress egg deposition in workers in early nests through physical and chemical forms of dominance. Later in the season, workers function less under the influence of the queen and begin to deposit eggs that will develop into males.

Bumblebees are highly defensive, and both queens and workers sting. When disturbed, they often produce a buzzing sound in the nest that is caused by vibrating flight muscles. This raises their body temperature and prepares them for a defensive flight. The buzzing sound may also function as a warning to intruders. Since the lancets of the venom apparatus do not support the large barbs found in honeybees, they do not lose their sting in the hide of a victim and thus they are capable of repeated stinging.

Honeybees

Honeybees are known to tolerate only a single queen in each colony (Kak, 1991; Wilson, 2004). Thus, they are despotic. If a queen is lost, workers (infertile females) produce a new queen brood by feeding already existing larvae a product from head glands called royal jelly. Upon emerging, potential queens demonstrate strong agonistic behavior toward one another (intolerance), usually ending in the death of subordinates. One female becomes the dominant (despotic) reproductive female (queen).

Honeybee societies become extremely large, workers generally numbering in the thousands. All workers are daughters of the queen. The lancet barbs in the venom apparatus are large and result in leaving the sting in the victim's hide, a process known as sting autotomy (Hermann, 1971).

In 1973, Karl von Frisch revealed the amazing abilities of honeybees to provide their colonies with information. "If a food source is less than 300 ft (100 m) from the hive, returning workers perform a simple round dance, outlining a circle in one direction on the vertical surface of the hive, then reversing the circle in the other direction." Worker bees observing this behavior know that food is close to the hive.

A waggle dance, carried out by a worker that has discovered a food source at a greater distance from the hive is carried out by wagging its abdomen and following a specific movement pattern. The waggle dance conveys information about the position of the food with relation to the position of the sun and the distance the food source is from the hive.

New colonies of honeybees involve the movement of a queen from her nest by a process referred to as absconding. This is similar to group founding in tropical social wasps. When a hive loses its queen, workers produce a new queen by feeding larvae with royal jelly.

Ants

Ants represent a large group of eusocial insects that are most closely related to wasps (Bolton, 1995; Hölldobler and Wilson, 1998; Ward, 2007). It is a diverse group of insects with a variety of species-specific behaviors. Most exhibit a despotic dominance in which the queen deposits eggs and the workers do not. In species with multiple queens (polygyny), dominance interactions do occur between females vying for the alpha position.

Other than a few groups of ants (which are monomorphic, all workers being of equal size), most ant species show some anatomical differences (polymorphism) that demonstrate not only a reproductive division of labor but an even greater demarcation of a division of labor among the worker force. Small workers may remain in the nest and care for the young, medium (intermediate) workers may be the prime foragers, and the largest workers may function most in colony defense.

These roles may show considerable overlap. Some major workers possess large mandibles, which they use in defense (especially well expressed in certain army ants, Subfamily Dorylinae). While polymorphic species may exhibit these different divisions of labor, their behavioral demarcation may not be as easily discernable as their anatomical features.

Eggs produced by the queen are cared for by workers who continue the care of the young until the completion of pupation and the subsequent emergence of adults. Workers indulge in trophallaxis between them, larvae, and reproductives, a social exchange of substances from the mouth.

Ant intelligence is difficult to discern because much of their behavior centers around chemical communication (e.g., pheromones). However, researchers have found that some determine directional problems by using celestial or other visual cues, as well as having an internal pedometer that allows them to remember the distance they have traveled from the nest.

Mollusks

Of the mollusks, cephalopods (such as cuttlefish, squid, and octopuses, which most probably evolved in the Late Cambrian from a monoplacophoran-like ancestor, with a curved, tapering shell, as found in gastropods, such as snails) are considered the most intelligent invertebrates and an important example of advanced cognitive evolution in animals (Hanlon and Messenger, 1996; Hochner et al., 2006; Nixon and Young, 2003; Wells, 1962). Cuttlefish and octopuses have the highest brain-to-body mass ratios of all invertebrates. However, their precise social behavior is unclear.

Unlike most other mollusks, almost all cephalopods are active predators. Their requirement to locate and capture their prey has been a probable driving force behind the development of their intelligence, which is uniquely advanced in their phylum. As an example, the Humboldt squid hunts schools of fish, showing extraordinary cooperation and communication in its hunting techniques.

Octopuses seek out lobster traps and steal the arthropods inside. They are also known to climb aboard fishing boats and hide in the containers that hold dead or dying crabs. Another example of cephalopod intelligence is the communication that takes place between the more social species of squid.

The octopus is one of the prime examples of an invertebrate animal that has repeatedly been shown to exhibit flexibility in its use of tools. Some species have been witnessed retrieving discarded coconut shells, manipulating them, transporting them some distance, and then reassembling them to use as a shelter.

Based on learning experiments, researchers have concluded that cephalopods have "short-term" and "long-term" memories. However, there are variations in such capabilities, the long-term memory capability of nautiluses being found to be much shorter than that of other cephalopods. Octopuses, on the other hand, can remember conditioning for weeks.

According to J.G. Boal (*Social Recognition: A Top Down View of Cephalopod Behavior*; 2006), evidence for or against the recognition of dominance, offspring, other kin, familiarity, mates, and individuals is largely lacking in cephalopods. Boal states that "it appears reasonable to expect cooperation in cephalopods that is directly selfish and that cooperation which requires recognition of kin or individual recognition is probably unlikely. Squids are clearly the most gregarious of the cephalopods."

It is often speculated that the three most famous attributes of cephalopods—complex nervous systems, sophisticated visual systems, and complex body patterning—could all serve to support complex, visual intraspecific communication. Cephalopod social behavior includes a sophisticated visual language, accomplished through rapid changes in body patterning, as expressed by Moynihan and Rodaniche in 1982.

Signals that contain information associated with reproduction, for example, species, sex, receptivity, and fitness advertisements, benefit the signaler to the extent that they are perceived and responded to by potential mates, and benefit the recipient to the extent that the recipient is also looking for mating opportunities. Such mating signals, which are widespread throughout the animal kingdom, have been documented in most cephalopods. All use chemical signals to coordinate reproductive behavior.

VERTEBRATES

As pointed out by Richard Fortey in *Horseshoe Crabs and Velvet Worms, The Story of Animals and Plants that Time has Left Behind* (2012), the ultimate beginnings of the line of organisms to which we belong go back to the Cambrian Period of the Paleozoic Era when the initial appearance of a tough connective tissue chord (the notochord) formed just beneath the dorsal hollow nerve cord of certain primitive invertebrates, giving rise to a new group called chordates (Table 5.1). This chordate line was spawning in Paleozoic seas approximately 500 million years ago. Early chordates were small invertebrate animals, possessing a primitive segmented body. They are represented in contemporary times by such tiny sea creatures as the larval stages of tunicates (the urochordates) and all stages of lancelets (the cephalochordates; Nielsen, 2012; Fig. 3.1).

Thus, it was not the tunicate adults that possessed chordate features. Adults are globate in body shape rather than elongate. Their larvae possess the notochord. Lancelets, on the other hand, possess a notochord in both the immature and adult forms, as though they were neotenic (retention of juvenile characteristics in adults) tunicates. The dorsal hollow nerve cord of these early chordates is primitive in structure, lacking a distinct brain.

Natural selective forces throughout the subsequent millions of years assisted chordates to pass through progressive steps during the Paleozoic, Mesozoic, and Cenozoic eras, including the types of animals and accompanying events found in the following phylogenetic sequence:

- ectothermic (unable to produce a stable body heat) jawless fish, such as hagfish and lampreys, with the development of vertebrae in certain members of the group, such as contemporary lampreys, but not in others, such as contemporary hagfish; Thus, vertebrates arose in this group;
- ectothermic cartilaginous fish, such as sharks, sawfish, ratfish, skates, and rays, with a subsequent progressive deposition of bone in selected regions of the endoskeleton;
- ectothermic bony fish, with the completion of bony deposits throughout the endoskeleton along with the development of lungs and lobe fins in certain species;
- ectothermic tetrapodomorphs (four limbed), lobe-finned fish, with gills, lungs, and walking appendages in various stages of evolution (Fig. 3.1);
- ectothermic lunged tetrapod creatures that developed in an aquatic environment but subsequently (millions of years later) invaded terrestrial environments and gave rise to the amphibians (Fig. 3.1);
- ectothermic reptiles that were more tolerant to drier conditions because of anatomical modifications (e.g., scales) and the development of an amniotic egg (Figs. 3.1 and 3.6); some reptiles, especially certain dinosaurs, may have developed endothermic features (ability to produce a stable body heat);
- finally, the planet was introduced to two new groups, the endothermic (able to produce their own body heat) birds and mammals (Fig. 3.1).

Fishes

Fishes represent the largest vertebrate group and the objects of considerable scientific attention. Sociality in fishes appears somewhat different from that in most other animals in terms of the concepts used to explain it (Wong, 2011).

As Hans Hofmann and Russell Fernald point out (2001), "fish species have evolved a remarkably diverse collection of behavioral adaptations," including, schooling. "Many species of small bony fishes school together, allowing them to effectively appear comparable to a large animal, a behavior which is said to discourage predators." They explain that individual fish have a lesser chance of being eaten by a predator when in a school than when alone. Schooling also functions in other ways, for example, facilitating the location of food sources, posing a hydrodynamic advantage and increasing reproductive success.

Spawning aggregations are schools that consist mainly of mature individuals that come together for the purpose of reproduction. Cod, for instance (family Gadidae), often form spawning schools.

Migrating schools of bony fish, which form along particular navigation routes, often transform into other types of schools, such as spawning schools. Salmon (family Salmonidae), for example, form migrating/spawning schools as they travel upstream to spawn.

Breeding seasons sometimes influence the development of what has been termed social behavior in certain fishes. Large predatory bony fishes, such as groupers (family Serranidae), for instance, often demonstrate asocial behavior except during breeding seasons.

Many species of small fishes, such as northern anchovies (*Engraulis mordax*), form other types of schools. As an example, schools may develop in feeding grounds, primarily due to the concentration of food organisms. Feeding schools can be made up of many different species of bony fishes, representing different developmental stages.

Seasonal changes may result in yet other forms of schools. Wintering schools, for instance, originate in wintering grounds. Various species may congregate in these schools during the winter months for survival purposes and disband when the weather changes.

As a group, bony fishes have sharply contrasting territorial behavior (Clarke, 1970). Even though damselfishes (family Pomacentridae) are relatively small, for instance, they are fearless in defending a territory. On the other hand, most large groupers (family Serranidae) will retreat from their territory if approached by another animal.

Hofmann and Fernald (2001) point out that "among the teleosts (most of the common fishes), cichlids have attracted particular attention because their explosive speciation has produced a striking diversity of behavioral systems and corresponding sensory adaptations." This group has also gotten the attention of certain researchers because of the complexity of their social behaviors.

As an example, they use the mouthbrooding cichlid (*Astatotilapia burtoni*) from the East African Lake Tanganyika to show how a species can teach us about social influences on the brain. Its complex but easily observable behaviors and the occurrence of two distinct classes of males, those with territories and those without, have facilitated insights into how physiological mechanisms interact with social behavior.

"Maturation of juveniles," for example, "is suppressed in the presence of adult fish." Furthermore, the hypothalamic–pituitary–gonadal axis (a hormonal channeling system) is shaped by social experience, as can be seen in the change of gonadotropin-releasing hormone and neuron size after a change in social status. These neural and endocrine modifications in adult animals are reversible, as every individual can change between dominant and subordinate states. In addition, body growth depends on social status and immediate social history.

"This phenomenon can be explained in the context of life-history theory as a differential allocation of resources toward growth or reproduction, depending on social status. Putative causal factors include the neuropeptide somatostatin and growth hormone. An understanding of the social control of physiological processes gained in *A. burtoni* will be useful in understanding the evolution and neurobiology of other social systems."

Amphibians

Amphibians are not particularly social. In *The Social Behavior of Anuran Amphibians*, Kentwood D. Wells (1977) says that "Temporal patterns of anuran (frog and toad) reproduction fall into two broad categories: prolonged breeding and explosive breeding." The spatial and temporal distribution of females determines the form of male–male competition. Males of explosive breeders in dense aggregations engage in "scramble competition," attempting amplexus (clutching of females by males to facilitate fertilization) with every individual and struggling among themselves for possession of females.

Males of prolonged breeders usually call from stationary positions to attract females and often maintain some sort of intermale spacing. Many aspects of vocal behavior and chorus organization can be viewed as consequences of intrasexual competition. Males of some prolonged breeders defend all-purpose territories, oviposition sites, or courtship areas against conspecific males.

"Males with high quality territories may enhance their attractiveness to females and obtain several mates in one season. The social organization of some species resembles the lek behavior (gathering of males for competitive breeding) of other vertebrates. Males or females of some tropical species care for eggs and tadpoles, but the evolution of parental care has not yet been studied in detail."

According to S. Boysen, *The Smartest Animals on the Planet* (2009), "many species of mammals and birds have been studied for their ability to show some understanding of numerical competence. As a comparison at a lower phyletic level, red-backed salamanders were found to have an ability to discriminate (within limits) between plates that had different amounts of food."

REPTILES

In *The Evolution of Reptilian Social Behavior*, B.H. Brattstrom (1974) points out, "reptiles display a diversity of behavior that is reflective of their evolutionary heritage from fish and amphibians and their ancestral contribution to the diversity found in birds and mammals. Much of the behavior observed in reptiles seems specific to the ecological setting within which they live. As a result, a diversity of behavior is found in each of the groups of modern reptiles."

Recent studies on the social behavior of lizards have proven certain of them to be capable of a variety of behavioral postures, sequences, and sociality that exceed that found in some mammals and birds. While many species of lizards are territorial, others are hierarchical and some have harems. For all those territorial species studied, crowding results in increased social interaction, increased aggression, and a switch to hierarchical behavior.

"While smell and sound may be important stimuli for social behavior in some reptiles, posture, actions, and especially color appear to be most important in diurnal lizards. Temperature and energy studies suggest that the large extinct dinosaurs probably fought considerably less than commonly portrayed in movies and stories, but were also probably much more brightly colored than commonly shown in reconstructions," since most of the species demonstrated diurnal activity.

Birds

Most researchers now support the view that birds represent a group of animals that arose from theropod dinosaurs during the Mesozoic era.[3] Birds and dinosaurs shared features such as hollow, pneumatized bones, digestive gastroliths, nest-building, and brooding behaviors. There also are genetic relationships as well. Only a few scientists still debate the dinosaurian origin of birds, suggesting descent from other types of archosaurian reptiles.

Birds vary in their degree of sociality, but the subject is currently undergoing extensive investigation in order to understand its significance to other social groups (Alcock, 2001). In short, the social biology of birds requires additional investigation.

In *Cognitive Adaptations of Social Bonding in Birds* (2007), Nathan J. Emery et al., point out that "while most monkeys and apes form stable groups, most birds are monogamous and only form large flocks outside of the breeding season (Emery et al., 2007). Some birds form lifelong pair bonds, and these species tend to have the largest brains relative to body size. Some of these species are known for their intellectual abilities (e.g., corvids and parrots), while others are not (e.g., geese and albatrosses)."

These authors present empirical evidence that "rook and jackdaw partnerships resemble primate and dolphin alliances." Although social interactions within a pair may seem simple on the surface, they argue that "cognition may play an important role in the maintenance of long-term relationships," something they refer to as "relationship intelligence."

Because of differences in the social nature of birds and mammals, Nathan Emery et al. found it necessary to further compare these two groups. In early discussions of the social intelligence hypothesis (SIH), "a number of biological, ecological, and behavioral preconditions were proposed as essential to the presence of sophisticated social processing in primates. These were a large brain, long developmental period before maturation, individualized social groups, and extended longevity. Indeed, in primate species with complex social systems…, these preconditions are fulfilled. In addition, primates with large brains (or more precisely, a large neocortex ratio when compared to the rest of the brain) tend to form larger social groups than species with a smaller neocortex ratio."

According to these authors, whether there is any relationship between the avian brain and sociality is unclear. They suggest that such inconclusiveness may be because avian social systems are flexible, depending on changing season or environmental conditions, and that earlier analyses were based on factors that may not be good indicators of social complexity, such as group size.

"Some birds have brains that, relatively speaking, are the same size as those of chimpanzees (after removing the effects of body size). Although not directly comparable to the neocortex, the most extensive dataset in birds for a more specific brain area is the size of the telencephalon (forebrain, the anterior part of the prosencephalon)." It appears that there is a strong relationship between social complexity and size of this structure in birds. However, as these authors have stated, "this analysis used a strange social category, 'transactional,' that included those species which demonstrated 'complex' forms of behavior (not necessarily social), such as ceremonial dancing, communal gatherings, fission–fusion societies, memory performance, food sharing, 'parliaments,' 'weddings,' aerial acrobatics, social play, milk-bottle opening, and other problem-solving."

They repeat that "the social organization of most birds tends to be very flexible, both temporally and spatially." As such, species may vary in a social system based on their geographical location and its different ecological pressures. A harsh environment in which food is scarce or difficult to locate could make raising healthy offspring more challenging. Long-term monogamy may be favored in these conditions, as parents that cooperate in raising their offspring year-in and year-out may gain an advantage that would not apply in less harsh environments.

"The avian brain may possess a similar social behavior/chemical network to the mammalian brain (Fig. 5.2). For example, an analysis of arginine vasotocin (the nonmammalian equivalent to arginine vasopressin and oxytocin) distribution in the brains of various species of birds in the family Estrildidae (grass finches and waxbills), which are all monogamous but which differ in species-typical group size, revealed patterns of receptor binding in the parts of the brain associated with sociality."

In their concluding remarks, these authors say that "although rooks (and possibly other large-brained birds) may have evolved similar sociocognitive abilities to primates, such forms of social knowledge appear to be used primarily within the context of the pair bond, rather than applied to a larger social network, such as [that] found in primates." Long-term monogamy depends on different forms of social information processing compared with polygyny (the most common mammalian mating system). For example, recognizing the subtle social signals produced by a partner and using such information to predict their future behavior suggests different social skills than remembering who did what to whom.

"Indeed, long-term pair-bonded species, including those which form cooperative breeding groups, appear to have the largest brains within birds. There are differences between long-term monogamous species, in both brain size and the complexity of their partnerships, something which is highlighted by the social relationships of young rooks. The question of whether these partnerships are cognitively sophisticated, particularly whether pairs have an advantage in behavior-reading (especially when their partner is the one providing the social cues), remains to be tested."

Boyson points out several studies that reveal bird intelligence. Woodpecker finches on Santa Cruz in the Galapagos Islands, for instance, "use sticks and cactus spines as probing tools in the more arid climates on the island, compared to finches living in evergreen forests that have abundant food."

New Caledonian crows have been found to "carve out sticks, twigs, and portions of leaves to help them obtain food." Boyson states that "New Caledonian crows have exceeded the skills demonstrated by all other nonhuman species because they seem to recognize the functional requirements of a tool for a specific search effort, and are also the first species to use hooks as tools."

Captive New Caledonian crows were found to be capable of using different types of tools to exploit different food sources. Boysen states that only the great apes—especially chimpanzees and humans—have shown similar flexibility and ingenuity in the use of tools.

Many species of birds are excellent vocal mimics. "They may also be able to imitate the actions of others, both in Simon-says-like copying games and when solving puzzles for gaining food."

European starlings can mimic the ring of a phone. Corvids (crows, ravens, and magpies) have larger brains than most birds for their body size. The crow, for instance, has the same relative brain size as the chimpanzee in terms of its brain/body ratio. According to Boyson, "birds completely lack a cortex and instead have a portion of brain called the hyperstriatum that allows their brains to perform a variety of functions." The hyperstriatum in corvids is larger than it is in other birds.

Mammals

Mammals, like birds, arose from the reptile line in the Triassic period, more than 225 million years ago (Fig. 3.1). They had differentiated teeth (incisors, canines, molars) and large brains, hair, and mammary glands. They developed into three main groups: the egg-laying monotremes, pouched marsupials, and eutherians (often referred to as placentals). By this time, the notochord, which initially was separate from the dorsal hollow nerve cord, had grown into a very complex segmental skeleton to become an integral part of the vertebral column, and the axial and appendicular skeletal systems functioned together.

As a group, mammals exhibit various degrees of sociality. Carnivores, such as cats, some bears, mustelids (weasel-like mammals), and viverrids (civets and mongooses), are more solitary than social. Their degree of sociality often depends on the various ways they have developed to maximize their success in survival strategies and reproductivity.

Most species of mammals considered to be social live in well-defined groups, especially when raising young. Thus, social groupings tend to vary with season, breeding, life cycles, food availability, and habitat.

Emery et al. (2007) point out that the SIH has primarily been discussed with respect to the evolution of primate intelligence. However, it subsequently has become clear that "other social animals, such as dolphins, hyenas, and elephants, also demonstrate many of the biological, ecological, and behavioral preconditions for intelligence; sophisticated cognitive skills in the laboratory and field; and many of the complex social skills found in monkeys and apes."

All mammals, however, do not fit this picture. In relationships between brain (neocortex) size and social group size, a significant relationship has been found to be apparent in primates, carnivores, insectivores, bats, and cetaceans, but no such relationship was found to exist in the ungulates (hoofed animals).

Coati Mundis

Coatis (several raccoon-like species in the genera *Nasua* and *Nasuella*) are highly gregarious and have a wide range of social behavior, including cooperative grooming, nursing, vigilance, and evident defensive behavior. Bands are usually slightly antagonistic to other bands when they meet. However, peaceful interactions also occur, and are occasionally characterized by intergroup grooming sessions. Adult males that approach bands other than their own are generally chased away by one or several band members, including both juveniles and subadults. Females often groom one another in an excited manner following such chases. During the breeding season, males are submissive to females and immatures, and occasionally groom them. Bands often break into subgroups for periods varying from several hours to 1–2 days.

Dogs and Wolves

Dogs (*Canis familiaris*) are either feral or domestic canines that have been selectively bred from the gray or timber wolf (*Canis lupus*). Within the past 2000 years, dogs have developed into hundreds of breeds by this method. Domestic dogs, however, have carried over some of the behaviors demonstrated by their wild counterparts. When existing as feral animals, for instance, they demonstrate strong social behavior, often hunting in packs.

Dominant wolves, like their domestic counterparts, walk in an erect and proud style, with stiff legs and a deliberate pace, their head, tail, and ears up. They stare freely and casually at others of their species. When in the presence of potential rivals, the pelts of wolves and some dogs become bristled, they curl their lips to demonstrate their teeth, and dominants obtain their choice of food, space, and mates.

Submissive displays have been shown to include adopting a posture that is physically lower than other (more dominant) dogs, such as crouching, rolling over on the back and exposing the abdomen, lowering the tail (sometimes to the point of tucking it between the legs), flattening of the ears, averting the gaze of dominants, nervously licking or swallowing, dribbling of urine, and freezing or fleeing when other dogs are encountered.

Recent research on wolves has indicated that dominant behaviors have been misinterpreted as personality traits that determine the individual's place in a linear hierarchy in the pack. In contrast, some researchers argue that packs are family units, and the "alpha" individual of a pack does not change through struggles for dominance. Rather, they claim that the family unit serves to raise the young, which subsequently disperse to pair up with other dispersed wolves to form breeding pairs and packs of their own.

Research on domestic dogs (*C. familiaris*) has also questioned whether dominance is a personality trait (Draper, 1995), and a list of 638 human personality traits by the Massachusetts Institute of Technology did not include dominance (MIT, 2015). In a behavioral study using 15,000 dogs of 164 breeds in an attempt to identify major personality traits, researchers performed a factor analysis of their data, resulting in the identification of five major traits: "playfulness," "curiosity/fearlessness," "chase-proneness," "sociability," and "aggressiveness." Two major factors were identified in social behavior: "confidence" and "aggression–dominance."

These studies suggest to some researchers that dominance, per se, may not be a personality trait. Rather, underlying personality traits, such as aggressiveness, confidence, and curiosity, may affect the prevalence of dog behavior that is viewed as dominance. However, refer to the original definition of aggression in Chapter 1 that was defined in terms of canine behavior.

Lions

Cats are generally considered to be solitary, nocturnal stalking mammals, existing together only as mating pairs or mothers with offspring. Lions are an exception (Barnett et al., 2006; Smuts, 1982). Currently existing in the wild only in Africa and India, although they have, at one time, been known to be native to Africa, India, Asia, and Europe. They exist in prides. They are territorial, the pride being dominated by an adult male or coalition of males.

Prides are long-term social groups. Individuals within a pride remain together for generations and consist of females and their offspring. Two or three unrelated males or four or five related males form a coalition and defend a territory against other male coalitions. Such males mate with the pride's females over a period of two or three years, at which time they may be driven out by another group of males.

Pride members often demonstrate cooperative hunting, fanning out to close in on their prey. While females are the hunters, males have first access to the prey.

Lions are known to communicate with one another through roars, determining potential threats because of how many roars they hear in or around their pride. According to Boysen, roaring serves "two purposes: first, communicating to potential intruders the numbers they are up against and second, summoning help."

Boysen also states that "most animal cooperative behavior, such as group hunting in lions, appears to be opportunistic rather than intentionally coordinated. However," she says, "there is some evidence from studies of primates that they do monitor each other's behavior when jointly working on a problem."

Elephants

Three living species of elephants are currently recognized: the African bush elephant, the African forest elephant, and the Indian or Asian elephant (Alter, 2004; Carrington, 1958; Debruyne, 2005). Some treat the two African species as one species, and still others recognize the existence of a fourth species in West Africa. While there are anatomical and geographical differences between the species, their behavior is generally similar. I will use the African bush elephant as an example of this group (Cranbrook et al., 2008; Fernando et al., 2003; Laursen and Bekoff, 1978; Shoshani, 2005; Shoshani and Eisenberg, 1982).

African elephants usually exist in groups of 6 to 12 animals under the influence of a matriarch (a female dominant), generally including her female offspring and their young. Larger families of 12 to 25 individuals are also common, and matriarchal societies may include sisters and their offspring. Such groups of related individuals are called bond groups, and bond groups may form larger units called clans.

Males born into the society may remain in the group but gradually grow more independent and spend more of their time on the outskirts of the group. They eventually leave the family, joining other males of varying ages in groups called bands of bulls.

Dominance behavior between males can appear very fierce, but typically it is not life-threatening.[4] Most of the bouts are in the form of deceptive aggressive displays. Generally, smaller, younger, and less-confident animals back away from agonistic encounters before encounters become threatening. Battles during the breeding season can get very aggressive, and elephants competing for dominance at that time are occasionally injured. During the breeding season, bulls approach and usually display agonistic behavior with any other males they encounter. Otherwise, they spend most of their time hovering around females to find a receptive mate.

As tool users, elephants have been found to modify branches and twigs to make switches that are used to keep flies at bay. They also cover themselves with mud, which protects their skins from the sun. It may also protect them to a degree from parasitic arthropods.

Female elephants are known to communicate over long distances through rumbles (infrasound) that are below the frequencies heard by humans. Elephant infrasound has been recorded over six miles (10 km) away from its source and covering a range of 38 square miles.

Boysen states that "infrasounds are also known to be produced by squids, lions, giraffes, whales, alligators, some species of birds, and the rhinoceros." Some species hear such sounds "through the ground or seismically while others detect it through the air or atmospherically." Still others, for example, whales, are able to process it through water.

Within recent years, it has been determined that elephants are capable of recognizing themselves in a mirror, a behavior that is also known for great apes and dolphins. Elephants have also been able to use mirrors to locate food items that are placed out of direct sight (Boyson, 2009).

Cetaceans

Dolphins and porpoises are social mammals that engage in cooperative and compassionate behavior (Mann et al., 1999; Mead and Brownell, 2005). Even cooperative defense is common. They have been known to remain with other members of their group that are injured and even charge attackers.

They travel in groups of variable size, which facilitates both their cooperative hunting and reproductive strategies (mating and care of the young). Their grouping behavior also serves as protection from predators. Herds vary in size, depending on species-specific hunting strategies and the amount of food that is available.

Mother–calf bonds form the base for their social behavior. Dolphins and porpoises are polygynous (one male has a number of females with which they breed), but males and females often swim in separate groups. In anticipation of mating, males approach females and often demonstrate aggressive behavior toward other males that seem interested in mating. Aggressive individuals present their body in a perpendicular fashion, expose their teeth, snap their jaws, jerk their head or tail rapidly, and swat or ram an animal that is the object of their aggression.

Both leadership within cetacean herds and their herding behavior are described by researchers as fluid. Dominant individuals appear to steer the course of a group from the rear of the herd.

Orcas, the largest dolphins (found throughout the world's oceans), are versatile and opportunistic predators, feeding mostly on fish, seals, sea lions, and some whales (Morell, 2011). They have a sophisticated social behavior, existing in small nuclear and extended families that are referred to as matrilines. The central figure in their pods, clans, and communities is the matriarch (older, dominant female). She and her offspring generally remain together throughout life.

Orca females are able to reproduce by 11 to 13 years of age. Maturing females care for their younger sisters (i.e., they are alloparents) until they have their own offspring. Males mature between 12 and 14 years of age.

Pods are extended family groups of related females that travel together. Occasionally, different pods cooperate in sharing the same area. A clan consists of different pods that share a common language. Interestingly, pods from separate clans often socialize with one another within a community in spite of language differences, cooperation, coordination, trust, and acceptance being clearly evident in their societies.

There are cases in which transient groups of seven to eight nonfamily orcas travel together. Overtly expressed aggressive behavior between either males or females has never been observed in orcas.

In all of these social species, dominance establishment is an important method of selecting those individuals that appear to have the qualities desired to provide progeny for the next generation. Thus, it is part of the process of natural selection.

Boysen points out that wild female dolphins of western Australia use sponges as tools to protect their beaks while foraging, the behavior being passed on to their daughters through observational learning.

Bottlenose dolphins are known to produce distinctive whistles underwater that emanate from their head. As the sounds bounce off objects, it provides them with a "sound picture" of what lies ahead. Each dolphin may have its own unique signature whistle, which may be used by other dolphins as a recognition sound.

Further testing of dolphin intelligence determined that they respond to mirror recognition just as humans and other apes do. In tests of cognitive capacities, it has been found that they understand gestured commands, including sequences of gestures that require complex behavioral responses.

These examples of social invertebrates and vertebrates illustrate that there are common features that relate to a social existence in many animals, but each group and species expresses its own characteristic form of sociality, dominance, and aggression, which often play a significant part in their lives. We find this is true of primates in the next chapter.

Note

Superscript numerals appearing in this chapter refer to additional text/explanation given in the appendix.

5

From Whence We Came: Primates

Non-human primate societies are primary groups, as apparently were human societies before the adoption of agriculture. Everyone operates in one or more primary groups, holding a position in one or more face-to-face dominance hierarchies. *Mazur, A., 2005.* Biosociology of Dominance and Deference

THE RISE OF PRIMATES AND THEIR PROSIMIAN BEGINNINGS

About 80 million years ago, prior to the mass extinction of reptilian dinosaurs and termination of the Mesozoic era, insect-eating mammals (insectivores) had diverged into many forms. Between 25 and 35 million years ago, about 30 million years after the close of the Mesozoic era, a group of these mammals had evolved certain characteristics that would influence further development toward the group currently referred to as primates (Chatterjee et al., 2000; Janečka et al., 2007; Pozzi et al., 2014) (Fig. 5.1).

Some of the major changes in mammalian features that led to the primates include an increase in size and complexity of the brain, a broader diet, binocular vision, opposable thumbs and toes, and the development in most cases of thin nails in place of claws. Over time, the brain of these animals continued to increase in size and change in complexity (Fig. 5.2). New types of nervous tissues even developed (Ramachandran, 2011; Simon, 1999; Thompson, 2000).

Their tails grew, some forming a fifth limb by being prehensile (New World monkeys) (Sussman, 2003). Others retained their tails but were unable to use them in a prehensile fashion (Old World monkeys). Some forms of primates later lost any external evidence of a tail (the Apes) but retained a tail remnant referred to as the coccyx (Postacchini and Massobrio, 1983).

While primates were destined to disperse to many tropical locations in both the New and Old World, distinction between tailed monkeys and apes (nontailed primates) actually occurred in the Old World, and divergence between apes continued for many millions of years until the rise of what Desmond Morris and others have referred to as *The Naked Ape* (1967). The entire story of the human animal's rise to stardom is a fascinating romp through natural selective changes that led them to become the most dominant animal of the Cenozoic Era and the most influential animal in geological history (Chatterjee et al., 2000; Pozzi et al., 2014).

Table 5.1 lists the major events and species that occurred during the appearance of chordates and what led to the rise of humans within the past few million years. As the reader see, our species, often referred to as modern humans (*Homo sapiens*) has existed on Earth for about 200,000 years.[1]

Our dominant status has been accompanied by the retention of old and development of new behavioral expressions, sometimes with a serious divergence of an aberrant nature. It is clear that we are not totally the animals we came from.[2] However, all organisms change over time. It is a characteristic feature of evolution which is never ending.

To thoroughly understand our significance as the primate elite, it is necessary to examine the entire primate group and discuss the various anatomical, behavioral, genetic, and phyletic (evolutionary) characteristics expressed by various members of the group (Fig. 5.1). After all, our behavior has been shown to be linked with that of certain other primates. The nature of the human animal is focused upon in Chapter 8, and the aberrant nature as influenced by our genetic make-up, and some forms of dominant behavior are discussed in Chapter 9.

The mammalian brain which eventually became associated with primates was undergoing rapid evolutionary changes (Keverne et al., 1996). In comparing plesiomorphic (old) and apomorphic (new) features of the mammalian

Dominance and Aggression in Humans and Other Animals
http://dx.doi.org/10.1016/B978-0-12-805372-0.00005-5

FIGURE 5.1 Primate phylogeny, including prosimians and simians (anthropoids). The earliest primates (or mammals approaching the primate line) that can be discerned in the fossil record have been mouse-sized creatures like *Purgatorius* (Bloch et al., 2006; Buckley, 1997; Clemens, 1974), a mammal that lived on insects, and the somewhat later *Plesiadapis*, a mammal that probably ate mostly leaves and fruit. As seen in the figure, chimpanzees are the closest relative of humans, sharing most of its DNA.

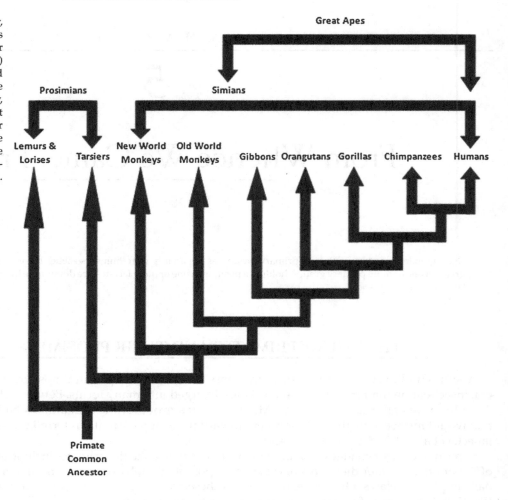

FIGURE 5.2 Comparison of animalian brains. While the cerebrum exists in all animals shown, it is larger in those that are phyletically advanced. In mammals, the cerebrum also takes on a convoluted appearance, showing well-defined giri and sulci. Convolutions become more extensive in the primates, especially humans. At the same time, there is a dwindling of the olfactory bulbs.

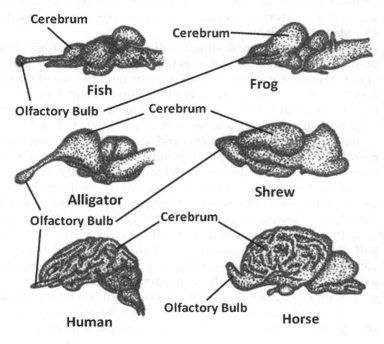

TABLE 5.1 Geological Time Scale, With Emphasis on the Cenozoic Era

Time Scale (Years BP)	Era	Period	Epoch	Events
	Cenozoic/ Quaternary recent			Industrial revolution, 1750–1850. Commencement of agriculture and a more sessile existence, 10–12TYA.
10,000			Pleistocene	Humans dispersed out of Africa about 100–50TYA. Earliest definite *Homo sapiens*, about 200TYA. Archaic *H. sapiens* was developing 400–250TYA. *Homo neanderthalensis* lived in Europe from 400–300TYA. *Homo heidelbergensis* lived from about 800 to 200TYA and may be the predecessor of *H. sapiens. Homo erectus* and *H. ergaster* were first to leave Africa and spread through Africa, Asia and Europe, about 1.8–1.3MYA. *H. erectus* and *H. ergaster* lived from about 1.8MYA to 70TYA. *Homo rudolfensis* and *H. georgicus* existed from about 1.9 to 1.6MYA.
2,000,000	Tertiary		Pliocene	*Homo habilis* existed from 2.5MYA until about 1.4MYA. *H. erectus* and *H. ergaster* first dispersed out of Africa about 2.5MYA. *Homo habilis* was probably the earliest member of the genus *Homo. Australopithecus africanus*, 3–1.7MYA. Oldest known tools used by humans (Oldowan stone tools), 3.4–2.6MYA. *Kenyanthropus platyops* (between *Australopithecus* and *Homo*, 3.5–3.2MYA. *Australopithecus anamensis* (earliest hominid with absolute upright bipedal features), 4.2–2.9MYA. *Ardipithecus ramidus*, 4.4MYA. *Ardipithecus kadabba*, several African localities, 5.8–5.2MYA. *Orrorin tugenensis*, North Kenya, 6MYA. Human/chimp split, 7–5MYA. First hominine, *Sahelanthropus tchadensis*, west Africa, 7MYA. Early hominid bipeds were *Sahelanthropos, Orrorin*, and *Ardipithecus*. Gorillas split off from human/chimp line, 8MYA. About 9–8MYA, descendants of dryopithecines in Africa diverged into two lines, one that led to gorillas and another to humans, chimpanzees, and bonobos. Several hominine genera in Eurasia and Africa, 13–9MYA.
12,000,000			Miocene	Less hospitable cooler conditions in northern hemisphere resulted in extinction of certain primates while others moved to Africa and South Asia. Ponginae (orangutans) diverged from Hominidae predecessor about 14MYA. The group that included our ancestors (likely members of the genus *Dryopithecus*) was adapting to life on the edges of expanding savannas in southern Europe about 15MYA. Great apes (Hominidae) diverged from hylobatids about 20–12MYA. South America reconnected with North America and invading placentals forced the extinction of most marsupials there. One of earliest of monkey-to-ape transitional primates (*Proconsul*) lived in African forests 21–14MYA. Numerous catarrhines appear about 22MYA. Earliest apes (proconsuloids) moved among tops of large branches in humid forests of east Africa in search of fruit, 23–16MYA. Apes apparently evolved from monkeys early in this epoch. Fossil prosimians and monkeys are comparatively rare from most of Miocene, but apes are common. First known Old World monkeys (catarrhines), dated at 24MYA. First mastodons, giant ground sloths, primitive dogs, horses, and antelopes.
26,000,000			Oligocene	First true carnivores, first cats. First true monkeys appeared about 34MYA.
36,000,000			Eocene	By end of Eocene, many prosimians had become extinct. India began crashing into Asia at a rate of 25–30cm/year about 50.5MYA. First rodents, rhinoceroses. Earliest primate fossils, 50–55MYA, about 10–15million years after nonavian dinosaurs became extinct. Early Eocene coincides with emergence of placental mammals. Placental mammals with larger bodies and bigger brains began to appear. Along with an increase in atmospheric oxygen, an abrupt global warming of 5–9°C lasted at least 200,000years.
			Paleocene	Transitional primate-like mammals were evolving by the end of the Mesozoic era. India was not yet part of Asia but was heading toward it at about 20cm/year. South America had drifted away from Africa and was not connected to North America after 80MYA. First hoofed mammals. First primates, the earliest primate fossils appearing about 55MYA.
65,000,000	Mesozoic	Cretaceous		Last nonavian dinosaurs. Last toothed birds. Primate evolution most likely began in the late Cretaceous. Early primates (prosimians) diverged from other mammals about 85MYA.

Continued

TABLE 5.1 Geological Time Scale, With Emphasis on the Cenozoic Era—cont'd

Time Scale (Years BP)	Era	Period	Epoch	Events
135,000,000		Jurassic		The supercontinent Pangaea began to break up about 100MYA. First flowering plants. Giant dinosaurs. First birds. First mammals (Hadrocodium wui, about 195MYA). Most mammal species were small, ranging in size from a mouse to size of a medium dog.
180,000,000		Triassic		Last seed ferns, first dinosaurs.
225,000,000	Paleozoic	Permian		Last trilobites. Last euryptids. First modern insects.
280,000,000		Pennsylvanian		First conifers. First primitive insects (wingless). First reptiles. Large coal swamps. The supercontinent Pangaea formed about 300MYA.
310,000,000		Mississippian		
355,000,000		Devonian		First ferns appeared about 360MYA. First fungi. First sea-bearing plants. First amphibians. First ray-finned fish, 390MYA.
405,000,000		Silurian		The oldest known bryophytes (mosses and their relatives) appeared in the Devonian (409-354MYA). First land animals. First cartilaginous fish, 409MYA. First lobe-finned fish, 412MYA.
440,000,000		Ordovician		First bryozoans. First vertebrates appear as jawless fishes. 500MYA.
500,000,000		Cambrian		The supercontinent, Panotia was breaking up about 541MYA. Cambrian explosion, resulting in great diversity in most groups of organisms.Tunicates (urochordates) existed in shallow marine waters about 555–548MYA.
600,000,000	Proterozoic			The supercontinent, Panotia, had developed by 600MYA years ago. Brachiopods, marine worms (trails and burrows).
	Archeozoic			Marine bacteria, marine algae.
3,500,000,000				Age of oldest photosynthetic bacteria on Earth (cyanobacteria).
4,000,000,000				Approximate time of first life on Earth.
4,500,000,000				Age of Earth.

The Time Scale has been constructed with the most recent events at the top. Material in this table was assembled with the literature cited in the text (*see review: Anonymous, 2016. Human Evolution. Wikipedia. https://en.wikipedia.org/wiki/Human_evolution*). While the table is not complete, it gives some idea of the most important forms of animals represented in the literature, the times they existed and some of the events that accompanied them. As is evident, the Miocene was the ape's heyday. All known human species have not been included, and times may vary with different authors and as the result of new investigations. *BP*, before the present; *MYA*, million years ago; *TYA*, 1000 years ago.

brain, there are elements of the primate brains that had been present in reptiles, namely the parts associated with automatic functions (Fig. 5.2).[3]

The medulla oblongata functions as a relay station between the spinal cord and other parts of the brain (Hughes, 2003). Control centers for heart rate, blood pressure, and breathing are located there, as well as reflexes that control such functions as vomiting and coughing. Our state of consciousness, nerves that control facial expressions, eye movement, olfaction, and vision are influenced by the medulla as well.

Nerves concerned with both fine and gross motor movements that extend from the cerebral cortex to the cerebellum pass through the pons. Movements (e.g., buttoning a shirt or walking) are examples of functions influenced by these nervous elements.

The cerebellum functions as a central control unit for motor activities, such as learning how to play a musical instrument, and helping to perfect such movements. Finally, it assists in making learned movements more automatic.

Humans and all other mammals have retained this reptilian brain, as well as adding new sections of nervous tissues to the brain over the subsequent millions of years (Fig. 5.2). The most outstanding new section of the human brain, the section that has increased most in size and complexity, is the forebrain (Dobbing and Sands, 1973). While there are a number of important functions of this section of the brain, the parts that are uniquely human are primarily parts of the cerebral hemispheres. These structures are subdivided into several separate lobes that are demarcated by deep fissures (Fig. 5.3).

The front of the human brain is associated with the integration of perception, a function that is coordinated with tissues at the back of the brain. The thalamus stands at the intersection between the two, helping us to understand our sensory perception (Fig. 5.3A).

As evolution progressed in primates, the brain continued to evolve at a rate so rapid that the process has come to be referred to as "the brain explosion" (Rilling and Insel, 1999; Striedter, 2004).[4] Much of the change in brain evolution has occurred in the cortex (the outer forebrain tissue), resulting in a quite new tissue referred to as the neocortex (Fig. 5.3), the tissue that encapsulates underlying parts of the brain. In humans, it is about 2.5 sq. ft in area and less than an eighth of an inch thick. In adults, it contains about 100 billion neurons that are supported and nurtured by billions of glial (nurse) cells.

Over millions of years, the brain of primates had increased tremendously in size, although brain size and complexity had changed at different rates in different species. While orangutans are somewhat larger than humans in

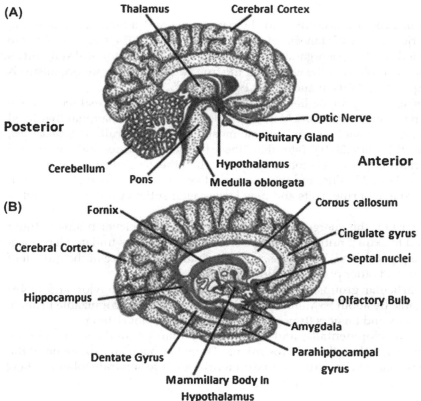

FIGURE 5.3 Sagittal views of human brain. (A) General, including the centrally located thalamus. The front of the human brain (on the right in the drawings) is associated with the integration of perception, a function that is coordinated with tissues at the back of the brain. The thalamus stands at the intersection between the two, helping humans to understand their sensory perception. (B) Parts of the limbic system, including cingulate gyrus, parahippocampal gyrus, hippocampus, dentate gyrus, amygdala, septal nuclei, mammillary bodies, olfactory bulbs, fornix, and associated tissues. The limbic system is sometimes referred to as the "emotional brain" because of its primary role in emotions, including affection, anger, docility, pain, and pleasure. It also functions in memory.

body size, for instance, their brains are 69% the size of the human brain. In addition, the surface of the human brain is more deeply fissured.

There are also several types of neurons (nerve cells) that appear to be unique to humans and higher primates. Examples are spindle cells which may integrate emotion with motor function and mirror neurons which may encode performance and ideas of action and abstract concepts. It was the continued development of the forebrain that led primates to higher cognitive levels, providing insight into new socially related survival strategies in the form of more complex group living (Ciochon and Fleagle, 1987).

All primates live social lives in one way or another, and thus many of them band together to cooperate in searching for food, for child care, and for defensive reasons. Their social significance is not as clear in some of the early primate forms. We must not lose sight of the fact that agonistic behavior need not be expressed on a continuous basis. Once dominance is established, dominance and defense may only occasionally require subtle clues in short spurts. Thus, we would expect less agonistic behavior in an organism's day once the initial stress of dominance establishment is complete.

While territorial disputes and agonistic behavior common to dominance establishment are often emphasized in characterizing primate social behavior, there are other behavioral features that may be just as important or possibly more important to understanding its true nature. Based on studies of 60 different primate species, 10% of their time was found to be spent in performing affiliative interactions (e.g., food sharing, grooming, playing, and huddling) and in establishing alliances (Jack, 2003). Only 1% of their time was spent in demonstrating aggressive interactions (fighting, visual and vocal threats, and subordinate behaviors). Thus, affiliative behavior has apparently increased along with brain size, being 5% higher in the great apes than in other primates (Kling and Steklis, 1976). While agonistic behaviors are important in establishing a dominance hierarchy and group harmony, affiliative interactions form a strong basis for social cohesion in all primates as well.

While development of the neocortex and frontal lobes have largely been responsible for the human animal's dominant status within the major ecosystems on the planet, genetic changes and social diversity have sometimes led to the development of abnormal behavioral expressions toward other of our male and female fellow humans, their offspring, and other biotic and abiotic components. Although these features of human behavior are discussed in Chapter 9, we must first take a comparative look at our predecessors.

Primate Rankings

Biologists rank the phylogenetic position of primates according to their primitive and advanced anatomical, behavioral, and genetic characteristics (Chatterjee et al., 2000; Janečka et al., 2007; Pozzi et al., 2014) (Fig. 5.1). Prior to the acquisition of knowledge about DNA, phyletic ranking of organisms largely depended on anatomical similarities and differences. However, convergent evolution complicated the issue, and the true nature of all extant organisms is found in their comparable genomes (DNA sequences) (Watson and Crick, 1953a,b).

The earliest primates (or mammals approaching the primate line) that can be discerned in the fossil record have been mouse-sized creatures like *Purgatorius* (Bloch et al., 2006; Buckley, 1997; Clemens, 1974), a mammal that lived on insects, and the somewhat later *Plesiadapis*, a mammal that probably ate mostly leaves and fruit. In *Neanderthal, Neanderthal Man and the Story of Human Origins*, P. Jordan (2001) states that "these [early insectivores and herbivores] were the first of the primate line to possess brains larger in relation to body size than the general run of animals of their time; they must have had a use for those enlarged brains, and that use most likely related to their cleverness in spotting and securing high quality food in the shape of ripe fruits and new leaves against the visual background of a noisy clutter of branches and foliage."

"High quality food," Jordan says, "could sustain the energetic nervous activity of their clever brains without requiring the elaborate and sizeable gut needed to extract nutrition from poorer food." Thus was initiated a promising evolutionary feedback mechanism involving good food, braininess, and a reduction in the size of the gut, all of which have played an important part in human and other primate evolution (Jordan, 2001).

Primates are generally divided into two particular groups, depending on their anatomy, behavior, and DNA: prosimians (ancestral to the simians) and simians. By about 55 million years ago, the first truly recognizable primate forms (the prosimians), resembling today's lemurs and tarsiers (types of prosimians), were flourishing.

Simians are more advanced behaviorally, developmentally, and anatomically than prosimians. The simian brain case, when compared to their bodies, is larger than that of prosimians. There has been an expansion of the brain's cortex in these higher primates, with increased fissuration and differentiation into separate lobes (Gilbert et al., 2005).

PROSIMIANS

Dating from the early Cenozoic era (about 50 million years ago), the group referred to as prosimians (Fig. 5.1) currently includes 77 known extant species of primates in several families that are primarily nocturnal arboreal forest dwellers, although there is considerable variation in their habitats. They communicate with one another through calls, scents, and visual displays to attract group members and mates, as well as functioning in repelling territorial intruders (Sussman, 2003).

As the most primitive assemblage of primates, contemporary prosimian species have retained some ancestral mammalian features. They generally have a long snout, and their small cranial cavities house brains that have retained significantly large olfactory and auditory organs (Fig. 5.2). Although they have binocular vision, their eyes are encapsulated in large bony orbits.

Prosimians have a short, well-defined breeding season, and they usually raise their young in nests or tree holes. Other than the primate features of stereoscopic vision and opposable thumb (along with a slightly larger brain), these and some of the following features are typical of nonprimate mammals.

Most prosimians have finger and toe nails rather than claws (Ashton, 1975). However, a digit or so in certain species have retained claws like those found in other mammals. The uterus of all prosimian females is divided into two bifurcating lobes like those of most other small- to medium-sized mammals (referred to as a bicornuate uterus), while the uterus of simians is undivided.

While menstruation is well developed in Old World Monkeys and apes, it is lacking in lemurs and lorises and poorly developed in tarsiers and New World monkeys (Fichtel and Kappeler, 2009). Some prosimian females have two to three pairs of teats associated with their mammary glands while all simian females have one pair.

Dominance is often expressed in prosimian females, appearing to be the rule in many Madagascan lemurs (Dunham, 2008). Some researchers have claimed that a combination of the following features has shaped the evolution of female dominance so group-living females can minimize energy expended during resource competition with males: altricial offspring (pattern of development in which the young are dependent upon their parents for their needs); a high degree of prenatal investment; a low basal metabolic rate; and rapid neonatal growth rate.

Researchers have found that captive male and female ring-tailed lemurs show different behavioral strategies, depending on their age, rank, and the dispersal of food to meet daily energetic demands and reproductive output. Highest-ranking (older) males, which are apparently deferent (subordinate) to females when they feed in small food patches, are successful in mating, possibly because deference to females results in low levels of aggression between females and high-ranking males (Dunham, 2008).

In the remainder of this chapter, we discuss selected species of primates by group, including both prosimians and simians. Humans are discussed more comprehensively in the following chapters. The following types of primates are prosimians.

Lorises

Eleven extant species of lorises are found in Africa, South Asia, and Southeast Asia (Fig. 5.1). All lorises are nocturnal and insect eaters. In the Smithsonian handbook, *Mammals*, J. Clutton-Brock and D. E. Wilson (2002) list the slender and slow loris as examples. They both demonstrate the characteristic slow movements of lorises, which help them to be cryptic and escape the notice of predators. The potto and angroantibos are also members of the family to which lorises belong. The potto, from West and Central Africa, feeds on fruits, leaves, sap, fungi, and small animals. Angwantibos, from West Africa, forage on insects (Brandon-Jones et al., 2004).

Galogos

Galogos are referred to as bushbabies because of their vocal demonstrations. Clutton-Brock compares the greater galogo, which grows to about 16 in., and the South African galogo, which grows to about 6.5 in. They both occur in Central, East, and South Africa. While the South African galogo makes kangaroo-like vertical leaps up branches, the greater galogo is unable to use its hind legs for landing (Groves, 2005).

Lemurs

There are 35 extant endemic species of lemurs in Madagscar (Godinot, 2006). They are quadrupeds, living in large groups of males and females. Their societies are female dominant. Sexual behavior is seasonal. Clutton-Brock discusses several lemur species. The fat-tailed dwarf lemur from West and South Madagascar lives in primary and secondary dry forest. During certain times of the year, it sometimes becomes torpid to overcome food shortages. Upon revival, it commences feeding at night on fruits, other vegetation, and insects.

The ring-tailed lemur is very sociable, often being found in groups of 5–25 and with a hierarchical group of adult females who dominate the males and defend their territory with loud calls. Brown lemurs, from Northern and Western Madagascar, are generally found in fluid groups. While their territories overlap, neighboring groups avoid contact. Black lemurs, from Northern Madagascar, occur in groups of 5–15, led by a single female. The ruffed lemur, from Eastern Madagascar, lives in groups of 2–20, with several dominant females that defend a common territory.

Bamboo lemurs from Northern and Eastern Madagascar, normally exist in groups of 3–5 (but sometimes as many as 40), led by a single dominant male. The weasel sportive lemur from Northeastern Madagascar is a leaping species but lacks an ability to grasp branches like other species.

Indris and Sifakas

Indris and sifakas belong to the same family (Wilson and Reeder, 2005). Verreau's sifaka, from Southern and Western Madagascar, is bipedal when walking on the ground, awkwardly hopping sideways with its arms held aloft. It lives in social groups of varying size, using vocal demonstrations in territorial disputes. The indri, from Eastern Madagascar, occurs in male–female pairs and their offspring. Males defend their territory, but females have first access to food.

Aye-Ayes

The aye-aye, from Northwestern and Eastern Madagascar, lives in forest habitats, as well as spiny desert and plantations. It has been dubbed the "primate woodpecker," because it finds and eats wood-boring insects within the wood of trees after detecting their minute sounds of activity (Groves, 2005).

Tarsiers

Tarsiers, prosimians from Southeast Asia, with four extant species, live in tropical rain forests (Clutton-Brock and Wilson, 2002) (Fig. 5.4A). While tarsiers are included in the prosimian group, some researchers view them as a link between prosimians and simians. Clutton-Brock uses the western tarsier as an example of the group. They are nocturnal and adept at climbing and grasping tree branches. It has a small, compact body, with slender fingers, padded toes, and sharp claws. It feeds chiefly on insects and usually sleeps on branches, rarely using nests for shelter.

SIMIANS

The rise of simians (also referred to as anthropoids) from prosimian stock took place in the early Cenozoic era (Table 5.1) (Figs. 5.1 and 5.4). The earliest simian fossils are about 30 million years old and include New and Old World monkeys and apes (Groves, 2005; Hartwig, 2011; Hoffstetter, 1974; Wilson and Reeder, 2005) (Fig. 5.4B–F).

Anatomical differences between simians (anthropoids) and prosimians are as follows. Simians are larger; they have fused mandibles, whereas prosimians are able to move them independently; simians have larger brains relative to their body weight; simians have both binocular and color vision, whereas prosimians do not have either (except for the tarsiers). Prosimians have a bicornuate uterus, with two separate uterine chambers. A bicornuate uterus is rather typical for most mammals, but in simians, the uterine chambers have fused. Prosimians usually have litters rather than single offspring, the latter of which is the norm in higher primates (Nowak, 1999).

Behaviorally, simians can be distinguished as arboreal, semiterrestrial, or terrestrial. Nonhuman simians may also be distinguished by their preference for savanna or forest. Most spend their lives in rather large social groups, the size depending on several factors. Semiterrestrial species, for instance, gain necessary protection from predators and are better able to protect scarce food resources by existing in large groups. Baboons are an example of this primate type.

On the other hand, leaf-eating primates, such as langurs and colobus monkeys, show less competition for food and thus tend to form smaller groups. Most simian primates are active during the day. The few that are nocturnal are relatively solitary hunters.

FIGURE 5.4 Primates. (A) Tarsier, a nocturnal prosimian that some biologists feel is a link between prosimians and simians. (B) Gibbon, a lesser ape, showing its long forelimbs that are used in brachiation. (C–F) Greater apes. (C) Orangutan. (D) Gorilla. (E) Chimpanzee, showing opposable big toe. (F) Bonobo.

Groups of simians other than humans occupy territories that keep them more-or-less isolated from other related groups. When they do encounter one another, altercations may ensue.

While females generally lack aggression, males are often aggressive. During what appear to be rare encounters between males of different groups, they are generally aggressive toward one another.

Chimpanzees seem to be a notable exception, but their attitude toward other groups often depends on the recognition of former friends and alliances (Minkel, 2006). While different chimpanzee troops may express themselves in a friendly and exciting manner upon coming together, agonistic behaviors may be expressed at other times in a violent fashion. Occasionally, it is during periods of group interactions when some females switch from one group to another for mating purposes.

Some forest-dwelling simian groups demonstrate hostile displays upon encountering one another. In gibbons, including the siamangs and howler monkeys, for instance, vocalizations are part of their territorial displays. Such displays are most often harmless and tend to facilitate the establishment of territorial boundaries without the danger of physical contact.

During encounters between males and females, bonds for the purpose of mating are formed. How long the male stays with a female and maintains family associations between him and the female and their offspring varies among simians. Primatologists often recognize six different patterns of family groupings, some of which are found in human populations (Strier, 2007):

- A family group with *single females and their offspring* (without the male) represents a common arrangement in many mammals but a rare one in primates. Nevertheless, this form of grouping has been found in some small nocturnal prosimians (galogos and mouse lemurs) and orangutans. Males generally have large territories that overlap with the territories of several females.
- *Monogamous family groups* include an adult male, an adult female, and their offspring, although there may be occasions in which promiscuous behavior is demonstrated. This is typical of most human societies in the world. Offspring remain with the family during their young life and leave their nuclear family when they are ready to mate. While this type of group is rare in nonhuman primates, it is found among prosimians, New World monkeys, and small Asian apes.
- *Polyandrous family groups* (one female with two or more males) are found in marmosets and tamarins, the smallest of New World monkeys. While a male and female may commence their relationship as a monogamous pair, a second male may join the pair and initially help in infant rearing. Both males may subsequently mate with the female. While polyandrous associations are rare in nonhuman primate societies, they do occur in certain human societies, such as in rural regions of Sri Lanka, India, Nepal, and Tibet.
- *Polygynous mating patterns*, involving one male and several females, are seen in geladas, langurs, howler monkeys, hamadryas baboons, gorillas, and many human societies. Polygynous associations have been culturally preferred in some Native American, South Asian, Arab, and African cultures.
- *Multimale–multifemale* associations are the most common societal associations in semiterrestrial primates (spending time in both arboreal and terrestrial settings). Dominance hierarchies are commonly established between both males and females. While agonistic behavior often determines dominance ranking, a dominant position is sometimes determined by the rank of an infant's mother.
- *Fission–fusion associations* are recognized in chimpanzee groups where social group size and composition change throughout the annual cycle, depending on different activities and situations. For instance, individual chimpanzees leave their communities from time to time. While males may leave to forage alone or join other males in a hunting party, it is the female, especially during periods of estrus, that casually changes membership from one group to another on occasion. Other simians are also known to change group membership from time to time. Adult rhesus macaque males, for instance, permanently leave their community to seek mates. Some langur and baboon species also change communities for the acquisition of food and mates.

These groups largely depend upon relatedness. Thus, kinship is an important consideration in the formation and maintenance of simian groups. How kinship functions in primate societies largely depends on the group being studied. Most nonhuman simians appear to depend on matrilineal (female) descent. In such cases, the strongest social unit is a mother and her offspring. Chimpanzees, for instance, maintain a bond with their female parent well into adulthood.

How the mother is ranked within the society sometimes determines how the offspring will be treated by other society members. The ranking of juvenile rhesus macaques, for instance, relates to their female parent's position of dominance (Maestripieri, 2007).

Sexual dimorphism is characteristic of semi-terrestrial monkeys and the great apes. When sexual differences are evident, the male is usually the largest and thus is physically dominant over the female.

New World Monkeys

New World Monkeys (from South and Central America) have flat noses and large, well-separated nostrils (Groves, 2005; Rylands and Mittermeier, 2009). Their tails are prehensile, they are arboreal, and they are almost always diurnal. Their family, territorial, and dominant behaviors are quite diverse. The following examples serve to provide information on New World monkey diversification.

Brown Capuchins

South American brown capuchins mingle with other species of monkeys to forage for food and gain group protection against potential predators. They use simple tools, like twigs and stones, to flush prey out of tree holes and to crack nuts.

According to Boyson, capuchins are known to adapt tool use in the rainforest for finding insects. Those living in other habitats have mastered the art of using rocks as hammers for cracking nuts. In tests to determine if capuchins could be taught to play a version of "Simon says," "they followed a learning procedure referred to as emulation. In other tests, they appeared to possess a sense of fairness in interactive play."

Bolivian Squirrel Monkeys

During courtship, male Bolivian squirrel monkeys compete with one another in an attempt to win the largest number of females. They move through their habitat in large troops of 40 to 200 individuals.

Golden Lion Tamarins

Golden lion tamarins usually exist in groups of 4–11. Sexual activities in subordinates are not suppressed by the dominant pair, but only the dominant pair produces offspring. Juveniles often assist in rearing the young (i.e., they are alloparents). Emperor tamarins often exist in groups with other closely related monkeys, each species responding to the calls of the other. Males often carry their young offspring, except when the young are suckled by the mother.

Pygmy Marmosets

Each group of 5–10 pygmy marmosets has one breeding pair and alloparents (helpers) that care for the young. The father cares for the young during the first few weeks after birth.

Humboldt's Woolly Monkey

Humboldt's woolly monkey from Brazil, Peru, and Bolivia, lives in mixed troops which sometimes split into subgroups for foraging. Their dominance hierarchies are based on age.

Chamek Spider Monkey

Chamek spider monkeys live in troops that occupy large territories. The social position of females and their survival is related to the dominant position of their mothers.

Black Howler Monkey

Mexican black howler monkeys demonstrate loud calls during crepuscular periods to establish their territory. Red howler monkeys are generally found in small groups, with an adult male, females, and young. Males that supersede an existing male may kill offspring (infanticide) that are not theirs so that all offspring will belong to them.

Other New World Monkeys

Just before dawn, dusky titis entwine their tails together and perform a sonorous choir to maintain social bonds and a territorial boundary. Like other marmosets, the silvery marmoset has troops with one breeding pair and siblings that assist in bringing up the young.

Old World Monkeys

Old World monkeys (from Africa, Southern Asia, and Southeast Asia) have noses that are more narrow than those of New World monkeys, and they point downward (Groves, 2005) (Fig. 5.1). They have tails that lack the prehensile feature, they are entirely diurnal, and there are both arboreal and terrestrial species.

The group referred to as Old World monkeys comes from primitive stock (prosimians) and eventually gave rise to the apes. According to P. Jordan (2001), "the lineage that runs from the first primates to the common ancestor of the great apes and humanity includes, at about 35 million years ago, the oldest known monkey-like creature called *Aegyptopithecus*, from the Faiyum Lake deposits of Egypt.

Between 31 and 35 million years ago, many genera of related monkey-like forms flourished in the Faiyum region. Their eye sockets indicate that they were no longer nocturnal primates [like their prosimian ancestors], and their teeth suggest that [at least] some of them were in fact closer to apes and men than to monkeys. Though they had larger brains per body size than the lower primates before them (prosimians), their eyes were not so perfectly forward facing as those of later higher primates, and their muzzles were longer. They were about the size of a modern gibbon, but they lacked the very long arms of tree-swinging apes of today."

Troops

Troops of extant Old World monkeys, for example, baboons, macaques, and vervets, are composed of different matrilineal families (lineage derived through the mother instead of the father), arranged in a linear dominance hierarchy. Even though they are matrilineal, there often is a central group of males who work together to dominate all others in their troops (so dominance establishment in males appears to have some despotic qualities). The females of certain matrilines in some troops of Old World monkeys may outrank females in other matrilines, representing a form of interfamilial dominance. Such coalitions are stable because of a high degree of association and relatedness between members of the matrilines.

Males that dominate in these societies exhibit certain types of overt behavior that are recognized by subordinate peers. Forms of behavior occur between individuals in rhesus monkey societies; for instance, in which the alpha male keeps his head and tail up, showing a deliberate form of regality in his movements, and he stares at other members of the group. When challenged, he shows his teeth and sometimes slaps the ground in anticipation of attacking. Its subordinates yield space, food, and estrous females to him.

Baboons

There are four extant baboon species, all living in social groups with multiple adult males and females (Bergman et al., 2003). Among the baboon species, the Hamadryas baboon behaves much like other baboons but is unique in that a harem is its basic social unit (it has a polygynous family unit). Dominant males keep several breeding females and their offspring under control by force.

Baboons are considered by Benis to exhibit both perfectionist and aggressive traits. "They play a vigorous game of dominance and submission based on the trait of aggression." Grooming is rooted in perfectionism. These animals do not exhibit smiles of recognition.

Guinea baboons generally forage in groups of about four individuals, but troops with as many as 200 individuals are not uncommon. Certain males within the group may have harems that they physically herd. Occasionally, females in the harems may step outside their group and mate with males from another group.

In troops of olive baboons, the young are tolerated until they begin to acquire adult coloration. At that time, males are driven off to battle their way into a new group, and females assume a position at the bottom of the social hierarchy.

Gelada baboons live in large troops with a dominant male and his harem. If he is dominated by a younger male, the younger male will kill the former male's offspring.

Other African simians provide us with a comparison of Old World social behaviors. Dominant mandrill males, for instance, may belong to large groups of up to 250 individuals and have a harem of up to 20 females. Patas monkeys live in troops of up to about 10 individuals. If attacked, the dominant male may function as a decoy, while his females take over other social functions.

Hierarchies develop in troops of red-capped mangabeys of up to 90 individuals, but subordinate males are allowed to freely mate with females. De Brazza's monkeys mark their territories with saliva and scents. While territorial, they attempt to avoid confrontation with intruders.

Celebes crested macaque monkeys form large mixed-sex troops of more than 100 individuals where males display very little aggression toward one another. Troops of up to 100 long-tailed macaques move around in noisy, quarrelsome groups, but hierarchies are poorly defined.

Single male guereza monkeys defend their territory with roars and spectacular jumping displays while leading a group that includes four or five females and their young. Social groups of hanuman langurs are considered flexible, sometimes peaceful troops with a number of males to troops with a single male.

Asian simians demonstrate a diverse array of behaviors as well. Proboscis monkey societies, for instance, are led by a single dominant male, who bares his teeth and uses loud nasal honks and penile erection to ward away territorial invaders. Large troops of several hundred golden snub-nosed monkeys split into smaller groups with a single male and several females to forage and breed.

Rhesus Macaques

Aside from humans, macaques are the most widespread primate genus, ranging from Japan to Afghanistan to North Africa (Groves, 2005; Zhang, 2003). According to D. Maestripieri, humans have behavior similar to rhesus macaques, monkeys that share with humans strong tendencies for nepotism and political maneuvering.

"After humans," he says in a book called *Macachiavellian Intelligence, How Rhesus Macaques and Humans Have Conquered the World* (2007), "rhesus macaques are one of the most successful primate species on our planet." In fact, he points out that it may be a similarity to our Machiavellian intelligence that fostered the success of both species.

Rhesus macaques live in complex societies with strong dominance hierarchies and long-lasting social bonds between related females. Individuals constantly compete for high social status and the power that comes with it, using ruthless aggression, nepotism, and complex political alliances. Sex can also be used for political purposes. The tactics these monkeys use to increase or maintain their power, according to Maestripieri, are not much different from those that political leaders employed during the Renaissance.

Alpha males, who rule the 50 or so macaques in a troop, use threats and violence to obtain the best food, the safest sleeping places, and access to the females in the group with whom they want to have sex. Like human dictators intent on holding power, dominant macaque monkeys use frequent and unpredictable aggression as an effective form of intimidation.

Less powerful members of the rhesus macaque group are marginalized and forced to live on the edges of the society's territory, where they are vulnerable to predator attacks. They must wait for the others to eat first and then have the leftovers, and they have sex only when the dominant monkeys are not looking.

"In rhesus society, dominants always travel in business class and subordinates in economy, and if the flight is overbooked, it's the subordinates who get bumped off the plane," Maestripieri said. "Social status can make the difference between life and death in human societies too," he pointed out. In the wake of Hurricane Katrina, for instance, the poorer members of the community accounted for most of the hurricane's death toll.

Male macaques form alliances with more powerful individuals, and take part in scapegoating on the lower end of the hierarchy, a Machiavellian strategy that a mid-ranking monkey can use when under attack from a higher-ranking one.

Altruism is rare and, in most cases, only a form of nepotistic behavior. Mothers help their daughters achieve a status similar to their own and to maintain it throughout their lives.

Female monkeys also act in Machiavellian ways when it comes to reproduction. They make sure they have lots of sex with the alpha male to increase the chances that he will protect their newborn infant from other monkeys six months later. "But while they have lots of sex with the alpha male and make him think he's going to be the father of their baby, the females also have sex with all the other males in the group behind the alpha male's back."

Struggles for power within a group sometimes culminate in a revolution, in which all members of a subordinate monkey's family suddenly attack the entire dominant family. These revolutions result in dramatic changes in the structure of power within rhesus societies, not unlike those that occur following human revolutions.

There is one situation, however, in which all of the well-established social structure evaporates: when a group of rhesus macaques confronts another group and monkey warfare begins. Rhesus macaques dislike strangers and will viciously attack their own image in a mirror, seeing their reflection as an intruder that is threatening them. When warfare begins, "Even a low-ranking rhesus loner becomes an instant patriot. Every drop of xenophobia (an unreasonable fear or hatred of foreigners or strangers or of that which is foreign or strange) in rhesus blood is transformed into fuel for battle."

Maestripieri says that, "What rhesus macaques and humans may have in common is that many of their psychological and behavioral dispositions have been shaped by intense competition between individuals and groups during the evolutionary history of these species." Rhesus groups can function like armies, and this may explain why they have been so successful in competition with other primates.

Based on these findings in successful primate groups, what can this all mean for humans? Pressure to find sinister solutions to social problems also may have facilitated the evolution of larger human brains and complex cognitive skills. While "our Machiavellian intelligence is not something we can be proud of, it may [actually] be the secret of our success. If it contributed to the evolution of our large brains and complex cognitive skills, it also contributed to the evolution of our ability to engage in noble spiritual and intellectual activities, including our love and compassion for other people."

Apes

Within this assemblage, apes diverged from a common ancestor with Old World monkeys about 20 million years ago (within the Cenozoic era) (Groves, 2005) (Figs. 5.1 and 5.4). Humans, chimpanzees, bonobos, one or two species of gorillas, and the organgutan make up the group referred to as modern great apes. Humans and other great apes split off from the lesser apes (gibbons) about 20 million years ago and are united by similar anatomical characteristics, including the lack of a tail, relatively large size, a relatively large brain and a barrel-shaped chest. Humans and other apes are capable of mating throughout the year, and nesting is not used by apes for raising the young.

Common Ancestor to Hominid Great Apes

A common ancestor of contemporary gorillas, chimpanzees, and bonobos existed about 15 million years ago (Fig. 5.1). The olfactory bulbs are reduced in size in the higher primates, while the cortex responsible for touch and vision is very large (Goodman et al., 1990; Groves, 2005).

Jordan (2001) states that *Proconsul* is the best known of the early apes of East Africa, representing a definite progress from *Aegyptopithecus* to the apes of today: muzzles were shorter; brains had become larger, both relative to body size and absolutely; and shoulder and elbow anatomy suggested the development of the suspensory, tree-swinging habits seen in modern apes.

At least some of the earlier monkey-like forms of apes appear to have had tails. *Proconsul* (hominoids of East Africa) probably did not.[5] *Proconsul* is an extinct genus of primates that existed from 23 to 25 million years ago during the Miocene epoch. Fossil remains are present in Eastern Africa, including Kenya and Uganda (Walker and Shipman, 2005). Four species have been classified to date: *Proconsul africanus*, *Proconsul gitongai*, *Proconsul major*, and *P. meswae*. The four species differ mainly in body size. Environmental reconstructions for the Early Miocene *Proconsul* sites are still tentative and range from forested environments to more open, arid grasslands.

Prehuman primates enjoyed a long run of evolutionary success, down until at least 14 million years ago and possibly millions of years later. Some of them show dental traits that point toward a similarity with living apes and humans (with indications that the diet could include hard nuts and tubers as well as soft fruit), but the long arms of most of today's apes (and of our earliest ancestors as well), were not yet in evidence.

About 15 million years ago, the northern latitudes had become colder, and the tropical and subtropical regions grew drier. The drier conditions put selective pressure on populations of primates in the Old World to change. African apes began to retreat into the tropical forests where they are found today. Due to continental drift and a closure between Africa, Europe, and Asia, apes began to disperse into the woodlands of southern Europe and Asia, as far as southwest China. Jordan (2001) believes that it was a relative or descendant of *Proconsul* that gave rise to the line of African hominines (more human-like apes) between 7 and 10 million years ago.

DNA is composed of long molecules, made up of repeating units called nucleotides. Each nucleotide consists of a phosphate group, a pentose sugar, and a chemical referred to as a nitrogenous base. There are four different nitrogenous bases in DNA. Thus, the differences in nucleotides depend on which of the four nitrogenous bases are present.

All organisms on Earth, including prokaryotic and eukaryotic forms, have DNA with the same basic chemical structure.[6] It is the arrangement of nucleotides (the sequences) and length of a DNA molecule that determines the differences between different forms of organisms.

DNA and its arrangement of nucleotides first and foremost represent a template or blueprint. Its sequence is used to make copies for itself (a process called DNA replication), as well as to make peptides (which ultimately become proteins). Proteins are the chemicals that run our bodies. Each species has particular sequences that are different from the sequences of other organisms, and genes, the important blueprints to make particular proteins, are represented by sections or portions of these sequences.

While we can show a relationship between our DNA and the DNA of many other life forms, similarities in their sequences increase the closer the relationship becomes. Because of the arrangement of these sequences, our genetic ties with other apes and all living organisms are evident. As stated by Matt Ridley in *Genome* (1999), we carry genes of all types from our past. As he points out, "there are genes that have not changed much since the very first single-celled creatures populated the primeval ooze."

There are genes that were developed as far back as when our ancestors were worm like. There are genes that must have first appeared when our ancestors were fish. There are genes that exist in their present form only because of recent epidemics of disease. Thus, while our DNA may change, depending on mutations that are sometimes selected due to environmental pressures, all organisms show links to their phylogenetic (evolutionary) past, including humans (Goodman et al., 1990).

Biologists agree that no two living organisms have arisen from another living organism. If they are closely related, they, instead, are considered to have had a common ancestor somewhere in the geological past. This thought can be brought forth here, that no two contemporary primates are descended from another contemporary primate, but common ancestors share characteristics that have been passed on to their descendants.

Within each of the 100 trillion cells in a human body (except the ova and spermatozoa in the ovaries and testes), there are 46 chromosomes (23 pairs) in each of the cell's nuclei under normal conditions, each set from the female and male parent containing 30,000 to 80,000 genes. Our closest relatives, the chimpanzees and bonobos (like gorillas and orangutans), have 48 chromosomes. The reduction in chromosome number in humans is apparently the result of fusion between two medium-sized chromosomes, now part of a large chromosome, Ridley (1999) suggests, is chromosome number two.

Based on our genome, we are considered to be between 94% and 98% related to chimpanzees and bonobos, these two great apes being more closely related to us than they are to gorillas. Comparing humans and chimpanzees, as

Ridley has pointed out, there is no bone, no known chemical, and no known part of the immune and digestive systems in chimpanzees that we do not share. We are less than 30,000 human generations from the time we split from our common ancestor with the group that later gave rise to the chimpanzees and bonobos.

We show some obvious contemporary differences in our anatomy. The shape of the skeleton has changed in humans to allow an upright posture and a bipedal method of walking, which, as Ridley states, "is well suited to long distances in even terrain; the knuckle-walking of other apes [partially supporting body weight on the middle knuckles of the hand] is better suited to shorter distances over rougher terrain."[7]

The skin has changed as well. It has become less hairy in humans, and it sweats profusely in the heat. "These features, together with the retention of a mat of hair to shade the head and the development of a radiator-shunt of veins in the scalp, suggest that our most immediate humanoid ancestors were no longer in the cloudy and shaded forest." They were beginning to walk in a relatively dry, open grassland environment (savanna).

Gibbons (the Lesser Apes)

Living in different geographical areas, the 11 extant species of gibbons are alike in their graceful brachiation (swinging) through trees and occasional short spans of bipedal walking (Groves, 2005) (Fig. 5.4). Gibbons are distinguished from the great apes by having little sexual dimorphism (sizes of males and females are similar), they are monogamous (males having a single female partner), and they are highly territorial. While they have some definite social characteristics, some (such as the siamangs, the largest of the gibbons) also possess asocial behavioral features in that they do not tolerate others of their species (Clarke, 1970).

As pointed out by Deborah Misotti (Talking Monkey's Project in Clewiston, Florida) (pers. comm.), one of the particular behaviors that is exhibited by gibbons is a unique form of vocalization, which they use to establish their respective territories, to call their family together and as a warning of danger. They use it also as they travel at great speeds over large areas of the arboreal forest to communicate with other gibbon families to ascertain their respective territories.

These vocalizations are expressed in gibbons more so than any other primate except the howler (a New World monkey) (Tenaza, 1984). The siamang, in particular, has a unique throat pouch which allows it to magnify and lengthen the time of its calls over a 3 mi area. The typical gibbon's call is heard up to 2 mi. It is a particularly individualistic behavior, used to establish territory and demonstrate their behaviors of aggression by the volume and length of their song.

Gibbons had a common ancestor with the great apes that split about 20 million years ago. There are gibbon-like fossils from the Oligocene and Miocene periods in Africa, the Miocene period in Europe, and the upper Pliocene and Pleistocene periods in Asia (Table 5.1).

While the siamang is the largest, darkest, and noisiest species of gibbon, they are nevertheless very small, lightweight, and arboreal when compared to other apes, spending most of their life in trees. They have a small head, arms that are longer than their legs, a short, slender body, and lightweight bones. As with other apes, there is no tail.

Siamangs are covered by long, dense, straggly black hair except on the face, fingers, palms, armpits, and soles. Their face is almost hairless, except for a slight mustache and beard. Their legs are dark, nostrils small, and their skin is jet-black.

Like other nonhuman simians, they have an opposable big toe. They have lost their opposable thumbs and thus brachiate without them. They are diurnal omnivores that live in small, stable family groups with a mated pair and their immature offspring. Grooming is important. Instead of making :sleeping nests" like other apes, they sleep in a sitting position in small groups in the forks of trees.

Siamangs live in very close-knit family groups, consisting of a dominant female, a male, and one or two offspring. Their home range is about 116 acres in size. Noisy territorial displays are generally heard for about half an hour every morning, their inflatable throat sac amplifying sound that is heard for as much as 2 mi away. Siamangs, like other gibbons, spend most of their time in trees, brachiating from branch to branch. On their rare trips to the ground, they demonstrate bipedal walking. They live about 35–40 years.

Mating pairs generally stay together for life. They can reproduce at between 5 and 7 years of age. Gestation is about 8 months, and they usually have a single offspring at a time. Siamang babies weigh about 6 ounces when born, and they have less hair than adults. Females wean their babies when they are about 1 year old, and the young remains as part of their mother's group for about 5–7 years, after which they start their own family group.

Orangutans

Orangutans are represented by two exclusively Asian species of extant great apes that are native to Indonesia and Malaysia (Cribb et al., 2014; Payne and Prundente, 2008) (Figs. 5.1 and 5.4C). They are currently found only in the rainforests of Borneo and Sumatra. While females demonstrate social traits in their care of offspring, early investigations had indicated that mature male orangutans are generally asocial except during a brief mating period. The authors above have shown that their apparent asocial behavior may be entirely different under varying environmental conditions.

Populations in Borneo and Sumatra, for instance, have been found to be different. Boysen reported that while the population in Borneo was fairly solitary, orangutans in the Sumatran population "were highly social and extremely tolerant of one another. At times, they would gather in groups as large as 100 animals." It was also in the Sumatran population that researchers also found orangutans using two types of sticks to acquire honey and insects deep inside tree holes and another type of stick when opening fruits. Such types of behavior have not been found in the Borneo population.[8]

About two-thirds the size of a gorilla, orangutans have a large and bulky body, a thick neck, short, bowed legs, and very long, strong arms. Their body is covered with long reddish-brown hair. They have a large head, and males have large cheek flaps. They have both opposable thumbs and big toes.

They are quite innovative, using leaves as umbrellas and water cups. Their daily activities are diurnal. They communicate with long, loud territorial, and sexual calls that carry over great distances. They can walk but most often brachiate through trees. They live about 50 years.

They can reproduce at 7–10 years of age, at which time males and females remain together for only a few days. Gestation is 8.5–9 months long. They have a single baby, and young are generally weaned when about 6 or 7 years of age. Male and female Bornean orangutan pairs remain together until the young is about 8 years old. Their home is defended with long calls and, if necessary, by fighting.

Gorillas

Gorillas are terrestrial omnivores which inhabit African forests (Dixson, 1981; Groves, 2002, 2005) (Figs. 5.1 and 5.4). According to some taxonomists, all gorillas belong to the same species. Some treat them as separate species with subspecific differences. According to some recent taxonomical reports, they are listed as two species, the western gorilla (*Gorilla gorilla*) and eastern gorilla (*Gorilla beringei*): (1) the western gorilla has two subspecies, the lowland gorilla (*Gorilla gorilla gorilla*) found at low altitudes in Cameroon, Equatorial Guinea, Central African Republic, Gabon, and Congo and the cross-river gorilla (*Gorilla gorilla diehli*); (2) the eastern lowland gorilla in eastern Zaire (*Gorilla beringei*) includes the mountain gorilla (*Gorilla beringei beringei*), found at high altitudes (5400–12,440 ft) in Zaire, Uganda, and Rwanda, and the eastern lowland gorilla (*Gorilla beringei graueri*).

They are the next closest living relative to humans, after the chimpanzee and bonobo (Groves, 2002). They are considered nonaggressive, exhibit a gingival smile, and chestslapping has been interpreted as a recognition display. They are narcissistic and perfectionists. Gorillas show tolerance toward their society members but do not share food. They vocalize with others of their species with such sounds as roars, growls, grunts, whines, chuckles, hoots, and high-pitched barks. They beat their chests, lunge, stare, lip-tuck, stick their tongue out, run sideways and, on particular occasions, raise up on two legs.

Dominant gorillas do not generally overtly assert their rank. When threatened by intruders, however, the male hoots, stands upright, beats his chest with cupped hands and throws vegetation. If his message of dominance fails, he charges with a huge roar and knocks the aggressor down with a massive hand swipe.

The gorilla body is relatively short and bulky, they have a wide chest, and their arms are longer than their legs. They are covered by brownish hair on most of their body (exceptions being on their fingers, palms, face, soles, and armpits). Adult males have a saddle-shaped patch of silverish hair on their backs, hence the name silverbacks. Their head is large, and their forehead is bulging. A sagittal crest on top of the head is largest in males. Both their hands and feet have an opposable thumb and big toe.

They are predominantly herbivores, eating leaves, fruit, seeds, tree bark, plant bulbs, tender plant shoots, and flowers. They occasionally supplement their plant diet with termites and ants.

They are considered shy diurnal polygynous primates that live in small family groups (bands) of six or seven individuals (a silverback, few females, and their young). Eastern gorillas are found in groups with ranges of 1000–2000 acres. The dominant silverback fathers most or all of the young in his group. Females groom their offspring, one another and the silverback. Females construct bowl-shaped nest-like areas with plant material for sleeping with their nursing infants.

They knuckle-walk, rarely stand only on their legs and remain on the ground most of the time. They become reproductive between 10 and 12 years of age. Gestation is 8–9.5 months. Young gorillas learn to crawl at about 2 months of age, and they walk before they are 9 months old.

Their closest relatives are chimpanzees and bonobos. They are not known to use tools.

Chimpanzees and Bonobos

The common ancestors of humans, chimpanzees and bonobos, are believed to have existed about 8 million years ago (Figs. 5.1 and 5.4E,F). Divergence between chimpanzees and bonobos came after the split with humans (De Waal, 2005; Groves, 2005).

Chimpanzees (*Pan troglodytes*) are primates native primarily to West and Central Africa. Biologists recognize four subspecies. Their closest relative, the bonobo (*Pan paniscus*), is separated from them by the Congo River. Chimpanzees and bonobos share a very close genetic relationship and are together the closest human relatives, both sharing between 94% and 98% identical DNA with us, but the social behavior between the two subhuman primates is extraordinarily different. As expressed by F. De Waal, we share some behavioral traits with both of these primates.

Although sometimes referred to as pygmy chimpanzees, bonobos generally cannot be distinguished from chimpanzees by size. Their features are different, however. Bonobos have a flatter, more open face, a higher forehead and long, fine, black head hair.

Both have arms that are longer than their legs. They are covered with black hair on most of their body (except on their fingers, palms, soles, and armpits). Young chimpanzees have pale skin in areas not covered by hair, and a white tuft of hair on the rump.

Chimpanzees have large ears, small nostrils, a slightly elongated snout, and a slight brow ridge. Their face is hairless except for a short white beard in both sexes. Baldness occurs in some adult females.

In addition to their opposable thumb, they have an opposable big toe, and thus grab with both hands and feet. There is a size difference in the sexes, males being larger than females.

They are omnivores. Besides leaves, fruits, seeds, tree bark, tender plant shoots, and flowers, they eat ants, termites, small nonprimate animals and young monkeys.

They use tools in the wild, such as sticks to obtain termites and ants, and they chew leaves so that they can soak up water. They are social and diurnal, living in small, stable communities or unit groups of about 40 to 50 individuals. Grooming is important. They construct "sleeping nests."

Sounds of communication are in the form of whining cries and loud barks. They are known to beg for food, hold hands, hug, and kiss. Lip puckering indicates a worried animal, and a smile indicates a relaxed, friendly one. Readiness for attack is indicated by lips that are tightly pressed together.

They knuckle-walk and occasionally walk upright. They are excellent climbers and spend much of their time in trees (including when sleeping). They live about 60 years.

They are able to reproduce between 12 and 13 years of age. Gestation is 8.5–9 months, and they usually have a single offspring. Twins are rare. Young animals remain with their mothers for about 7 years.

Chimpanzees demonstrate considerable behavioral variability. Some chimpanzees can be narcissistic and aggressive, while others have these qualities and are perfectionists as well. Male dominance is especially important in their world and is often associated with violent behavior. Some members exhibit a gingival smile. Thus, there is a distinct behavioral variability among individuals.

Chimpanzees have been known to become excited when encountering certain other populations, often spending time to renew friendships. However, they may demonstrate aggressive behavior if they feel they or their territories are threatened. At the same time, they usually show a degree of tolerance toward other members of their own group, rarely fighting over food. They even share on occasion.

As A. Mazur points out in *Biosociology of Dominance and Deference* (2005), chimpanzees can be murderous apes at times, especially to individuals that do not belong to their group. Jane Goodall states that chimpanzee communities do not mix, and males of different communities often engage in lethal battles.

James Case (2007) states that wild bands of chimpanzees "routinely seek to expand their territories by dispatching war parties against their neighbors. Ordinarily, they are the most vocal of creatures. However, chimps on the warpath steal silently through the forest to surprise the intended foe, picking up sticks and stones along the way for use as weapons."

Chimps live in groups of 15–120 individuals. Parties of adult males sometimes attack and kill intruding males. They may cooperate in killing and eating animal prey, such as monkeys, small antelopes, and birds.

Bonobos, on the other hand, are sometimes found in troops of up to 80. While very similar anatomically to chimpanzees, bonobos lack the violence exhibited by their close relatives (they are nonaggressive), and their society is influenced by female dominance. Instead of violence, some of their most outstanding behavior is associated with sexual expressions and social cohesiveness.

Their diet seems to include very little animal protein, and they have not been found to hunt monkeys. Their tool use is undeveloped.

Unlike chimpanzees, male bonobos usually limit dominance displays to brief runs, while dragging branches behind them. Male attempts at dominance establishment are generally ignored by females.

Chimpanzees almost never adopt a face-to-face position during copulation, but bonobos use this position most of the time. Increased female sexual receptivity is characteristic of bonobos. Instead of being receptive for a few days of her cycle, she is almost continually sexually attractive and active.

In addition to copulation, females demonstrate genito-genital rubbing. Males also engage in pseudo-copulation and a behavior called penis-fencing (rubbing their erect penises together). Sexual diversity in bonobos also includes sporadic oral sex, massaging another's genitals, and intense tongue kissing.

Comparing the sex lives of chimpanzees and bonobos, de Waal points to reconciliations of hugging and kissing between chimpanzees after fighting. Bonobos typically use sex rather than aggression. Bonobos are extremely tolerant of one another and have been considered by Mazur as being second to humans in their degree of civility in status interactions.

As a species, bonobos are best characterized as female centered and egalitarian. Sex is substituted for aggression. According to de Waal, "bonobos engage in sex in virtually every partner combination," sexual interactions occurring "more often among bonobos than among any other nonhuman primates." In the wild, reproduction, that is, sex for procreation, is about the same as in chimpanzees.

De Waal (2005) believes that sexual activity in bonobos is used in lieu of conflicts. He states that "anything (not just food) that arouses the interest of more than one bonobo at a time tends to result in sexual contact." Mothers who may lunge at another female that hits her young may subsequently engage in genital rubbing. Males who chase another male away from a female may subsequently engage in scrotal rubbing.

Both chimpanzees and bonobos live in what de Waal refers to as fission–fusion societies. They "move alone or in small parties of a few individuals at a time." If such a bonobo group meets another, one individual from a group may wander off with the second group.

Bonobo societies are both female centered and female dominated. While chimpanzee males may use threatening charges over a food item as a form of dominating others, female bonobos disregard male threats. If a male bonobo attempts to harass a female, females bond together to chase him away.

According to the Bonobo Conservation Initiative, bonobos are classified as endangered, estimates of their population size ranging from 10,000 to 20,000 bonobos left in the wild. While it has been difficult to collect reliable population estimates on bonobos, the population, which is already small, has become fragmented and is decreasing.

The collective threats impacting wild bonobos include poaching, civil unrest, habitat degradation, and a lack of information about the species. Disease transmission also poses a silent threat; many gorillas and chimpanzees have (already) succumbed to the Ebola virus.

Bushmeat hunting is the greatest threat to wild bonobo populations. Traditional taboos, which once protected bonobos in many areas, are breaking down in the face of economic desperation and human population pressure. In a region where more than 90% of residents can only afford to eat one meal each day, people are increasingly turning to wild sources for meat, both for sustenance and for profit in the commercial bushmeat trade. Due to years of war and insufficient infrastructure, the journey to marketplaces is long and arduous; smoked meat is one of the only commodities durable enough to withstand the trek. Because bonobos only bear offspring every 4–5 years, the population is slow to regenerate.

Subsistence agriculture in the Congo region relies on slash and burn farming, which is the practice of cutting and burning trees and vegetation to clear plots for planting crops. This practice quickly depletes the soil of its natural nutrients and requires the clearing of new plots every few planting cycles, driving agricultural activity deeper into the rainforest and encroaching upon bonobo habitat. Slash-burn agriculture is a method of producing plant food in primitive cultures around the world, but it ends in complete forest destruction and nutrient-poor, lateritic soils.

As the Congo is achieving greater political stability, large-scale industrial agriculture is also posing a larger threat. Industrial agriculture requires vast amounts of land and resources and could come into conflict with conservation aims:

> Despite a government moratorium on industrial logging concessions in the Congo, logging continues, contributing to the degradation and destruction of bonobo habitat. It also allows hunters to enter previously inaccessible areas of the forest via logging roads, perpetuating the bushmeat trade.

Concluding Remarks About Prehuman Primates

As indicated in the foregoing statements, there is a commonality of anatomy in primates, especially in the higher forms. Yet, they have each developed distinct forms of behavior which best suit their particular circumstances. Even in those that are closely related, their social, dominance, and aggressive forms of behavior are often widely divergent and distinctly their own (as in chimpanzees and bonobos).

On the other hand, we sometimes see common expressions of dominant and aggressive forms of behavior in distantly related primates; for example, rhesus macaques and humans, two species which have extensive population ranges in the contemporary world. It is clear that their similar behaviors have developed independently according to their particular way of life.

Since chimpanzee and human genetics show a close relationship between the two and their defensive and aggressive forms of behavior are similar, we tend to use chimpanzees as a model from which human behavior has arisen. However, there are millions of years of separation between the two. Recognizing the wide array of behavior seen throughout the primate group and the significant differences in the behavior of the two most closely related primates (chimpanzees and bonobos), it may be more logical to assume that the behaviors of chimpanzees and humans have developed independently and followed parallel paths in the two species.

Since it has been common practice to assume that human behavior has arisen from our common ancestor with the chimpanzee/bonobo line or more directly from chimpanzees, this approach is evident in the following treatment of humans and their most immediate predecessors.

Humans

Human and chimpanzee/bonobo groups diverged from a common ancestor about 4–6 million years ago (Fig. 5.1). Either *Sahelanthropus* or *Orrorin* may be our last shared ancestor with them (Heng, 2009; Patterson et al., 2006; Stringer, 2003a,b). Moving through the early bipeds, hominids eventually evolved into australopithecines that later gave rise to the genus *Homo* (see Table 5.1).

The earliest documented members of the genus *Homo* are in the species *Homo habilis* (the earliest species for which there is positive evidence of use of stone tools) which evolved around 2.3 million years ago. The brains of these early hominines were about the same size as that of a chimpanzee, although it has been suggested that this was the time in which natural selection provided a more rapid wiring of the frontal cortex in the line leading to *Homo* (Fig. 5.1). It was during the next million years that a process of increased encephalization began, and with the arrival of *Homo erectus* in the fossil record, cranial capacity had doubled to 850 cm^3. The increase in human brain size during that period is equivalent to every generation having an additional 125,000 neurons more than their parents.

Homo erectus and *Homo ergaster* were the first of the hominines to leave Africa, spreading through that continent, Asia and Europe between 1.3 and 1.8 million years ago. It is believed that these species were the first to use fire and complex tools. According to the Recent African Ancestry theory, modern humans (*Homo sapiens*) may have evolved in Africa possibly from *Homo heidelbergensis*, *Homo rhodesiensis*, or *Homo antecessor* and subsequently dispersed out of the continent at some time between 50,000 and 100,000 years ago, replacing local populations of humans as they increased their range, for example, *Homo erectus*, *Homo denisova*, *Homo floresiensis* and *Homo neanderthalensis* (Fig. 8.2).

An archaic form of *Homo sapiens*, the forerunner of anatomically modern humans, evolved between 400,000 and 250,000 years ago. Recent DNA evidence suggests that several haplotypes (groups of genes within an organism that have been inherited together from a single parent) of Neanderthal origin are present among all non-African populations, and Neanderthals and other hominids may have had limited interbreeding between them and other species.

Anatomically, the archaic form of modern humans (*Homo sapiens*) evolved in the Middle Paleolithic, about 200,000 years ago. According to many anthropologists, the transition to behavioral modernity with the development of a symbolic culture, language, and specialized lithic technology seems to have occurred around 50,000 years ago, although some researchers suggest a more gradual change in behavior over a longer time span.[9]

Chapter 6 more specifically describes the nature of this modern hominine, *Homo sapiens*, and the world in which it lives. This and subsequent chapters delve into features that make it stand out as a different kind of animal.

Note

Superscript numerals appearing in this chapter refer to additional text/explanation given in the appendix.

CHAPTER

6

The Human Animal

Since the beginning of history, people have struggled to unravel the mystery of human nature and find out exactly what makes us so special. *Robert Winston and Don E. Wilson, Editorial Consultants, Human, 2004, Smithonian Institution*

Based on the accomplishments of *Homo sapiens* in recent years, it is unquestionably a very remarkable species of social animal, cognitively superior, bipedal, extremely innovative, dominant, and aggressive, arising as the primate elite from a highly evolved group of prehuman predecessors referred to collectively by biologists as hominids (Aiello and Dean, 1990; Boyd and Silk, 2003; Cochran and Harpending, 2009).[1] The group includes other human-like forms (all of which are now extinct) and a number of other not-so-human animals we refer to collectively as great apes (as shown in the previous chapter).

Included in this privileged great ape group are humans, chimpanzees, bonobos, gorillas, and orangutans. Males are, on average, larger and physically stronger than females in all great apes, although the degree of sexual dimorphism varies greatly among species.

Most extant species of apes, like their lower primate predecessors, are predominantly quadrupedal, using their front limbs to walk on their knuckles, and in addition to humans, some species spend at least part of their time in an erect, bipedal position (Stokstad, 2000). They are all able to use their highly manipulative, extensively innervated, and oppositional hands (hands with opposable thumbs) for gathering food or nesting materials, extensive tool use, devices of defense, and, in some cases, for highly perceptive sensors.

Anatomical features used to characterize humans and other primates are sometimes found in other animals as well. Binocular vision, for instance, is generally a characteristic feature of predators while animals with laterally positioned eyes are generally preyed upon. Although opposable thumbs are considered a signature feature of primates, a number of birds and nonprimate mammals also have opposable digits. Some primates, such as gibbons, have lost their opposable thumbs. While certain other animals have developed large brains, the ratio between body size and brain and the brain's comparative complexity are particularly significant in humans, other primates, and certain other mammals (e.g., cetaceans: whales, dolphins, and porpoises).

Humans are rather unique primates, having a brain that has evolved to a point of being capable of what appears to be endless cognitive thoughts. While opposable thumbs, an erect posture and bipedalism have been emphasized as important features leading to human superiority on Earth, it has been primarily the power of the human brain that has led to the development of numerous traits that more adequately distinguish it from all other animals (see Chapter 7).

WHERE DID CONTEMPORARY HUMANS COME FROM?

Most of the evolution of humans and their predecessors occurred in Africa, especially eastern Africa (Heng, 2009; Tyson, 2008). Before and during the rise of *H. sapiens*, millions upon millions of other types of organisms had evolved on Earth as well, some of which formed relationships with humans, which we tentatively and collectively refer to as quasi-coexistences.

Human evolution has occurred over the last few million years (Table 5.1), stemming from a common ancestor with other apes, such as chimpanzees and bonobos. Over the past 7 million years, becoming human involved moving from a forest habitat into a savanna-like habitat; a significant change in its acquisition of and type of food; a series of anatomical modifications, including some on the skull and teeth, as well as changes in the axial and appendicular skeleton for walking and running upright; a loss of much of its extensive terminal body hair; an increase in brain size and complexity; the accompanying acquisition of ever-increasing technological skills; and an uncanny ability for locating, using, and manipulating planetary resources. While there are a number of features that are distinctly human, they have retained numerous features from their predecessors that are distinctly animalistic.

HUMANS AND THE EARTH THEY LIVE UPON

The important relationship between humans, their environment, and Earth had developed and continued for most of their 200,000 years in a fashion similar to other animal relationships, revealing that Earth has always been and still is an important provider of resources for supporting the lives of humans and many other animal species during its 4.5-billion-year existence. However, life on Earth has not always been pleasant for many life forms (Buttel and Humphrey, 2002; Dunlap and Michelson, 2002; Foster, 2000; Redclift and Woodgate, 1997). Periodic shortages of food and other resources and dramatic environmental changes have sometimes put intolerable stress on populations (humans included), causing them to either change or become extinct.[2] In the process, little damage was usually incurred by Earth from its ever-changing biota other than minor local alterations, which, for the most part, were insignificant.

Throughout Earth's history, it has been quite unusual for a single species, or even a group of species, to influence cosmopolitan changes on Earth and its diverse populations. However, in recent years, Earth has suffered greater-than-average abuse from our species, causing it and its nonhuman faction to change considerably. Many organisms that have encountered *H. sapiens* in its rise to dominance have not survived.

Some of Earth's physical changes due to human activity (indicating its dominant, strongly influential position in the world) have been so cataclysmic that they are evident in satellite pictures taken from space (Winter, 2014). And while we cannot see feral population changes from outer space, we can be sure that when cataclysmic physical changes are evident from a celestial view, nonhuman populations have undergone overwhelming modifications because of the human species and its activities.

Upon admitting that human dominance on Earth has seriously altered other populations, we may wonder if any comparable circumstances have ever resulted in catastrophic changes during the 4-billion-year period of life before humans arrived on the scene. Looking back into geological time, we will find that planetary changes due to human presence on Earth actually do not represent the first time a group of organisms has been responsible for massive, cataclysmic changes to the planet, resulting in catastrophic modifications in populations. It happened at least once before on a grand scale, long before humans, other multicellular organisms, and even unicellular eukaryotic organisms came on the scene, commencing about 3.5 billion years ago, when unicellular photosynthetic organisms (organisms capable of making their own food from sunlight, water, and carbon dioxide) arose and dramatically changed Earth's atmosphere and the course of evolution by producing an unusual, almost nonexistent gas (at the time) referred to as oxygen.

This gas, so important to humans and most other current forms of life on the planet, led to a gradual but highly significant change in the atmosphere that is affectionately referred to by geologists and biologists as "The Great Oxygen Event." The major portion of this event seems to have occurred 2.3 billion years ago (Biello, 2009).

In the process of altering the Earth's atmosphere from an anaerobic to an aerobic one, increasing amounts of oxygen reacted with methane and carbon dioxide (both of which are greenhouse gases), resulting in a period of extreme cold referred to as "Snowball Earth," commencing about three quarters of a billion years ago. Since these times, Earth has undergone repeated periods of global warming and cooling, causing tremendous changes in sea level and climatic conditions which in turn put stress on all of Earth's populations.

Even during much shorter periods of time, climatic conditions have always changed tremendously, fluctuating from one extreme to another, and populations of organisms have changed as well. As pointed out in an article at PhysicalGeography.net called *Earth's Climatic History*, "The period from 2,000,000 to 14,000 BP (before present) is known as the Pleistocene or Ice Age. During this period, large glacial ice sheets covered much of North America, Europe, and Asia for extended periods of time."

However, the extent of glacial ice during the Pleistocene was not static. The Pleistocene had interglacial periods as well, during which glaciers retreated because of warmer temperatures, and periodic returns to colder periods during which glaciers advanced. During the coldest periods of the Ice Age (a period which commenced about 2.6 million years ago), average global temperatures were probably 4–5°C colder than they are today.

The article points out that "the most recent glacial retreat is still going on. We call the temporal period of this retreat the Holocene epoch. This warming of the Earth and subsequent glacial retreat began about 14,000 years ago

(12,000 BC). The warming was shortly interrupted by a sudden cooling, known as the Younger-Dryas, at about 10,000 to 8500 BC. Scientists speculate that this cooling may have been caused by the release of fresh water trapped behind ice on North America into the North Atlantic Ocean. The release altered vertical currents in the ocean which exchange heat energy with the atmosphere."

During the undulatory nature of planetary changes, "the warming resumed by 8500 BC. By 5000 to 3000 BC, average global temperatures reached their maximum level during the Holocene and were 1–2°C warmer than they are today. Climatologists call this period the 'Climatic Optimum' (Koshkarova and Koshkarova, 2004). Just prior to and during the Climatic Optimum, many of the Earth's great ancient human civilizations began and flourished. As an example of improving conditions for a flourishing human population in Africa, the Nile River had three times its present volume [of water]," providing abundant fresh water to growing sessile human populations.

Climatic conditions again changed, and "from 3000 to 2000 BC, a cooling trend occurred. This cooling caused large drops in sea level and the emergence of many islands (e.g., in the Bahamas) and coastal areas that are still above sea level today. A short warming trend took place from 2000 to 1500 BC, followed once again by colder conditions. Colder temperatures from 1500 to 750 BC caused renewed ice growth in continental glaciers and alpine glaciers, and a sea level drop of between 2 and 3m below present day levels."

Continuing with climatic fluctuations, "the period from 750 BC to 800 AD experienced a warming trend." Temperatures, however, did not get as warm as the Climatic Optimum. During the time of the Roman Empire (150 BC to 300 AD), a cooling began that lasted until about 900 AD. At its height, the cooling caused the Nile River (829 AD) and the Black Sea (800–801 AD) to freeze.

"The period 900–1200 AD has been called the 'Little Climatic Optimum.' It represents the warmest climate since the Climatic Optimum. During this period, the Vikings established settlements on Greenland and Iceland. The snow line in the Rocky Mountains was about 370m above current levels. A period of cool and more extreme weather followed the Little Climatic Optimum. Florida, currently just above sea level, was twice as wide as it is now. Due to more recent rises in sea level, archeologists and anthropologists have to dive into Gulf waters to locate artifacts from locations that were once terrestrial Native American villages. A great drought in the American southwest occurred between 1276 and 1299. There are records of floods, great droughts, and extreme seasonal climatic fluctuations up to the 1400s."

During the period 1550 to 1850, "global temperatures were at their coldest since the beginning of the Holocene. Scientists call this period the 'Little Ice Age'," during which "the average annual temperature of the Northern Hemisphere was about 1.0°C lower than today. During the period 1580 to 1600, the western United States experienced one of its longest and most severe droughts in the last 500 years. Cold weather in Iceland from 1753 to 1759 caused 25% of the population to die from crop failure and famine. Newspapers in New England were calling 1816 the year without a summer."

From 1850 to present the climate experienced a period of general warming. "However, beginning in 1935, positive anomalies became more common, and from 1980 to 2006 most of the anomalies were between 0.20 and 0.63°C higher than the normal period (1951–1980) average."

Under current conditions, in which global warming is slowly occurring, growing water demands, driven by population growth and foreign land and water acquisitions, are straining the Nile's natural limits. According to the World Wildlife Organization, "many of the water systems that keep ecosystems thriving and feed a growing human population have become stressed. Rivers, lakes, and aquifers are drying up or becoming too polluted to use. More than half the world's wetlands have disappeared. Agriculture consumes more water than any other source and wastes much of that through inefficiencies."

"Climate change is altering patterns of weather and water around the world, causing shortages and droughts in some areas and floods in others.

At the current consumption rate, this situation will only get worse." Authors of this chapter claim that "by 2025, two-thirds of the world's population may face water shortages, and ecosystems around the world will suffer even more."

Such changes in climate, although seemingly minor, can result in tremendous organismic fluctuations and reductions, even extinction for many forms of biota. On the other hand, they sometimes represent opportunities for organisms that find the new conditions encouraging for survival and reproduction.

We may notice that the nonscientific community often pays little attention to these environmental changes, many of which humans are actually responsible for. Unless changes are cataclysmic, they generally take long periods of time to express themselves, and we may not recognize them as threatening. Perhaps we think that demands by the human population are temporary or not as serious as they may seem. Perhaps we conclude that people who bring these statistics up are alarmists and we will somehow solve such problems as the demands become greater.

Why should we be concerned about change now, when the Earth, except for what may seem to be minor climatic changes, appears to be more-or-less settled? Maybe we are going through a temporary cyclic phase that will end up in a homeostatic end. After all, many organisms have successfully progressed through millions upon millions of years of evolution that have resulted in an astounding array of multicellular forms, including the human animal which in many ways appears to be experiencing the heyday of its existence, at least in developed countries. Yet, current appearances may be deceiving.

The answers to these questions are hinted at in geological history, the undulating conditions that have prevailed through time and the realization that nothing is static. As indicated in the foregoing paragraphs, climatic shifts of great magnitude have occurred in the past, and it is likely that they will occur in the future. Along with them, all populations will react to the changes in one way or another, influencing food chains and food webs around the planet and the possibility of extinction of many of the planet's life forms.

Along with climatological changes, food abundance, predation, parasitism, and other ecological changes, depending on population size and health, opened the door for opportunists and closed the door for organisms that cannot cope with the change. In addition, whether we like it or not, fluctuations in climate and subhuman populations affect human existence as well, in spite of our personal sense of being different and independent from other organisms on the planet.

Survival of the organisms experiencing these changes depends on their preadaptive nature when faced with a change. Humans are diverse, intelligent creatures, easily adapting to minor climatological conditions and fluctuations of other organismic populations. While many humans may suffer from global changes, such changes probably will not cause human extinction.

But what about major changes, ones that affect not only humans but all other organisms on Earth? While many species may suffer serious cut-backs in their presence and even become extinct under adverse conditions, a portion of human diversity and the likelihood of its survival appear to be partially hidden in its incredibly enlarged brain and the intelligence that is associated with it, but in actuality, it is more than that. Humans (do I dare say it again) are different in certain key ways from all other animals that have ever existed on the planet.

Humans are highly evolved, intelligent beings. As already stated, they currently represent the most dominant form of life on Earth, and they have adjusted the environment and living conditions to make life more pleasurable for themselves. Yet, they retain many animalistic traits which, together with innovation, have been instrumental in producing maladaptive fluctuations, sometimes even adverse environmental conditions and resource shortages that, to say the least, are not self-perpetuating.

Many of the species that have occupied Earth throughout the Paleozoic, Mesozoic, and Cenozoic eras (a period of over 450 million years), especially the dominant ones, have enjoyed life-sustaining conditions (interspersed with threatening climatological and ecological changes) for long periods of time, but, Earth is constantly changing, and populations, no matter who or what they are, have never lasted indefinitely.

DEVELOPMENT AND EVOLUTION OF THE EARTH

The extraordinary story of life on Earth, the third terrestrial planet from the sun, with an average equatorial diameter of 7926 miles (12,756 km), required preliminary life-giving and life-sustaining events to get it started. From its inception, it has undergone local and planetary changes that have challenged its life forms.

Since most people have lost their connection with Mother Earth and her biota, it is worth briefly mentioning the most outstanding events Earth has progressed through in order to understand and appreciate what humans have inherited from the past and enjoy today. Ideally, learning to appreciate Earth and its delicate relationship with its biota and the worldwide changes that have come about at the hands of the human animal, we may be able to see why our population must experience a significant reduction in order to become homeostatic:

> We are lucky, on Earth. We are lucky because we—as complex and self-aware organisms—are here. We are sustained, given air to breathe, and water and food, by a very ancient planet, a planet past its midpoint, a planet that is nearer death than birth. *J. Zalasiewicz and M. Williams*, The Goldilocks Planet

Its Anaerobic Beginning

In the beginning, during the Earth's formative years, it was not a fit planet for humans or any other forms of life.[3] Following its abiotic beginning, it took time, billions of years in fact, for the evolution of both Earth and its life forms to be ready for the likes of humans.

Earth, the only planet known at this time that can be called home for this highly evolved primate, is and always has been a fascinating cosmological entity, its abiotic birth occurring approximately 4.5 billion years ago from the coalescence of interstellar gases and galactic debris, following the formation of a remarkable star that we call our sun. Earth, like the sun, was initially a molten, anoxic (oxygen-free) sphere that eventually cooled and formed a lithospheric crust (Prothero and Dott, 2003). But what is it that makes our star so accommodating to life forms on a celestial body like Earth?

Comparing the sun with other high-energy bodies, it is, as pointed out by Stephen Hawking in *A Brief History of Time* (1996), a good star with medium size (one referred to as a main sequence white dwarf star) that emits a wide array of electromagnetic rays (from ultra-short gamma waves to ultra-long radio waves). We say that it is a good star because its distance from Earth and size, gravity and density are such that it burns at the right intensity to help warm Earth and supplies it with visible light and other forms of radiant energy. It became a star when its matter coalesced and its increasing gravity and density reached a high-enough level to be able to produce helium from fusing hydrogen molecules. Yet, while it is a star, it is not so large and dense that it burns at a higher temperature and consequently will not burn out at a faster rate. It and Earth fit very well into a category we might call a Goldilocks Zone, a zone in space that is suitable for life (Zalasiewicz and Williams, 2012).

As indicated, some of the electromagnetic rays it produces during the hydrogen–helium reaction are extremely important in assisting the heating of Earth, as well as providing valuable wavelengths of light and a steady supply of energy for Earth's food chains and livable environment.

But food chains did not exist during Earth's initial stage of development. It took half a billion years for Earth to become cooler and develop an environment that could support life.

Once the Earth cooled to the point of solidifying peripherally, the surface formed a number of crustal (tectonic) plates that make up Earth's continents and their shelves (Fig. 6.1). These plates are in constant motion, moving ever-so-slowly around Earth, sliding by, colliding with, and separating from one another, sometimes being enlarged or torn apart, reshaped, and resized, depending on Earth's inner forces (Kious and Tilling, 2001). At their edges and sometimes within their centers, volcanic and earthquake activities occur, adding new land masses and tearing apart old ones. Such movements are driven by convection currents within the planet from several sources, for example, the sun's electromagnetic rays, and formative and radioactive heat.

From its violent inception, Earth has maintained a very close relationship with the sun. To this day, in fact, the Earth and most life that exists on it are entirely dependent on certain of the sun's electromagnetic waves for a constant source of energy.[4] Without the sun's energy, humans and other forms of life would perish.

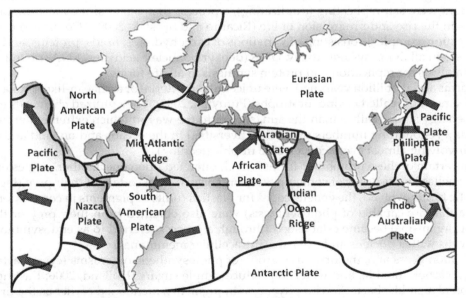

FIGURE 6.1 Approximate shape and position of some of the tectonic plates on Earth and the direction of their movement. Tectonic plate movement results from convection currents within the Earth and are thus tied to Earth's inner heat. A major point of pressure results from the activity of volcanic activity along the mid-Atlantic Ridge, forcing North America and South America to move away from Africa. When plates collide, subduction zones occur in which one tectonic plate may move over or beneath another, causing the formation of a mountain range (see Fig. 1.5). Such movement happened between an Antarctic plate that moved east and then north, eventually slamming into the Asian continent and creating the Himalayan Mountains.

With what we know about Earth, we can understand that its size and position with respect to the sun, moon, and other cosmological and atmospheric features make it unique in our solar system for the support of life (Kasting, 1993). And while we, like the early Greek philosophers (Socrates, Plato, and Aristotle), may think of it as a relatively stable body that has minor variations, it actually is a very dynamic system of tremendous physical and chemical change.[5]

Due to phenomena such as energy and magnetic shifts, its orbit around the sun, its circling satellite (the moon), rotation, wobble, tectonic activity, and relationship with other celestial bodies within and outside the solar system, Earth is an ever-changing host to atoms and molecules that constantly combine, disintegrate, and recombine to form nearly 55 million different known chemicals. Thus, Earth and its important diversity of chemicals undergo constant transformations, and together with other changes, Earth fluctuates back and forth in energy consumption and radiation to express itself in a vast array of chemical and physical variations. Internal convection currents, colliding and separating tectonic plates, volcanism, earthquakes, the formation of new rocks and mountains, the breakdown and erosion of land masses, and a variety of chemical cycles that are constantly going on in the environment are but a few expressions of Earth's very dramatic and ever-changing nature (Fig. 6.1).

While we can read about these events that Earth goes through, we may question whether they are actually occurring. Many changes are hidden and take eons of time to take place, but they go on nevertheless. As an example, volcanic activity at the bottom of the Atlantic Ocean is spewing out molten rock to form the largest mountain chain on Earth, and the outward pressure created is pushing the tectonic plates represented by North America and Africa apart at a rate of 2–4 cm a year (Condie, 1997).

While these plates were part of a supercontinent (Pangaea) during the Paleozoic era (commencing about 300 million years ago; Zhao et al., 2004), they began moving apart about 100 million years ago and continue to move today (Fig. 6.1). Eventually, they will collide with other tectonic plates to form new land masses and mountain chains. We may not see it happening, but geologists have been studying it for years and can make predictions about how they move and when they will collide.

LIFE ON EARTH

During Earth's enormous modification over time, carbon (an element that resulted from the fusion of helium atoms within the core of a star) has dominated as the basic building block for living systems, partly because of its ability to form complex polymer (long chain) molecules (Pace, 2001; Fig. 6.2). At some time between 3.5 and 4 billion years ago, carbon, hydrogen, oxygen, nitrogen, and other assorted chemicals combined to form complex protobiotic (early-life) entities that were self-replicating (duplicating), a process that is one of the most important, mysterious, and amazing events in the rise and progression of life (Ricardo and Szostak, 2009).[6] Continued evolution produced organisms that were more complex, consisting of variations of carbohydrates, lipids, proteins, and nucleic acids, the same chemicals we see in all living systems today. DNA (deoxyribonucleic acid) and RNA (ribonucleic acid) became the templates and catalysts for replication and protein synthesis in all life forms.

The primal life forms about 4 billion years ago were unicellular (bacteria- or archaean-like) prokaryotes (one-celled organisms that lack a nucleus) called chemoautotrophs (Futuyma, 2005). They utilized chemicals (e.g., sulfur) in the environment as an energy source rather than the sun, similar to the way in which extreme thermophilic archaeans produce energy today. Astounding numbers of species diversified in the oceans and on land and eventually established food chains involving almost every conceivable trophic (feeding).

It is believed that certain of these protobiotic forms, which often fed upon one another, sequestered mitochondria (the respiratory organelles in most cells of contemporary organisms) from certain of their prey (Fig. 6.3). As bacteria and other prokaryotic forms evolved, the genetic codes for plastids (units of pigments in contemporary plants, algae, and certain bacteria that are capable of photosynthesis) were also obtained from their prey or their prey's DNA. These structures (or organisms) became established (through a process referred to as endosymbiosis) in eukaryotic organisms (having cells with a nucleus and extensive intracellular membranes).[7]

About 2.7 to 3.5 billion years ago, the first large group of photosynthetic organisms (cyanobacteria) arose, utilizing the sun's energy, carbon dioxide, and water to produce simple sugars (Holland, 2006; Kasting, 1993; Fig. 3.3).[8] Cyanobacteria (colonial alga-like bacteria with phycocyanin pigments, mentioned earlier as the first photosynthetic organisms) were derived from simpler prokaryotic forms, becoming quite different in how they fed and in the energy they produced. By using the sun's energy, CO_2, their pigments, and water, their construction of simple sugars and starches (polysaccharides that represent stored energy) resulted in the production of oxygen as a waste gas (Figs. 3.3 and 6.4). Over the subsequent millions of years, they and their photosynthetic descendants created an environment around, on, and within Earth that was progressively more aerobic.

Carbon Cycle

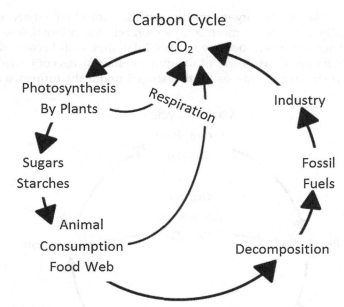

FIGURE 6.2 Carbon cycle in the contemporary world, with photosynthesis and respiration continuing since the first photosynthetic cyanobacteria commenced using the sun's energy and CO_2 and the more modern breakdown of fossil fuels and belching of toxic fumes from industrial plants. This is but one of several important chemical cycles that constantly occur on Earth. Carbon dioxide, like water vapor, methane, ozone, chlorofluorocarbons, and nitrous oxide, is a greenhouse gas. Greenhouse gases are gases that absorb and emit infrared radiation in the wavelength range emitted by Earth. Cumulative anthropogenic (i.e., human-emitted) emissions of CO_2 and other such gases are a major cause of global warming.

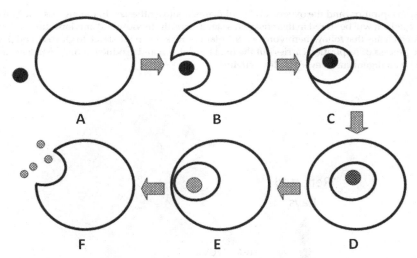

FIGURE 6.3 Processes of endocytosis and exocytosis, showing how a food item (*solid black sphere* in A) is consumed by the plasma membrane by invaginating and surrounding the food with part of the plasma membrane. Typically, food that enters the cell undergoes digestion by lysosomes inside the food and digestion vacuoles, and the waste materials are cast out in exocytosis. At some point in the early evolution of prokaryotes, organisms housing mitochondria and plastids were taken in by endocytosis, but there was no or very little digestion. Instead, the organism, or parts of them, became established as organelles within the organism that consumed it. The acquisition of these new organelles (the process which is referred to as endosymbiosis) represented major changes in ancestral prokaryotes. All plant cells house both chloroplasts and mitochondria. Of these two organelles, animal cells house only mitochondria. The rise of chloroplast-laden cells led to the release of oxygen as a waste gas and eventually changed the atmosphere from an anaerobic one to an aerobic one. The rise of mitochondria-laden cells led to more efficient energy production and organisms that are able to produce their own body heat.

Simultaneously, the increasing amount of oxygen around Earth reacted with high-energy ultraviolet light rays (ultraviolet C, which is part of the electromagnetic energy given off by the sun), forming an ozone screen around Earth, which hinders short-wave, high-energy ultraviolet C wavelengths from getting through to Earth but allows longer wavelength ultraviolet A and some ultraviolet B to reach Earth (Fig. 6.5). Without filtering out the more life-threatening ultraviolet C, it would be difficult for most life on Earth to exist.

Over millions of years, unicellular aerobic organisms diversified, some of which became the dominant life forms, and those populations eventually gave rise to more complex unicellular, colonial, and multicellular forms (Grosberg and Strathmann, 2007). When organisms become multicellular, their cells become differentiated to accomplish specific tasks to maintain body homeostasis. Many of the multicellular forms continued the trend toward increasing complexity in behavior and structure. Thus, as Earth changed under the influence of abiotic pressures, so did

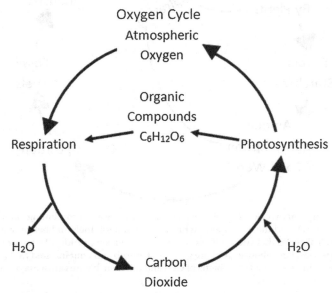

FIGURE 6.4 Photosynthesis, respiration, and the oxygen cycle. All plants photosynthesize. In the process, they take in the sun's energy, CO_2, and water and produce glucose, which will be stored in their tissues as starch (mostly amylose and amylopectin), and oxygen, which will be given off as a waste product. In the food chains that follow, herbivores eat the plants, break down the starch to glucose, and pass the breakdown products to the mitochondria where the process of respiration carries out the final breakdown and production of ATP (adenosine triphosphate). Energy in ATP's chemical bonds is used by all organisms in every bodily activity.

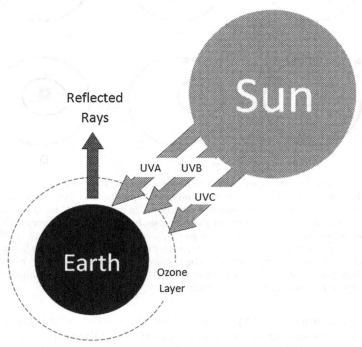

FIGURE 6.5 Numerous electromagnetic rays from the sun travel toward the Earth. Ultraviolet waves are examples. UVA waves reach the surface of Earth, but the amount of UVB and UVC are dramatically reduced because of an ozone layer surrounding the planet. UVC wavelengths are the shortest but most energetic and are threatening to life on Earth.

the populations of organisms, and this increase in complexity became the path followed by many organisms in the evolution of life.

MUTATIONS, DIVERSIFICATION, AND NATURAL SELECTION

Yet, with all its magnificence, the biological world was never a perfect world. During self-replication, chemical mistakes (which we refer to as mutations) occurred in all groups of organisms, from unicellular to multicellular forms, resulting in variability in their populations.[9] With repeated mutations, the acquisition of foreign DNA from the environment and epigenetic changes (changes arising from the rejuvenation of old genes under environmental influence) resulted in a constant source of variation.

With constant variation within populations, along with the isolation of population segments, populations changed and continued to change to become different from one another (they diversified, often forming new species). While I refer to mutations as mistakes, they added a new, very important dimension to population dynamics, collectively becoming the driving force behind biological diversity and thus a form of genetic variability that keeps many species from becoming extinct.

Diversification has continued over the following millions of years, providing Earth with an incredibly wide array of organisms of varying complexity. Many of the simplest types of multicellular life forms that we know today or the groups to which they belong arose in the oceans as far back as 600 million to a billion years ago, coinciding with the formation of a supercontinent called Panotia (Hoffmann, 1999).

In the subsequent Cambrian Explosion (a major event of great diversification at the beginning of the Paleozoic era, which commenced after the breakup of an earlier supercontinent called Panotia), they continued to evolve through the three major geologic eras of the Phanerozoic eon: the Paleozoic Era, which commenced 542 million years ago and lasted until about 251 million years ago; the Mesozoic Era, which commenced 251 million years ago and ended 65 million years ago; and the Cenozoic Era, which commenced 65 million years ago and continues to the present time (see geologic time scales, Fig. 1.7 and Table 5.1).

Throughout the span of time since life on Earth was represented solely by unicellular forms, organisms have repeatedly undergone mutations, variation has been present in all populations, and populations have undergone various degrees of isolation. Facing environmental threats, individuals that possessed the right combination of features in their genetic makeup survived environmental changes and passed their attributes on to the following generations, and those that did not succumbed. Thus, nature had formed a stunning system of natural selection that has a built-in mechanism for preserving individuals that are the fittest members of their population and passing their genes to succeeding generations (Fig. 6.6).

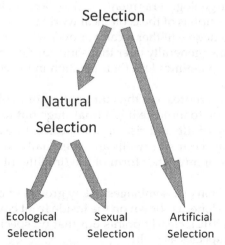

FIGURE 6.6   Process of natural selection. Populations naturally develop variations over time. We call them clines, variations from one end of a population to the other. Variations can be behavioral, physiological, anatomical, and genetic. Organisms best suited to their environment have an edge over organisms that are not suited to their environment and will be the most reproductively successful. Artificial selection is seen in the production of ideal bloodlines of domestic animals, for example, dogs and cats, in which humans have selectively bred animals to bring out desired features.

We can imagine that surviving these millions of years on Earth was very difficult for most organisms. They (including humans) had to contend not only with such negative biotic factors as predation, parasitism, inadequate food supplies, and aggression between different species and within their own species, but with devastating environmental changes as well (Fig. 6.5).[10] As a consequence, most species that have arisen over time have become extinct, and contemporary species are not the same species as those that existed at earlier times (Alroy, 2008). Everything has changed and continues to change. Change and natural selection are and always have been two of the most outstanding features of biotic forms on Earth.

Occasionally, environmental changes in the past have reached catastrophic levels. Researchers recognize at least five major periods of environmental crisis for life forms in which the Earth changed dramatically (each resulting in phenomenal mass extinctions), including an Ordovician–Silurian crisis (445 million years ago, MYA), a Frasnian–Farmenian (Late Devonian) crisis (375 MYA), a Permian–Triassic crisis (251 MYA), a Triassic–Jurassic crisis (200 MYA), and a Cretaceous–Tertiary crisis (65 MYA). In addition, each species, as well as its phylogenetic ancestors and descendants, have had to contend with individual crises, which periodically threatened their existence (Raup and Sepkowski, 1982; see Table 5.1).

Hypotheses for the causes of these mass extinction events include: (1) impact events involving sufficiently large asteroids or comets, which could result in tsunamis, global forest fires, and a reduction of impinging sunlight; (2) climate change, although it has been suggested that recent cycles of ice ages have had only mild impacts on biodiversity; (3) volcanism, especially extensive activity in short durations; and (4) plate tectonics, in which the opening and closing of seaways and land bridges may play a role in extinction events as previously isolated populations are brought into contact and new dynamics are established in ecosystems (Fig. 6.6). While gamma ray bursts have also been implicated, researchers state that they are not possible in metal-rich galaxies like the one Earth belongs to.

While a vast number of species have arisen throughout the millions of years that life has been on Earth, many have become extinct due to these environmental events that sometimes caused the annihilation of most life forms that existed at that time. As an example, 96% of oceanic species and 70% of terrestrial vertebrates were lost in the Permian–Triassic crisis alone.

Over the subsequent millions of years, life rebounded, populations changed, and new life forms arose until a new environmental crisis (millions of years later) wiped many of them out as well. Thus, life on Earth has been a series of constant environmental changes and recycling of periods of mass speciation (diversification), followed by mass extinction. Between and during the major and many minor crises, speciation and extinction were naturally contrasting events in an ongoing process of natural selection (Fig. 6.6).

RISE AND FALL OF DOMINANT FORMS

As life forms evolved during the three geologic eras mentioned earlier (Table 5.1), certain groups of organisms rose to dominance. Dominant organisms, the alphas of the ecological world (as in the climax trees mentioned in Chapters 1 and 2), control their environments to a degree (either locally or over wide areas), influencing what other organisms will be found in their presence. Over time (generally over thousands or millions of years), when major environmental changes occur, dominant organisms sometimes lose their position in the ecological hierarchy, and new groups of organisms become the dominant forms.

Our species, *H. sapiens*, in some ways, represents the culmination of evolutionary changes over these millions of years. I use the term "culmination" not to indicate it is a final stage but to point out what arrangement we have currently arrived at since the beginning of life and rise of our species. Life and life forms will exist and continue to change whether we are here or not until our sun swells and eventually collapses about 5 billion years from now, but this does not mean that our species or any other form of life (intelligent or otherwise) will be present through this time.

A glance at the phylogenetic (evolutionary) assemblages of key groups of organisms throughout the span of time that life has existed in the Paleozoic era alone, as shown on the inside front cover of Sebastien Steyer and Alain Benateau's book, *Earth Before the Dinosaurs* (2012), clearly points out that no species has reigned in a dominant position forever. Will this be true for the human species as well?

Our most threatening forms of population destruction can have both anthropogenic (caused by humans) and nonanthropogenic properties. Some contemporary anthropogenic sources of risk are found in such events as the rise of artificial intelligence, nanotechnology, biotechnology, warfare and mass destruction, global warming and cooling, ecological disasters, world population and agricultural crises, and experimental technology accidents. Those of a nonanthropogenic nature are said to include global pandemics, critical temperature changes, volcanism, megatsunamis, geomagnetic reversals, asteroid impacts, and cosmic threats. While pandemics and temperature changes are listed here as nonanthropogenic, the growing human population (overcrowding), overuse of certain medications, and global warming may have anthropogenic relevance, which could certainly influence their outcome.

We, like any other species, can and will most likely vanish from the scene at some point. It has happened many times before to the planetary alphas and others. However, human extinction is not a pleasant thought, and we are quite different from other animal forms in certain ways. Could there be a possibility that we will not perish from Earth like other species have done? Based on the evidence in this and other books about our dominant and aggressive nature and our disregard for the impact of overpopulation, we can clearly see that we are and have been for a while a threat to ourselves, other species, and even the planet itself.

With the current rate of vanishing species on Earth, some scientists are comparing contemporary extinction rates with those that led to the great mass extinctions of the past. Some scientists are even calling the time period we exist in the Anthropocene (an unofficial designation of present time, with respect to the influence humans have had on planet Earth) to point out how important we are as a dominant species and how we are changing the world.

Could these life-threatening changes and human influence actually cause our species to vanish? If so, there must be a reason or possibly many reasons. One of the differences between our species and the numerous other dominants that have existed on Earth is that their rise and fall to and from dominance was almost always the result of powerful environmental changes, while we have evolved to a point of ultra-domination that has the potential of resulting in our own annihilation, as well as the annihilation of many of Earth's other creatures.

Can this actually happen? Are we so out of control that we cannot recognize and repair the damage we have caused to Earth and its biota? Are we a species that must fade away, like all the other dominant forms of the past, or can we prolong our existence on the planet that took so long to support intelligent forms?

The journey taken by humans from their inception to the present is an extremely interesting story about an extraordinary animal, filled with events that reveal how we became what we are and how diverse (sometimes strange) our behavior has become, because of and also in spite of our animalistic past and current super intelligence.

In our presentation of earlier and the following chapters, we must realize that in a biological world, we are nothing more than a biological entity. We are animals, and as animals, we express a variety of animalistic virtues. We have arisen from a common ancestor with another life form (just as all species do), and we must entertain biological thoughts in order to survive, reproduce, and avoid extinction.

And yet, we are very unusual animals, highly social and brainy ones, equipped with extreme intelligence and extraordinary innovative skills, features that enhance our dominant status on Earth. It is the combination of these qualities that has provided an extremely equitable life for many of us on Earth. At the same time, they may have put us in an exceedingly tenuous position, one that no other animal has ever experienced in the 4 billion years of evolution of life on Earth.

Other than in an anoxic–aerobic shift, like the one that occurred during The Great Oxygen Event, it is difficult to believe that one dominant species can change Earth to such a degree that it can actually become uninhabitable for them and many other forms of life. Yet, that is precisely what we have done and continue to do, and we have reached a point at which we must carefully evaluate our dominant role in all the things we do:

> With every technological change, we instantly mutate into a new—and for the ecosystem—an exotic kind of creature. Like other exotics, we are a paradox, a problem for both our environment and ourselves. We search for explanations in terms of external controls or worldly problems, but it may be more profitable to concentrate on our own confusion. *Neil Evernden* The Natural Alien, *Humankind and Environment*

Note:

Superscript numerals appearing in this chapter refer to additional text/explanation given in the appendix.

7

Similarities Between Humans and Other Living Organisms

A balanced perspective cannot be acquired by studying disciplines in pieces but through pursuit of the consilience among them.
E.O. Wilson, Consilience, The Great Branches of Learning

HUMANS AS SOCIAL ANIMALS

Recall from earlier chapters that humans, their primate relatives, and certain other animals are social, that is, they live in groups and interact in various social ways, often demonstrating dominant, cooperative, competitive, and aggressive forms of behavior within and outside of their groups. Such forms of behavior are expressed in diverse ways, from what appears to be egalitarian cohesiveness to the establishment of well-defined dominance hierarchies, despotic tendencies, and aberrant behaviors referred to as antisocial disorders.

At times, depending on the personality and abilities of its group members, dominance and aggression in humans (as in other animals) function in establishing and maintaining population and environmental homeostasis. At other times, when dominant and aggressive behaviors extend beyond the boundaries of tolerance and morality, they create population and environmental instability or disorder, entropy if you will.

When population and environmental instability reach a point at which the existence of all life forms and the Earth itself are threatened, as they often have become under human domination, it is time to step back and critically examine the causes of such instability and contemplate making a change in the system which has resulted in the dilemma. To do this, the human animal's attributes and influence on the world must be scrutinized from various perspectives.

We find that compensating for maladjusted approaches to living in a modern world is an ongoing event for the human species (Moore, 1993; Wallerstein, 2011). When we approach problematic situations, we analyze them and sometimes attempt to correct them to bring them back to homeostasis. However, we often are reluctant to correct them until they become a serious threat to humanity, the environment, and the world.

Within the past few years, an increasing number of excellent books by leading scientists and environmentalists have appeared which describe the nature of Earth and its inhabitants, most of which issue warnings about the seriousness of Earth's destruction at the hand of the human species. In addition, human fragility in terms of pathogens that are quick to evolve into new forms that can wreak havoc in crowded animal populations around the world, especially under the influence of human-generated drugs, represents a serious threat to the human species, which is growing to a more ominous level.

As Laurie Garrett in *The Coming Plague, Newly Emerging Diseases in a World Out of Balance* (1994) states, understanding *Homo sapiens* requires a new paradigm in thoughts about its relationship with the entire biosphere and its many interrelated components. "Preparedness demands understanding. Perspectives must be forged that meld such disparate fields as medicine, environmentalism, public health, basic ecology, primate biology, human behavior, economic development, cultural anthropology, human rights laws, entomology, parasitology, virology, bacteriology, evolutionary biology, and epidemiology." Under human behavior, we should list psychology and sociobiology since they are important in understanding who we are.

Many contemporary humans ignore signs of environmental disruption, but as a population, we should not pretend that concerned scientists are crying wolf when they speak of environmental problems. The data have been presented by leaders in the scientific community, and we should treat the matter as serious enough to warrant our attention. Was it crying wolf when Rachel Carson and others exposed the threat of ecological destruction in the 1960s? Was it crying wolf when virologists and tropical medicine specialists first reached a conclusion in 1988 that it was time to sound an alarm on the importance of emerging microbes?

Nature is not benign. The bottom line is: the units of natural selection—DNA, sometimes RNA elements—are by no means neatly packaged in discrete organisms. They all share the entire biosphere. The survival of the human species is not a preordained evolutionary program. Abundant sources of genetic variation exist for viruses to learn new tricks, not necessarily confined to what happens routinely, or even frequently. *Joshua Lederberg, Nobel Laureate Opening Speech, National Infectious Diseases, The Fogarty International Center, Rockefeller University, 1989*

In spite of the credentials of these writers and the importance of what they say, it is often the case that their comments fall on deaf ears. And when action is eventually taken, it often treats the superficial symptoms rather than get to the root of the problem. We will find that the root at the base of most human problems lies in the presence of some rather paradoxical qualities that humans possess and the overabundance of people who make up our international population:

We have created a Star Wars civilization, with Stone Age emotions, medieval institutions, and god-like technology. *E.O. Wilson, The Human Condition, The Social Conquest of Earth*

We may wonder why humans in different walks of life sometimes do not agree on the seriousness of the human animal's destructive nature. Much of the controversy arises among our political leaders because of differences in the forms of education and personality among contemporary humans and their understanding and misunderstanding of science and its findings. Many people avoid science in school and thus are not exposed to or interested in the importance of other-than-human biota or even the populations to which they themselves belong.

Some of these people ignore what is around them unless it pertains directly to them, holding their position of dominance in society or providing them with personal status and comforts. Others may have even more selfish reasons for avoiding or even condemning scientific findings. We belong to societies that are designed for power and profit. Still others assume that the problems will take care of themselves, possibly with the help of God, that humans have coped with threatening situations many times and have survived them. Really? Has the loss of millions upon millions of human lives in wars been a way of taking care of our problems? Has overpopulation created a positive step toward planetary homeostasis?

To understand the true nature of a human approach to life and the games they play, it is up to the reader to consider the scientific implications and evaluate the human condition. It is also the reader's responsibility to have an open mind and be honest about how they interpret the literature.

FOSSILS AND DATING

Fossils and radiometric dating have been important for estimating relatedness and understanding populations and lineages of organisms that existed in earlier geologic history (Boltwood, 1907). In spite of the many fossils that are being found these days, however, fossilization is actually a relatively uncommon occurrence in nature, usually requiring the presence of hard parts and death in an extremely dry area or near a site in which sediments are being rapidly deposited, providing quick burial and an anaerobic environment. Much about the evolution of humans and many other animals is based on hard work and expertise by knowledgeable people who report on fossil remains and archeological artifacts, since DNA may be (and most often is) lost in the process of fossilization.

Deterioration of hard parts has a lot to do with climatic conditions. Much of the history of musical instruments and other artifacts in the southeastern United States has been lost due to a high moisture content of the soil, whereas artifacts have been preserved in the Southwest where moisture content does not reach a critical level (Hermann, 2011). Even wood, which is difficult to break down, may last no more than two years in soil with a high moisture content as a result of bacterial and fungal presence.

Evidence of organisms prior to the development of hard body parts, for example, shells, bones, and teeth, is especially scarce. Fortunately, certain soft-bodied fossils do occasionally exist in the form of ancient microfossils and impressions, usually due to rapid burial under sediments, such as volcanic ash.

In lieu of fossils, Richard Fortey points out in *Horseshoe Crabs and Velvet Worms* (2012) that Earth has other sources of organisms that have originated in early geologic periods while retaining primitive phylogenetic traits:

Evolution has not obliterated its tracts as more advanced animals and plants have appeared through geological time. There are, scattered over the globe, organisms and ecologies which still survive from earlier times. They speak to us of seminal events in the history of life. They range from humble algal mats to hardy musk oxen that linger on in the tundra as last vestiges of the Ice Age. The history of life can be approached through the fossil record, a narrative of forms that have vanished from the Earth. But it can also be understood through its survivors, the animals and plants that time has left behind.

COMPARING HOMOLOGOUS STRUCTURES AND THEIR CHEMISTRY

In addition to anatomical studies of groups of animals that show homologous structural features (features stemming from the same origin), vestigial structures, and embryological (ontological) development, comparisons in contemporary species offer contributing factors toward an understanding of common descent. Universal biochemical organization and molecular variance patterns in all organisms also show a direct correlation with common descent.

BIOGEOGRAPHY

Further evidence comes from the field of biogeography because isolation mechanisms often provide a thorough explanation of the geographical distribution of organisms across the world. This has become especially important in terms of island biogeography, as shown by R.H. MacArthur and E.O. Wilson (1967) in *The Theory of Island Biogeography* and by Darwin himself. Combining what we know about species diversification with a knowledge of plate tectonics and radioactive dating provides a logically consistent scenario of how the distribution of living organisms has changed over time.

POPULATION CHANGE

The development and spread of antibiotic-resistant bacteria, along with pesticide resistance in plants and insects and numerous investigations of population change, provide contemporary evidence that evolution due to natural selection and epigenetic changes are ever-continuing processes in the natural world. This is true for all organisms, humans as well.

In 1491, *New Revelations of the Americas before Columbus*, Charles Mann (2011) points out that unlike A.R. Holmberg's view in *Nomads of the Long Bow: The* Sirionó *of Eastern Bolivia* (1969), the aboriginal Sirionó people "existed almost without change in a landscape unmarred by their presence." Thus, it first appears that "indigenous peoples of the Americas floated changelessly through the millennia until 1492." In truth, human populations constantly change, although not at the same rate and in the same direction, no matter what their cultures are like. In addition, the process of change in relation to population isolation mechanisms and subsequent speciation of organisms in general have been observed directly and indirectly in both nature and under artificial conditions.

Perhaps one of the best examples of rapid evolution that has been brought to our attention within recent years involves populations of African cichlid fish that underwent extraordinary diversification from just a few species to as many as 500 new species during a period lasting 100,000 years (McMahan et al., 2013). Current genomic techniques are helping to integrate empirical and theoretical studies by identifying the genes that underlie the phenotypic differences among species.

There are certainly numerous anatomical differences between people around the world, differences that have come about through isolation from other groups. Those who dwell in cold climates, for example, generally have a stockier build. Those who evolved in equatorial areas often have darker skin.

To understand human populations, we must understand the basic biological concepts involving mutations, variability, isolation, and speciation. Mutations occur in all groups of organisms (Hastings et al., 2009). Mutations result in population variability. Over time, the process of natural selection molds a group of organisms to best comply with environmental stresses.

With isolation between groups, mutations cause the populations to not only vary within their group but also to develop greater differences between isolated groups. Many of the differences are superficial, for example, anatomical differences between different ethnic groups. However, the longer the groups are isolated from one another, the greater the differences will become. Eventually, when differences involve dramatic changes in the groups' genome, they become genetically incompatible. The ultimate result of continued mutations, variation, and isolation is speciation (formation of separate species).

All humans (without considering sexual differences) are 99.9% genetically identical (Sanchez-Mazas, 2008). While they established different groups during their emigrations within Africa and dispersal out of Africa, they never reached a point of developing more than superficial tribal traits. They all remain genetically compatible and differ most in superficial anatomical features and cultural traits.

All contemporary humans have evolved from African stock, and as a species, they have lived in Africa throughout most of their evolution. Thus, there is more human genetic diversity in Africa than anywhere else on Earth (Pennisi, 2007). Starting with 14 initial ancestral population clusters, certain humans from these clusters emigrated and dispersed to various African sites over time. Human genetic diversity changes in native populations with emigrational and dispersal distance from their original sites, their differences being the result of bottlenecks (evolutionarily difficult periods and stresses) during human movements.

Only a small part of Africa's population dispersed out of the continent, and thus only part of the original African genetic compliment went with them. There are certain African populations that harbor genetic alleles that are not found anywhere else in the world. In addition, all the common alleles found in populations outside of Africa are found in the African continent.

Most human biological variation (especially evident in anatomical features) is based on clines (variations from one side of a population to another) in which features blend gradually from one area to another (Begon, 2006). This is also the case with any widely distributed animal species. Each subpopulation on Earth which has anatomical differences shares a different clinal distribution pattern.

Human adaptability in different environments varies both from person to person and from subpopulation to subpopulation. The most efficient adaptive responses are found in geographical subpopulations where environmental stimuli are the strongest (e.g., high amounts of melanin in the skin being associated with peoples who spend a lot of time in the sun or Tibetans being highly adapted to high altitudes).

As with populations of other animals, if populations of humans would have remained isolated long enough, their differences would have resulted in speciation (the rise of new species). However, understanding clinal geographical genetic variation is further complicated by movements and mixing between human populations, which has been occurring since prehistoric times, deemphasizing group differences. As an example, contemporary humans travel around the globe, and they often share a complex mixing of genetic material and thus do not demonstrate the same patterns of variation through geography.

There is a statistical correlation between particular features in a population of humans because of a higher degree of sharing of genes, but different features are not expressed or inherited together. Thus, genes that code for superficial physical traits, for example, skin color, hair color, or height, represent a minuscule and insignificant portion of the human genome and may not correlate with genetic affinity. Dark-skinned populations that are found in Africa, Australia, and South Asia are not especially closely related to each other. Even within the same region, physical phenotype (genetic features we can see) is not related to genetic affinity: dark-skinned Ethiopians, for instance, are more closely related to light-skinned Armenians than to dark-skinned Bantu populations (Isichei, 1997).

Despite pygmy populations of Southeast Asia (Andamanese) having similar physical features with African pygmy populations, such as short stature, dark skin, and curly hair, they are not genetically closely related to these populations (Cavalli-Sforza et al., 1994). Genetic variants affecting superficial anatomical features (such as skin color), from a genetic perspective, are essentially meaningless; they involve a few hundred of the billions of nucleotides in a person's DNA. Individuals with the same anatomy do not necessarily cluster with each other by lineage, and a given lineage does not include only individuals with the same trait complex.

Due to practices of group endogamy (the practice of marrying within a specific ethnic group, class, or social group), allele frequencies cluster locally around kin groups and lineages, or by national, ethnic, cultural, and linguistic boundaries, giving a detailed degree of correlation between genetic clusters and population groups when considering many alleles simultaneously (Sarich and Miele, 2004). Despite this, there are no genetic boundaries around local populations that biologically demarcate any discrete groups of humans. Although we recognize different phenotypic anatomical features in certain humans, human variation is actually continuous, with no clear points of demarcation. In contemporary times, there are no large clusters of relatively homogeneous people, and almost every individual has genetic alleles from several ancestral groups.

BASIC BIOLOGICAL FUNDAMENTALS AND THEIR ORIGINS

The fundamental biological inertia for all life that is generally expressed in universal traits, from the tiniest microorganism to the largest and most dominant multicellular life forms, is found in their need to survive and reproduce (Freeman, 1977; Voigts et al., 2014). Interestingly, the dynamics of survival and reproduction for humans follow most of the same rules that influence other animal populations, although they may at first appear different. Also, an increase of complexity in the lives of humans has added additional stresses that may eventually affect our survival and the survival of many organisms around us (Leslie, 1996; Rees, 2003).

SIGNIFICANCE OF THE BRAIN

Reasons for the most important behavioral differences between humans and other animals on Earth lie in the size and complex nature of the human brain (Andrews, 2001; Kandel et al., 2000; Ramachandran, 2011; Thompson, 2000). Animal brains have been increasing in size since the onset of simple ganglionic development over half-a-billion years ago. Tunicate larvae and lancelets, representing two of the first invertebrate types of organisms with chordate features (presence of a notochord, pharyngeal slits, dorsal hollow nerve cord, muscular somites, and a postanal tail), lack a distinctive brain, although a dorsal nerve cord is present (Zimmer, 2000; Fig. 8.1).

Even though an actual brain is almost nonexistent in these simplest of chordates (the group to which humans belong), it has consistently increased its size and complexity throughout the millions of years of craniate (brainy) evolution to become an extremely multifarious organ (Striedter, 2004; Fig. 5.2). After having passed through periodic stages of exceptionally rapid growth in the distant past, the contemporary human brain belongs to an animal who can do just about anything.

While the human brain is large and complex when compared to that of other mammals (having upwards of 86 billion neurons in adults), primates in general (both prosimians and simians) have unusually large and complex brains for their body size as well. Scientists have long suspected that this is because primates have an unusually complex social life and need a sophisticated brain to cope with it.

In a 2013 Scientific American Mind Volume article (2013), anthropologist John Hawks stated, "For the first two-thirds of our history, the size of our ancestors' brains was within the range of those of other apes living today." The species of the famous Lucy fossil, *Australopithecus afarensis*, had skulls with internal volumes of between 400 and 550 mL, whereas chimpanzee skulls hold around 400 mL and gorillas between 500 and 700 mL. During this time, *australopithecine* brains started to show subtle changes in structure and shape as compared with other ape-like humans. For instance, the cortex (the outside covering of the brain that is related to most cognitive thoughts) had begun to expand, reorganizing its functions away from visual processing toward other regions of the brain (Fig. 5.2).

Hawks stated that "the final third of our evolution saw nearly all the action in brain size. *Homo habilis*, the first of our genus *Homo* who appeared 1.9 million years ago, saw a modest hop in brain size, including an expansion of a language-connected part of the frontal lobe called Broca's area. The first fossil skulls of *Homo erectus*, 1.8 million years ago, had brains averaging a bit larger than 600 mL.

All of our brain development was not in hops and spurts. "From here the species embarked on a slow upward march, reaching more than 1000 mL by 500,000 years ago." Early *H. sapiens* had brains within the range of people today, averaging 1200 mL or more. As our cultural and linguistic complexity, dietary needs, and technological prowess took a significant leap forward at this stage, our brains grew to accommodate the changes. As the shape changes, we see accentuated areas that belong to the regions related to depth of planning, communication, problem solving, and other more advanced cognitive functions.

However, development of our brain over time was not always uphill. "With some evolutionary irony, the past 10,000 years of human existence actually shrank our brains. Limited nutrition in agricultural populations may have been an important driver of this trend. Industrial societies in the past 100 years, however, have seen brain size rebound, as childhood nutrition increased and disease declined. Although the past does not predict future evolution, a greater integration with technology and genetic engineering may catapult the human brain into the unknown."

Based on comparative brain data for primates, R.J.M. Dunbar and S. Shultz (2007a,b) (*Understanding Primate Brain Evolution*) have summarized the currently known facts about brain size and its possible correlates with sociality in humans. As they point out, the human brain's growth has been primarily in the development of the neocortex, a six-layered brain surface tissue that contains an enormous number of neural cells (Fig. 5.2).

In order to evolve a large neocortex, they say, a species must first evolve a large brain in general to support it. This, in turn, requires an adjustment in diet (to provide the energy needed, such as in the eating of more meat instead of only vegetation) and life history (to allow sufficient time for both brain growth and software programming). It has been shown that any type of meat provides more calories than virtually any type of vegetable, as long as the serving sizes weigh the same. As an example, a 3-ounce serving of lean steak has 180 calories, while a 3-ounce serving of green-leaf lettuce has only 13 calories. Calories, expressed as small calories (gram calories) and large calories (kilogram calories or dietary calories) can be thought of as units of energy. A small calorie (equal to 1/1000 of a large calorie) is the approximate amount of energy required to raise the temperature of 1 g of water by one degree under a specific atmospheric pressure.

The human brain uses more energy than any other organ in the body, accounting for up to 20% of the body's total. Likewise, oxygen consumption is directly related to the needs of brain activity. According to M.E. Raichle and D.A. Gusnard (2014) in an article called *Appraising the Brain's Energy Budget*, the average adult human brain represents only about 2% of the body weight but accounts for about 20% of the oxygen used. Two-thirds of the brain's energy consumption is used to assist neuron transmission. The remaining third is apparently used by neuroglial activities, for example, maintaining homeostasis in nervous tissue.

The brain of animals (including humans) is not just an organ to handle routine activities. It must handle automatic reactions to the environment and a wide assortment of cognitive processes involving daily events in their lives (Jones, 2012). The more complex the life is, the more the cognitive response required. Thus, what at first appears to be a brain and neocortex size correlation is actually a much broader concept, involving a variety of phenomena which all relate to social complexity.

Dunbar and Shultz (2000a,b) state that "a clear distinction can be drawn between two versions of the cognitive challenge that humans have faced which underpins what has been referred to as the 'social intelligence hypothesis'." One view focuses on the social bonding of groups as the critical issue. The other assumes that feeding and nutrient flow is the critical constraint, making the social learning of efficient foraging strategies the principal selection pressure for the evolution of sociocognitive skills.

Species with large brains have, on average, higher metabolic rates (when corrected for body size), larger bodies, longer life spans, and longer juvenile periods. Life history traits can permit species to support larger brains (metabolically and developmentally), but the overall architecture of the brain is not itself tied to—or necessarily constrained by—life history characteristics:

> Overall, the consistent relationship between slow life history characteristics and higher-than-predicted metabolic rates indicates that a complex suite of life history characteristics is necessary to support the development of large brains. While total brain size correlates with the conventional gestation/lactation measure of parental investment, relative neocortex volume correlates better with the length of the juvenile period (weaning to first reproduction), suggesting an important role for the learning of sociocognitive skills. Thus, extended life histories may be necessary both to allow the laying down of large quantities of brain tissue and to allow neural tissue to be finely tuned (a learning process) through social and other more conventional processes.

Based on Dunbar and Shultz's report, it appears that both ecology and brain size are strongly associated with social group size. Individuals must solve ecological problems in order to function in large groups. Yet, although the social brain hypothesis has often been formulated in terms of group size, it is more correct to think of it in terms of the complexity of social relationships.

The size of grooming cliques (coalitions), for instance, correlates significantly with a neocortex ratio (neocortex volume divided by the volume of the rest of the brain) across primates. As pointed out by Robert Trivers in *The Folly of Fools* (2011), frequencies of tactical deception, whereby animals appear to deliberately mislead other individuals, are also important, and it appears that there are both developmental and energetic constraints on final brain size. With a neocortex volume–sociality correlation, a deep relationship with the suite of traits described (an adaptive complex), and constraints or consequences that arise in the context of life history, natural selection has resulted in the development of both a larger brain and an increasingly sophisticated social life.

It has become obvious that humans have retained the sequence of embryological development of its predecessors but its brain has undergone further divergence that is exclusively human (Schoenemann, 2006). It is with this large and complex brain, other anatomical features, scientific curiosity, and technological capabilities that humans have become the most dominant animals on Earth, ever.

DEVELOPMENT OF THE HUMAN BRAIN

Embryonic development of the human brain gives us important clues about how the brain has evolved since the first ganglionic mass began directing animal development and behavior. With reference to life in the Paleozoic era, Michael Sweeney (2014) states that as ocean life diversified into the ancestors of today's animals, neural networks in early brains began to diversify as well. "Some connections began to specialize in vision, and others in hearing, tasting, and smelling." Olfaction became extremely important in early vertebrates. By the time organisms evolved to cope with a terrestrial environment, their brains were already laying down new neural tissues on top of the earlier neural networks, and neurons competed in their own dominance struggles in a process that has been referred to as neural Darwinism.

Biologists have found that the phyletic stages developed in the evolution of the brain over millions of years are evident in embryonic and fetal development of higher organisms. A few days after conception in humans, for instance, embryonic cells begin to specialize. "Some form a simple neural plate, which changes into a groove and then a tube. The rather large cerebral cortex that distinguishes the human brain develops last, in the final months before birth, in a sequence like the one it followed in its evolution."

"Human brains evolved by adding new tissue over tissue that already existed. These sets of tissue represent three stages in paleontological change, as well as the sequence in which they develop in an embryo. In vertebrates, the brain developed by reptiles is often referred to as the R-complex" (MacLean, 1985; Fig. 5.2). It "oversees sleeping and waking, breathing and heartbeat, temperature regulation, and automatic muscle movements. It also plays a crucial role in the processing of sensory signals from the peripheral nervous system." This brain provides "sufficient mental power to find food, compete with other animals for survival, and pass along genes of the dominant, most fit individuals to succeeding generations." Thus, the seat of dominant struggles lie in the R-Complex.

The second region of the brain is the limbic or paleomammalian system. "It gives rise to emotions and simple memory, as well as rudimentary social behaviors." While dominance struggles had already developed by this time, they became honed in this second stage.

The third region of the brain is the cerebral cortex, a region that is most highly developed in humans. "This region, sometimes referred to as the neomammalian brain, gives humanity its capacity for language, culture, memory of the past, anticipation of the future, empathy, and ability to see the world through the eyes of others." With the development of the neomammalian brain and further enlargement of the neocortex, human dominance behavior has taken on its most complex form.

The human embryo passes through its early stages to the blastocyst (Fig. 3.4) prior to its first 14 days of development, and the development of its three primordial tissues, development of neuroblasts, and commencement of movement and connecting to other cells by about week three. A neural plate, neural groove, dividing of neural tissues into hemispheres, and closure of the groove occurs by the fourth week.

By weeks four through eight, the brain and spinal cord develops, and the brain forms its three primary lobes, the prosencephalon (forebrain), mesencephalon (midbrain), and rhombencephalon (hindbrain; Fig. 7.1). Further development into distinct cerebral hemispheres, cerebellum, pons, and medulla is complete by 11 weeks (Fig. 5.2).

Cerebral hemispheres enlarge to dwarf other brain components by five months, but the gyri and sulci that characterize a mature human brain are not yet present. It is during the latter six to nine months of fetal life that the brain develops these structures.

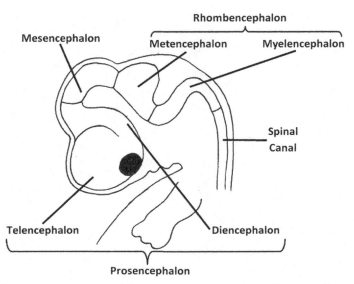

FIGURE 7.1 Formation of the human brain from primordial ectoderm. Following an invagination to form a neural groove and eventually a dorsal hollow nerve cord, cephalic neural tissue first segments into a prosencephalon, mesencephalon, and rhombencephalon. The prosencephalon subsequently divides into a telencephalon and diencephalon. The rhombencephalon subsequently divides into a metencephalon and myelencephalon. The seat of most endocrine production in the human body is in the diencephalon, which further divides into a hypothalamus, thalamus, and epithalamus. The hypothalamus produces nine different hormones, many of which influence the pituitary gland to produce seven hormones that function throughout the human body, including hormones that control the reproductive process.

Sweeney points out the damaging effects of poor nutrition, alcohol, smoking, and other drug use, especially in the early stages of pregnancy when development is most critical, to an unborn baby. It is during this stage that a human embryo undergoes its most important physical and physiological development, and both anatomical and physiological anomalies may result because of it. As we will see, how a person turns out in life depends on: (1) inherited factors that have been passed down from parents; (2) the environment the embryo and fetus exist in during development; and (3) how the individual is treated during postnatal life, especially during the early formative and adolescent years.

THE ANIMAL BECOMES HUMAN

Thus, we have seen in this chapter that humans have retained much of the DNA of its distant past, clearly shown in its retention of the sequence of embryological development of its predecessors, going back in time to the earliest of multicellular animals, and the rise of the earliest cephalochordates over 450 million years ago. And from its prochordate beginning, its nervous system has steadily evolved from a simple brainless dorsal hollow nerve cord to a large and complicated brain, which gave it powers to depart from its lower animalistic past.

In the following chapter, we notice that *H. sapiens* has retained other anatomical and behavioral links from its predecessors as well, but it also developed distinct human characteristics that complemented the development of its highly cognitive brain. It was the combination of animalistic and humanistic features that propelled humans into an elite status among planetary biota in which it could take charge of its capability to increase its population size and to dominate the Earth like no other animal.

8

Human Nature

Human nature is still an elusive concept because our understanding of the epigenetic rules composing it is rudimentary. *E.O. Wilson,* *1999 Consilience, The Fitness of Human Nature*

As we would expect from a highly evolved, very cognizant animal in an extremely stressful contemporary environment, human behavior is quite complex, its expressions differing widely in individuals with different personality traits and different stresses (Levi, 1987). A general examination of human behavior is found scattered in the various chapters of this book. This chapter focuses on the basic features of anatomy and behavior as it relates to humans and other animals and elaborates on certain expressions of human nature, especially personality, which is the cornerstone of human behavior. Concepts discussed here will prepare the reader for Chapter 9, Alternate Human Behavior.

Before getting into human nature per se, it may be worth repeating a thought that is briefly brought out at the end of Chapter 7 that the human approach to coping in the world represents a combination of both ancient animalistic features and new, more human, ones. As one of the outstanding features of an animalistic existence, animals, including humans, characteristically overproduce (Champman, 1928). Under natural conditions, populations are brought into homeostasis through the presence of community dynamics, including predation, parasitism, and food availability (Arrow et al., 1995) (Fig. 6.5). Subhuman animals do not consider concepts such as overpopulation or their impact on other organisms in their environment. Their populations are strictly run by a survival and reproduction approach. If their populations get overly large, they suffer the consequences, with population decline and natural selective modifications of the remaining population. There are population fluctuations, and at times, a species that cannot adjust to ecological pressures will become extinct in the process. Thus, since humans do not have a natural population stabilizer, a similar approach by humans toward their environment and other biota is an ever-growing threat to ecological stability.

But as we notice in Chapter 7, humans are cognizant creatures with great intelligence and innovative ways, many of them being able to think outside the box, and care dearly about retaining a homeostatic relationship with the environment and its biotic and abiotic components. Some humans understand that when their populations become too large, it puts undue pressure on other (nonhuman) populations, as well as their own, such as through dwindling resources, habitat destruction, and the introduction of exotic species (Woodward, 1997).

While the mind of humans is capable of retrospective and predictive thinking and has been a driving force in the development of extraordinary technological achievements in the contemporary world, it appears that rising signs of intelligence have been accompanied by both the retention of old survival traits and a vast array of new anatomical, behavioral, and emotional changes. All such traits combined define the nature of the dominant contemporary human (Goldstein, 2010) (Fig. 8.1).

Therefore, it will help the reader understand who this human animal is by examining some of these features within the human population and comparing them with similar features in the populations of its closest relatives. This prepares us for the extraordinary human features described in later chapters.

Before going into anatomical and behavioral differences, however, I would like to present a brief taxonomic view of the contemporary human (Bjorklund and Pellegrini, 2002). This will set the stage for understanding where we have arisen as a species and how we fit into the biological framework of a modern world (Table 5.1).

Domain Eukarya

- Humans and other organisms that have a well-defined nucleus in their cells are called eukaryotes, all of which belong to the domain Eukarya. This separates eukaryotes from the domains Bacteria and Archaea, which include

Dominance and Aggression in Humans and Other Animals
http://dx.doi.org/10.1016/B978-0-12-805372-0.00008-0

FIGURE 8.1 The Vitruvian Man, a modern depiction of a drawing in pen and ink on paper by Leonardo da Vinci around 1490, is based on the correlations of ideal human proportions with geometry, described by the ancient Roman architect Vitruvius in Book III of his treatise *De Architectura*. He described the human figure as being the principal source of proportion among the classical orders of architecture. He determined that the ideal body should be eight heads high. Da Vinci's drawing is traditionally named in honor of the architect. The drawing provides an example of da Vinci's deep understanding of human proportion and represents a cornerstone of his attempts to relate man to nature. He believed the workings of the human body to be an analogy for the workings of the universe, and indeed humans have worked hard during the last few hundred years to understand the universe and determine what place they will play as the universe becomes clearer.

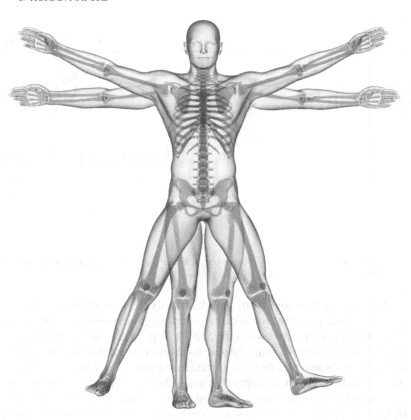

organisms that do not have a nucleus, groups that biologists refer to as prokaryotes. Thus, all other organisms except Bacteria and Archaea are eukaryotes.

Kingdom: Animalia

- This taxonomic category includes multicellular organisms that are made up of cells without cell walls or organelles such as contractile vacuoles and plastids (usually). While humans have as many as 200 types of cells in their adult body, they are very similar in origin, function, and development to those of other animals.

Phylum: Chordata

- This taxonomic category includes animals with a cartilaginous notochord or a more elaborate cartilaginous or bony endoskeleton, which has been derived from the notochord, pharyngeal slits and bars, a subpharyngeal endostyle, dorsal hollow nerve chord, and a postanal tail. While the embryonic human has a notochord-like arrangement, a postanal tail, and pharyngeal slits during its early development, the postanal tail is reduced through apoptotic events to a short os coccyx, the pharyngeal slits and surrounding tissues go into the makeup of other structures and tissues, and the endoskeleton becomes a complex system of cartilage and ossified hydroxyapatite (bone).[1]

Class: Mammalia

- This taxonomic category includes chordates with hair and mammary glands. Adult human hair is composed of the protein keratin but varies in structure and texture among mammals. In humans, it varies between the sexes and changes somewhat following birth and during adolescence. Women typically have more vellus (fine) hair, and men sport more terminal (longer and thicker) hairs, these features depending on hormonal influence.[2] Fetal humans have fine hair called lanugo, which is later replaced by vellus or terminal hair.
  The number of teats varies among mammals, but humans and other simian primates have two. Although mammary glands are characteristic of mammals, certain other animals are known to produce milk-like products as well.[3]

Order: Primates

- This taxonomic category includes mammals such as prosimians (early primates) and simians (monkeys and apes), all characterized as having binocular vision and opposable thumbs (discussed in other chapters) (Wilson and Reeder, 2005). Prosimians include lemurs, bush babies, tarsiers, and several other types (suborder Prosimii).

Simians include the New World monkeys (five families of primates that are found in Central and South America: Callitrichidae, Cebidae, Aotidae, Pitheciidae, and Atelidae); Old World monkeys (superfamily Cercopithecoidea), which are native to Africa and Asia today, and the apes (including Hylobatidae, which has four genera and 16 species of gibbon, including the lar gibbon and the siamang, commonly referred to as the lesser apes); and hominids (the great apes), consisting of orangutans, gorillas, chimpanzees, bonobos, and humans.

Family: Hominidae (commonly referred to as hominids)

- Members of this taxonomic category are known as great apes, including four extant genera: chimpanzees and bonobos (*Pan*), gorillas (*Gorilla*), humans (*Homo*), and orangutans (*Pongo*). This group does not include the New World monkeys, Old World monkeys, or the lesser apes.
  Subfamily: Homininae (commonly referred to as hominines or hominins)
- This taxonomic category includes all forms of humans, gorillas, chimpanzees, including all extinct hominids that arose after the split from an orangutan-like common ancestor (Ponginae). Based on the finding of fossils, the common ancestor between humans and members of the genus *Pan* existed about seven million years ago. More human-like apes included species in the genera *Sinapithecus*, *Sahelanthropus*, *Ororrin*, *Ardipithicus*, *Australopithecus*, *Paranthropus*, and *Homo*.

Genus: *Homo*

- This genus includes modern humans and extinct species closely related to them, for example, *Homo habilis*, *Homo gautengensis*, *Homo heidelbergensis*, *Homo antecessor*, *Homo cepranensis*, *Homo erectus*, *Homo ergaster*, *Homo floresiensis*, *Homo neanderthalensis*, *Homo rhodesiensis*, and *Homo rudolfensis*.[4]

Species: *sapiens*

- This is the only extant species of human. Thus, all groups of contemporary humans belong to the same species, and like other animal species, the human species demonstrates a considerable amount of anatomical and behavioral variation.

Subspecies: *sapiens*

- This name forms a trinomen (third name, *Homo sapiens sapiens*) in the scientific name, indicating that it bred with other human species during its rise to stardom, for example, *Homo sapiens neanderthalensis* when the two coexisted in Europe. However, many biologists use the binomen *Homo sapiens* which indicates that breeding between the two is not evident or that we are talking about these organisms on the species level. Contemporary taxonomic procedure supports the use of subspecific names (trinomens) for these two homonines. *H. s. sapiens* is the only extant subspecies of *H. sapiens* and thus, I will not use the subspecies designation unless necessary.

Thus, contemporary humans (*H. sapiens* Linnaeus 1758) are animals in the Kingdom Animalia, phylum Chordata, class Mammalia, order Primates, family Hominidae, and subfamily Homininae. The name following the species name, "Linnaeus," and the date 1758 mean that the taxonomist Carolus Linnaeus described our species in 1758.

Domain: Eukarya
Kingdom: Animalia
Phylum: Chordata
Class: Mammalia
Order: Primates
Family: Hominidae (common name hominids)
Subfamily: Homininae (common name hominines or hominins)
Genus: *Homo*
Species: *sapiens*
Subspecies: *sapiens*
Scientific Name: *Homo sapiens sapiens*

THE RISE OF ADVANCED HOMININES

Based on the rapidly growing number of fossils being found, the main radiation of advanced hominines (beyond the nonhuman great apes) occurred in the Miocene from hominids (including members of the genus *Sivapithecus*), which lived in African forests alongside stretches of grasslands, probably feeding on seeds and roots (Table 5.1) (Fig. 1.8)

(Teaford and Walker, 2005). If they did eat meat, it was undoubtedly scavenged because they were not yet hunters. Their tool use had not extended much beyond an animalistic level (Andrews, 1983).

The earliest forms of human-like primates (including species similar to contemporary chimpanzees) are aged from 7 to 17 million years ago (Chen and Li, 2001). Hominines were bipedal as early as 5.2 million years ago. The human lineage diverged from a common ancestor with chimpanzees and bonobos at some time between 5 and 7 million years ago. Since that time, humans have evolved many unique characteristics that have assisted in their rise to dominance.

The oldest known human-like (hominine) fossil discovered is *Sahelanthropus tchadensis* (Guy et al., 2005), which lived about 6–7 million years ago. It was chimp-like in having a small body size, relatively small brain, and large brow ridges. Its most evident human features were a short, flattened face, reduced canines, and enlarged cheek teeth (premolars and molars) with a heavy crown. It also walked upright. *Ororrin tugenensis* (5.7–5.9 million years ago) (Guy et al., 2005) and *Ardipithicus ramidus* (discovered in the Afar Depression of Ethiopia and dated at 4.4 million years old) (White et al., 2009) both walked upright as well. Since that time, hominine diversity has increased dramatically, especially between 2 and 4 million years ago, at which time a group of hominines called australopiths emerged.

Australopithecus anamensis (Leakey et al., 1998), the first and oldest known member of this genus, was found in rocks in Kenya that were 3.9–4.2 million years old (during the Pleistocene epoch). A more advanced *Australopithecus afarensis* (Kimbel et al., 1994) (affectionately referred to as Lucy) was first found in Tanzanian rock and aged at 3.0–3.4 million years old. Both of these species were bipedal but not as erect as *Homo sapiens*. They had large canines and incisors and a large overhung jaw. Some *A. afarensis* features were carried over to a more advanced species called *Australopithecus africanus* (Strait et al., 2009).

Australopithecus africanus, found in East Africa and referred to as the gracile southern ape because of its relatively dainty jaw, its relatively small cheek teeth, absence of skull crest, and the presence of a centrally located cranial opening (foramen magnum), lived about 2.5–3 million years ago, coexisting with species of *Homo* (the genus to which contemporary humans belong). It also existed with two other more robust australopithecines, *Australopithecus robustus* and *Australopithecus boisei*, about 1.5–2.0 million years ago (Skelton and McHenry, 1992). Some investigators put these latter two species in the genus *Paranthropus*.

At some point, late prehuman and early human-like species entered savannas and commenced walking long distances, and an accompanying reduction in body hair kept the body cooler. Dark skin color may have been selected to protect their skin against UV-induced folic acid loss, which is required for reproductive viability and fetal development. Also, dark-skinned humans in equatorial Africa received sufficient UV radiation to synthesize vitamin D, while in areas with less sunlight, human skin is believed to have become lighter through natural selection to sequester additional light energy.

Homo habilis (Tobias, 1987) existed 1.75 million years ago at Olduvai in Kenya. *Homo erectus*, which existed primarily between 0.25 and 1 million years ago, was first found in Pleistocene deposits in China. Specimens found on the western shore of Lake Turkana were dated at 1.6 million years old. *Homo erectus* or a common ancestor with another human form walked fully erect and may have given rise to *H. sapiens* about 200,000 years ago.

Evidence for the close relationship between *H. erectus* and *H. sapiens* comes from anatomical similarities, as well as the presence of tools, use of fire, construction of stone and wood shelters, and the arrangement of their shelters into communities (Gibbons, 1998).

Homo sapiens arose in equatorial (tropical) Africa (an area that is limited in latitude by the Tropic of Cancer in the northern hemisphere at about 23°26′ (23.5 degree) N and the Tropic of Capricorn in the southern hemisphere at 23°26′ (23.5 degree) S), the warmer part of the continent (Fig. 8.2). Some of the first major movements of species of *Homo* involved *H. erectus* dispersing into China and *H. sapiens* dispersing to areas now recognized as England and Germany (Figs. 1.1 and 8.2). It first appeared in the Middle East over 90,000 years ago and in Europe 45,000 years ago (Fig. 1.2). Two closely related species (some treat them as subspecies) of the genus *Homo* (*H. sapiens* and *H. neanderthalensis*) coexisted with one another in the area now considered Europe for about 10,000 years.

We may wonder what the world was like when *H. sapiens* was arising as a species? Climatic conditions were periodically very harsh. The evolutionary surge that led to the rise of *H. habilis* (Tobias, 1987) began during the transition between the Pliocene and Pleistocene Epochs around 2.5 million years ago. At that time, climates were becoming cooler and drier.

All subsequent species of the genus *Homo* evolved during the Pleistocene (Rightmire, 1988). This was generally a time of more extreme world cooling and recurrent glaciations (ice ages). During the coldest periods, global temperatures dropped by about 9°F (5°C), and long-lasting ice sheets spread out from the poles and high mountains. Between the four or more major glaciations of the Pleistocene, there were interglacial warming periods with temperatures similar to current conditions. Both the glacials and the interglacials lasted tens of thousands of years.

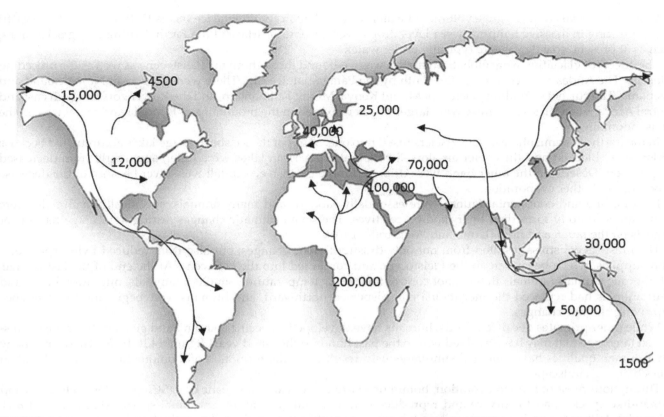

FIGURE 8.2 Movement of *Homo sapiens* from its point of origin in Africa to other parts of the world. In parts of its range, it overlapped with the ranges of *Homo neanderthalensis* (in Europe) and *Homo erectus* (Africa, parts of Europe, and Asia). While *H. sapiens* is about 200,000 years old, its dispersal out of Africa and major movements around the world have been during the last 100,000 years. *H. sapiens* lived a totally animalistic life until the Neolithic Revolution, about 12,000 years. Compare this figure with Figs. 1.1 and 1.2.

Continents of the northern hemisphere were more affected by glaciations than more southern areas, which generally remained mostly tropical and subtropical. The coldest regions of the world became arctic deserts. However, the great hot deserts of North Africa and western North America were mostly vast grasslands with large permanent lakes and abundant large game animals during the Pleistocene ice ages.

Sea levels were an estimated 450 ft (137 m) lower than today during the coldest periods, and a substantial volume of the world's water in terrestrial environments was locked in 1–2 mile (about 1.5–3 km) thick glacial sheets covering thousands of square miles. As a consequence, vast areas that are now shallow seas and ocean bottoms were exposed for thousands of years. Twice during the last ice age, lowered sea levels resulted in Siberia being connected to Alaska by a 1200–1300-mile (1900–2100-km)-wide corridor. Some Asian hunters may have used this route to disperse into the western hemisphere to become the first Native Americans (Hermann, 2011; Wells and Read, 2002).

Human evolution was very likely strongly affected by the dramatic climate swings that existed during the Pleistocene, the changes no doubt presenting powerful new natural selective pressures. Many animal species were driven to extinction by the advancing and retreating ice ages. Humanity no doubt survived primarily by its innovative capabilities and intelligence.

One of the greatest problems in the cold regions would have been the relative scarcity of plant foods, conditions that may have confined them to the more hospitable areas of Africa and the areas that they successfully dispersed to. Changes in subsistence patterns were no doubt essential for human survival. During such ice ages, those species that were not driven to extinction by the cold commonly evolved larger, more massive bodies as a means of producing and retaining more heat. This was especially true of mammals, including humans, in the northern hemisphere.

Thus, the climate during the Pleistocene when *H. sapiens* arrived on the scene was marked by repeated glacial cycles in which continental glaciers pushed to the 40th parallel in some locations (Van Andel, 2002). At maximum glacial development, 30% of the Earth's surface was covered by ice, a zone of permafrost stretching southward from the edge of the glacial sheet for hundreds of kilometers. The mean annual temperature at the edge of the ice was −6°C (21°F).

With each glacial advance, huge volumes of water were tied up in continental ice sheets 1500–3000 m (4900–9800 ft) thick, resulting in massive temporary sea-level drops over the entire surface of the Earth. During interglacial times, such as at present, coastlines were typically under water.

The effects of glaciation were global. Antarctica was ice-bound throughout the Pleistocene as well as the preceding Pliocene. The Andes were covered in the south by the Patagonian ice cap. There were glaciers in New Zealand and Tasmania. The current subsiding glaciers of Mount Kenya, Mount Kilimanjaro, and the Ruwenzori Range in east and central Africa where *H. sapiens* arose were larger. Glaciers existed in the mountains of Ethiopia and to the west in the Atlas mountains.

In the northern hemisphere, many glaciers fused into one. South of the ice sheets, large lakes accumulated because outlets were blocked and the cooler air slowed evaporation. African lakes were fuller, apparently from decreased evaporation. Deserts on the other hand were drier and more extensive. Rainfall was lower because of the decrease in oceanic and other evaporation.

Both marine and continental faunas were essentially modern and many animals, specifically mammals, were much larger in body form than their modern relatives. The severe climatic changes during the ice age had major impacts on the fauna and flora (Lamb and Woodroffe, 1970).

The most severe stress resulted from not only drastic climatic changes, but also from reduced living space and food supplies, which began late in the Pleistocene and continued into the Holocene. At the end of the last ice age, ectothermic animals (animals that cannot control their body temperature), smaller mammals, migratory birds, and swift animals had replaced the megafauna and dispersed northward, and humans were beginning to experience improving climatic changes.

While other primates vocalize sounds, humans have developed a specific anatomy and genetics for more sophisticated language that may have evolved within the past hundred thousand years (Humboldt, 1988). These and many other human qualities have emerged simultaneously, resulting in the formation of a unique human animal and an ever-changing landscape.

Throughout most of human evolution, hominine existence was an animalistic one (Stringer, 1992). Humans and human-like species had to survive and reproduce in a feral state like all other animal species (Figs. 8.3 and 8.4). Otherwise, failure to survive and reproduce as a species would have terminated in extinction. Some of the following characteristics were inherited from predecessors:

- Bipedalism, leading to quicker travel and a higher degree of defensiveness.
- Increasingly efficient binocular vision, leading to greater focusing ability for the acquisition of resources.
- An opposable thumb, leading to greater manipulation of tools. At the same time, a condition with less of an opposable toe and the development of a foot that facilitated bipedalism, migration, emigration, and dispersal occurred.
- A steadily increasing brain size and complexity, leading to advanced cognitive abilities.
- Greater differentiation of brain tissue, leading to an increase in the size and complexity of the forebrain's surface area (neocortex).
- Development of more diverse dominance hierarchies, leading to greater communication, more complex forms of deception, increase in the complexity of selfishness, and increased defensive and food-gathering abilities.
- Demarcation of territorial boundaries, leading to isolation and development of intergroup differences and greater need for territorial expressions.
- Development of sedentary behavior (Fig. 1.2), leading to greater communication, increased reasons for protecting territorial resources and extended benefits of a defensive nature.

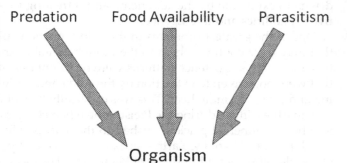

FIGURE 8.3 Population pressures typical of feral organisms. Population growth is generally kept in check by predation, parasitism, and food availability. Humans have been responsible for altering populations by killing off predators or bringing in exotic ones, changing food availability by forest destruction, and releasing exotic species that compete with indigenous organisms. Humans have destroyed entire ecosystems through such activities as habit destruction to construct pastureland, erect buildings, and lay down roads.

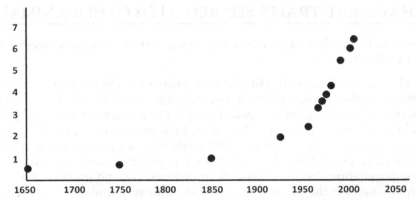

FIGURE 8.4 Approximate human population growth since 1650, showing a rapid upturn from the mid-1900s, increasing its rate exponentially around 1950. Vertical axis shows the global number of people in billions. At the dawn of agriculture (about 8000 BC), the population of the world has been estimated to be approximately 5 million. By 1 AD, it grew to an estimated 200–600 million. According to Worldometers (http://www.worldometers.info/world-population/#table-historical), based on fertility estimates by the United Nations Department of Economic and Social Affairs Population Division, the world population was 7.349 billion as of July 1, 2015. Predictions for further growth vary, but based on current estimates, world population growth is slowing and will reach 10 billion persons in the year 2056 unless attention is put on cultivating a population decline. The world's urban population is currently growing by 60 million people a year, about three times that of the rural population. By 2030, it is expected that nearly 5 billion (61%) of the world's 8.1 billion people will live in cities.

As with any animal species, variations within isolated human populations found on the planet allowed natural selection to play its role in the preservation and extinction of individuals and species that existed before contemporary humans. It has been only in the last few thousand years of our evolution that we have distinguished ourselves as especially brainy dominant animals with extraordinary technological potential. But along with this status, our populations have increased almost exponentially, leading to a threat to all other life forms and to ourselves (Fig. 8.4).

Early humans most likely were scavengers, with supplements of seeds, roots, and fruits (Cashdan, 1994). They were social animals, existing mostly as family groups. Their life at that time was probably much like that of contemporary nonhuman apes.

About 50,000 years ago, humans began to elaborate on tool-making, which allowed them to hunt with more efficiency (Boesch and Boesch, 1990; Boesch and Boesch-Achermann, 2000; Van Schaic et al., 1996). Their degree of dispersal increased as well, and it was at that time that humans reached Australia and other isolated islands. Dispersal from Asia to North America occurred as late as 15000 years ago. By the time they reached North America, they were well-established hunters and on the verge of establishing increasingly sedentary habits.

An agricultural way of life developed in Africa, Eurasia and the Americas about 10 to 15000 years ago, leading to the rise of more complex tools and a more sedentary way of life (Kerridge, 2006). Sedentary living led to the development of villages and a greater attraction between humans with similar interests. As we will see, it also resulted in greater conflicts between different human groups because of their differences and territorial complications.

MAMMALIAN ANATOMICAL FEATURES FOUND IN PRIMATES

In addition to our improved brain, we owe our existence to adaptability, similar to that which is shown by ever-changing populations of any animals that have developed traits important for survival and reproduction (Hill and Kaplan, 1999). As mammals, we are characterized as having hair and mammary glands, but it was the rise of certain primate features that allowed us to go beyond the typical mammalian state and occupy a position of such dominance and influence over other life forms on our planet.

As we know, evolution is a continuous process. Change is inevitable. The fossils that have been unearthed may represent separate species, but any changes that have come about in the rise of any species will ultimately show overlap with a previous one. As fossils are found, we will learn more about how these overlapping populations were related to one another.

BEHAVIORAL TRAITS SHARED WITH OTHER ANIMALS

Some of the predecessor traits that humans have carried along with them as they developed through the throws of human evolution are as follows:

- Humans have retained an animalistic propensity for living up to their biotic potential. Biotic potential is defined as the maximum number of individuals a species can produce (Fig. 8.5). As with other organisms, this is and always has been a survival strategy against food deprivation, predation, and parasitism (Fig. 8.3). Under natural conditions, animals that overproduce have their population reduced by inadequate food supplies, parasitism, and predation (Smart et al., 2006). Since food supplies have been adequate (for the most part in industrialized countries) for a thriving human population, contemporary humans do not have predators to keep their populations in check (other than themselves), and parasites have been eliminated or severely cut back in many parts of the developed world, the human population is increasing almost at an exponential rate and growing to a dangerous level (Cohen, 1995; Ehrlich and Erhlich, 1991; Gallant, 1990; Gore, 1992).[4]

  Actually, there are many signs, as elaborated upon throughout the chapters of this book, that we have long surpassed the environmental (ecological) carrying capacity (the maximum number of individuals that the environment can tolerate) (Fig. 8.5). Further, we are stretching the limits of our food production by creating new, larger, and more abundant supplies of fruits and vegetables, using pesticides and genetically modified approaches to keep insects from feeding on our food plants, taking increasing amounts of forest for developing pastures and feeding most of our water and grain to grow food animals, using fossil fuels to make plant fertilizers and paying little attention to decreasing our populations. And while we are in the process of engulfing genetically designed foods, we are putting ourselves in an extremely vulnerable position, since we are not aware of what these foods are doing to us (Black, 2012; Engdahl, 2007).

- Like other social creatures, humans have retained a propensity for developing agonistic behaviors and dominance hierarchies that affect every part of human life (Pusey and Packer, 1997). Also, as with other social species, such hierarchies tend to give structure to human society, facilitating the development of a societal division of labor. At the same time, they contribute to a waning tolerance and intrahuman conflict, for example, aggression and wartime behavior, at many levels of human society.

- Paradoxically, humans, like other social animals, express territorial behavior, another survival strategy that provides dominant individuals and populations with adequate food, mates, status, and protection from predators and conspecific intruders (Altman, 1975). However, territoriality, combined with an enlarged brain, a dominant status, technological skills, a noncaring attitude toward our cohabitants, and an overpopulated world, is resulting in an imbalance in species diversity and an increased tension among different groups of humans. To avoid conflict within their various subpopulations, humans have established a system of world trade, a social mechanism that helps distribute resources to individuals who are unable to provide such resources for themselves and those at the edge of human existence.

FIGURE 8.5 Population curve, showing a line for exponential growth or overproduction, as carried out for all organisms, and a more typical curve (the horizontal portion to the right) that is established by environmental pressures (e.g., abundance of food, predation, and parasitism). All organisms overproduce, and their populations are generally brought back to a stable level by environmental pressures. Without environmental pressure, an animal population will realize exponential growth but at some time will suffer from population overabundance.

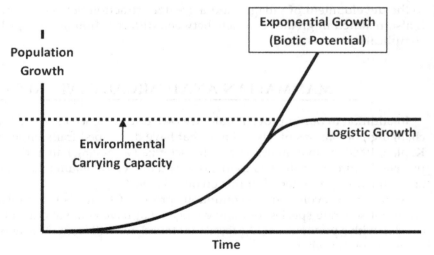

- Humans and other social animals often express altruistic behavior in terms of relatedness (Ferriere and Michod, 2011). Thus, as with other animals, humans are usually more altruistic toward related individuals and the groups they belong to than toward individuals and groups they do not know or belong to.
- Humans and other social creatures share a reciprocally altruistic nature (reciprocity), allowing them to engage in social endeavors that provide the most abundant advantages to more than one party (Falk and Fischbacher, 2006). This has become especially important in business and politics where individuals or companies can benefit one another, and it is best demonstrated in humans because of their cognizant brain and ability to contemplate such actions.
- Selfishness is a behavior that has been absolutely necessary for survival throughout the four billion years of evolution for all organisms (Estlund, 2011). Strangely, selfishness and an extremely dominant attitude seem to counter altruistic and reciprocally altruistic acts. However, all of these behaviors can be selfish in certain respects. Altruism and reciprocal altruism (reciprocity), for instance, are ways for a species or an individual to acquire the most for its colony and species. And since they are directed more toward close relatives, they benefit their own family unit and species but possibly ignore or destroy the benefits of other groups or species.
- Deceitfulness (deception), another selfish survival strategy, provides the needs of individuals who require them (Hinton, 1973). There are many examples of deceit in all animal populations, and they are used for many reasons. Deception appears to be especially valuable in the trials of life and conflict.
- Toleration is a behavior that tends to detoxify overly expressed dominance in social groups of animals (Tan, 2000). As pointed out earlier with respect to social wasps, toleration, like expressions of dominance, varies considerably within a population of social animals, including humans. Unfortunately, toleration is a behavioral expression that is often lacking in human society unless it benefits the person or group that expresses it.

ANATOMICAL TRAITS SHARED BETWEEN HUMANS AND OTHER PRIMATES

As expressed in earlier chapters and beginning of this chapter, humans have evolved through a long chain of events over millions of years from nonprimate mammals to prosimians, through a series of simians to a variety of hominines (Fig. 3.1). Since anatomical and behavioral changes have evolved together to make humans what they are today, humans have carried over and thus share both anatomical and behavioral characteristics that are common with other members of the primate group (Patterson et al., 2006). Some anatomical traits that humans share with their primate relatives and which are distinct from their earlier mammalian predecessors are (Fig. 8.6):

- While the clavicle (collarbone) is prominent in all primates, it is absent or reduced in many nonprimate mammalian groups. In humanoids, it is elongated. The clavicle is a bone in the shoulder girdle on each side of the body that serves to link the scapula (shoulder blade) and sternum (breastbone).
- The primate shoulder joint, involving the scapula, clavicle, and proximal end of the humerus, allows a high degree of limb movement in all directions.
- The primate elbow, involving the distal end of the humerus, ulna, and radius, allows the joint to undergo considerable rotation.
- The five-digit forelimb (chiridian limb) is constructed with an opposable thumb. While the large toe of humans is larger than the other toes, it is not opposable as it is in other primates.
- Claws generally found in mammals lower than primates are found developed into flattened nails in most primates. There are some prosimians that have claws on certain digits. Claws and nails share a common chemical nature, both being composed of the protein keratin.
- There are sensitive tactile pads on the terminal portions of the digits in primates.
- All primates share a complex binocular vision, capable of visual acuity and color perception, although the degree of binocular vision varies somewhat from the prosimians to the higher simians.
- Primates share an enlarged brain, relative to body size, especially well developed in the cerebral cortex. While all primates have large brains as compared to lower mammals, the human brain:body ratio is greater than in other primates.
- Mammary (milk-producing) glands characterize mammals. Prosimians show affinities with other mammals in having a higher number of teats, while the number has been reduced to two in simians.
- Female humans and females of other simians have a uterus that is not forked. Prosimians possess branched (bicornate) uteri like nonprimate mammals.
- Humans and other simians generally have one offspring per pregnancy, and the offspring is required to have a prolonged postnatal relationship with its parents.

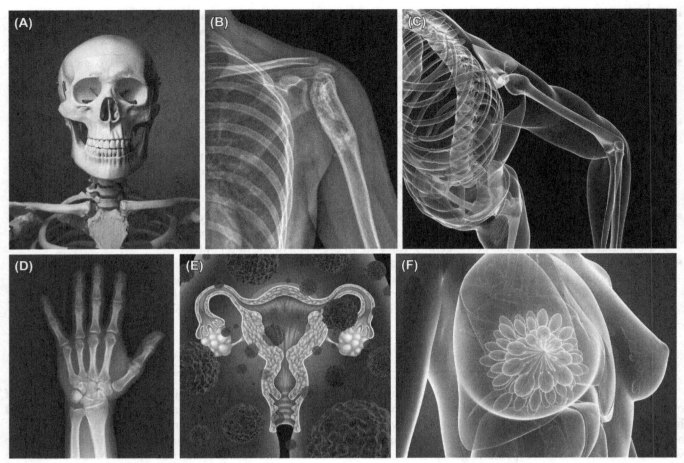

FIGURE 8.6 Anatomical features humans share with other simians. While the clavicle (collarbone) (A) is prominent in all primates, it is absent or reduced in many nonprimate mammalian groups. In humanoids, it is elongated. The primate shoulder joint (B), involving the scapula, clavicle, and proximal end of the humerus, allows a high degree of limb movement in all directions. The primate elbow (C), involving the distal end of the humerus, ulna, and radius, allows the joint to undergo considerable rotation. The five-digit forelimb (chiridian limb) (D) is constructed with an opposable thumb. While the large toe of humans is larger than the other toes, it is not opposable as it is in other primates. Claws generally found in mammals lower than primates are found developed into flattened nails in most primates. There are some prosimians that have claws on certain digits. Claws and nails share a common chemical nature, both being composed of the protein keratin. There are sensitive tactile pads on the terminal portions of the digits in primates. All primates share a complex binocular vision (A), capable of visual acuity and color perception, although the degree of binocular vision varies somewhat from the prosimians to the higher simians. Primates share an enlarged brain, relative to body size, especially well developed in the cerebral cortex. While all primates have large brains as compared to lower mammals, the human brain:body ratio is greater than in other primates. Female humans and females of other simians have a uterus that is not forked (E). Prosimians possess branched (bicornate) uteri like nonprimate mammals. Humans and other simians generally have one offspring per pregnancy, and the offspring is required to have a prolonged postnatal relationship with its parents. Mammary (milk-producing) glands characterize mammals. Prosimians show affinities with other mammals in having a higher number of teats, while the number has been reduced to two in simians (F).

DISTINCTLY HUMAN ANATOMICAL TRAITS

Since humans have separated from other simian stock, they have developed anatomical features of their own (Aiello and Dean, 2006). The following characteristics distinguish them from other primates (Fig. 8.7):

- Humans typically have an upright posture, partially due to the vertebral column that forms an S-curved shape. These curves are due to postnatal changes that occur when a baby raises his or her head and crawls (cervical curve) and when upright walking is accomplished (lumbar curve). This upright condition and accompanying bipedal locomotion are two of the most crucial anatomical characteristics of humans. While other animals occasionally demonstrate a vertical posture and bipedal locomotion, humans typically do.
- The joint for the neck vertebrae (the atlanto-occipital joint) is in the center of the skull base, another feature that accompanied the human's upright posture. The first cervical vertebra (the atlas) articulates with two occipital

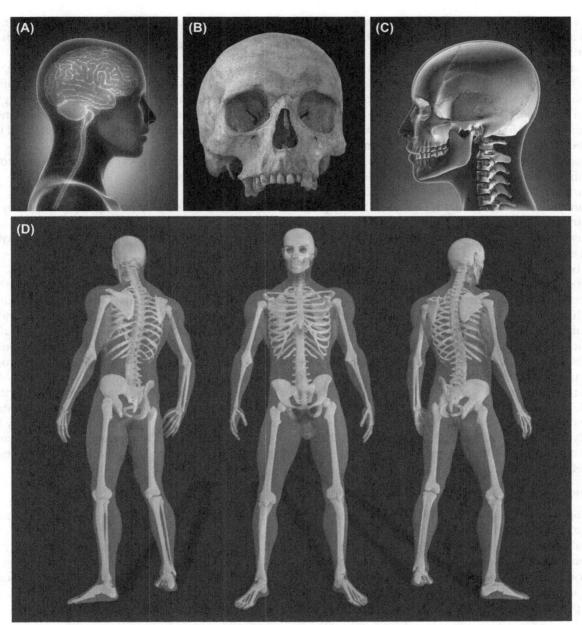

FIGURE 8.7 Human anatomy. The brain (A) is uniquely large as compared to body size, and the cerebrum is especially large and complex. The face (B) is short and located beneath the frontal section of the brain. The jaw is short, and the dental arch is round. The canines (B) are generally no longer than the premolars, and there are usually no gaps between them and other teeth. The first human premolar is similar to the second one, and tooth structure is generally distinctive, indicating that it is associated with diverse food types. The atlanto-occipital joint (C) is in the center of the skull, and the spine has taken on a curvature that protects it (to some degree) from becoming injured. The body is upright and bipedal (D). The legs are typically longer than the arms, putting emphasis on walking upright and running rather than knuckle-walking and brachiating. The big toe is not opposable like it is in other apes. The thumb is opposable and highly manipulative. Hair on the body is terminal and vellus, but reduced from the condition found in other apes. Humans live considerably longer than their simian relatives, including the nearest hominid relatives.

condyles, which are to each side of the foramen magnum. And it swivels around the dens of the second cervical vertebra (the axis) in the atlanto-axial joint. The entire spinal cord assumes curves directly inferior to these first two bones to absorb tension created as a result of an upright posture.

- A human's legs are typically longer than the arms.
- Human toes are short; the first (big) toe is usually the longest, but it actually has fewer phalanges than the other toes and is not divergent (opposable) as in other primates.

- The human's thumb is opposable and thus hands are prehensile, as they are in other primates. However, human hands are highly manipulative. Combined with bipedalism and a highly developed brain, this condition has led humans to new technological heights.
- The human body is primarily bare or with sparse, inconspicuous hair (vellus or reduced terminal hairs), except on the head, by the eyes (eye lashes and eyebrows), in the underarms, and in the pubic region. These shifts in anatomical design accompanied movements from a forest-like environment to open savanna. Although human males sometimes show significant body hair, it is never as extensive as in other primates under normal circumstances. Along with the change in hairiness, the number of sweat glands has increased, certain of them (eccrine sweat glands) functioning in thermoregulation.
- The human brain is uniquely large as compared to body size, and the cerebrum is especially large and complex.
- The human face is short and located beneath the frontal section of the brain.
- The human jaw is short, and the dental arch is round.
- Human canines are generally no longer than the premolars, and there are usually no gaps between them and the other teeth.
- The first human premolar is similar to the second one, and tooth structure is generally distinctive, indicating that it is associated with diverse food types.
- It has been found that humans live considerably longer than their simian relatives, including the nearest hominid relatives, the chimpanzees. In a 2013 Scientific American article, *Long Live the Humans*, Heather Pringle pointed out that chimpanzees (with a genome about 99% identical to that of humans) have a life expectancy at birth of about 13 years while human babies born in the United States in 2009 possessed a life expectancy at birth of 78.5 years. In a 2010 article in the Proceedings of the National Academy of Sciences, *Evolution of the Human Lifespan and Diseases of Aging: Roles of Infection, Inflammation and Nutrition*, Caleb Finch pointed out that the reason for the distinction between human and other simian life spans may relate to the human immune system, especially immune responses related to APOE e4 DNA sequences. As investigators research DNA's involvement in human longevity, we may recognize other features of our genetics that influence longevity. Pringle contemplates the question about whether we should expect the trend toward longer lives to continue, that "some scientists have predicted that many babies born after 2000 in countries where life expectancy had already been high will live to 100 years of age," although complications related to increasing obesity in modern populations, unpredicted climate changes, and problems that arise due to overpopulation may affect longevity in deleterious ways.

Behavioral Traits Shared Between Humans and Other Primates:

- Their opposable thumbs (Fig. 8.6) allow all primates to manipulate articles better than other mammals and to engage in higher forms of behavior involving cognition.
- Binocular vision in all primates (Fig. 8.6) provides greater depth perception, resulting in more intimate relationships with their immediate environment.
- The brains of all primates are similar, although the human brain is larger and more developed than the ones found in lower primates, resulting in more complex forms of behavior.
- Some primates other than humans and some nonprimates are tool users.

DISTINCTLY HUMAN BEHAVIORAL TRAITS

As with anatomy, humans have developed behavioral traits that characterize them as different from other simians (McGue and Bouchard, 1998). The following list points out some of these traits:

- Humans have extraordinary curiosity, attention, memory, and imagination. While some animals may have some of these qualities, they are far more developed in humans.
- Humans have the capability of improving their adaptive nature through rational thought, although they still are threatened by their animalistic past.
- Humans use and make tools of tremendous variety and complexity. Some such tools are for improvements in the way of living while others have been invented for defensive and destructive purposes.
- Humans are self-conscious or retrospective, reflecting on the past, present, and future, including such thoughts as life and death and behavioral consequences.
- Humans are capable of making mental abstractions and symbolisms.

- Humans have developed complex languages.
- Humans have not only a sense of beauty but are capable of altering the environment to complement this sense.
- Humans have developed a belief in supernatural spirits and superstitions, leading to the rise of religions and metaphysical emotions.
- Humans have gone to the nth degree in developing cultures based on a social existence, and they have carried forth biological concepts that have helped these traits develop.
- Humans are emotional animals. Emotion is poorly understood in nonhuman animals. While it certainly exists in our closest simian relatives and in certain other animals, it has become extremely complex in humans.
- Humans have an extraordinary array of social expressions we refer to as personalities. Personalities define the nature of the human animal.
- Since humans developed their earliest civilizations, there has been a movement among certain individuals in positions of power to attempt to improve the species. As carried out by eugenicists, such attempts have been expressed in both positive and negative ways, suggesting breeding the upper echelons of society and eliminating the lower echelons (Black, 2012; Engdahl, 2007).

As we may realize, most of the behavioral differences between humans and other primates, as well as the accomplishments of the species, are related to the development of a bigger and better brain. Both anatomical and behavioral changes occurred along with the increased size and complexity of this organ.

EMOTIONS

Emotion is a complex reaction that involves both our brains and other parts of our body when we are exposed to an individual, concept or event that results in awareness, physical arousal, and a reaction (Izard, 1978). Emotional triggers are registered in the brain's thalamus which reacts by signaling the adrenal glands by way of the amygdala (Fig. 5.3). Adrenaline is released into the blood, resulting in an increased heart rate and oxygen consumption for possible quick reaction. Signals from the thalamus also stimulate the brain's prefrontal cortex, which determines the significance of the original signal and the body's response.

Some human responses to individuals and environmental events relate to what psychologists refer to as the big six universal emotions: joy, surprise, fear, anger, distress, and disgust. We also feel and show satisfaction, ecstasy, bliss, gratitude, pride, love, irritation, shame, exasperation, guilt, and a wide assortment of other emotions. Some of these responses may overlap and result in a combination of more complex emotions.

Dominance establishment, for instance, involves a number of emotions, especially fear, anger, distress, and others that are associated with aggression between individuals. When properly established within a society, a dominance hierarchy functions in establishing a division of labor and order within groups. Dominance establishment by despotic humans, who are generally pathological individuals with extreme selfishness and a lack of tolerance, often results in a loss of societal homeostasis and generally involves an assortment of immoral personality traits.

PERSONALITY

How humans react to other individuals, their environment, and events have a lot to do with whether they develop a personality that fits within the "norm" (Cialdini, 2003; Sherif, 1936). Personality, which is partly innate (temperament) and partly learned, incorporates the totality of an individual's behavioral and emotional tendencies.

The "norm" for personality is generally characterized by behaviors that are accepted by a particular society and usually involves what that society designates as moral behavior. While most humans develop a personality that fits into an acceptable range, certain members of human society express behaviors that are outside the limits of acceptability. We comment on this later in this chapter, and Chapter 9 identifies and elaborates upon this segment of our population.

In *The Normal Personality, A New Way of Thinking about People* (2009), Steven Reiss compares accepted views about human personality development, stating that "The shadow of Freudian analysis looms over modern psychopathy, driving many psychologists to try to understand their clients' personal troubles and personalities using

Freudian-related constructs to study mental illness." Using this approach, Reiss believes many psychologists find it problematic in distinguishing between individuality and abnormality, causing them to sometimes have difficulty in separating problematic behavior from genuine disorder.

While many psychologists believe that all or most dark, unconscious mental processes found in people originate in problems that arise during the formative years of an individual's life, Reiss shows how "normal" motives drawn from later life experiences may underlie many personality and relationship dysfunctions, for example, divorce, infidelity, combativeness, workaholism, loneliness, authoritarianism, weak leadership styles, perfectionism, under-achievement, arrogance, extravagance, stuffed shirt-ism, disloyalty, disorganization, and overanxiety. Yet, these "dysfunctions" may be influenced by both innate and earlier formative experiences. According to Reiss, all brain regions that function in personality construction, along with genetic influence and learning experiences, contribute to the development of a fundamental part of a normal personality called a self-image (self, ego). Self-image becomes increasingly complex as a person's life unfolds.

A person's self-image may be analogized to a mental picture that is quite resistant to change which depicts not only details that are potentially available to objective investigation by others (height, weight, hair color, gender, IQ score, and so on), but also details that have been learned by that person about themselves, either from personal experiences or by internalizing the judgments of others. A simple definition of a person's self-image is their answer to the question: What do you believe people think about you?

Social and cognitive psychologists use the term self-schema to describe self-image (Valentino et al., 2008). Like any schema (a pattern of thought or behavior that organizes categories of information and the relationships among them), self-schemas store information and influence the way we think and remember. Self-schemas are also considered traits with which people define themselves, drawing information about the self into what they feel is a coherent scheme.

Self-image has been defined by Aaron T. Beck (1975) as consisting of (1) core beliefs, (2) rules and predictions, and (3) automatic thoughts about one's self. These categories change and become more complex as learning increases. Beyond self-image, a description of personality generally consists of what others recognize as traits in a person.

One approach to understanding and describing human personality has been a five-dimension approach (the Big Five Model) that consists of the following categories (Judge et al., 2006): openness or open-mindedness, conscientiousness, extraversion, agreeableness, and neuroticism. Let us pursue the categories individually.

- *Open-mindedness*: Open-mindedness has been described as the readiness to consider different points of view. Open-mindedness and closed-mindedness refer to attitudes held on topics usually considered to arouse emotion, such as positions on political, social, and economic views. Persons with "open minds" will tend to weigh evidence critically; those with "closed minds" will tend to refuse to consider dissenting evidence. Topics such as religion, political ideology, sex/gender differences, racial differences, or status quo may be more likely to result in closed-mindedness in certain individuals.
- *Conscientiousness*: This is a distinguishing feature of mental life, variously characterized as (1) a state of awareness as well as the content of the mind, that is, the ever-changing stream of immediate experience, comprising perceptions, feelings, sensations, images, and ideas; (2) a content effect of neural reception; (3) a relation of self to environment; and (4) the totality of an individual's experiences at any given moment.
- *Extroversion*: Also spelled extraversion, this trait is "a tendency to direct interests and energies toward the outer world of people and things rather than the inner world of subjective experience." By comparison, introversion is "a preoccupation with the self and personal thought, feelings, and fantasies rather than with the outer world of people and things." Certain researchers have considered this orientation to be the basis of a distinct personality type which they characterize as contemplative, reserved, sensitive, and somewhat aloof. Others have found that anxiety states, obsessive reactions, and depressive disorders tend to be associated with this pattern.
- *Agreeableness*: This personality trait reflects individual differences in cooperation and social harmony. People who score high on tests designed to recognize the agreeable nature of humans tend to see most people as honest, decent, and trustworthy. However, people scoring low on such tests are generally considered to be less concerned with others' well-being and have less empathy. Such individuals appear to be less likely to go out of their way to help others. People who show a low degree of agreeableness are often associated with skepticism about other people's motives, resulting in suspicion and unfriendliness. They also appear to have a tendency to be manipulative in their social relationships and are also more likely to compete with others than to cooperate.
- *Neuroticism*: Neurotic personality is described as a pattern of traits and tendencies that increase susceptibility to neurosis of one type or another. Neurosis is defined as a class of functional mental disorders that involve distress but not delusions nor hallucinations, whereby behavior is not outside of socially acceptable norms.

A persistently tense, apprehensive, insecure individual may be prone to anxiety neurosis; the overly orderly, cautious, meticulous person is more likely to develop an obsessive-compulsive disorder; a tendency toward fixed irrational fears may develop into a phobic disorder; and a susceptibility to bodily complaints may lead to a hysterical (conversion) disorder.

In searching for a common thread in all personal growth models around the world, Steve Pavlina developed three categories that overlap the Big Five model devised by Aaron Beck (Rose, 2013): truth, love, and power. These categories, in turn, touch on other principles, for example, oneness, authority, courage, and intelligence.

"Truth," Reiss says, "has a lot to do with a highly developed self-image." Power in the context of Pavlina's model is not comparable to the power repeatedly mentioned in this book in association with a dominant personality but rather the power one has with being in control or self-control in the Big Five model mentioned earlier:

> Oneness combines love and truth, enabling one to feel not only personal feelings but the feelings and emotions of others as well. Authority connotes true knowledge of one's self, the world, and others. Accordingly, with power and truth, a person will become an authority in whatever field they are specialized in. This authority allows a person to become more successful in reaching goals.

Courage develops as an offshoot of love and power, for example, the courage to look beyond short-term goals. It also relates to the absence of fears and the presence of sufficient power. "Love," Reiss states, "is required to commit one's self to others and their well-being." See the discussion about "love" in Chapter 9.

Reiss points out that studies repeatedly show "that a deviant and unhealthy upbringing, full of emotional and physical abuse, changes the brain in an abnormal way." Unfortunately, people who undergo such abuse "simply cannot undo their basic and rigid styles of reacting to specific triggers." In *Braintrust: What Neuroscience Tells Us about Morality* (2011), Patricia Churchland stated, as Reiss has also commented upon, that evidence points out that the development of "a normal personality is strongly related to ethical behavior."

Personality types depicting behaviors such as aggression, narcissism (the pursuit of gratification from vanity or egotistic admiration of one's own attributes), and perfectionism (a personality trait characterized by a person's striving for flawlessness and setting excessively high performance standards, accompanied by overly critical self-evaluations and concerns regarding others' evaluations) point out the diverse nature of a group of primates who live in an extremely complex society (Fig. 8.8).

Over the years, there have been many other attempts to characterize human personalities (personality assessment), based on personality traits, attitudes, behavior patterns, physique, or other outstanding characteristics (Eysenck, 2013; Guntrip, 1995; McCrae and Costa, 1997). A personological approach is one in which "personality is studied from a holistic point of view, based on the [concept] that an individual's actions and reactions, thoughts and feelings, as well as their personal and social functioning, can be understood only in terms of the whole person."

It is hoped that individuals who undertake the chore of characterizing personality from descriptions of human behavior remain conscious of using a scientific approach because some such descriptions resemble those of contemporary western astrology, which is often associated with systems of horoscopes that purport to explain aspects of a person's personality and predict significant events in their lives based on the positions of celestial objects. We can use this approach, as long as we keep in mind that thoughts about personality have actually been researched and are modified if the approach is incorrect.

Most people are familiar with A and B personality types, which have contrasting attributes (Holbrook and Hirschman, 1982). Type A individuals show dominant features and are considered ambitious, rigidly organized, highly status-conscious, sensitive, and impatient. They have a tendency to take on more than they can handle, they want other people to get to the point, and they are anxious, proactive, and concerned with time management. People with Type A personalities are often high-achieving "workaholics," they push themselves with deadlines and hate both delays and ambivalence.

In his 1996 book dealing with extreme Type A behavior (*Type A Behavior: Its Diagnosis and Treatment*), M. Friedman (1996) suggests that Type A behavior can be dangerous when expressed through its three major symptoms: (1) free-floating hostility, which can sometimes be triggered by even minor incidents; (2) time urgency and impatience, which causes irritation and exasperation usually described as being "short-fused"; and (3) a competitive drive, which causes stress and an achievement-driven mentality. The first of these three symptoms is believed to be covert and therefore less observable, while the other two are more overt:

> "Type B personality individuals live at lower stress levels." They typically work steadily, and may enjoy achievement, although they have a greater tendency to disregard physical or mental stress when they do not achieve. When faced with competition, they may focus less on winning or losing than individuals with a Type A personality and may focus more on enjoying the game regardless of winning or

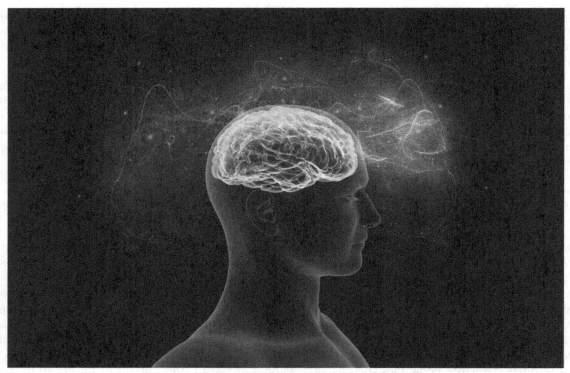

FIGURE 8.8   The cognizant human, the most dominant animal to ever exist on planet Earth, an elite status that has come about because of its brain. Humans have extraordinary curiosity, attention, memory, and imagination. They have the capability of improving their adaptive nature through rational thought. They use and make tools of tremendous variety and complexity. They are self-conscious or retrospective, reflecting on the past, present, and future, including such thoughts as life and death and behavioral consequences. They are capable of making mental abstractions and symbolisms. They have developed complex languages. They not only have a sense of beauty but are also capable of altering the environment to complement this sense. They have developed a belief in supernatural spirits and superstitions, leading to the rise of religions and metaphysical emotions. They have gone to the nth degree in developing cultures based on a social existence, and they have carried forth biological concepts that have helped these traits develop. They are emotional animals. They have an extraordinary array of social expressions we refer to as personalities.

losing. Unlike the Type A personality's rhythm of multitasked careers, Type B individuals are sometimes attracted to careers of creativity, for example, writer, counselor, therapist, actor, or actress. However, network and computer systems managers, professors, and judges are more likely to be Type B individuals as well. Their personal character may enjoy exploring ideas and concepts. They are often reflective, and think of the "outer and the inner world."

According to Friedman et al. (1986), limitations of the original study comprise the inclusion of only middle-aged men and the lack of information regarding the diets of those subjects. While the latter could serve as a confounding variable, the former calls into question whether the findings can be generalized to the remaining male population or to the female population as a whole.

These researchers conducted a randomized controlled trial on 862 male and female post-myocardial infarction patients, ruling out (by probabilistic equivalence) diet and other confounds. Subjects in the control group received group cardiac counseling, and subjects in the treatment group received cardiac counseling and Type A counseling. The recurrence rate was 28% in the control group and 13% in the treatment group, which was regarded as a strong and statistically significant finding.

According to the *APA Dictionary of Psychology*, personality assessment includes (1) observational methods that use behavior sampling, interviews, and rating scales; (2) personality inventories, such as the Minnesota Multiphasic Personality Inventory; and (3) projective techniques, such as the Rorschach Ink-blot Test and Thematic Apperception Test (Morgan, 2002; Schacter et al., 2009). "The uses of personality assessment are manifold, for example, in clinical evaluation of children and adults; in educational and vocational counseling; in industry and other organizational settings; and in rehabilitation."

In 1928, William Marston (*The Emotions of Normal People*) developed a DISC personality model, which appears to be utilized by psychologists today. DISC is an acronym for four personality types (Fig. 8.7): (1) dominance (D) relates to control, power, and assertiveness; (2) influence (I) relates to social situations and communication; (3) steadiness (S) relates to patience, persistence, and thoughtfulness; and (4) compliance (C) relates to structure and organization.

According to Marston, D-personality people (also known as high D) enjoy dealing with problems and challenges. They are demanding, forceful, egocentric, strong-willed, determined, aggressive, ambitious, and pioneering. I-type people influence others through talking and activity. They are more interested in people than carrying out tasks.

S-personality people like steady paces and security. They are calm, relaxed, patient, predictable, deliberate, stable, consistent, rather unemotional, and poker-faced. C-personality people are cautious and conscientious. They adhere to rules, regulations, and structure.

As Marston suggested, additional personalities may be recognized by combining the four personalities (as in the outgoing, reserved, task-oriented, and people-oriented categories in Fig. 8.9). Whether the DISC system is useful in truly characterizing a person's personality, we must remember that personality, like dominance and aggression, is fluid among members of a population. A positive attribute of the DISC system is bringing out the terms that describe how people behave. Combining descriptive terms that suit the personalities of people allows us to form a personal reference catalog, which describes a particular person.

Determining a person's personality is important to understand precisely what a person is like, that is, the person's nature. As stated in the APA Dictionary of Psychology, "Personality is generally viewed as a complex, dynamic integration or totality, shaped by many forces, including: hereditary and constitutional tendencies, physical maturation, early training, identification with significant individuals and groups, culturally conditioned values and roles, and critical experiences and relationships."

In business, Robert Rohm (Personality Insights, 2016) states, the starting point of understanding people is to realize that everyone is different. Psychologists, he says, apply the DISC personality model with four approaches in mind to (1) highlight and encourage a person in his or her strengths; (2) address a person's blind-spots without assuming that a weakness exists; (3) recognize that each person has a unique blend of all the major personality traits to a greater or lesser extent; and (4) recognize that behavioral patterns are fluid and dynamic as a person adapts to his or her environment.

Since we are delving into the concept of dominance, it is worth pursuing this DISC approach further but with emphasis on the dominance quadrant. According to Marston, a person who fits into the dominance quadrant ("normally") "tends to be direct and decisive." They would prefer to lead than follow, and their personality leans toward leadership and management positions. They tend to have high self-confidence and are risk-takers and problem-solvers, which enables others to look to them for decisions and direction (Fig. 8.9):

They think about big picture goals and tangible results. In team play, they are bottom-line organizers that can lead an entire group in one direction. They place great value on time frames and seeing results. They may change the status quo and think in a very innovative way. On the other hand, they tend to overstep authority because they prefer to be in charge themselves. At times, they can be augmentative and not listen to the reasoning of others. They tend to dislike repetition and routine and may ignore the details and minutiae of a situation, even if it is important. They may attempt too much at one time, hoping to see quick results. Since they like to be in control, they may fear the idea of being taken advantage of by others.

Dominants will "likely be very autocratic managers in a team environment and rise to the top during crisis moments. They will provide direction and leadership, push groups toward decision making, will maintain focus on the goals, and will push for directness and lack of social interest around others. They are generally optimistic thinkers, but may have personality conflicts with others they perceive as negative. They function well with heavy workloads and when under stress, and welcome new challenges and risks without fear":

They may be perceived as always speaking and not listening to others." In order to create a homeostatic environment, "the dominant may need to strive to listen more actively, be attentive to other team members' ideas, and to strive for consensus instead of making decisions alone. Dominants can be controlling and domineering at times. Unless they work at it, it may be difficult for them to be friendly and approachable.

Dominant people can be recognized by certain behaviors. "They will make gestures and postulate themselves to increase their size. They may frown, purse their lips, maintain direct eye contact, and hold their head still. They may lead from the front, often going through doors before others and walking ahead of the crowd. They may freely touch other people in a nonemotional way but not accept this behavior in return. Alpha behavior may be displayed when meeting someone for the first time. They often firmly place their hand on top of someone else's when greeting someone. Their grip will be firm and sustained until they choose to end the greeting. Dominants are verbal, reinforcing their body language, controlling the conversations."

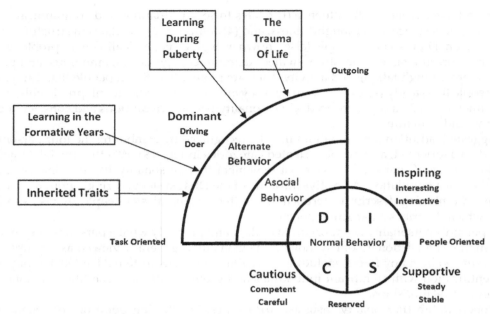

FIGURE 8.9 Modification of the DISC personality system, as developed by Marston (1929), a system which was designed to depict "normal" human behavior. Modification includes the introduction of two additional outer circles on the dominant side, representing asocial and alternate behavior, as well as the influence on personality development by inherited genes, experiences in the formative years, during puberty, and the trauma one experiences during life. What is not depicted are the numerous variations within each type of behavior when the categories are combined, potentially, forming a multitude of personality types and tremendous variation within each category. While attempts are made to characterize people and their personalities, such differences between personalities make it difficult to place them in precise categories. Nevertheless, the DISC method conveys important information about human personality that can be used to understand human nature.

Carney et al. (2010) state that two types of dominance are useful to distinguish (1) intrinsic dominance, which is based on an individual's own ability to use force and (2) derived dominance, based on a learned ability to fight and the formation of coalitions and alliances. The combined effect of these forms of dominance "may be expressed in visual or vocal threats, physical attack, other forms of aggressive display, verbal threat or verbal aggression, including forms of commanding, ordering, ridiculing, teasing, threatening, accusing, blaming, and criticizing. Expression of dominance may also include individual power, collective physical power, or technological power from weapons or other devices."

In social situations, Carney et al. state that, power can be exercised for the following reasons:

- Competition for resources, including dyadic behavior, resource-specific coalitions, or a criminal approach, for example, theft, burglary, and so on.
- Competition for mates, including male–male competition, male aggression toward females, female–female competition, and female aggression toward males.
- Competition for associates, including dyadic and polyadic displays for a position in the dominance hierarchy.
- Protection of other humans, including kin, a sexual or social partner, and a friend, or altruistic and reciprocal behavior toward an unknown.
- Individual and cooperative defense.
- Play-derived aggression.
- Redirection of aggression.
- Psychopathic crimes.
- Infanticide.
- Dissocial (unprovoked) aggression for gaining praise or approval of others, for example, in street gangs.
- Intergroup aggression, including xenophobia (joining together to threaten or attack members of another group) and territorial defense (threatening potential trespassers) and intergroup aggression.
- Group invasion.
- Warfare.
- Intergroup alliances.

- Violent demonstrations.
- Terrorism.

The distribution of power among individuals is determined by (1) dominance orders; (2) revolutionary alliances (forming a group to overthrow the leader); (3) conservative alliances (forming a group to maintain the stability of the leader); and (4) protective alliances (forming a group to protect the leader) and intergroup structures.

Psychological processes for learning and propagating dominance depend on (1) individual learning, including material reinforcement to gain resources and social reinforcement to increase stature; (2) modeling influences, including social facilitation and observational learning; (3) instructional control; and (4) self-reinforcement, including moral justification, displacement of responsibility, dehumanization of victims, and attributing blame to victims.

Rohm points out that intelligence quotient (IQ) measures a person's intelligence and personality quotient (PQ) refers to a person's ability to understand themselves. "Studies have shown that technical skill, beginning with intelligence and developed through education and experience, accounts for 15% of success in the workplace. The other 85% of workplace success comes from people skills, which are developed through learning better ways to behave and interact." Rohm also relates to methods used to recognize different managerial behavior patterns at work, which involves personality traits:

- *Managerial autocratic behavior*, which includes a strong assertive and confidential tone. It communicates the message: "I am a strong, competent, and knowledgeable person on whom you can rely for effective guidance and leadership." This approach generally requires other people to be submissive, obedient, and respectful. Extreme forms of behavior in this group "may include a dictatorial and dogmatic approach."
- *Responsible-hyper-normal managerial behavior*, which includes affectionate, friendly, dominant, and supportive expressions. It communicates the message: "I am strong, competent, and emphatic person on whom you can depend for emotional support."
- *Cooperative-excessively comfortable behavior* indicates that there is a striving for compromise and harmony. It communicates the message: "I am an exceedingly friendly, agreeable, unchallenged person who would like you to like me."
- *Obedient-dependent managerial behavior* is generally submissive and friendly. It conveys the message: "I am a weak and helpless person in need of your help and support."
- *Self-deprecatory-masochistic behavior* is characterized by dependent and submissive expressions, which are influenced by aggression and hostility. It tends to communicate the message: "I am a weak, deficient, unworthy person, just deserving of your domination, rejection, and contempt." This type of behavior stimulates others to be arrogant and cruel, and to exploit the person.
- *Rebellious-suspicious behavior* is characterized by a combination of hostility and antisocial behavior. It communicates the message: "I reject and mistrust you, for you are certain to become unworthy of my affection and esteem." It evokes in others a punishing and chastising attitude and rejection.
- *Aggressive-sadistic behavior* is a direct expression of hostility and dominance at the verbal and nonverbal level. This type of behavior communicates the message: "I am a dangerous and threatening person and you are a suitable target for my wrath." This type of behavior evokes in others feelings of fear and guilt.

In determining personality, a person's culture may provide special challenges. Culture refers to the cumulative deposit of knowledge, experience, beliefs, values, attitudes, meanings, hierarchies, religion, notions of time, roles, spatial relationships, concepts of the universe, and material objects and possessions acquired by a group of people in the course of generations through individual and group determination. According to G. Hofstede (1997), "the position that the ideas, meanings, beliefs, and values people learn as members of society determines human nature (cultural determinism). People are what they learn," he says. He points out that there is no universal "right way" of being human. The right way in one society "almost never corresponds to the right way in other cultures. Proper attitude of an informed human being could only be that of tolerance."

Hofstede also points out that "different cultural groups think, feel, and act differently." Studying differences in culture among groups and societies presupposes a position of cultural relativism. Ethnocentrism is the belief that one's own culture is superior to that of other cultures. Cultural differences manifest themselves in different ways.

Other popular ways to characterize personality type are seen in the introversion–extraversion distinction, which has already been discussed earlier, and Carl Jung and Erich Fromm's character types (Burston, 1991), such as the exploitative orientation and market orientation. The latter approach, explained at a website called *About Health*, includes (1) a receptive character type, (2) an exploitative type, (3) a hoarding character type, (4) a marketing character type, and (5) a productive character type.

- *Receptive type*: This type is characterized by a need for constant support from others. They tend to be passive, needy, and totally dependent upon others. They require constant support from family, friends, and others, but they do not reciprocate this support. Receptive types also tend to lack confidence in their own abilities and have a difficult time making their own decisions. Individuals who grow up in households that are overbearing and controlling often tend to have this personality orientation.
- *Exploitative type*: This type is willing to lie, cheat, and manipulate others in order to get what they need. In order to fulfill their need to belong, they might seek out people who have low self-esteem or lie about loving someone they really do not care about. These types take what they need either through force or deception and exploit other people to meet their own selfish needs.
- *Hoarding type*: This type copes with insecurity by never parting with anything. They often collect massive amounts of possessions and often seem to care more about their material possessions than they do about people.
- *Marketing type*: This type looks at relationships in terms of what they can gain from the exchange (reciprocity). They might focus on marrying someone for money or social status and tend to have shallow and anxious personalities. These types tend to be opportunistic and change their beliefs and values depending on what they think will get them ahead.
- *Productive type*: This type is a person who takes their negative feelings and channels the energy into productive work. They focus on building loving, nurturing, and meaningful relationships with other people. This applies not only to romantic relationships, but also to other familial relationships, friendships, and social relationships. They are often described as a good spouse, parent, friend, coworker, and employee. Of the five character types described by Fromm (Burston, 1991), the productive type is the only healthy approach to dealing with the anxiety that results from the conflict between the need for freedom and the need to belong.

It is hoped that information provided here to describe human personality will be helpful in understanding the tremendous variation in human behavior and how difficult it is to categorize people under specific labels. In addition, we have primarily been discussing what is considered "normal" behavior and methods of characterizing it. It becomes more difficult to characterize alternate (e.g., antisocial) behavior. It is here that people differ from the "norm" and extend their behavior further into the dominant realm of immoral behavior.

THE BETTERMENT OF HUMANKIND

Judging from the totality of human behaviors mentioned earlier and throughout this book, human personality runs a gradient from despotic dominance to complete subordination. Dominant personalities rule the society, and subordinate personalities follow the rules that dominants make. At times, dominants possess antisocial personalities and take advantage of subordinate followers. Fig. 8.9, a modification of the DISC personality system devised by Marston (1929), points out the influence of personality by inherited and learned behavior, as well as how the dominant quadrant can be expanded to incorporate asocial and alternate behavior.

From the times of early civilization, certain dominants (eugenicists) have concerned themselves with the betterment of humankind, and many secret societies have formed throughout human civilizations for that express purpose. Eugenics is defined as the study of or belief in the possibility of improving the qualities of the human species or a human population, especially by such means as discouraging reproduction by persons having genetic defects or presumed to have undesirable inheritable traits (negative eugenics), or encouraging reproduction by persons presumed to have desirable inheritable traits (positive eugenics) (Black, 2012; Engdahl, 2007).

According to Edwin Black in *War Against the Weak, Eugenics and America's Campaign to Create a Master Race* (2012), undesirables include people such as criminals, the sick and weak, alcoholics, anatomical deviants, people with mental ailments, and entire cultures that appear to be different. In attempts to improve the human species and rid it of such objectionableness, groups of dominants have devised and used the concept of eugenics. The entire concept illustrates how far humans will go to dominate and exterminate their fellow humans. This concept is brought out further in later chapters.

HUMANS IN A COMPLEX WORLD

Unlike the societies of most lower primates, most human societies demonstrate behaviors that are not routine replications of the previous day of their lives (Henshilwood et al., 2002). Other than in aboriginal human groups, events in contemporary human society can change on a daily basis. And depending on the diverse genetic makeup of humans and the stresses of life, extremes of these personalities are often expressed by individuals who may have suffered traumatic experiences at some time in their life.

Unlike personality, temperament only concerns those aspects of an individual's personality that are regarded as innate rather than learned. Thus, personality has both innate and learned components. However, some treat personality and temperament as being similar. It is said that a person's temperament is established by age five. Personality, on the other hand, may be influenced by the numerous trials and tribulations one experiences through life, although the time of primary importance is during the formative years and adolescence. Once established, temperament and personality traits generally remain with humans for their entire life.

EVOLUTION OF PHILOSOPHICAL THINKING IN HUMANS

While humans are biological animals, they possess an extremely diverse assortment of personality traits, emotions, and temperaments that make them more complex behaviorally than other animals (Benjamin, 2008). This is true even for aboriginal cultures.

Understanding the human animal and its dominant nature is steeped in controversial issues. Since a human in a modern environment is able to contemplate his or her existence and, as a group, should be capable of understanding themselves as animals that are governed by biological forces, human behavior as a whole is complicated by incorporating natural (animalistic), innovative, and metaphysical roots. To further complicate our diverse behavioral repertoire, we are victims of an ever-increasing complexity of societal stresses, often due to a more complex and increasingly stressful life style and overpopulation (Suedfeld, 2003).

There are other, more cortical ways of characterizing the complex human animal, and early thinkers have sought a characterization by examining its philosophical nature. Over the years, philosophical thought has led to a more scientific approach, and thus logic and biological evidence has altered the way we, as contemporary humans, think.

Philosophers since Plato, Socrates, and Aristotle have been concerned with defining human nature through concepts such as a(n):

- *Philosophical naturalistic approach*, whereby humans behave by natural laws, just as any animal would. Accordingly, humans, like all animals, are products of evolution by way of random mutations, environmental stresses, and natural selection. Humans that ideally fit this category do not believe in supernatural beings or an afterlife.
- *Abrahamic approach*, whereby humans are spiritual beings that have been created by a single god in his image.
- *Polytheistic approach*, whereby humans live in a world populated by a number of spiritual or mythological beings.
- *Holistic approach*, whereby humans exist within god, or as part of a divine cosmos.

From a biological standpoint, a more contemporary characterization of humans would be a modification of the first approach: that humans, like all living creatures, are products of evolution who have reached a mental capacity to question their existence and have often sought an explanation through metaphysical channels.

In a religious context, humans have selected themselves to flaunt a god-given right to be dominant over all other life on Earth (Chapter 12), and they are directed to live by certain moral codes (Miller, 1994). In terms of morality, humans are capable of imposing a wide variety of moral codes on themselves, such as:

- *Moral realism*, in which moral codes are believed to exist outside of human opinion, that is, things may be right or wrong, regardless of human opinion.
- *Moral relativism*, in which moral codes are thought to be a function of human values and social phenomena.
- *Moral absolutism*, in which certain acts may be right or wrong, regardless of their context.
- *Moral universalism*, which is a compromise between moral absolutism and moral relativism, favoring the thought that there is a common universal core of morality.

Contemporary biological thinkers accept that most humans would typically live by a moral code whether they are influenced by religious or other doctrines, but no matter what their religious or nonreligious approach to life, there will always be humans who display aberrant behavior that does not follow a moral code (Simon, 2008). Human populations establish dominance hierarchies at all levels of their existence, and in spite of moral codes, many individuals that have a dominant nature abuse the moral code to obtain the things in life that they desire, including those in both a religious and nonreligious context. Not all dominants function in this way, but many do. Antisocial individuals, for instance, live outside of moral codes, and many religious individuals live outside the codes they profess to obey. Chapter 9 points out the nature of such individuals and their lack of interest in moral codes.

NATURE VERSUS NURTURE

Controversy about philosophical and biological approaches to understanding humans exists between such fields as sociology, psychology, and biology in that similar phenomena are often explained in terms of one field without regard for the others (Hofstede, 2001). In addition, even though we live in a contemporary world and understand the genetic and learned phenomena known to take a part in the formation of a human personality, behaviorists sometimes are not settled on what behavior is genetically based and what is learned. In short, the controversy of nature versus nurture sometimes persists in the literature.

Early philosophers leaned toward the nurture concept. John Locke, for instance, identified human nature as a tabula rasa in which the mind is a "blank slate" at birth (Anstey, 2003). Information, he said, is acquired during life, and rules for processing such information are formed by sensory experiences. Later investigators, such as Konrad Lorenz, brought out the innate nature of certain animal qualities (Krebs et al., 1992). Contemporary investigators have learned that both nature and nurture are important in determining what a human and any animal become behaviorally (as in the formation of personality).

HUMAN BEHAVIORAL FEATURES

There are many common behavioral features in human populations (Borgerhoff and Schacht, 2012; Cronk, 1991; Cronk et al., 1999; Hames, 2001; Winterhalder and Smith, 2000):

- Humans, in general, have similar smiles.
- Humans use their eyes in similar ways to convey cognition or flirtatiousness.
- Facial attractiveness is perceived in a similar way.

In *Human Universals* (1991), D. Brown identifies approximately 400 different behaviors that are considered invariant among all humans. Some may have connections to genetic roots while others may be learned.

Many behaviors, cues, and postures are recognized that are used to determine a person's temperament, personality, and position within the various dominance hierarchies that we face in life. This is a subject covered in Chapter 11, *Dominance and Aggression in the Workplace*.

Thus, behavior clearly has both innate and learned components. Using identical twins as examples, their behavior would most likely be similar if behavior was all innate. When raised in separate environments, however, even identical twins demonstrate differences in behavior. Likewise, individuals who are not related may have similar behaviors if raised under the same conditions.

On the other hand, some behaviors are clearly innate and thus are derived from their predecessors (Mandal, 2010; Maslow, 1954). For instance, biological concepts, such as preservation, reproduction, overproduction of offspring, territoriality, and dominance (agonistic) behavior, present in all animal species, are clearly innate.

There are numerous other traits that are difficult to place as innate or learned. How a group of animals accomplishes or expresses these actions and the degree to which they are expressed is influenced by learning. The process of learning, as already established in this and other chapters, generally relies on the trials and tribulations experienced in life, especially in the formative and adolescent years and in instances of traumatic events.

In addition, like all other organisms, humans demonstrate a tremendous amount of superficial anatomical, physiological, and behavioral variation within their population. It is variation that allows the process of natural selection to preserve the fittest among organisms. It also allows humans as a whole to have a wide variety of personalities and thus enables them to express agonistic behavior in various ways so that they may occupy different positions within a wide array of dominance hierarchies.

VARIATION, THE HALLMARK OF HUMAN CULTURE

It is clear that variation is the hallmark of human culture and that there are many ways to express this variation. In addition to features that we have already used to characterize humans, Mazur (*Biosociality of Dominance and Deference*) (2005) points to additional characteristics that make humans stand out as different from all other primates, such as their indulgence in athletic sports, body adornment, cleanliness training, cooking, a particular form of courtship, music, decorative art, division of labor, organized games, specific greetings, joking, personal names, religion, sexual restrictions, shelters, elaborate tool making, and use of fire. Many of these features are associated with dominance interactions that determine status and rank.

We define humans in terms of these traits, but only with reference to our populations as a whole. If we would select any one of these features associated with humanity, we would find individuals in our population who are exceptions to the rule.

When differences between people become widely separated from what we consider the "norm," we often apply the terms alternate or aberrant to confirm it as strangely unusual (Cialdini et al., 1991). Often, such aberrant behaviors do not represent a threat to other societal members. Individuals who express nonthreatening aberrant behaviors are said to be asocial (Fig. 8.7).

On occasion, such behaviors become extreme divergences from our moral codes, and when they approach a point at which they begin to exert negative influence on other members of our population, we say that the behaviors are antisocial (Coie et al., 1998). As we will later see, antisocial people often express dominant behavior, and their dominance is coupled with an amoral approach to life and other people around them.

Dominant people are considered to be those who direct other people, often telling them what to do. Looking at them in a positive light, we may say that they are generally considered to be respected, influential, and of leadership quality. Submissive or subordinate people are considered to be noninfluential or nonassertive, and they are generally directed by others (the dominants).

However, because of the presence of degrees and types of dominance expressed by humans and other animals, degrees of dominance and deference (subordinance) overlap. Those who share traits of dominance may be overly tolerant, moderately tolerant, slightly tolerant, slightly intolerant, moderately intolerant, or despotic, and an overly dominant person (as in some of the social wasps discussed) may not be respected at all. Lack of assertion or influence, likewise, depends on the degree of dominance and subordinance expressed.

One of the rewards for being dominant is status (Hollingshead, 2011). Individuals who occupy high-status positions generally show it. In humans, presenting particular facial expressions and postures, having special types of vehicles and other material possessions of value, wearing fashionable and expensive clothing and jewelry, having a pleasant-to-look-at and much younger spouse or one with wealth, prestige, power, or inherited virtues are signals of high status.

Mazur (2015) divides status into (1) constant signals that are not controlled by the individual expressing them and (2) controlled signals that an individual can change. The reputation of an individual or the individual's family or clan are the result of constant signals. Gestures (as in showing fear, anger, sadness, disgust, happiness, surprise, joy, and anxiety) and actions (such as flaunting expensive clothes, jewelry, cars, and houses) are examples of controlled signals.

Another controlled form of signals is found in certain facial expressions, although we often may not be accustomed to thinking about them (Fridlund, 1994). Dominant faces in men (especially Nordic men), for instance, often are thought to appear handsome and muscular, with prominent chins, heavy brow ridges, and deep-set eyes. Submissive faces are considered to be oval or rectangular, with close-set eyes.

In the following chapter, we look at a variety of human societies and recognize that "normal" and "aberrant" behaviors are often defined by the societies that express them. We also recognize that aboriginal human societies demonstrate a higher degree of animalistic behaviors that keep them grounded and closer to Mother Earth.

In *The Dictionary of Psychology* (1999), R.J. Corsini described the traits that we previously discussed: neuroticism, extroversion, openness, agreeableness, and conscientiousness. These traits may be variously expressed in personality types, such as "A" in which a person may be achievement-oriented or "B" in which a person may be the opposite. While "A" and "B" personalities are recognized in certain individuals, there is not actually such a sharp line of demarcation between people. Behaviors expressing different types of personalities vary tremendously and may be recognized as existing in overlapping behavioral hierarchies rather than strictly defined types.

Human personality has been examined in various other ways. An approach discussed by A.M. Benis (*A Model of Human Personality Traits Based on Mendelian Genetics*) (1986) is to examine certain traits from a genetic perspective, such as aggression, narcissism, and perfectionism. Humans, as a group, exhibit all such traits while their primate relatives may exhibit some but not others. On an individual basis, these personality types vary tremendously throughout a population and are largely influenced by experience.

According to Benis, human acts associated with aggression include body posturing, eye contact, and gestures of intimidation and deference (dominance and subordinance). In effect, dominance establishment, as it is used here, relates to the striving for power by an individual over one's environment and other individuals within it. Thus, in humans, it may be thought of as a component of ambition.

It is here that we must use a bit of caution because where simple ambition stops and more aberrant dominance behavior starts is sometimes difficult to discern. Both ambition and the striving of power are expressed in gradations from one behavior to the other rather than as a distinct demarcation between personality types. In psychiatric events, dominance may reveal itself in many forms, sometimes in subtle, harmless ways and at other times in sadistic (deriving pleasure or sexual gratification from extreme cruelty) or sadomasochistic (the deriving of pleasure, especially sexual gratification, from inflicting or submitting to physical or emotional abuse) ways (Berner et al., 2003).

Benis (1986) goes on to say that "the facial complexion [of humans expressing aggression] is nonsanguine, that is, tending toward sallowness or pallor in individuals with light skin. During the expression of rage, the facial complexion of pallor is accentuated."

Humans with a narcissistic trait (love of one's own body) exhibit the flaunting of "body posturing, expansive arm gestures, bowing, colorful self-adornment, and a natural attraction to the limelight of personal recognition." This trait signifies a striving for glory in one's environment, and thus it can be thought of as another component of ambition.

There is an innate facial expression referred to as a gingival smile (broadly exposing teeth and gums), accompanied by facial complexion, which usually is blood red or ruddy in individuals with a light skin color. During rage, the sanguine complexion becomes more florid.

Perfectionism (striving to be perfect, especially in terms of moral character) is not considered a basic drive and is not associated with rage. It has been considered a mediator between aggression and narcissism. Acts associated with this trait are obsessiveness, compulsiveness, repetition, and maintenance of neatness.

Characterizing personalities is sometimes made more difficult by the various, often incongruent, approaches used by different investigators to study human nature, such as from biological, psychological, or sociological viewpoints. In addition, the use of terms that are employed to describe certain behaviors, such as dominance, social, asocial, anti-social, sociopathy, and psychopathy, sometimes vary or overlap in their meaning, depending on who is speaking of them. Thus, it is only through a multitude of approaches and a comparative examination of other animal species that we will truly begin to understand the complexity of the human species.

Note

Superscript numerals appearing in this chapter refer to additional text/explanation given in the appendix.

CHAPTER

9

Alternate Human Behavior

We've all got both light and dark inside us. What matters is the part we choose to act on. That's who we really are. **Sirius Black**, J.K.
Rowling's Harry Potter, Order of the Phoenix

Dominance interactions are found in all walks of life and at all levels of human existence. They may be powerful expressions that influence society in a positive or negative fashion (e.g., a political leader who genuinely cares about his or her constituents, versus a political psychopath who does as he or she pleases, sometimes destroying the lives of those who get in the way) (Meloy and Shiva, 2007). On the other hand, they may be less-significant expressions; for example, simply winning at a competitive sport or game of cards. People who have an antisocial personality typically express their form of dominance in a more aggressive, sometimes violent fashion.

The development of a sedentary lifestyle in humans led to the formation of larger groups and increased interrelationships, including individuals who were extra-familial (those who have extended their society to include people outside their family group). Most urban and suburban human societies continue in this fashion, while aboriginal societies either live in multifamily groups or have retained the earlier family-oriented groupings. As R. Winston and D.E. Wilson (2004) stated in *Human*, our predecessors "learned that living in a large group was a more effective way of meeting material and spiritual needs, and it functioned as a better system of defense against threats." Unfortunately, opposing groups became larger as well, and their differences often led to altercations and mass killings.

As pointed out in the initial chapters of this book, existing in groups comes with certain responsibilities. Rules must be made (generally based on religious, moral, and political beliefs, Chapters 12 and 13), and members of such a group are expected to become familiar with and practice behaviors that depict a social "norm" (Fig. 9.1). While social norms vary, depending on the society to which an individual belongs, some members of a society are willing to abide by social rules while others are not. Those who do abide by the rules are considered part of the "norm," while those who do not are said to express alternate behavior (Cialdini et al., 1991).

Thus, human society is composed of a tremendous variety of individuals at both extremes and others who fit somewhere in between. They occupy different behavioral categories and different rungs within the linear and despotic dominance hierarchies that exist.

It has already been said that it is often dominant individuals who influence the making of societal rules, and it is often dominant individuals who break them. In rule making, we desire individuals who are morally selfless and possibly sometimes even altruistic. While this may happen on occasion, many individuals who make the rules have more materialistic or sadistic reasons for domination (see Chapter 13). Such individuals are accustomed to using deceit to influence their followers, and their deceitful mannerisms are often difficult to recognize (Trivers, 2011).[1]

In rule making, dominant individuals (especially those who have an antisocial personality) often make rules that are not self-demanding and are often beneficial for the rule maker (Fig. 9.1). Thus, it is a selfish attitude. In rule breaking, dominant individuals generally demonstrate an even more selfish approach as well. This chapter is about those people, people with personalities who desire to dominate others by aberrant means and do not care about what the consequences of their behavior are to others. In the minds of most of society, they are the social outcasts, bullies, burglars, corporate and identity thieves, abusers, murderers, and rapists that every society must contend with.

Many people who demonstrate behaviors other than the "norm" do not precisely fit descriptions of antisocial personalities, and there are no definite lines of demarcation separating one group from another. Rather, there is a gradation of personalities from one end of the scale to the other, a bell curve of personalities, if you will.

Dominance and Aggression in Humans and Other Animals
http://dx.doi.org/10.1016/B978-0-12-805372-0.00009-2

139

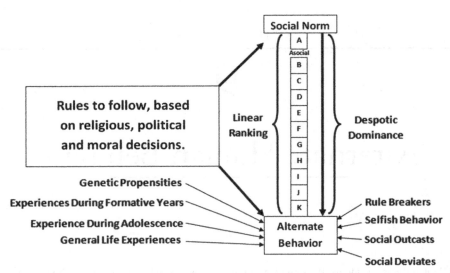

FIGURE 9.1 Rules are generated primarily by dominant personalities, generally based on religious, political, and moral decisions. As seen in this figure, people who follow the rules form a "social norm." Those who do not follow the rules demonstrate alternate (antisocial) behavior. Between these two groups are a wide array of personalities (shown by categories A through K). The array forms a more-or-less linear ranking of individuals. Ranking may lose its linear makeup, depending on the presence or absence of an alpha, as well as the degree of dominance the alpha individual expresses. Asocial behavior exists just outside the social norm, and the individual may appear more strongly asocial as it moves closer to the alternate side. Despotic dominants bypass the ranking system and demonstrate alternate behavior so strong that a linear hierarchy does not exist (shown by the vertical arrow on the right). Alternate behavior is expressed by individuals who have developed their personality from genetic propensities, experience during their formative years, experience during adolescence, and general experiences through life. People who demonstrate alternate behavior are dominant individuals, for example, rule breakers, selfish individuals, social outcasts, social deviates, and anyone else who exists outside of moral limits.

While many people are perfectly comfortable in following societal rules, some people choose to isolate themselves from society by not becoming involved, staying away from other individuals and societal functions. They practice what has been described as an asocial way of life, the prefix "a" designating a lack of social qualities.

Webster's New Collegiate Dictionary defines asocial as "rejecting or lacking the capacity of social interaction" (Fig. 9.1). However, whatever behavioral problems they may have, people in this category generally do not represent a particular threat to society. In fact, they may influence society in positive ways by contributing thoughts and ideas which positively influence others in one way or another. Many writers and others who enjoy working alone fit into this category.

Individuals within human society who exist as antisocial individuals live a more isolated and aberrant lifestyle, a dominant, egotistical existence that drives them to get what they want in spite of societal rules. Webster's New Collegiate Dictionary defines antisocial as "hostile or harmful to organized society, especially being marked by behavior deviating sharply from the social norm or adverse to the society of others."

Social norms mean very little or nothing to the people who reside outside of normality. In order to understand such individuals, let us reexamine certain characteristic aspects of human nature.

One of the most interesting features of human behavior discussed in Chapter 8 is personality because it tells us precisely what type of person an individual is. With both genetic and learned components, personality thus refers to the way a person reacts to both genetic drives and experiences through life.

Since personality develops as a consequence of both nature and nurture, the combination of features derived from them may determine whether a person becomes a productive member of society or an individual with antisocial personality disorders (ASPDs). For instance, much of our potential personality variation may come down to us genetically from our parents and their predecessors. How we are nurtured through our formative years (the first few years of life), events that we experience throughout our subsequent years, how we react to our peers, and the culture we are born into also make us who we are.

We must keep in mind that in spite of their genetic constitution and degree of retrospective thinking, how we treat a young person will influence the way that they respond to their environment and other humans. In most cases, for instance, happy and content babies generally tend to be easy to wean, put to bed, and bathe. Being unhappy because of abuse can project the infant into a realm of mixed emotions and potentially violent behavior later in life.

We have already seen how the human brain develops. In *A User's Guide to the Brain* (2002), J. Ratey describes early human development, pointing out the importance of moral treatment in the guidance of young people. On the day an infant is conceived, he or she begins to perceive the external world and also becomes aware of its own internal states, such as hunger, tiredness, and discomfort. As the baby enters postnatal life and develops, he or she initially begins to pay attention to light, then to voices, and progressively to more external and internal stimuli.

As children grow and learn, they begin to develop a conscious understanding of what they are perceiving. Along the way, the two hemispheres of their brains, their male or female characteristics, and their major brain functions (e.g., movement, memory, language, emotion, and social ability) mature. Ultimately, every child becomes an individual with a unique identity, a special set of behaviors, and his or her own problems, which are mostly mild. However, in some cases, problems that young people encounter are serious enough to qualify as mental disorders.

We must remember that when they meet other children, they generally enter some form of dominance hierarchy. As in adult life, some children develop dominant personalities, some develop subordinate personalities, and others fit somewhere in between. Bullying plays a role in the development of a personality in both the bully and the target, and if these forms of behavior continue, they influence who the individuals become.

By age 5, nature and nurture have generally played their roles in determining a child's preemptive personality, which is often similar to what it will be in adulthood (Bandura and Walters, 1963). While early life after 5 years of age can certainly influence who a person becomes, another particular round of influence comes during adolescence, when hormones and brain chemistry play an important role in who the person becomes. Once personality is established, some people react impulsively to events while others are more cautious.

There are many other differences between people, many of which we are not familiar with. Some people are more dependent on other people (they hold a more dependent or subordinate position) while others are more independent (they lean toward the more dominant side).

While we have admitted that there is a genetic basis for different levels of dominance and aggression, children express various degrees of aggression in the home for a number of reasons. These factors, together with the pain a child experiences and any aberrant brain chemistry that they have, may influence the temperament and degree of aggression an individual expresses in life. Since brain chemistry may cause individuals to react to their environment in different ways, some alternate, chemical-dependent behaviors may be corrected by medication.[2]

Psychologists sometimes use terms such as sociopathy and psychopathy for individuals who demonstrate aberrant (antisocial) behavior. These terms are sometimes confusing because of the various ways they are defined in the fields that may use them. For instance, sociopathy is defined as a loosely applied term that is represented by a personality disorder characterized by a lack of social responsibility and failure to adapt to ethical and social standards of the community. To many researchers in the field of psychology, sociopathy is an outdated term that has been replaced by Antisocial Personality Disorder (*Diagnostic and Statistical Manual of Mental Disorders, Fourth Edition,* 1994). However, the term sociopathy, like psychopathy, may be useful to indicate certain forms of antisocial behavior.

According to D.T. Lykken (*The Antisocial Personalities,* 1995), sociopaths are influenced by ineffective parenting and inadequate socialization during childhood and adolescence. Lykken recognizes four types of sociopaths (Fig. 9.2)[3]:

1. *A Disaffiliated Type* in which individuals develop antisocial traits and have an inability to relate emotionally to others, which affects relationships on a global level;
2. *A Disempathetic Type* in which individuals are capable of demonstrating affection and attachment to relatives, friends, or spouses, but they are prone to relate to others as objects. This serves a protective function due to childhood experiences of trauma, which can be viewed as being dissociative in nature and a form of desensitization;

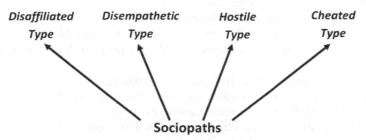

FIGURE 9.2 Four types of sociopaths, based on the work of D.T. Lykken (1995), which are influenced by ineffective parenting and inadequate socialization during childhood and adolescence. While the terms "sociopathy" and "psychopathy" are still in use, these terms have largely been replaced by "antisocial personality disorder" (*Diagnostic and Statistical Manual of Mental Disorders,* 1994).

3. *A Hostile Type* in which the individual is an angry, resentful, and aggressive person who purposefully rejects the social norms and mores of society and displays antisocial and traditional psychopathic traits as a result of their hostile beliefs;

4. *A Cheated Type* which is much like the hostile type, being hostile, antisocial, and rejecting the norms and mores of society, but for different reasons. These individuals feel rejected by society due to real or perceived inadequacies, most likely learned through experiences with an abusive parent. In later life, they create a personality in which rules do not apply to them because they have been wronged by others.

Psychopathy is defined as a mental (antisocial) disorder in which an individual manifests amoral and antisocial behavior, shows a lack of ability to love or establish meaningful personal relationships, expresses extreme egocentricity, and demonstrates a failure to learn from experience and other behaviors associated with the condition. Psychologists argue that the terms ASPD and psychopathy are not synonymous. ASPD refers to broad behavioral patterns based on clinical observation, whereas psychopathy refers not only to specific behavioral patterns but also to measurable cognitive, emotional, and neuropsychological differences. In other words, psychopathy assesses character as well as behavior.

ASPD is described by the American Psychiatric Association's *Diagnostic and Statistical Manual* as a personality disorder characterized by "...a pervasive pattern of disregard for, and violation of, the rights of others that begins in childhood or early adolescence and continues into adulthood." Psychopathy is considered by many psychologists today as a synonym of sociopathy, and both terms have largely been replaced by the term ASPD.

In *Human*, R. Winston and D. Wilson (2004) point out that although nature and nurture work together during childhood to mold individuals into the adults they become, they still have some ability to override natural instincts. We may be capable of change: that is, we may possess attributes for change, but whether we have the ability to utilize this capability is another question. Most literature on psychopathy indicates that antisocial behavior, once established, is very difficult to override.

MENTAL ILLNESS

In *50 Signs of Mental Illness*, J.W. Hicks (2005) points out that everyone experiences some form of mental health trauma in their life. In certain circumstances, mental trauma may lead to alternate behavioral problems. Psychiatrists refer to particular forms of alternate behavior as disorders, forms of mental illness.

The cause of mental illness is not well understood. Some illnesses are genetically based, but no genes have been isolated that are known to be responsible for their cause. Individuals with a genetic predisposition to mental illness may show symptoms of disorders early in life while others may not experience overt symptoms until later. Substance abuse (e.g., alcoholism, drug use, or other behavior employed to handle forms of craving) may complicate disorder issues and help them surface (see Chapter 10, *The Chemical, Physical, and Genetic Nature of Dominance, Aggression, and Aberrant Behavior*).

Learned behavior may play a significant role in mental health. Unpleasant experiences in life, such as the death of family members or friends, and mental and physical abuse may result in the acquisition of a mental disorder, especially in individuals who are genetically predisposed for certain disorders.[4]

Hicks (2005) lists several major categories of disorders and subsequently proceeds to discuss each of the categories in detail in the remainder of his book. I repeat his listing simply to give the reader a glimpse of what psychiatrists have recognized as disorders in humans:

- adjustment disorders (temporary emotional reactions to stress);
- anxiety disorders (phobias, panic attacks, and disabling worries);
- depression (which affects mood, sleep, appetite, sexual desire, and energy level);
- bipolar disorder, formerly known as manic-depressive illness (periods of depression alternating with elevated mood and hyperactivity);
- schizophrenia (hallucinations, delusions, and disorganized thinking);
- obsessive-compulsive disorder (intrusive thoughts and repetitive behaviors);
- posttraumatic stress disorders (reactions to life-threatening events);
- personality disorders (persistent and extreme character styles that often lead to problems in relating to others);
- drug and alcohol disorders (intoxication, addiction, and withdrawal);
- physical complaints and worries (can reflect psychological difficulties);
- sexual disorders (performance problems and unwanted urges and preoccupations);

- autism, mental retardation, hyperactivity, and other learning disorders emerging in childhood;
- dementia and delirium (memory loss and confusion, most common in the elderly).

Hicks points out that "symptoms of psychiatric illness frequently overlap and are easily misdiagnosed. Also, overlapping symptoms indicate a more complicated disorder." This will be evident in the brief treatment of specific disorders discussed later.

Anger, nervousness, and sadness are what Hicks states are typical responses to some forms of stress. "When we feel threatened, harmed, obstructed, betrayed, or disrespected, our bodies prepare to fight or escape" (the fight or flight reaction). Anger is an instinctive reaction to such stress, but the extent to which a person expresses their anger depends on their attitude, which in turn depends on their personality.[5]

Like personality, attitude is partially genetically based and appears to be "the most genuine and characteristic emotion of concern in antisocial personality disorders." Hicks further points out that most domestic violence and violence against strangers is committed by individuals with ASPDs.

Some personality disorders that may be associated with antisocial behavior are[6]:

- cravings (desiring something that you know is bad for you, especially if it involves drugs or sex);
- deceitfulness (lying and cheating);
- grandiosity (an exaggerated conviction that you are a special and important person);
- histrionics (feeling unattractive or worthless);
- psychosis (out of touch with reality);
- sexual preoccupation (having excessive sexual preoccupations and uncontrolled sexual urges, especially in connection with behaviors such as pedophilia and rape).

PSYCHOPATHIC BEHAVIOR

People with ASPDs (expressed here as equivalent to psychopathic behavior, as expressed by some psychologists) are dominant individuals who will do what they want to do, in spite of whether it hurts someone else. In *Without Conscience* (1993), R.D. Hare discusses antisocial personalities of certain humans, stating that their hallmark "is a stunning lack of conscience, their game being self-gratification at another person's expense. Many spend time in prison, but many do not. All take far more than they give from society and fellow humans."

Some of the obvious characteristics of psychopathy involve a flagrant criminal violation of society's rules. They are "self-centered, callous, and remorseless" people who are "profoundly lacking in empathy and an ability to form warm emotional relationships with others." They are people who function without the restraints of conscience, and they lack the qualities that allow humans to live in social harmony.

Many people considered to be psychopathic are criminals, but there are many who are not imprisoned, people who use "their charm and chameleon-like ability to cut a wide swath through society and leave a wake of ruined lives behind." See Chapter 11 (*Dominance and Aggression in the Workplace*), Chapter 13 (*Dominance in Politics*), and Chapter 14 (*Human Aggression, Killing, and Abuse*).

According to Hare, psychopaths make up a significant portion of the population described by the media as thieves, swindlers, con men, wife beaters, white-collar criminals, hype-prone stock promoters, boiler-room operators, child abusers, gang members, drug barons, professional gamblers, members of organized crime, doctors who have lost their licenses, lawyers who have been disbarred, cult leaders, mercenaries, unscrupulous business people, serial killers, and rapists.

Hare lists a number of individuals who have stood out as some of human society's most antisocial members:

- *Kenneth Bianchi*, one of the "Hillside Stranglers" who raped, tortured, and murdered a dozen women in the Los Angeles area in the late 1970s, turned in his cousin and accomplice and fooled some experts into believing that he was a victim of a multiple personality disorder and that the crimes had been committed by "Steve."
- *David Birkowitz*, the "Son of Sam" killer who preyed on young couples in parked cars.
- *William Bradfield*, a smooth-talking classics teacher convicted of killing a colleague and her two children.
- *Ted Bundy*, the "All-American" serial killer who was responsible for the murders of several dozen young women in the mid-1970s, who claimed that he had read too much pornography and that a "malignant entity" had taken over his consciousness.
- *John Gacy*, a Des Plaines, Illinois, contractor and Junior Chamber of Commerce "Man of the Year" who entertained children as "Pogo the Clown," had his picture taken with President Carter's wife and murdered 32 young men in the 1970s, burying most of the bodies in the crawl space under his house.

- *Ed Gein*, a psychotic killer who skinned and ate his victims.
- *Jeffrey Dahmer*, the "Milwaukee Monster," who pleaded guilty to torturing, killing, and mutilating 15 men and boys and was sentenced to 15 life terms.
- *Joe Hunt*, a fast-talking manipulator who masterminded a rich-kids' phony investment scheme (popularly known as the Billionaire Boys Club) in Los Angeles in the early 1980s, conned wealthy people into parting with their money and was involved in two murders.
- *Edmund Kemper*, the coed killer, sexual sadist, and necrophiliac, who mutilated and dismembered his victims.
- *Jeffrey MacDonald*, a physician with the Green Berets who murdered his wife and two children in the 1970s, claimed that "acid heads" had committed the crimes, became the focus of a great deal of media attention, and was the subject of the book and movie called *Fatal Vision*.
- *Ken McElroy*, who for years robbed, raped, burned, shot, and maimed the citizens of Skidmore, Missouri, without conscience or remorse until he was finally shot dead in 1981 as 45 people watched.
- *Colin Pitchfork*, an English flasher, rapist, and murderer who was the first killer to be convicted on the basis of DNA evidence.
- *Clifford Olson*, a Canadian serial murderer who persuaded the government to pay him $100,000 to show the authorities where he buried his young victims and does everything he can to remain in the spotlight.
- *Constantine Paspalakis and Deidre Hunt*, who videotaped their torture and murder of a young man.
- *Robert Ramirez*, a Satan-worshipping serial killer known as the "Night Stalker," who proudly described himself as evil, was convicted in 1987 of 13 murders and 30 other felonies, including robbery, burglary, rape, sodomy, oral copulation, and attempted murder.
- *Charles Sobhraj*, a French citizen born in Saigon who was described by his father as a destructor, became an international confidence man, smuggler, gambler, and murderer who left a trail of empty wallets, bewildered women, drugged tourists, and dead bodies across much of Southeast Asia in the 1970s.
- *Kenneth Taylor*, a philandering New Jersey dentist who abandoned his first wife, tried to kill his second wife, savagely beat his third wife on their honeymoon in 1983, battered her to death the next year, hid her body in the trunk of his car while he visited his parents and his second wife, and later claimed he had killed his wife in self-defense when she attacked him following his discovery that she was sexually abusing their infant child.
- *Gary Tison*, a convicted murderer who masterfully manipulated the criminal justice system, used his three sons to help him escape from an Arizona prison in 1978 and went on a vicious killing spree that took the lives of six people.

The list of such individuals is exhaustive (Stone, 1998). These and other psychopathic killers are examples of the individuals who have stood out in the media, but as Hare mentions, "the majority of psychopaths manage to ply their trade without murdering people." He states that victims of most psychopaths are far more likely to lose their life savings to an oily-tongued swindler than their life to a steely eyed killer. Within recent years, computer scam artists have become examples of this group.

According to Hare, a frightful and perplexing theme that runs through the case histories of all psychopaths is "a deeply disturbing inability to care about the pain and suffering experienced by others."

Why are these people like this? It is sometimes difficult to understand how a person becomes a psychopath. Hare states that while it is true that the childhoods of some psychopaths are characterized by material and emotional deprivation and physical abuse, "for every adult psychopath with a troubled background, there is another whose family life apparently was warm and nurturing and whose siblings are normal, conscientious people with the ability to care deeply for others." In addition, Hare points out that "most people who had horrible childhoods do not become psychopaths or callous killers."

These realizations seem to point either to a genetic base for violence, the crowd an individual becomes associated with, the boredom an individual experiences in life (possibly associated with IQ), abuse that an individual receives, or a combination of these conditions. For whatever reason they become psychopaths, individuals who develop antisocial personalities appear to know that what they are doing does not fit in with the norm, and they often exert some energy in hiding their inner feelings.

Hare lists other characteristics of psychopaths that help to conceal or identify them in society:

- They are often witty and articulate.
- They can be amusing and entertaining conversationalists, with a quick and clever comeback.
- They can tell unlikely but convincing stories that cast themselves in a good light.
- They can be very effective in presenting themselves well and are often very likeable and charming.
- To some people, they seem too slick and smooth, too obviously insincere, and superficial.

- Astute observers often have the impression that they are pretending or play-acting.
- They may ramble and tell stories that seem unlikely in light of what is known about them.
- They typically attempt to appear familiar with sociology, psychiatry, medicine, psychology, philosophy, poetry, literature, art, or law.
- A signpost to psychopathy is a smooth lack of concern at being found out.
- They have an extreme narcissistic and grossly inflated view of their self-worth and importance, a truly astounding egocentricity and sense of entitlement and see themselves as the center of the universe, as superior beings who are justified in living according to their own rules.
- They often come across as arrogant, shameless braggarts who are self-assured, opinionated, domineering, and cocky.
- They love to dominate others and seem unable to believe that other people have valid opinions different from theirs.
- They appear charismatic or electrifying to some people.
- They are seldom embarrassed about their legal, financial, or personal problems. They see them as temporary setbacks, the results of bad luck, unfaithful friends, or an unfair and incompetent system.
- They do not comprehend how to achieve goals.
- They often think big but usually with someone else's money.
- They may talk fast but contradict themselves from one sentence to the next.
- Prison records show they have an inability to learn from past experience.
- They sometimes verbalize remorse but contradict themselves in words and actions.
- They lack remorse or guilt and have a remarkable ability to rationalize their behavior and shrug off personal responsibility for actions that cause shock and disappointment to family, friends, associates, and others who have played by their rules.
- They frequently see themselves as the real victims.
- They view people as little more than objects to be used for their own gratification.
- Lying, deceiving, and manipulation are natural talents for them.
- They frequently are successful in talking their way out of trouble.
- They usually do not get along with other psychopaths.
- They usually exhibit serious behavioral problems at an early age, including stealing, taking drugs, cutting school, having early sexual experiences, and demonstrating cruelty to animals and other children. Their antisocial behaviors as adults are continuations of behavior patterns that first appeared in childhood.
- They show no loyalty to groups, codes, or principles other than to look out for themselves.

While Hare (1993) states that not all rapists are psychopaths, he estimates that perhaps half of the repeat or serial rapists are. Their acts are the result of a potent mixture: uninhibited expression of sexual drives and fantasies, desire for power and control, and perception of the victims as objects of pleasure or satisfaction. Thus, dominance over a subject or group in general is important to a psychopath, either in the context of women victims or in situations with same-sex victims.

While there is some controversy about the parts nature (genetics) and nurture (learned behavior) play in psychopathy, the presence of psychopathic behavior most certainly appears to be influenced by both. The disorder remains little understood, and no effective treatment has been found. Psychopaths do not seek help on their own, and they detest being in a subordinate position. They frequently dominate both individual and group therapy sessions, imposing their own views and interpretations on others.

SCALE OF EVIL

Michael Stone, Columbia University psychiatrist, has developed a twenty-two-category scale for what has been referred to as a "scale of evil" (1998, 2009). It is based on an examination of hundreds of criminals and demonstrates the wide variation of behavioral expression in antisocial disorders. The list begins with individuals who are the least apt to be psychopathic and proceeds to those who possess most psychopathic traits:

- People who kill in self-defense and do not show psychopathic tendencies.
- Jealous lovers who, although egocentric or immature, are not psychopathic.
- Willing companions of killers: those with aberrant personalities—probably impulse-driven, with antisocial traits.
- People who kill in self-defense but who have been extremely provocative toward the victim.

- Traumatized, desperate people who kill abusive relatives and others but lack significant traits. Such people are genuinely remorseful.
- Impetuous, hot-headed murderers, but without marked psychopathic features.
- Highly narcissistic but not distinctly psychopathic people with a psychotic core who kill people close to them (e.g., in jealousy or with an underlying motive).
- Nonpsychopathic people with a smoldering rage who kill when rage is ignited.
- Jealous lovers with psychopathic features.
- Killers of people who are "in the way" or who kill, for example, witnesses (they are egocentric but not distinctly psychopathic).
- Psychopathic killers of people "in the way."
- Power-hungry psychopaths who kill when they are "cornered."
- Killers with inadequate, raging personalities who "snap."
- Ruthlessly self-centered psychopathic schemers.
- Psychopathic "cold-blooded" spree or multiple-murder individuals.
- Psychopaths committing multiple vicious acts.
- Sexually perverse serial murderers and torture-murderers.
- Torture-murderers with murder as the primary motive.
- Psychopaths driven to terrorism, subjugation, intimidation, and rape (short of murder).
- Torture-murderers with torture as the primary motive but with psychotic personalities.
- Psychopathic torture-murderers, with torture as their primary motive.

The aforementioned traits of individuals with antisocial personalities are expressions involving dominance and aggression, which are found in individuals in many walks of life and in populations around the world. In the following chapters, many of these traits are elaborated upon in terms of the genetic and chemical nature of dominance and aggression, bullies, despotic leaders, in the context of religion, and in the process of various forms of killing:

> If the entire human species were a single individual, that person would long ago have been declared mad. The insanity would not lie in the anger and darkness of the human mind—though it can be a black and raging place indeed. And it certainly wouldn't lie in the transcendent goodness of that mind—one so sublime, we fold it into a larger "soul." The madness would lie instead in the fact that both of these qualities, the savage and the splendid, can exist in one creature, one person, often in one instant. *Jeffrey Kluger, Time Magazine, What Makes Us Moral*

Indeed, human thoughts may border on the edge of or beyond what we define as "normal," and there is no better realm of thought that demonstrates fluidity in human behavior as in their sexual fantasies. These love/hate relationships are evident in all forms of human behavior.

LOVE

Love is generally described as an intense feeling of deep affection for or a deep romantic or sexual attachment to someone. The causes and effects of love are difficult to explain because there are different kinds of love, and the ultimate form of love is affected by the diverse array of human personalities.

R.J. Sternberg's *Triangular Theory of Love* (1988) discusses three different types of love: commitment, passion, and intimacy. A relationship based on only one of these elements is less likely to survive than one based on two or more. Each of these is linked with other influences. Intimacy and passion but no commitment, for instance, is a form of romantic love. Consummate love includes all three features. Sternberg reveals the following categories in his explanation of this difficult-to-understand condition:

- An *intimate liking* of someone characterizes a true friendship, in which a person feels a bondedness, warmth, and closeness with another individual but not intense passion or long-term commitment.
- *Infatuated love* is often what is felt as "love at first sight." But without the intimacy and the commitment components of love; infatuated love may disappear suddenly.
- *Empty love*: At times, a stronger love deteriorates into empty love, in which the commitment remains, but the intimacy and passion have died.
  In cultures in which arranged marriages are common, relationships often begin as empty love.
- *Romantic love*: Romantic lovers are bonded emotionally (as in liking) and physically through passionate arousal.

- *Companionate love* is often found in marriages in which the passion has died out of the relationship, but a deep affection and commitment remain. Companionate love is generally a personal relation and that two individuals build to share their life with one another, but with no sexual or physical desire. It is stronger than friendship because of the extra element of commitment. The love ideally shared between family members is a form of companionate love, as is the love between deep friends or those who spend a lot of time together in any nonsexual but friendly relationship.
- *Fatuous love* can be exemplified by a whirlwind courtship and marriage in which a commitment is motivated largely by passion, without the stabilizing influence of intimacy.
- *Consummate love* is a complete form of love, representing the ideal relationship toward which many people strive but which apparently very few actually achieve. Sternberg cautions that maintaining a consummate love may be even harder than achieving it. He stresses the importance of translating the components of love into action. "Without expression," he warns, "even the greatest of loves can die."
Consummate love may not be permanent. For example, if passion is lost over time, it may change into companionate love. The balance among Sternberg's three aspects of love is likely to shift through the course of a relationship. A strong dose of all three components found in consummate love typifies, for many of us, an ideal relationship. However, time alone does not cause intimacy, passion, and commitment to occur and grow.

RESPECT

According to an anonymous article in TwoOfUs.org called *Showing Respect to Your Partner*, "self-help books and talk shows are replete with references to 'respect' and how foundational it is to a healthy relationship." However, the word is rarely defined in practical terms. Individuals in a partnership often focus on what they should be "getting" from their partner in terms of respect. But respect has a giving component as well. The following advice is recommended for a stable relationship. Whether the advice is followed depends on the personality of individuals in the relationship:

- *Choosing words carefully*: Words come out quickly and can be hard to take back. Thus, before a verbal tirade is launched against a partner, the desired outcome of the words should be considered. Does one really want to "punish" their partner—or is one simply longing for him/her to be more considerate of their needs? If so, a diplomatic approach is more likely to achieve this goal.
- *Acknowledging contributions*: Partners let one another down on occasion but may simultaneously make some positive contributions to the relationship. Advice from the writer is to be sure to affirm these qualities, even amid other frustrations. Doing so will help a partner lower his/her defenses and lead to a more constructive partnership.
- *Honoring boundaries*: Understanding and respecting a partner's personal boundaries regarding time together/apart, physical contact, and so on is recommended.
- *Being willing to compromise*: Being respected does not mean an individual's needs always take priority over those of a partner. Compromise provides a relationship with the flexibility it needs to keep from ripping it apart.
- *Showing consideration*: Helping with the housework is a good gesture, as well as giving sincere compliments and being thoughtful toward a partner.
- *Being strong enough to admit being wrong*: When a partner is confident in his or her self-worth, apologizing should not make either of them feel threatened. We all make mistakes; admitting so when it happens allows a relationship to move forward, rather than backward.
- *Protecting a partner*: A partner's physical or emotional well-being should never be compromised. If tempers are out of control, immediate professional help is recommended.
- *Understanding one's worth*: Self-esteem is not about thinking one person is better than the other. However, a partner should have an unshakeable conviction that their thoughts, feelings, and body warrant respect. If a person is truly convinced that they are worthy of respect, others are unlikely to doubt it.
- *Acting honorably*: While our fundamental human dignity calls for respect, being a person of character makes it easier for people to respect someone. People who act with integrity rarely do anything to harm another person; accordingly, such people are more likely to be respected by others.
- *Setting and upholding boundaries*: When a partner loves someone, it is easy to let certain things slide. "He or she did not really mean it." "It was just that one time." "I know, but he/she has been working really hard lately." If a person regularly makes excuses for their partner, he or she may be taking advantage of the person making excuses. It is up to each individual to protect their worth and boundaries from anyone who would undermine them.

- *Being a man or woman of one's word*: When an individual lies to their partner or breaks promises, they weaken the trust in the relationship. Lack of trust often leads to a lack of respect.
- *Showing respect*: To truly be respected, we must also respect. If an individual cannot find anything in their partner worth respecting, that person should consider why they are even in the relationship. If just a few of a partner's actions or attitudes are causing mistrust or resentment, those issues should be actively addressed. Taking a marriage/relationship education workshop can help resolve conflict more efficiently and respectfully.

Respect means recognizing one's own worth—and the worth of others. When we respect our partner, we are able to rise above pettiness, jealousy, and cruelty. When we respect ourselves, we are able to transcend insecurity, defensiveness, and fear. And respecting both ourselves and our partners enables us to build strong, lasting, and mutually supportive relationships.

The points about respect pertain to what two individuals should consider to maintain a homeostatic relationship. However, they do not consider differences in human personality, especially in relationships in which an antisocial personality is involved.

FALLING OUT OF LOVE

Previous comments are made about how feelings may wane in a loving relationship. Individuals in a partnership who do not possess the features needed for love or who do not express themselves in the categories listed may find themselves falling out of love. Falling out of love by one of the partners and not the other may cause unpleasant feelings and lead to unhappy circumstances. Mix such circumstances with a personality disorder, and one or both of the individuals may experience extreme displeasure as a result, stalking or a life-threatening consequence.

STALKING

When an individual falls out of love with someone, they may threaten the ego of their former mate, which may result in the mate simply leaving or commencing a stalking mode (Mullen et al., 2000). Stalkers (dominants) may think of the individual they are stalking as a possession, and the loss of that possession threatens the stability of the stalker.

The National Center for Victims of Crime defines stalking as "virtually any unwanted contact between two people that directly or indirectly communicates a threat or places the victim in fear." However, the legal standard in practice is usually somewhat more strict. Wikipedia defines stalking as "unwanted or obsessive attention by an individual or group toward another person." Stalking behaviors are related to harassment and intimidation and may include following the victim in person or monitoring them.

In *Stalkers and their Victims*, Mullen et al. (2000) state that individuals who stalk are often grouped into two categories: psychotic and nonpsychotic. While some stalkers may have preexisting psychotic disorders, for example, delusional disorder, schizoaffective disorder, or schizophrenia, most stalkers are nonpsychotic and may exhibit disorders or neuroses such as major depression, adjustment disorder, or substance dependence, as well as a variety of personality disorders (such as antisocial, borderline, dependent, narcissistic, or paranoid).

"Some of the symptoms of 'obsessing' over a person may be characteristic of obsessive-compulsive personality disorder. The nonpsychotic stalkers' pursuit of victims can be influenced by various psychological factors, including anger, hostility, projection of blame, obsession, dependency, minimization, denial, and jealousy. Conversely, as is more commonly the case, the stalker has no antipathic feelings toward the victim, but simply a longing that cannot be fulfilled due to deficiencies either in their personality or their society's norms."

According to D.V. James and F.R. Farnham in *Stalking and Serious Violence* (2003), a characteristic of the behavior in stalking is that it occasionally causes fear or apprehension in the victim. "Dramatic incidents of stalker violence brought stalking to public prominence and were the impetus to the introduction of antistalking legislation that, in most jurisdictions, is framed in terms of a behavior that places a person in fear of physical harm. However, most stalkers are not violent; rates for violent behavior range between 30 and 40% in most reported series."

Violence infrequently results in serious physical injury, with most victims being grabbed, punched, slapped, or fondled by the stalker. Serious violence is rare. It has been suggested that the homicide rate in stalking is probably less than two percent, but an analysis of prevalence rates of stalking and homicide illustrates that this percentage is a gross overestimation.

Two major studies, based on the psychiatric examination of series of stalkers, involved large enough samples to provide statistically significant associations between stalking and violence. Harmon et al. (1998) examined the records of 175 stalkers, collected over a 10-year period; of these, 81 exhibited violent behavior. Significant associations with such behavior were previous relationship to victim, threats of violence, and substance abuse. Mullen et al. (2000) examined 145 stalkers, of whom 52 (36%) were assaultive.

Significant associations were found between assault and previous threats, previous convictions, substance abuse, and the authors' stalking typology. Only previous convictions remained significant when all variables were considered in a regression analysis. Stalkers of different motivational types evidenced different frequencies of assault.

Two further important studies found significant associations with violence. These studies were based primarily on victim reports of stalking violence, rather than examination of stalkers. The study of 223 cases from the Los Angeles Police Departments Threat Management Unit (TMU) found a significant relationship between violence and former intimacy and between a history of general violence and violence committed during the stalking campaign.

The association with former intimacy was positively influenced by the suspect's level of proximity to the victim and by threats to the victim and property. A study based on the interview of 187 former intimate victims of stalking found a significant correlation between verbal threats and subsequent violence. There were also significant relationships between drug and alcohol use and the occurrence of violence resulting in physical injury.

In their review of the literature, Mullen et al. (2000) summarized the personal characteristics of stalkers likely to be associated with a higher risk of assault. "Principal among these were substance abuse; a history of criminal offenses, particularly violent or sexual offenses; male gender; threatening the victim; presence of a personality disorder; pursuit of an exintimate; unemployment; and social isolation." Less easy to measure were the presence of high levels of anger at the victim, an intense sense of entitlement, and fantasies about assaulting the victim.

There is an assumption that risk factors for violence in stalking samples are likely to have much in common with risk factors in nonstalking samples. Mullen et al. (2000) point to other studies and conclude that "the very limited predictive research to date has produced three variables which significantly and strongly predict personal and/or property violence among stalkers: prior criminal convictions, substance abuse, and prior sexual intimacy with the victim."

"It is assumed that all violence, to persons or property, is a homogeneous entity, with one set of predictive factors. The violence in the major studies of stalkers appears to have been general violence, which was predominantly minor in nature." Harmon et al. (1998) defined a stalker as violent if the stalker physically assaulted the target or an associate, or attacked or damaged the property of the target, including any physical contact—that is, the defendant banging on the door repeatedly; or making physical contact with the target, someone close to the target, or a surrogate for the target.

This definition fails to differentiate between repeatedly banging on a door and homicide, of which there were two instances in their study. Mullen et al. (2000) restricted their definition to physical assault on the person. Among the incidents involving 52 assaultive stalkers, there was one fractured jaw, one stab wound, six indecent assaults, and eight attempted or accomplished rapes. However, overall, "physical injuries were largely confined to bruises or abrasions." In studies based largely on victims' reports, these authors' definition of reported violence ranged from "pushing, slapping, kicking, and biting, to rape, assault with a weapon, and so on." Physical injury ranged from "small cuts and bruises" to "broken bones."

Palarea et al. (1999) separated violence against the person from violence against property, but did not offer definitions of these categories. "Violent behavior is not homogeneous, and the predictors of one form of violence may be quite different from the predictors of other types."

"There is no reason to assume that the associations of minor assault in stalking should, for instance, be the same as those of homicide." Yet, there are no results in the stalking studies just cited to indicate whether different degrees of violence in stalking may have different associations.

TOLERANCE

According to L. Namka in *Conflict Over Values* (1995a,b), and N.S. Jacobson in *Behavioral Couple Therapy: A New Beginning. Behavior Therapy* (1992), "when two people join their lives to be together in a relationship, they bring along their deepest values as to what they hold dear in the world." As we have learned throughout this book, there are fundamental differences in the values of people.

Individuals are different, for instance, "due to their upbringings, the needs of their central nervous systems, and how they see the world." People have different needs and different desires. Some values are a strong part of the person's personality and are resistant to change. Some couples are even attracted to each other because at first they value their partner's different ways of looking at the world.

Following the honeymoon phase of the relationship, conflict comes up when there are noticeable differences in values (between the individuals) of the couple. Part of the problem is often due to unrealistic expectations for the relationship. It is a sad fact that Americans have set up such high standards for their relationships. In our society, some people expect to get all of their emotional needs met within the relationship. The unrealistic expectation that one's partner (can) meet every need for intimacy is wrecking many marriages. Spiritual people seem to have even higher expectations than others. With rising expectations for the partner lies the danger for lack of personal happiness and longevity of the relationship. High expectations and focus on the partner's weaknesses rather than their strengths result in personal unhappiness and a threat to the relationship.

Demand for change of the partner comes from what Virginia Satir says in *The Big Game of Life* (Rubin, 2008) is found in statements like "I have the right to tell you what to do." "Couples who play The Big Game polarize their demands. Each sits on an opposite pole caught in their own pain and shout at each other, demanding change but ending up in [a stalemate]. They believe that not only the other person can change, but also that they should, and [immediately]."

"Permanent behavioral change does not happen, especially if there has been a demand for change." Intolerance in a relationship, coupled with angry, coercive behaviors, always causes distancing between the two people in the relationship. Thus, human beings have a need for control in their life. However, when personal needs spill over into demanding changes from the partner, it backfires and prevents intimacy from developing.

This author states that "What we all want is to be loved and love others in return."

A life without love, that's…terrible. Love is like oxygen! Love is a many splendored thing. Love lifts us up where we belong. All you need is love! **Christian,** *Moulin Rouge*

"But some people get caught up in destructive, intrusive behaviors that are used to try to obtain love. Their agenda becomes: 'The quicker I can get you to agree with me and change, the better.' Not knowing other ways to get what they want, [such people] continue their demands for change." Their purpose is to obtain love for themselves, but they criticize their partner in the process. They are often "well intentioned in trying to get change to bring about more happiness for themselves, but it does not work. They do what they know how to do and do it louder and longer, but it still does not work." Demanding the other person clean up their act is considered a coercion model of change.

A circular loop of anger based on [the phrase] "I have the right to tell you what to do" keeps the couple caught in countless arguments. The first scenario is when the first person is angry and demands change in the partner. The partner may comply with [behavior that will] preserve peace but resents it. The behavior changes but then the partner slides back into old actions. The first person is angry and blames the person for not trying hard enough. He or she spends time justifying how bad the partner is for not meeting his or her needs and tries harder to change the partner. People who demand change from their partners dig themselves in deeper and deeper into an impasse with their partner as they cycle around the loop of destructive communication. The second scenario is when the second partner responds by blaming and insisting on changes that he wants in the first person—tit for tat, but it ends up with constant fault-finding.

Attempts to change the other person, even though well intentioned, may become an even bigger problem. For some individuals, the constant (attempt) to change a partner is the biggest problem they run up against, even bigger than the behavior in their partner in which they are trying to change.

Some people [remain] angry and upset through their attachment to their belief that they cannot be happy unless the partner changes. The result is poor self-esteem, isolation, and continual frustration. Blaming, fights that no one wins, and distancing characterize the relationship.

Later, the behaviors that seemed initially so stabilizing or freeing in their partner seem to wear thin. What they formerly valued in their partner becomes an irritant. The Big Game of Life begins in earnest. Then the war begins to try to make the partner over to be a carbon copy of their own values. Then the Hardening of the Categories begins— once I believe something bad about your behavior, I look for data to confirm it. Instead of seeing their own need for control and insistence on having things their way, the other person becomes the enemy. Psychologist Neil Jacobson et al. have identified the most common differences between couples that set up most of the fights (1992). These differences stem from basic values:

• I need closeness and intimacy with you/I need my space, don't bother me; 50% of fights are related to this statement.
• We must act like free spirits and be spontaneous/We must be logical and rational.

- We can spend money and get what we want; let's put it on plastic/We must save money.
- I must have control over you/Don't tell me what to do; I must be an independent person.
- We must keep the secrets (strong silent type)/Let's be open and talk about everything.
- Let us have sex all the time/Sex is okay once in a while.
- Conflict and anger should be expressed and accepted/Don't get angry.
- I do not want to clean the house/We must live in a neat, clean house.
- You must have a successful career/I just want to live a laid-back kind of life.
- Children should be raised; strictly/children should be allowed to be free and spontaneous.
- My work over everything/You must be here for the family or you don't love me.
- I withdraw when we fight to take care of myself/I'll go on the offensive and go after you.
- Our marriage must be monogamous/I want to experience different sexual partners.
- My parents, friends, or the kids come before you/Our partnership comes first.

These differences in values reflect different ways of looking at the world and living one's life. Some couples see their partner's flaws but have deep commitment to remaining in the relationship. Some marriages last with happiness despite great differences in values. Their commitment is that they learn to accept [or tolerate] their differences rather than make them sources of arguments. Talking about the problem as if it is the enemy, not the partner's behavior, and seeking common solutions ends up being the solution to the problem.

Some couples agree to stay together and take on daily challenges despite widely divergent needs and desires. It is as if each agrees, "Yes, I love and accept him or her, despite those flaws," or "Yes, there are big value differences between us, and we can stay together," or "No matter what, we can work it out."

How do couples weather out their basic differences to stay together to achieve a happy relationship? Some learn to become more tolerant of their partner's behavior by giving up the struggle for change on demand. They agree to stop pushing the buttons that set their partner off. They are committed to staying together despite their differences and put the interests of the partnership over fighting about who is right or wrong. They accept that each partner is different, and that differences bring variety and challenges to the relationship. With this acceptance comes the desire to acclimate more to their partner's wishes. True personal change involves openness, nonblaming, and nondefensive communication about conflict.

It is a funny thing about human values—they often become stuck when another person demands change; they loosen when there is closeness, acceptance, and seeing the partner through sympathetic eyes. When both begin to see the pros and cons of each position and see their partner with compassion [and respect], they begin to change naturally as they no longer have to defend their stance. Closeness comes about when they see the problem, not the partner, as the enemy.

At times, there are differences between the couple that are hard to reconcile. Some destructive behaviors, such as the three A's (severe, ongoing abuse, adultery, and addictions), are impossible to tolerate or accept. Over time, these destructive behaviors grow into fundamental differences that cannot be tolerated.

The greater the number of differences or the intensity of the beliefs between the partners, the more difficult time they will have in achieving stability in their relationship. One or both partners can become highly sensitive to their partner's behaviors and become greatly critical and angry over small issues. Commitment to the partner and the relationship starts to wane.

Subsequently, "communication breaks down and depression may set in." One or both partners may start to seek outside people to verify their beliefs about how bad their partner is. They seek interests outside the relationship and begin to invest more energy away from their partner. They may seek out a transitional person as a friend or lover to get them through this critical time. If the transitional people they choose support them in their belief on how bad the partner is, then they are more likely to consider leaving.

Breaking up happens when one or both members of the couple decide that there are fundamental differences in their partner that they cannot stand. The decision to leave the relationship happens when one or both decide that there is a fatal flaw in their partner that they no longer can live with.

Chains do not hold a marriage together (Namka, 1995a,b). It is the threads, hundreds of tiny threads, which sew people together through the years. A couple's commitment to living the C words sews the relationship together: cooperation, connection, compassion, clear conscience, and ongoing communication.

The best advice for happiness in a relationship is to keep expectations of a partner modest and gratitude high. One of the highest level skills that keeps a relationship together is to feel appreciation for the partner's positive traits. Acceptance of things as they are and stopping the struggle to change the partner is a way to achieve happiness in the relationship. Desirable changes happen when communication strategies are learned and the goal is to become more open, noncritical, and nondefensive, and to listen to accommodate the other. Often one partner's focusing on the goal of acceptance leads to the other partner starting to make change. The achievement of a good marriage and being loving with a partner in the lean years, as well as the good, is a fine art to be learned and practiced.

"Couples who commit to stay together do make changes in their values across time. In an atmosphere of openness and trust, the need to hold dogmatically to one's own position and make the other partner wrong decreases. Values and differences start to soften and fade when there is mutual agreement to see things from the other partner's point of view. The couple takes mutual responsibility for their problems. They know each must change themself for the betterment of the relationship. As they put their commitment to being a couple above their individual needs, they begin to grow more like each other in their values. Couples who want to stay together develop the ability to accept the unresolvability of some of the problems and enjoy each other anyway. They learn to turn their differences into a source of strength and enrichment of the relationship."

ADULTERY

Adultery, also referred to as infidelity, promiscuity, cheating, or having an affair, pertains to having extramarital sex. In most contemporary societies, it is considered objectionable. What exactly constitutes adultery is defined in a cultural context. While adulterers have suffered capital punishment, mutilation, and torture as a consequence of their promiscuousness, adultery is not a criminal offense in most contemporary society. On the other hand, adultery can be punishable by death in certain societies.

Adultery is commonly thought of as a breach of trust by the person who suffers from an adulterous relationship. It can result in feelings of jealousy and guilt, and if the adulterous affair continues over a long period of time, the third person in such a relationship may carry feelings of deceit and may encourage divorce. Couples in which adultery has been carried out may simply withdraw from one another with ongoing feelings of guilt, carry on an obsession with their lover, may choose to reveal the affair, or in rare cases commit violence or other crimes toward one another.

HATE AND HATRED

In a marital situation in which there are differences between people who cannot seem to learn to tolerate their partner, a hatred can develop. Likewise, hatred can develop outside the confines of a marital situation, where tolerance is not practiced toward other humans for one reason or another.

In its simple definition, hatred (or hate) can be thought of as a deep and emotional extreme dislike. It can be directed against individuals, entities, objects, or ideas. At the individual level and within the context of our present subject, hatred may develop in a defunct marital situation, as described earlier.

In the 2002 edition of the *Penguin Dictionary of Psychology*, Reber and Reber define hate as a "deep, enduring, intense emotion expressing animosity, anger, and hostility towards a person, group, or object." Hatred is often associated with feelings of anger and a disposition toward hostility. While hate can develop out of a partnership, as indicated in the foregoing paragraphs, the origin of most hate is found outside of a marital context.

In *Ethnolinguistics and Cultural Concepts: Truth, Love, Hate and War*, J.W. Underhill (2012) stresses that love and hate are social and are culturally constructed. From a psychoanalytic view, hate can be thought of as an ego state that wishes to destroy the source of its unhappiness. Because hatred is believed to be long lasting, many contemporary psychologists consider it to be more of an attitude or disposition than a temporary emotional state.

From a neurological standpoint, S. Zeki and J.P. Romaya (2008) point out in Neural Correlates of Hate that hate has been investigated with an fMRI procedure, in which people had their brains scanned while viewing pictures of people they hated. The results showed increased activity in the middle frontal gyrus, right putamen, bilaterally in the premotor cortex, in the frontal pole, and bilaterally in the medial insular cortex of the human brain (Fig. 5.3).

According to a report by R. Stotzer in 2007, hate (bias-motivated) crime generally refers to a criminal act that is believed to have been motivated by hate. Those who commit hate crimes target victims because of their perceived membership in a certain social group, usually defined by race, gender, religion, sexual orientation, disability, class, ethnicity, nationality, age, gender identity, or political affiliation. Incidents may involve physical assault, destruction of property, bullying, harassment, verbal abuse or insults, and offensive graffiti or letters (hate mail).

Hate speech is speech perceived to disparage a person or group of people based on their social or ethnic group, such as race, sex, age, ethnicity, nationality, religion, sexual orientation, gender identity, disability, language ability, ideology, social class, occupation, appearance (height, weight, skin color, and so on), mental capacity, and any other distinction that might be considered by some as a liability. The term covers written as well as oral communication and some forms of behaviors in a public setting. It is also sometimes called antilocution and is the first point on Allport's scale that measures prejudice in a society.

In many countries, deliberate use of hate speech is a criminal offence prohibited under incitement to hatred legislation. It is often alleged that the criminalization of hate speech is sometimes used to discourage legitimate discussion of negative aspects of voluntary behavior (such as political persuasion, religious adherence, and philosophical allegiance). There is also some question as to whether or not hate speech falls under the protection of freedom of speech in some countries.

Both of these classifications have sparked debate, with counterarguments, such as, but not limited to, a difficulty in distinguishing motive and intent for crimes, as well as philosophical debate on the validity of valuing targeted hatred as a greater crime than general misanthropy and contempt for humanity being a potentially equal crime in and of itself.

HATRED AND HATE GROUPS

Darkness cannot drive out darkness: only light can do that. Hate cannot drive out hate: only love can do that. **Martin Luther King Jr.**, *The Testament of Hope: The Essential Writings and Speeches*

With a population of over seven billion people in the world, tension between people and groups is growing, especially when food, jobs, and other resources are becoming scarce. Such tension generally builds in groups of people who are emotionally or chemically unstable, suffering from an ASPD, maladjusted because of abuse or tired of being an underdog in a world that consists of difficult-to-scale dominance hierarchies.

Emotions are described by psychologists as various bodily sensations that respond to certain behavioral traits, such as mood, temperament, personality, disposition, and motivation. They also respond to the presence of certain hormones and neurotransmitters, for example, dopamine, noradrenaline, serotonin, oxytocin, and cortisol (see Chapter 10). Emotions are often the driving force behind motivation, whether they are positive or negative.

Cognition (a group of mental processes that includes attention, memory, producing and understanding language, solving problems, and making decisions) is an important aspect of emotion, particularly with relationship to the interpretation of events. As an example, fear is generally recognized as a feeling that usually occurs in response to a threat. The cognition of danger and subsequent arousal of the nervous system (e.g., rapid heartbeat and heavy breathing, sweating, and muscle tension) are integral components of subsequent interpretations and labeling of that arousal as an emotional state.

Hatred is a form of emotion which the Merriam Webster dictionary defines as (1) an intense hostility and aversion that is usually derived from fear, anger, or a sense of injury; (2) extreme dislike or antipathy. As a noun, it is synonymous with words such as abhorrence, detestation, odium, revulsion, disgust, and extreme dislike. As a verb, it is synonymous with words such as abhor, detest, loathe, cannot stand, find insufferable, cannot bear, and be repulsed by. An antonym is love.

While hatred is common in human society, either toward individuals or certain groups of people, it is generally considered a futile emotion that serves no discernible purpose to the human species apart from causing disequilibrium in its population. Hatred is considered a learned emotion, a by-product of traits such as greed, fear, and envy. Children generally learn hatred from their parents, other guardians, teachers, influential members of their community, or peers. When these emotions become strong enough in people, they result in hatred.

Misanthropy, a form of hatred, is the general intense dislike, mistrust, or disdain of the human species or human nature. A misanthrope, or misanthropist, is someone who holds such a view or feeling.

There are many misanthropists on Earth, often expressing hatred toward people in terms of their race, religion, gender, and politics of sexual orientation. To them, they are the misfits. As such, they feel they have wreaked havoc on world homeostasis for centuries (see examples in Chapter 13). Unfortunately, hate groups appear to be on an increase, and because of their nature, they most likely will continue wreaking havoc for many years into the future. Increases in population and shortage of resources will no doubt feed discontentment and the rise of hate groups around the world.

As mentioned earlier, hatred appears to be associated mostly with learned rather than innate behavior, although a predisposition to antisocial behavior may influence the development of hatred. It is one of the classical emotions that also includes love, anger, and fear. Classical emotions are expressions of paleocircuits of the mammalian brain, for example, modules of the cingulated gyrus, which also facilitate the care, feeding, and grooming of offspring.

According to the most accepted neurological model (the Direction Model), hatred, like anger, activates the left pre-frontal cortex (Fig. 5.3) from regions generated in the senses before passing to the paleocircuits. It is often expressed as prejudice, violence, and discrimination and may lead to heinous moral behavioral errors in human judgment.

Hate groups are organized groups or movements that advocate and practice forms of intolerance, for example, hatred, hostility, or violence, toward members of a race, ethnicity, religion, gender, sexual orientation, or other sectors of human society. Two organizations in the United States monitor such groups: the Anti-Defamation League (ADL) (League, 2001) and the Southern Poverty Law Center (SPLC) (Bauer and Reynolds, 2009). Well-known hate groups that are known to practice violent behavior may be tracked by the FBI as well.

A list of what these organizations deem to be hate groups includes: supremacist groups, anti-Semitic, antigovernment, and extremist groups that have actually committed hate crimes. While inclusion of a group in the list does not imply that a group actually advocates or engages in violence or other criminal activity, the Southern Poverty Law Center has noticed a 50% increase in hate-group activity from 2000 to 2008, behavior that has the potential to lead to violent crimes (Barnes and Ephross, 1994). A report by this organization stated that the center recorded 1018 hate groups operating in 2011.

There is a dispute about the significance of hate talk. While some individuals who represent antihate groups believe people have a right to think and say what they want, others recognize hate talk as a step toward acting out hate crimes. The California Association for Human Relations Organizations (CAHRO) asserts that hate groups, such as the Ku Klux Klan (KKK) and White Aryan Resistance, clearly preach violence against racial, religious, sexual, and other minorities in the United States (Kaplan, 2010).

In the past, hate groups traditionally recruited members and spread extremist messages by word of mouth or through the distribution of flyers and pamphlets. However, the popularity of the Internet in the mid-1990s brought new international exposure to many organizations, including groups with hate-group beliefs, for example, white supremacy, neo-Nazi, homophobia, Holocaust denial, and Islamophobia. Websites dedicated to attacking their perceived enemies were founded by several white supremacist groups. In 1996, Internet access providers were asked to adopt a code of ethics that would prevent extremists from publishing their ideas online, and a Consultative Commission on Racism and Xenophobia (CRAX) had the task of investigating and stamping out the current wave of racism on the Internet (Bradbury and Williams, 2006; Genn and Lerman, 2010).

Since the web allows members all over the world to engage in conversations, the Internet has been a boon for hate groups in terms of promotion, recruitment, and expanding their base to include audiences of all ages. Even Facebook has hate page/group creators who choose their target, set up a site, and then recruit members. Findings indicate that Facebook fans build on historical stereotypes and cultural narratives to frame celebrities (e.g., the Obama family) negatively. Other groups focus on attacking the president's politics.

In a 2009 iReport, the Simon Wiesenthal Center (SWC) (Lennings et al., 2010) identified more than 10,000 problematic hate and terrorist websites and other Internet postings. The report includes hate websites, social networks, blogs, newsgroups, YouTube, and other video sites. The findings illustrate that as the Internet continues to grow, extremists find new ways to seek validation for their hateful agendas and to recruit members. As a result, the face of contemporary hate continues to transform.

The Southern Poverty Law Center has designated several Christian groups as hate groups including American Family Association, Family Research Council, Abiding Truth Ministries, American Vision, Chalcedon Foundation, Dove World Outreach Center, and Traditional Values Coalition. In addition, many hate organizations are influenced by religious belief, such as Nation of Islam (NOI) (Moore, 2009). The white supremacist Creativity Movement (Wong et al., 2015), formerly known as the World Church of the Creator (Michael, 2007), is associated with violence and bigotry. Aryan Nations (Kaplan, 2003) is another religion-based white supremacist hate group. The Westboro Baptist Church (Baker, 2015) is considered a hate group for its provocative stance against homosexuality and America.

According to a 2003 FBI Law Enforcement bulletin (Foremski, 2010), a hate group, if unimpeded, passes through seven successive stages. In the first four stages, hate groups vocalize their beliefs and in the last three stages, they act on their beliefs. The report points to a transition period that exists between verbal violence and acting that violence out, separating hardcore haters from rhetorical haters. Thus, hate speech is seen as a prerequisite of hate crimes and as a condition of their possibility.

Individuals who belong to hate groups, like those who are members of gangs, find solace in belonging to a group whose members think like they do and show compassion for their fellow members. They learned to hate and will likely teach their children to hate as well. The question becomes: Why do people hate other people and groups and what does their hate provide in living their lives?

Hate groups, like any formal human groups with a particular purpose, are arranged into a dominance hierarchy. Dominants pass the rules to their underlings, and the underlings perform their duties. This is also true for gangs.

WHO ARE HATED?

It appears that many people who hate are intolerant of others who are not like themselves. Yet, there are specific groups that are hated most.

According to W.B. Rubenstein (2004), the federal government has collected data on hate crimes reported throughout the United States since 1990. "To date, the conventional account of that data has simply been to report that racial hate crimes are the most frequently reported type, followed by religious hate crimes and sexual orientation hate crimes." While this conventional story is not technically wrong, Rubenstein argues in the article that it is not the real story the data tell.

Undertaking the first comprehensive empirical analysis of this data, this article develops a new account of hate crimes in the United States. First, the article pierces the neutral categories (race, religion, and sexual orientation) to demonstrate that three subgroups —blacks, Jewish people, and gay people—report, by far, the most hate crimes. Second, [he] adjusts the raw data to account for the differing population sizes of targeted groups: per capita, gay people report the greatest number of hate crimes, followed by Jewish people and blacks, these three groups reporting hate crimes at greater per capita rates than all other groups. Third, gay people are especially likely to report personal, as opposed to property-based, hate crimes.

A final section of the article presents the first scholarly analysis of the staggering growth of anti-Islamic and anti-Arab hate crimes after September 11, 2001. The methodology of this article enables a per capita perspective on this increase, showing that Muslims and Arabs reported hate crimes in 2001 at rates even greater than those at which gay people, Jewish people, and blacks have reported hate crimes over the past half-decade. While this post-9/11 spike leveled off in 2002, Muslims and Arabs are still reporting hate crimes at very high rates.

"As Congress intended hate crimes data to assist in designing public policy initiatives, the article concludes by calling on Congress to respond to what the data actually demonstrate."

GANGS

We see signs of...brutality and oppression all around us, from the streets of Los Angeles and Brooklyn to the hills of Bosnia and the forests of Rwanda. Rather than resolving the problems of intergroup hostility, we merely appear to stumble from viciousness to viciousness. Why? *J. Sidanmus and F. Pratto, Social Dominance, An Intergroup Theory of Social Hierarchy and Oppression*

Gangs are social groups of individuals who come together as an organization with a hierarchical base, dominated by an alpha individual who directs other members of the gang (Decker et al., 2013; Joe, 1997). They live in and protect a territory and individually or collectively engage in some form of violent or illegal behavior.

Street gangs are self-formed associations of peers, united by common interests. They are involved in many types of street crimes, including extortion, smuggling, robbery, kidnapping, arm and drug trafficking, and murder. While psychopathic individuals generally work alone when expressing criminal and violent behavior, gangs generally function as cooperative groups to accomplish the same tasks.

The formation of gangs is not just a modern phenomenon, and they are not just found locally. While the United States has numerous contemporary gangs, many are found internationally. Gangs that were responsible for numerous deaths in India, for instance, were known to exist between the years of 1740 and 1840. Gangs in Victorian London were well known for picking pockets.

Other well-known international gangs have been: the Bowery Boys, Chasers, American Old West outlaw gangs, Dead Rabbits, Adam the Leper's gang, The Order of Assassins, Indian Thugs, Penny Mobs, Catford Massif, Snakehead, Chinese Triads, Italian Mafia, Jewish Mafia, and Pancho Villa's Villistas. The best-known criminal gang in the United States has probably been Cosa Nostra (the Mafia).

The first street gang in the United States, known as the 40 Thieves (Gray, 2015), commenced their activities in New York City in the late 1820s. More than 200 gang wars occurred in New York City in 1850. Chicago is known to have had over a thousand gangs in the 1920s. That was an interesting time in the United States, a time which included prohibition.

Within recent years, gangs have become much larger and more abundant (Decker et al., 2013), especially in large cities and certain countries. There were about 785,000 active street gang members in the United States in 2006 (Cartwright et al., 1970). According to the 2012 *Gang Book* (Decker et al., 2013), Chicago currently has 150,000 gang members. Los Angeles county alone has 120,000 gang members. These are interesting figures, especially since many reports are making statements about dwindling signs of criminal activity.

Across the United States in 2007, there were about 30,000 gangs and 800,000 active gang members. By 2009, there were about 900,000 gang members, 147,000 of which were in prisons or jails.

Internationally, Mexican drug cartels have as many as 100,000 members (Decker et al., 2013). Over 1800 gangs are operating in the United Kingdom. There are at least 102,400 known members in Japan. Hong Kong's Triad had about 300,000 members in the 1950s.

The most obvious question is: Why would anyone join a gang, an activity that completely contradicts the moral codes humans are supposed to follow? Authorities say that most gang-related troubles begin in unstable neighborhoods where there are broken homes, violent role models, and access to drugs (Komro et al., 1999). As many people appear to understand, we are not spending enough time with our young people and allowing them to experience a stable family life. Youths living with troubled adults, single-parent families, youths with low academic achievement scores, youths with delinquent friends, and youths using marijuana are considered more likely to join gangs.

These seem to be some of the same reasons that psychopathic killers develop into the persons they become, which include abuse and inappropriate surroundings. With proper treatment in their formative years and a likelihood of a more promising future for young people, some authorities claim that gangs would most likely have less members.

The most common reason given for joining a gang appears to be unemployment. Other reasons that were mentioned earlier have to do with social disorganization in the community and disintegration of social institutions, for example, family, school, and personal safety.

In addition, there are definite motives for gang members. Lacking any hope for a prosperous future, they see a way of obtaining power, respect, money, and protection in a gang. Dominance hierarchies develop and aggression rises. Many gang members feel ostracized from the community they belong to and feel that joining a gang is the only way to gain status and success.

According to a Heritage Foundation article called *North American Youth Gangs: Patterns and Remedies* (2005), Stephen Johnson states that "youth gangs have flourished wherever there have been population shifts and unstable neighborhoods." Recent growth in the number of gangs has resulted in global affiliations. Gangs that flourished in Los Angeles during the 1960s, for instance, now have fraternal links to thousands of members in Mexico and Central America, as well as connections to gangs that are expanding across the United States.

California-based gangs are particularly strong. The 18th Street or Calle 18 gang "coalesced among Mexican migrants in Los Angeles in the 1980s." The Mara Salvatrucha 13 or MS-13 gang formed in the same neighborhoods in the 1980s but initially consisted of Salvadoran migrants (Cruz, 2010).

It appears that struggling economies and weak justice systems are the primary reasons for the increasing sizes of these gangs. As an example, nearly a million young Mexican adults have joined the Mexican labor force each year to find only 200,000 available jobs.

Johnson says that "transnational gangs will continue to flourish as a by-product of larger social and economic processes, e.g., the growth of transient, unstable neighborhoods and the expanding numbers of undocumented migrants who live on the margins of society." Like other forms of crime, he says, "youth gang activities can be reduced but never eliminated."

Could this be a problem we must deal with globally, a problem that is directly associated with the number of people in our populations? Could it have anything to do with the cage-of-rats example mentioned in Chapter 14?

TERRORISTS

Recall the previous paragraphs that gangs are social groups of individuals who come together as organizations with a hierarchical base, dominated by alpha individuals who direct other members of the gangs. While terroristic groups represent a giant step beyond simple gangs, the basic concept is similar, and terrorist groups generally incorporate some of the principles of both gangs and hate groups (Canham, 2010; Ganor, 2002).

In its simplest definition, terrorism is the use of violent acts to frighten people in an area as a way of trying to achieve a political goal (Kossowska et al., 2010). A more precise definition of terrorism is difficult to form because it is politically loaded and emotionally charged. According to an *Encyclopædia Britannica* definition and Charles Ruby (*The definition of Terrorism*) (Ganor, 2002), terrorism is "the systematic use of violence (terror) as a means of coercion for political purposes." It includes violent acts that are intended to create fear (terror), and they are perpetrated for religious, political, or ideological goals. In performing terroristic acts, the terrorist faction deliberately targets or disregards the safety of noncombatants (civilians).

A terrorist is a radical individual (person who has radical ideas or opinions) who employs terror as a political weapon, usually organizes his or her activities with other terrorists in small cells, and often uses religion as a cover for terrorist activities. A radical cell or terrorist cell is a cell made up of terrorists (often three to five members) who often do not know each other or the identity of their leadership in order to insure operational security.

A cyber-terrorist, cyberpunk, or hacker is a programmer who breaks into computer systems in order to steal or change or destroy information as a form of cyberterrorism. A sleeper is a spy or saboteur or terrorist planted in an enemy country who lives there as a law-abiding citizen until they are activated by a prearranged signal. A suicide bomber is a terrorist who blows himself up in order to kill or injure other people (Brym and Araj, 2012).

This chapter has reviewed both "normal" and alternate forms of behavior in humans, showing that because of their cognitive abilities, humans are extremely complex animals who sometimes have genetic and behavioral flaws. Many individuals who display behavioral flaws may have an ASPD while others take a more subordinate stance and are influenced by antisocial leaders for one reason or another. With tremendous diversity in personality types, along with variations in dominance and aggressive behavior, antisocial individuals and their followers demonstrate behavior that leads to less harmony in the human population. Harmony is further unstable due to other factors, for example, chemical alterations that occur in individuals.

Note

Superscript numerals appearing in this chapter refer to additional text/explanation given in the appendix.

10

The Chemical, Physical, and Genetic Nature of Dominance

The determination of the sequence of the human genome and the assignment of function to these genes is having a dramatic effect on our understanding of the role of genetics in human behavior. Similarly, developments in imaging techniques, allowing changes in neuronal activity to be correlated with thought processes, are affecting our thinking about relationships between the functioning of the mind and chemical activity in the brain. *Cashmore, A., 2010.* The Lucretian Swerve: The Biological Basis of Human Behavior and the Criminal Justice System. Proceedings of the National Academy of Sciences.

It was suggested in Chapter 2 that morality within a society involves a suite of interrelated behaviors that cultivate and regulate complex interactions to provide survival and/or reproductive benefits. Each human culture has some sort of a moral code that it (as a population) more-or-less follows, in which various forms of behavior can be labeled as right or wrong.

In Chapter 9, it was mentioned that while members of a society generally follow a moral code, certain individuals behave in opposition to such codes. Whether individuals within a society choose to follow the moral code and practice tolerance and understanding toward their fellow humans or partners depends on the presence or absence of a poorly understood homeostatic condition within the mind and body of each member of the society. Disruption of homeostatic mechanisms may be rooted in chemical or behavioral trauma, as a result of both genetic and learned input.

What is understood is that mental and behavioral homeostasis, under "normal" conditions, is dependent upon a proper genetic make-up and proper learning environment, especially during the formative and adolescent years (Christie and Viner, 2005; Dorn and Biro, 2011). We also know that the human body, like that of all organisms, consists of chemicals that are important in the way they function. A homeostatic chemical environment within the body of each individual generally stems from processes that relate to a "normal" genetic predisposition and good nutritional habits, as well as a lack of developmental aberrations and brain trauma.

Life is all about chemistry (Holmes, 1987; Voet and Voet, 2011). We and everything in the universe are made of chemicals, organic and inorganic ones, small and extremely large (macromolecular) ones. Abiotic components (e.g., water, phosphorus, nitrogen, and carbon) often pass through environmental cycles in which chemicals are repeatedly built up and destroyed. The bodies of all organisms, including humans, are dependent upon their chemical make-up which must undergo internal cycles as well, involving various metabolic processes (both anabolic and catabolic). A homeostatic internal environment is dependent upon the chemicals which enter and depart from it.[1]

Our bodies rely on the intake of specific elements and the anabolic processes utilized to construct important biological molecules from the four basic chemical groups: carbohydrates, lipids, proteins, and nucleic acids. Just as proteins are the workhorses of chemical reactions in our bodies and DNA is the blueprint for making them, certain chemicals are particularly important in causing the body and mind to function in certain ways. Collectively, they may work as a system to keep the internal environment homeostatic as long as their quantities are properly adjusted for the jobs that must be accomplished and the physiological mechanisms are operating in a regular manner (Gutteridge and Thornton, 2005; Kent, 2009).

However, if either of these mechanisms is out of kilt, especially those that operate in our brain, we are sometimes faced with a situation that results in unusual behavioral modifications. The degree of dominance and aggression associated with these modifications depends on what chemicals are involved (Clarke and Faulkes, 1998).

Dominance and Aggression in Humans and Other Animals
http://dx.doi.org/10.1016/B978-0-12-805372-0.00010-9

While there have been a significant number of reports on the effects of certain chemicals within the body that influence dominance and aggressive behavior, the extraordinary number of variables that are inherent in subjects who express these forms of behavior have provided confusing results. Two of the problems in identifying the nature of chemicals in the brain are that different chemicals may overlap in their functions, and certain chemicals may have multiple functions. Genetic influence is also not well understood, although recent investigations and reports have shed promising light on the subject. Excellent summaries of what is known and the possible mechanisms behind chemical imbalances may be found in the reports of A. Mazur et al. (Mazur, 2005, 2015; Mazur and Booth, 1998).

RELATIONSHIP BETWEEN BRAIN STRUCTURE AND FUNCTION

In an earlier chapter, aggression was defined as a behavior intended to inflict physical injury and/or psychological pain on another individual. Not all authorities agree with this definition, because personal and societal defense is often initially expressed as warning behaviors that allow a species to avoid dangerous altercations (Hermann et al., 2017). In such cases, there would be no intention to inflict physical injury or psychological pain. Examples are in the warning behaviors of social wasps, the showing of teeth by certain mammals, the spreading of wings in birds and some insects and aposematic coloration in species that otherwise may have a means of injuring an aggressor. Humans often use similar tactics to avoid confrontation (e.g., military threats from an already powerful country, demonstrations of smart bombs or other futuristic weapons, and the wearing of red or otherwise aposematic uniforms, along with an arrogant or dominant attitude).

In spite of mechanisms for avoiding conflict, certain forms of aggression can lead to serious problems in both the aggressor and recipient of an aggressive act if a homeostatic condition is not maintained. To understand these concepts, researchers are investigating aggression from a variety of directions, including medical, psychological, chemical, and biological perspectives.

ANATOMICAL VARIATIONS IN THE BRAIN

Single photon-emission computerized tomography (SPECT) has been used to identify regions of the brain that are associated with aggression (Weir and Green, 1994). Color variations that appear in the brain reveal abnormalities and regions that have been damaged. Brain activity that is associated with aggression is generally found in three particular areas:

- Aggressive thoughts appear to reside primarily in the left temporal lobe.
- Repeated thoughts and the degree of attention given to a particular thought process appear to reside primarily in the anterior cingulated gyrus.
- Impulse is controlled mostly by the prefrontal cortex.

Regions of the brain, which are particularly associated with aggression, are the hypothalamus and the midbrain's periaqueductal gray matter. These areas are connected to the brain stem that controls responses and with structures like the amygdala and prefrontal cortex (Fig. 5.3). It has been suggested that the amygdala controls aggression, although its role in primates is difficult to ascertain (Antai-otong, 2009).[2]

GENETICS AND AGGRESSION

As pointed out a number of times in previous chapters and by Caitlin Jones in *Genetic and Environmental Influences on Criminal Behavior* (2005), both genes and environment play a role in the development of personality in an individual. Having a genetic predisposition for criminal behavior, for instance, may not totally determine the actions of an individual, but if they are exposed to the right (or wrong) environment, their chances are greater for engaging in criminal or antisocial behavior.

Also stated earlier, environmental changes of most importance in a developing personality occur primarily during the early developmental years and adolescence, although postadolescence experience generally is important as well. Children who are genetically predisposed to violence and suffer childhood abuse are generally more prone to developing personality disorders (Lewis and Christopher, 1989). Personality disorders in young people often lead to more serious disorders as they grow older. Changes in adolescent individuals, along with body and brain development, may further promote behavioral changes that may influence the development of an antisocial personality (Johnson et al., 2000). Traumatic experiences, for example, the death of relatives or friends, failures in marriage and friendships, or a combination of all of these, may affect who a person becomes (Brewin et al., 2000). These statements are supported by studies discussed in the following paragraphs.

Mutations

Since genes are passed down to offspring from both male and female parents, parental genes generally affect their offspring in one way or another; in addition, genes of grandparents and even more distant predecessors may affect them as well. While genes may be mostly associated with routine bodily functions, good health and "normal" brain function, offspring may receive a combination of genes that may result in normal development or cause malfunctions in anatomy, behavior, and chemical reactions (Charlesworth et al., 1993). Malfunctions can also result from mutations, as well as chemical and physical trauma during the embryological stage.

Mutations are changes in a genomic sequence in the DNA molecule. They are caused by such things as radiation, viruses, transposons (DNA sequences that have been transposed) and mutagenic chemicals (especially during the embryological stage), as well as errors that occur during meiosis (formation of spermatozoa and ova) and DNA replication (Kondrashov, 1988). They can also be induced by the organism itself through cellular processes, for example, hypermutation (an immune response to foreign elements) (Odegard and Schatz, 2006; Oprea, 1999; Dunn-Walters et al., 1998).[3]

There are literally thousands of genetic disorders in humans. While some are the result of mutations in individuals who have the disorder, most are inherited from family members. Some are common whereas many are rare. Some types of genetic disorders are as follows:

- *Autosomal Dominant Genetic Disorders*: Caused by inheriting a defective gene from a single parent (Griffiths et al., 2012). This defective gene belongs to an autosome. An autosome is a chromosome that is not an allosome (i.e., not a sex chromosome). This condition is also known as autosomal dominant pattern of inheritance.
- *Autosomal Recessive Genetic Disorders*: Disorders that manifest themselves only when an individual has two defective alleles of the same gene, one from each parent (Barroso et al., 1999). These disorders are inherited via the autosomal recessive pattern of inheritance.
- *Sex-Linked Disorders*: These are disorders related to sex chromosomes (X and Y) or genes in them.
- *Multifactorial Genetic Disorders*: The result of genetic and environmental factors (Bodmer and Bonilla, 2008).

Such defects can be caused by the following mechanisms:

- *Mutations*: Sudden inheritable changes in the nucleotide sequence of a gene.
- *Aneuploidy*: A condition that is caused when there are abnormal number of chromosomes in an organism, due to loss of a chromosome (monosomy) or the presence of an extra copy of a chromosome (e.g., trisomy and tetrasomy).
- *Deletions*: Loss of a part of chromosome, as in the case of Jacobsen syndrome.
- *Duplications*: Duplication of a portion of a chromosome that results in extra genetic material.
- *Inversions*: A condition in which the nucleotide sequence is altered because a portion of a chromosome has broken off, got inverted, and became reattached at the original location of the chromosome.
- *Translocations*: A condition in which there is an interchange between chromosome segments. In some cases, a portion of a chromosome may simply get attached to another chromosome.

Monoamine Oxidase

Faulty genes appear to be associated with some expressions of aggression, for example, recessive genes on the X chromosome that are referred to as MAOA and MAOB (Cases et al., 1995). They both produce enzymes called monoamine oxidase collectively. Monoamine oxidase is an enzyme that has been shown to be associated with antisocial behavior. Norepinephrine, serotonin, and dopamine are metabolized by both MAOA and MAOB. Among their catalytic functions, these particular enzymes are vital to the inactivation of monoaminergic neurotransmitters.

Nitric Oxide

In mice, a gene in males that influences the production of a neurotransmitter called nitric oxide causes them to be extremely aggressive physically and sexually (Beckman and Koppenol, 1996). Normally aggressive females with brains that are not producing nitric oxide are not aggressive. However, there are other considerations. It is believed that the presence of nitric oxide alone is not the sole cause of aggressive behavior. Also, rats lacking known gene mutations show differences in the degrees of aggression between males and females. While genetic ties to aggression are not well understood, many of the chemicals in forthcoming sections of this chapter are produced in structures that nevertheless have a genetic base.

Stimulation of the Amygdala and Hippocampus

Stimulation of the amygdala in hamsters has resulted in aggressive behavior. On the other hand, lesions of homologous structures in lizards have been found to reduce aggression. Being associated with aggression, dominance behavior may be affected as well. Rhesus monkeys that had neonatal lesions in the amygdala or hippocampus were found to demonstrate a reduced expression of social dominance (Bauman et al., 2006).

Chemical Influence and Stress

Under chemical influence, nonhuman primates appear to establish and maintain their status in hierarchies through a series of short face-to-face (agonistic) competitions between members of the group (Miller, 1965). An exchange of threats or attacks is seen as an attempt by each animal to out-stress or intimidate the other by inducing fear, anxiety, or other discomfort. Thus, the chemistry of each organism affects changes in the chemistry of the competitor. Stress is experienced as both a feeling of discomfort and a syndrome of neurological responses. The animal that out-stresses its adversary is the winner.

Behavioral Influence and Stress

Certain forms of behavior also lead to changes in stress (Bell, 1977). Staring, which is a stress-inducing behavior, is a dominant sign associated with high status. Eye aversion is a deferential sign associated with low status. In other words, a dominant act (staring) elicits stress in a recipient; a submissive act (eye aversion) relieves stress in an actor. When stress is too great for one, a subject generally switches from dominant to deferential actions, thereby relieving stress and simultaneously signaling an acceptance of lower rank.[4]

CHEMISTRY AND AGGRESSION

According to the FBI's Uniform Crime Reporting Program, violent crimes rose 2.3% in 2005 and 1.9% in 2006, which is reported as the first steady increase since 1993. Why there was an increase could relate to stress and other factors that deal with modern human life. Stress is often associated with chemical imbalances.

Most biochemical processes in the human body strive to maintain equilibrium (homeostasis), a steady, balanced state that influences cell, tissue, and chemical reactions. Both internal and external stimuli continually disrupt homeostasis, causing alterations in the organism's optimal condition for living. Factors causing alterations in homeostasis can be interpreted as stress (Antonovsky, 1979). Also, a life-threatening situation, such as physical trauma or prolonged starvation, can greatly disrupt homeostasis. An organism's attempt at restoring conditions back to or near homeostasis can also be interpreted as stress.

Inadequate and Improper Body Chemicals

According to Mark Sircus in *Transdermal Magnesium Therapy* (2007), it was Linus Pauling who first highlighted the supremacy of nutrition in correcting abnormalities in the chemical environment of the brain. Thus, there is truth in the statement, "we are what we eat."

Nutrients such as ascorbic acid, thiamine, niacinamide (vitamin B), pyridoxine, vitamin B_{12}, folic acid, magnesium, glutamic acid, and tryptophane were presented by Dr. Pauling as "intimately linked to brain function and mental illness." Sircus further states that "deficiencies in certain necessary nutrients lead to psychotic symptoms and depression while supplementation of other nutrients help attenuate and improve the symptoms of mental illness."

In the following paragraphs, chemical deficiencies and their outcomes are discussed in both children and adults. The reasons for deficient amounts of required chemicals is often unknown, but the information points out their importance to the maintenance of behavioral homeostasis, especially in individuals who are not eating properly. Genetic differences may also affect the amount and type of chemicals required in the diets of certain individuals. Studies show that understanding the relationship between chemicals and behavior is still in its infancy.

Behavioral variations may come about as a result of chemicals in the foods we eat and in toxic chemicals we put into our bodies. A study by R E Kleinman et al. (*Hunger in Children in the United States: Potential Behavioral and Emotional Correlates*) (1998) points out that an estimated four million American children "experience prolonged periodic food insufficiency and hunger each year, representing 8% of the children under the age of 12 in this country." The same studies show that "an additional 10 million children are at risk of hunger."

That study examined the relationship between hunger as defined by food insufficiency attributable to constrained resources and variables reflecting the psychosocial functioning of low-income, school-aged children. It showed that "children from families that report multiple experiences of food insufficiency and hunger are more likely to show behavioral, emotional, and academic problems on a standardized measure of psychosocial dysfunction than children from the same low-income communities whose families do not report experiences of hunger."

Elements and Simple Chemicals in Foods and Their Influence on Behavior

Vitamins: Vitamins are organic or similar compounds that are required as vital nutrients in small amounts by an organism. They cannot be synthesized in sufficient quantities by an organism and thus must be obtained as dietary supplements.

In a comprehensive review on nutrition and aggressiveness, M.R.Werbach (*Nutritional Influences on Aggressive Behavior*) (Werbach, 1995) points out that "deficiencies of several vitamins are known to be associated with irritability. These include niacin, pantothenic acid, thiamine, vitamin B_6, and vitamin C." In industrialized societies, classic vitamin deficiency diseases are rare, although marginal vitamin nutriture due either to inadequate intake or to vitamin dependency appears to be fairly common. Moreover, under laboratory conditions, adverse behavioral changes precede specific clinical findings in a number of vitamin deficiencies.

"It is not known how frequently overaggressive behaviors are a manifestation of marginal vitamin nutriture," he says. While little has been published to prove a relationship between the aggressive behavioral syndrome in humans and marginal vitamin nutriture, Lonsdale and Shamberger reported on 20 people eating "junk food" diets who were found to have biochemical evidence of marginal thiamine deficiency. Their subjects, particularly the adolescents, were impulsive, highly irritable, aggressive, and sensitive to criticism:

Following thiamine supplementation, their behavior improved concurrent with laboratory evidence of improved thiamine nutriture, suggesting that marginal thiamine deficiency may have contributed to their aggressive behavioral syndrome.

Werbach states that "the acquisition of chemicals known as elements and trace elements is sometimes important in both the presence and absence of aggressive behavior. Elements are forms of chemicals in their most basic form which are required in food to maintain body homeostasis. Trace elements are elements which are required in minute amounts."

Beyond the elements mentioned in the following paragraphs, the amounts of more complex chemicals may often be associated with behavioral problems. At times, insufficient amounts of these chemicals are at the root of aggression, while an overabundance is responsible in other cases. They will be discussed in due process.

Iron: Iron is the most common nutritional deficiency in industrialized societies (Hider and Kong, 2013); 10% of American males and 3% of American females are overtly iron deficient. A deficiency of iron is known to interfere with proper brain function. Dopamine, a chemical in which iron is important for its proper function, is a major neurotransmitter in the brain. Animal studies have shown that iron deficiency may cause learning deficits and consequent behavioral impairment by diminishing dopamine neurotransmission. Iron is also needed as a cofactor for enzymes that metabolize not only dopamine, but also serotonin and norepinephrine, which also have a potent influence on behavior. These chemicals are discussed further in the following paragraphs:

Evidence is now emerging that iron deficiency may be an important contributor to the aggressive behavioral syndrome. Among adolescent males, iron deficiency has been shown to be directly associated with aggressive behavior (a conduct disorder). Moreover, in a population of incarcerated adolescents, the prevalence of iron deficiency was nearly twice that found in their non-incarcerated peers.

Lithium: There is considerable evidence that pharmacologic doses of lithium, which has no known essential function, can reduce abnormal aggressive behaviors including self-mutilation. Lithium has been used successfully with treatment-resistant hospitalized children with diagnoses of a certain form of aggressive conduct disorder, as well as with brain-injured and mentally retarded patients with aggressive, combative or self-destructive behavior.

Werbach (1995) states that While often effective, lithium at pharmacologic doses (generally 900,000 µg or more daily) has serious limitations. It suffers from many potential side effects, some of which are common. Because it becomes toxic at a serum level which is not much higher than the therapeutic range, serum lithium levels must be tested periodically. For these reasons, patients must be under medical supervision so long as they are taking the drug.

It is possible that lithium may exert a powerful effect on overaggressive behaviors at far lower levels of intake. Using data from 27 Texas counties, Schrauzer and Shrestha found that the incidences of suicide, homicide, and rape were significantly higher in counties whose drinking water supplies contained little or no lithium than in counties with higher water lithium levels, even after correcting for population density.

Corresponding associations with the incidences of robbery, burglary, and theft were also significant, as were associations with the incidences of arrests for possession of opium, cocaine and their derivatives. Only the incidences of arrests for possession of marijuana, driving under the influence of alcohol and drunkenness failed to correlate with the water lithium level.

While the effect of low-dose lithium supplementation on overaggressive behaviors has not been reported, results of an uncontrolled study suggest that low-dose lithium derived from vegetable concentrates may have a powerful effect on mental state and behavior. Thirteen depressed patients with bipolar disorder were treated with natural lithium derived from vegetable concentrates. All improved in about 10 days and there were no adverse effects. After 6 weeks, they were taken off of lithium and all regressed to their former depressed state within 3 days. Two days after lithium was resupplied, their depressions lifted again:

> If we assume that a person consumes about one liter of water daily from municipal supplies, it is striking that the level of lithium provided from the vegetable concentrates approximates that consumed by residents of the Texas counties with higher lithium levels: "Natural" lithium dosage was 150 µg daily; lithium level of drinking water in the Texas counties with higher levels was 70–170 µg per liter.

It has been suggested by certain individuals (e.g., Mark Sircus in *Transdermal Magnesium Therapy*) (2007) that while drugs (e.g., lithium) prescribed by physicians may appear to solve health and mental problems, it does not get to the root of the problems. Rather, such problems may originate from other deficient chemicals that are required for proper health. It has been suggested that supplying such chemicals often may adjust behavior back toward "normality" so that mind-altering drugs are not required.

Magnesium: Studies on rodents suggest that magnesium has a complex relationship with aggressive behavior (Sircus, 2007). Magnesium deficiency reduces offensive aggressive behavior but increases defensive aggressive behavior. Lower levels of magnesium supplementation increase the number of attacks on intruders while higher levels have the opposite effect:

> In humans, magnesium deficiency, which enhances catecholamine secretion and sensitivity to stress, may promote aggressive behavior. Increased catecholamines, in turn, induce intracellular magnesium losses and, eventually, increase urinary losses of magnesium. It has been suggested that the Type A behavior pattern which is associated with chronic stress and aggressive behavior may both cause and be caused by magnesium deficiency. Also, suicide attempts, which are violently aggressive acts against the self, have been correlated with lowered magnesium levels in cerebrospinal fluid.

Sircus states that "magnesium deficiency can result in disruptive behavior. Even a mild deficiency can cause sensitiveness to noise, nervousness, irritability, mental depression, confusion, twitching, trembling, apprehension, and insomnia." Sircus further states that "magnesium deficiency or imbalance plays a crucial role in the symptoms of mood disorders. Observational and experimental studies have shown an association between low magnesium and aggression, anxiety, attention deficit hyperactivity disorder (ADHD), bipolar disorder, depression, and schizophrenia."

Manganese: Massive overexposure to manganese (a trace element) produces a condition called manganese madness, "which may initially be marked by violence, criminal acts and a state of mental excitement (Masters et al., 1998; Raine, 2002); later, neurological impairment slowly develops, with signs and symptoms that resemble Parkinson's disease."

The behavioral effects of marginal levels of manganese toxicity have not been described. Recently, researchers have found elevated hair manganese in a population of violent male offenders, suggesting that marginal manganese toxicity may be associated with violent criminal behavior. Compared to hair manganese levels which they found, people exposed to levels of manganese pollution, which are known to be toxic possess hair values that are two to six times higher.

Elevated hair manganese levels have also been reported in hyperkinetic children and men with a history of childhood hyperactivity or those who have an increased rate of antisocial and drug-use disorders. In rats, chronic manganese exposure initially produces hyperactivity with an increased tendency to fight. While any hypothesis concerning the behavioral effects of marginal manganese toxicity in humans is highly speculative, these findings suggest that marginal manganese toxicity may promote overaggressive behaviors in adults.

Toxic metals: "Brain damage due to toxic metal exposure may promote aggressive, antisocial, and violent behaviors. Lead exposure," for instance, "is known to cause learning and behavioral problems, problems which are found in a substantial portion of juvenile delinquents" (Pounds et al., 1991).

The strongest evidence to date that lead exposure increases the frequency of aggressive behaviors comes from the Edinburgh Lead Study which included over 500 children between the ages of 6 and 9 years. After taking 30 possible confounding variables into account, the investigators still found a significant relationship between the log of blood lead levels and teachers' ratings of the childrens' behavior on an "aggressive/antisocial" scale and on a "hyperactive" scale, but not on a "neurotic" scale. As in other studies on the relationship between lead exposure and brain damage, a dose–response relationship was found between blood lead and behavior ratings, with no evidence of a threshold:

> In a study of the relationship of lead exposure and violent behavior in adults, hair lead levels from 19 violent criminals were found to be elevated as compared with those of 10 nonviolent criminals. This study was repeated 8 years later by the same research team with essentially the same results. However, their results were contradicted by those of the recent Gottschalk study on hair manganese levels; in that study, no significant differences were found between hair lead levels of 104 violent criminals, prison guards and local townspeople.

Cadmium: Studies comparing hair cadmium levels of violent male offenders to matched controls have had conflicting results (Stacey et al., 1980). "One study published in the *Journal of Learning Disabilities* looked at hair cadmium levels of 40 apparently normal young men entering US. Navy recruit training and found a highly significant relationship between hair cadmium levels and the number of demerits each recruit had received. Moreover, the three recruits who had the highest cadmium levels all displayed serious behavior difficulties during training" (Werbach, 1992).

Aluminum: Exposure to aluminum may also contribute to overaggressive behaviors (Zatta and Alfrey, 1998). Hair aluminum levels of a group of 22 juvenile offenders, as well as of another group of 10 severely delinquent, psychotic or prepsychotic adolescent boys, were elevated. However, both studies compared aluminum levels to laboratory norms rather than to matched controls; thus, other differences between the groups could account for the findings.

Serotonin: Serotonin, which is a major neurotransmitter, "has been found to play an important role in modulating aggressive behavior" (Cases et al., 1995). "Impulsive, violent and suicidal behaviors have repeatedly been shown to be associated with a reduction in serotonergic activity in the central nervous system." Further discussion of serotonin as a modulation chemical will be found later in this chapter.

Tryptophan: Tryptophan, which is an essential amino acid, is the dietary precursor to serotonin (Shea et al., 1990). Several lines of evidence have suggested that the amount of tryptophan in the diet relates closely to aggressive behavior. For example, rats given a diet almost lacking in tryptophan developed aggressive behavior toward mice.

"Tryptophan must compete with other large neutral amino acids to cross the blood–brain barrier; therefore, the ratio of the amount of tryptophan to the amount of competing amino acids (the tryptophan ratio) may provide a rough indication of the availability of tryptophan in the brain for conversion into serotonin." Researchers have calculated the dietary tryptophan ratio for 18 European countries in an attempt to relate it to homicide rates. While initially no correlation was found between low tryptophan ratios and homicide, once social and cultural differences were controlled for, low tryptophan ratios were found to be associated with high homicide rates.

"A more direct method of examining the relationship between the tryptophan ratio and aggression is by measuring the actual ratio in blood plasma. When a group of depressed alcoholics was evaluated in this manner, those with a history of aggression, including suicide attempts, also had the lowest tryptophan ratios." Werbach asks, "If a low ratio of tryptophan to competing amino acids is associated with aggressive behavior, will tryptophan supplementation reduce that behavior?"

Dietary tryptophan was manipulated in social groups of vervet monkeys by providing them with amino acid mixtures that were tryptophan-free, nutritionally balanced or excessively high in tryptophan. These mixtures were shown to have a marked effect on plasma tryptophan levels. During spontaneous activity, the only effect of the different mixtures was increased aggression in males on the tryptophan-free mixture. During competition for food, however, while the tryptophan-free mixture continued to increase male aggression, the high-tryptophan mixture reduced aggression in both males and females. These data suggest that tryptophan supplementation may be most effective in reducing aggression during times of stress.

When hospitalized male schizophrenics were given tryptophan, only those patients with high levels of hostility and a high lifetime frequency of aggressive incidents benefited; these patients showed a lessening of hostility and depression, a reduction in ward incidents and improvement on a standardized psychiatric rating scale. In another study of 20 aggressive patients, 6 g of tryptophan daily for 1 month failed to reduce the number of violent incidents, although it significantly reduced the need for potent medications to control violent or agitated behavior.

The rate of firing of serotonergic neurons in the brain increases as the level of behavioral arousal increases; thus, increased serotonin levels would be more likely to influence brain function at higher levels of arousal. Indeed, this fact probably explains why vervet monkeys only responded to tryptophan supplementation when they were put under competitive stress. It also may explain why altered tryptophan levels failed to affect aggression in a study of normal human males, while overaroused, hostile, and aggressive psychiatric patients responded well:

> In the conversion of tryptophan to serotonin, the intermediate step is its conversion to 5-hydroxytryptophan. Surprisingly, supplementation with 5-hydroxytryptophan may increase aggressive behavior, apparently because while tryptophan appears to enhance the serotonergic system exclusively, 5-hydroxytryptophan also appears to enhance the catecholaminergic system.

Glucose: Glucose is an extremely important chemical of value in the process of cellular respiration. In this chemical process, glucose is broken down to pyruvic acid which subsequently enters mitochondria for further breakdown and the making of ATP (adenosine triphosphate). All reactions in the body utilize energy from ATP. Complications arise when either insufficient amounts or an overabundance of glucose circulates in the blood. Imbalances may be due to liver or pancreas malfunction or to the amount of sugar intake in the food.

"There is early evidence that hypoglycemia (low blood sugar) during glucose tolerance testing is related to hostile, aggressive behavior such as that seen in habitually violent and impulsive criminals" (Benton, 1988). "Scientists have found that a group of habitually violent adult criminals had lower basal glucose levels during glucose tolerance testing than controls. Even in the normal population, there is evidence of a relationship between hypoglycemic tendencies and both frustration and hostility."

Assuming that there is an association between hypoglycemia and the aggressive behavioral syndrome, the question of whether hypoglycemia causes the syndrome remains [unanswered]. One method of investigating the issue of causality is by changing the amount of sugar in the diet and examining the behavioral effects. Since dietary sugar provokes insulin production that may cause a reactive hypoglycemia, a change in behavior following a change in sugar intake would be consistent with the hypothesis that dietary sugar influences that behavior.

In a series of increasingly sophisticated double-blind studies, Schoenthaler (1983) addressed this question by reducing the sugar intake of thousands of incarcerated juvenile offenders in different locations around the United States. Compared to offenders on a placebo diet, he found a significant reduction in various forms of antisocial behavior (such as assaultiveness, fighting, self-injury, and suicide attempts) in offenders restricted to a minimal amount of sugar in their diet—but only for males:

> While Schoenthaler's work suggests that dietary sugar may influence behavior, he did not examine blood sugar levels; it thus fails to address the role of reactive hypoglycemia in the aggressive behavioral syndrome. The finding that only males responded may either be because males are more likely to engage in aggressive behaviors, or because males are more sensitive to nutritional influences on aggression.

Indirect Effects of Chemicals

Many chemicals other than the ones mentioned above do not influence the brain directly, but may, along with certain activities, indirectly affect the release and function of other chemicals that are more directly involved with brain function. Some examples of how chemicals and lifestyle are correlated are as follows:

- Fatty acid imbalances and chemical sensitivities may result from modern food processing.
- High sugar and carbohydrates and a low-fat diet leads to higher amounts of insulin and prostaglandin imbalance.
- Lack of exercise leads to decreased serotonin and dopamine production.
- Lack of sleep, poor nutrition, and deionized air leads to reduced serotonin levels.
- Boring classes, boring jobs, and a lack of activity leads to reduced dopamine and norepinephrine.

Based on data on chemical imbalances, it is obvious that the body may respond to such imbalances in a variety of ways. The chemical nature of the brain and internal systems that are associated with it may influence not only the behavioral repertoire of an individual, for example, the degree of aggressiveness and position within a dominance hierarchy, but also the individual's vulnerability to depression and susceptibility to immune-related illnesses (Hecht, 1967):

> Though it is a complex matrix of causes that cuts across physical, emotional, mental, and spiritual levels of being—it's arguable that a significant portion of the blame for violence and depression can be laid on nutritional causes. **Mark Sircus** *Transdermal Magnesium Therapy*

More Complex Chemicals in Foods and Their Influence on Behavior

Common Foods: According to Werbach (1992), "it appears that overaggressive behaviors can be provoked by a reaction to common foods. Reactions range from irritability to a psychotic aggressive reaction. Children who improved after food eliminations had previously been irritable, fretful, quarrelsome and could not get along with others. Often they had to be taken out of school as they upset the classes and were considered incorrigible. After food eliminations, however, their personalities dramatically changed, and they became happy and social."

"A study reported in [a publication called] the *Lancet* suggests that food sensitivities may be quite common among behaviorally-disturbed children. Eighty-one out of a group of 140 children with behavior disorders (almost two-thirds) experienced significant improvement following the elimination of certain foods along with food additives. When they were challenged with the specific foods which had been eliminated, their behavior problems returned. Moreover, 75% of these children reacted to a double-blind challenge with salicylates but not to placebo." The following report illustrates how food sensitivities may affect aggressive behavior (Werbach, 1995):

- When he was 5 years and 1-month old, G.L. was seen because of uncontrollable temper tantrums. He was believed (to be) aphasic because of poor speech development and was too uncomfortable to do initial IQ testing. The EEG showed 14-per-second spikes, large amounts of sharp activity in the motor leads, temporal single, polyphasic sharp waves and a long run of sharp waves in the right temporal area. Allergy tests revealed strong reactions to milk, chocolate and yeast.
- He was placed on a diet free of milk, chocolate, and cola drinks. Seven and one half months later, his EEG was normal. 6 months after the repeat EEG, he was learning better and his behavior was much improved. He was challenged again with the suspected foods for one week, during which time his behavior again became uncontrollable.
- The EEG now showed two and one half to six per second activity on the right, greater in the mid-temporal and parietal leads, accentuated by drowsiness. Light cerebral dysfunction was diagnosed.

Adults may also display overaggressive behaviors due to food sensitivities. For example, MacKarness has written of a woman who had been hospitalized 13 times for violent behavior and depression; after common foods were eliminated from her diet, she no longer became violent or depressed. Instead she felt fine and obtained a regular job.

While the research literature suggests that any commonly ingested food or food additive may be responsible for provoking pathological, psychological, and other behavioral reactions, milk may be a special case. Schauss et al. (1979) found that chronic juvenile delinquents consumed much more milk than matched controls without a history of delinquency. Male offenders consumed an average of a gallon of milk daily compared to a little less than a quart a day for the controls, and females showed similar differences:

> Schauss and Simonsen believe that overconsumption of milk causes antisocial behavior. He has reported that when several Michigan detention centers reduced their inmates' milk consumption, the incidence of antisocial behavior declined; when they permitted milk consumption to increase again, antisocial behavior also increased.

Werbach (1995) concludes his article by saying that "the literature offers numerous clues, but little scientific verification," which is consistent with the hypothesis that the aggressive behavioral syndrome can be prevented and treated by manipulating nutritional factors. Epidemiological studies have repeatedly found associations between overaggressive behaviors and deficiencies of several essential nutrients: niacin, pantothenic acid, thiamine, vitamin B6, vitamin C, iron, magnesium, and tryptophan. While repletion of frank deficiencies is likely to be beneficial, the benefit of correcting marginal deficiencies remains to be proven.

While not an essential nutrient, lithium (mentioned earlier in this chapter) has been "proven effective in reducing overaggressive behaviors when provided at massive pharmacologic dosages. Moreover, even the relatively tiny daily lithium intake from municipal water supplies has been found to be negatively correlated with measures of the aggressive behavioral syndrome. In an open trial, supplementation with such natural levels of lithium appeared to be effective in treating bipolar depression. These findings suggest that natural lithium supplementation may be effective in the management of the aggressive behavioral syndrome, a hypothesis which remains to be explored experimentally."

Some evidence suggests that "overaggressive behaviors may be promoted by the toxic effects of aluminum, cadmium, and lead." Exposures to these elements (especially cadmium and lead) should be avoided; it is unknown whether treatments designed to chelate these metals in order to remove them from the brain are effective in reducing overaggressive behaviors:

> Reactive hypoglycemia may be more common among people displaying the aggressive behavioral syndrome and, in an open study, reducing sugar consumption was followed by a reduction in antisocial behavior. Whether treating documented reactive hypoglycemia

reduces overaggressive behaviors remains unknown. Finally, sensitivities to foods and food additives appear capable of inducing overaggressive behaviors. Most of the evidence remains anecdotal; however, salicylates have been shown to provoke behavioral disturbances under double-blind conditions.

In spite of the relative paucity of scientific evidence from controlled studies, "clues from case reports, open trials, observational (correlational) studies, and animal studies suggest that attention to nutritional factors may reduce overaggressive behaviors and the devastation resulting from them. Those clues, plus the safety of most nutritional interventions, argue that a nutritional approach should be considered in the treatment of the aggressive behavioral syndrome."

Brain Chemicals

As indicated in the foregoing statements, certain brain chemicals (many of which have been mentioned repeatedly here) have also been found to be very important in the development, expressions and lack of aggression. In fact, they may play a key role in increased levels of such behavioral expressions as school shootings, muggings, murder and road rage.[5]

Serotonin: A report by R. J. Larsen and D. M. Buss called *Psychology: Domains of Knowledge about Human Nature* (Larsen and Buss, 2008), pointed out that serotonin is a neurochemical that plays an important role in certain personality traits, for example, depression, anxiety, and bipolar disorder. Investigations indicate that serotonin is one of the most important central neurotransmitters underlying the modulation of impulsive aggression.

As shown earlier in this chapter, high serotonin levels are associated with a low degree of aggressive behavior. It has been shown that serotonin levels in animals vary with changes in daily activities, decreasing, for instance, as a result of animals experiencing repeated victorious episodes of aggression.

Low serotonin levels are most readily detected in people who engage in impulsive and violent forms of aggressive behavior. Introduction of supplemental serotonin has been found to restore behavior to a normal function, suppressing aggressive behavior.

Thus, serotonin deficiency appears to be related to pathological, violent forms of aggressiveness, not including the aggressive behavior that is important in what we may refer to as normal, everyday survival. However, everyday survival may add stress to our lives and stress results in reduced levels of serotonin.

Strangely, serotonin levels seem to affect dominance status in an opposite fashion (Young and Levton, 2002). High status, as established in forming dominance hierarchies, may be influenced by elevated levels of serotonin. Primates with high social rank are found to have higher serotonin levels than subordinates, and they are less likely to experience impulsive or depressive behavior. Nonhuman primates are known to rise in status when serotonin is administered to them. Thus, high serotonin levels and dominant rank are positively correlated, while low serotonin levels and antisocial behavior are positively related.

Reduced serotonin levels are connected with a variety of other conditions, for example, attention deficit disorder (ADD), irritability, depression, aggression, anxiety, lack of concentration, chronic pain, restlessness, fatigue, nausea, obsessive-compulsive disorder, weight loss, fibromyalgia, arthritis, chronic fatigue syndrome, heat intolerance and other syndromes (Frost et al., 2011). High sugar intake can result in food deprivation, poor concentration, and depression. Insulin affects serotonin levels and many other body systems. Poor nutrition (especially involving lower levels of vitamins B_6, C, and E—the stress vitamins) can cause lower serotonin levels.

It should be obvious at this point that the association between aggression and levels of brain chemicals is real but complex. Further complexity becomes evident when considering the involvement of other chemicals, for example, arginine–vasopressin (AVP), oxytocin, dopamine, norepinephrine, melatonin, insulin, prostaglandins, and testosterone; these, in turn, appear to be strongly influenced by stress, diet, exercise, sunlight, sleep, and other factors that influence an individual's life.

Arginine–vasopressin: Release of AVP in the vicinity of the amygdala appears to regulate maternal aggression toward female intruders, an apparent attempt by the body toward insuring the survival of offspring. In males, however, it has been implicated in certain types of aggression and other social behavior, especially in relation to pairbonding (Caldwell and Young, 2006).

Oxytocin: Oxytocin, a neurotransmitter that is released in the brain during birth and the process of breastfeeding, reduces anxiety and fear. This is believed to enable new mothers to face threat-potential intruders with more defensive aggression. To complicate the issue, AVP and oxytocin have shown signs of being interdependent.

Dopamine/norepinephrine: Dopamine, another brain chemical, may indirectly affect aggressiveness (Yanowitch and Coccaro, 2011). It is associated with pleasure, and according to a paper by F. A. Elliot called *A Neurotological Perspective of Violent Behavior* (Elliott, 1988), "activation of both affective (emotionally driven) and predatory aggression are accomplished by dopamine. Reduced dopamine and norepinephrine manifests as ADD, impulsivity, lack of concentration, restlessness, depression and loss of pleasure."

Melatonin—Melatonin, 5-methoxy-*N*-acetyltryptamine, is a hormone found in many life forms, including humans, at levels that vary in daily cycles. Its connection with biological systems is quite diverse. Many biological effects of melatonin are produced through the activation of melatonin receptors, while others are due to its role as a pervasive and powerful antioxidant, with a particular role in the protection of nuclear and mitochondrial DNA. It has been shown to have connections with circadian rhythm, light dependence, the immune system, dreaming, pediatrics, medical complications, learning, memory, Alzheimer's, delirium, stimulants, headaches, gall stones, amyotrophic lateral sclerosis, radiation, and tinnitus.

In connection with circadian rhythms, melatonin has a regulatory function. Being indoors and thus lacking sunlight lead to reduced melatonin levels. Reduced melatonin may disrupt the sleep cycle and result in seasonal depression (SAD).

Insulin: Insulin is an animal hormone. Its presence informs the body's cells that the animal is well fed, causing liver and muscle cells to take in glucose and store it in the form of glycogen. Insulin also causes fat cells to take in blood lipids and turn them into triglycerides, and it has several other anabolic effects throughout the body. High sugar intake can result in food deprivation, poor concentration, and depression (Wender and Solanto, 1991). Insulin also affects serotonin levels and many other body chemical systems.

Prostaglandins: Prostaglandins are hormones that belong to a group of lipid compounds that are derived enzymatically from fatty acids and have a number of important functions in the animal body. Although they may not directly affect expressions of aggressiveness, they function as mediators and have a variety of strong physiological effects (Matsuoka et al., 2005).

Noradrenaline: Brain levels of the neurotransmitter noradrenaline have been shown to be lower in the hippocampus of dominant mice but higher in the central amygdala of submissive mice. Noradrenaline may help mediate the effects of stress on the body (Otten et al., 1999).

Cortisol: Children generally have been found to have higher cortisol levels in anticipation of entering a challenging or stressful situation than did children who did not anticipate entering one, but those with conduct disorders do not. Low cortisol in the saliva of boys has been correlated with antisocial and aggressive behavior (McBurnett et al., 2000).

Copper and zinc: Aggressive criminals have been found to have abnormal ratios of serum copper and plasma zinc when compared to nonaggressive individuals who are not criminals (Tokdemir et al., 2003).

Testosterone: Testosterone is produced in the testes of males and in the adrenal cortex of both males and females. While it appears to have a higher correlation with aggressive behavior than most chemicals, aggressive tendencies in the absence of high levels of testosterone indicates that testosterone is not the only factor responsible for aggressive behavior (Archer, 1991).

In evaluating the effects of testosterone and dominance in men, A. Mazur and A. Booth (1998) point to an exhaustive number of studies that approach the subject from various directions. With reference to its relationship to aggression and dominance, they make the following statements:

- Hens injected with testosterone become aggressive, each rising in dominance status, some to the alpha position.
- Testosterone affects human males importantly but differently at three stages of life: perinatally (in utero and shortly after birth), during puberty and in adulthood.
- Boys who committed the most violent crimes had slightly but not significantly higher testosterone levels than boys who committed only property crimes.
- Testosterone may provide a social stimulus during pubertal development, which may help to explain sexual and antisocial behavior among adolescent boys (Reynolds et al., 2007). For instance, there appears to be little or no relationship between the level of serum testosterone and problematic behavior among young boys. However, maturation, which is under the influence of testosterone, produces profound social effects on adolescents. Testosterone affects adolescent behavior mostly through indirect social responses rather than through direct activation of target receptors by testosterone in the bloodstream.
- The level of testosterone circulating in the bloodstream may affect dominating or aggressive behavior by activating receptors in organs of the nervous system.
- Prisoners with a prior record of violent and aggressive crimes have significantly higher testosterone levels than those without such a history.
- Testosterone levels are not significantly different between aggressive and dominant groups, but both have significantly higher testosterone levels than groups that are neither aggressive nor dominant.
- Rapists who are most violent in the act have higher testosterone levels.

- Testosterone levels relate to the violence of crimes, as well as peer ratings of toughness in the person committing the crimes.
- Testosterone levels correlate with assertive and dominant behavior.
- Testosterone levels correlate with dominant or aggressive behavior and antisocial norm-breaking.
- Women convicted of unprovoked violence had higher testosterone levels than other prisoners.
- Men with higher levels of testosterone are more likely to be arrested for offenses other than traffic violations, to buy and sell stolen property, incur bad debts, and use a weapon in fights.
- Men who were delinquent as juveniles were more likely to commit crimes as adults if they had higher levels of testosterone.

In *Biosociology of Dominance and Deference*, Mazur (2005) concludes that "while good evidence suggests that testosterone encourages dominant behavior, there is no strong indication that the hormone in either natural or synthetic form produces violence per se. Yet, young men are more likely than any other age-sex category to engage in overt dominance contests. Men are more violent than women even when dominance is not involved. Men more frequently kill themselves. Males, young and old, are persistently more likely than females to hunt, to play violently and to be interested in weapons." Add to this the antisocial nature of some men, as pointed out by R.D.Hare in *Without Conscience* (1993), whether testosterone-induced or not, and we have an explosive assortment of traits that often form a deadly alliance.

Other hormones: In *Hormones and Animal Social Behavior*, E. Adkins-Regan (2005) points out that "social behavior has been linked to many hormones, the last established connections being to certain steroid hormones, pituitary prolactin and a few peptide neurohormones." According to this author, biology has revealed astounding similarity in genomes and biochemical-signaling pathways across long-separated lineages of the animal kingdom differing in ecology, morphology, and behavior. In general, these chemicals help adjust behavior to circumstances and contexts, for example, those which are physical, social, and developmental.

"Social behavior," she says, "is often age-related, and hormones are a mechanism that can ramp up the behavior at the appropriate age. In animals with indeterminate growth, size rather than age may be the trigger for the onset of adult social behavior, and hormones can be a messenger between size and behavior."

The following hormones have been found to be particularly associated with aggression and dominance behavior. A more thorough understanding of the importance of hormones can be found on page 83 of Adkins-Regan's book.

Dehydroepiandrosterone (DHEA): This is an abundant androgen hormone that may function in territorial behavior in birds, rats and hamsters. It has also been implicated in aggression in humans.

Glucocorticoid hormones: These hormones have been shown to be important in regulating aggressive behavior in rats. Injections promote aggressive behavior, and corticosterone reduction decreases aggression. They also affect the establishment of social hierarchies, those with low corticosterone levels being dominant.

Pheromones: Pheromones are nonhormonal chemicals that are produced by an individual and elicit a behavioral response from another member of the same species. They are generally released from the body of one animal and have a volatile nature, being received by the other animal through some form of chemoreceptors. Studies in mice have shown that certain urinary proteins (Mups) promote innate aggression in males. Similar pheromones have not been identified in humans.

It is sometimes difficult to understand precisely how alterations of chemicals in the body affect behavioral changes. Iguanas, for instance, show variable degrees of aggression, due to chemical changes in their body. Interactions between alpha and subordinate males who usually dominate others may affect tolerance levels rather than alterations in aggressiveness.

It is clear that chemistry plays an important role in determining aggression and dominance in humans and other animals. It is also clear that further research is needed to understand the precise mechanisms behind the development and expressions of such behaviors.

PHYSICAL AND CONGENITAL ANOMALIES

In humans, abnormalities in the brain's frontal cortex have been associated with aggression (Brower and Price, 2001). Other organs and chemical imbalances also affect aggressive tendencies. The amygdala, hippocampus, hypothalamus, and pituitary gland may be involved.

Congenital anomalies in the brain and its supporting organs (e.g., the endocrine system) and acquired defects (e.g., those resulting from physical or chemical injuries) can add to the difficulty of understanding aggressiveness in humans and other animals. Such defects can result from early interactions between neonates or juveniles, while others may be due to the use of foreign chemicals.

SUBSTANCE ABUSE

In *Illegal Drugs, A Complete Guide to Their History, Chemistry, Use and Abuse*, Paul Gahlinger (2004) points out that humans and many other animals, "both domesticated and wild, will seek out [and use] intoxicating foods, such as fermented fruits and psychoactive plants." Drugs were used by Neanderthal people long ago, "as shown by archeological remains discovered in the Shanidar cave in Iraq, dated to 50,000 years ago. Whether they are used for medicine, pleasure, religion, or curiosity, drugs have become an integral part of human life":

> Humans are inquisitive creatures. If something can be done—no matter how bizarre, silly, or dangerous—somebody, somewhere will try to do it. If a substance can possibly be eaten, it is certain that somebody will have tried to eat it. Why, for example, would anyone lick the slime off an ugly toad? And why, when it was found that swallowing this slime caused horrendous sickness and nightmare visions, would someone then decide to smoke the secretions? *Paul Gahlinger.* Illegal Drugs, A Complete Guide to their History, Chemistry, Use and Abuse

Gahlinger states that psychoactive drugs have always been closely associated with certain religions. "In early belief systems, any substance with the ability to prevent or cure disease was considered sacred. The ancient texts from many lands—India, China, Egypt, Greece, and the Americas—are filled with references to drugs. The sacred scriptures of India, for example, contain over 1000 hymns in praise of Soma, the psychedelic mushroom. Manna of the Exodus," he says, "may have been a hallucinogenic mushroom."

Drugs that appeared to have healing qualities were associated with God, and drugs that appeared to be associated with evil were considered the work of the Devil. "In every [early] society, beneficial drugs became the specialty of healers, mind-altering drugs became the specialty of shamans or priests, and poisons became the specialty of sorcerers and witches."

One must weigh the effects of drugs. Most are dangerous for a variety of reasons, and if they cause societal problems, they are listed as illegal. "Illegal drugs are not necessarily those that are most poisonous, but those that result in other hazards, such as addiction or inappropriate behavior. If a drug causes someone to act in ways that are offensive to others, it presents a threat not only to the individual but to the whole society."[6]

Substance Abuse and the Brain

After our discussion of chemicals that affect humans and other animal behavior, it should be clear that the human brain is an amazingly complex organ that is influenced by the chemicals that pass through it. Chemical influence is important both during embryological development and in postnatal life.

By the time a baby is born, it has an average of about 100 billion neurons (nerve cells) (Parent and Carpenter, 1995). By 10 years of age, each neuron has established about 100 connections with other neurons. Some have as many as 20,000 connections, total interconnections numbering about 100 trillion.

Signals move between the various neurons by a chemical pathway called neurotransmission. At least 80 different neurotransmitters have been described. Some common examples are acetylcholine, dopamine, epinephrine, norepinephrine, endorphins, and serotonin.

Acetylcholine stimulates muscles and glands. In high concentrations, dopamine may influence the development of a mental condition referred to as schizophrenia (Van Os and Kapur, 2009). Epinephrine stimulates the entire body, preparing it to react to exciting and threatening situations. Norepinephrine, like epinephrine, gives a person the feeling of alertness. Endorphins, which are typically released during exercise, result in feelings of peace and well-being. Serotonin has a number of functions, excessive amounts causing hallucinations.

Modern human society, especially in the United States, is a drug-using society (Gahlinger, 2004). In addition to drug use for medicinal purposes, the United States has 4% of the world's population but consumes 65% of the world's supply of hard drugs. Many people take them in all walks of life for a variety of reasons.

Drugs are defined by Gahlinger as chemical substances used in the treatment, cure, prevention, or diagnosis of disease or used to otherwise enhance physical or mental well-being. Those that reach the brain function in different ways, often being influenced by age, ethnicity, heredity, sex, health, allergic responses, and psychiatric state. Combinations of drugs may also influence one another and the way the brain reacts to them. Those which alter the function of the brain are referred to as psychoactive.

Psychoactive drugs, which are generally simple and small molecules, interfere with neurotransmission. When drugs enter the brain, they flood the space between neurons and overwhelm the usual communication between cells. This may result in a stepped up transmission of signals, a block of signals or an overload of neurotransmitters, with effects that may be dramatic and unpredictable. Among the many negative results of drug use, overstimulation of neurotransmitters can result in pleasurable feelings and euphoria.

Certain drugs may help the body in some ways and thus are referred to as medicines. If they harm the body, they are referred to as poisons. All drugs become harmful in high concentrations, but they may be useful at low doses.

Each time a drug is used, there is a danger of a bad reaction, which may result from impurities in the drug, infection, overdose, or other problems. A bad reaction may be expressed within minutes, hours, days or even years from the time they are used. The most common long-term effect of drugs is a loss of interest and motivation in life. The greatest risk in taking drugs is addiction.

While many drugs used by people are illegal, alcohol and tobacco are two that are not, and in certain ways, they are considerably more dangerous. For instance, it has been said that alcohol kills about 200,000 people each year in the United States, and it is claimed that tobacco kills about 430,000 each year.

Taking psychoactive drugs has been a human endeavor at least since 4000 BC in Mesopotamia. Commencing with opiates, modifications of the structure of the original chemicals and invention of new ones has led to abuse and addiction in human society throughout the world.

Gahlinger (2004) states that 178 substances have been declared illegal in the United States. He also states that there are numerous psychoactive substances, which are legal because they have practical uses, are naturally found in nature but have not been declared illegal or they have been accepted by society.

People, he says, generally start using drugs because of curiosity, encouragement by friends or a desire to find relief from boredom, pain, anxiety, or depression. Use, he says, blends into abuse when drug activity starts to cause problems in the person's life.

Illegal drugs are restricted because they are a threat not only to the health of the individuals using them, but also to their families and to the rest of society. This is especially so of addictive drugs. The greatest danger of these drugs is not to the physical health of addicts but to their mental state. Addiction results in unpredictable or compulsive behavior that can bring enormous suffering to an addict's family, friends and coworkers. Drugs that change thoughts or behavior can make people less respectful of regulations and traditions.

Most illegal drugs "cause people to withdraw. Opiates and depressants produce an isolating numbness, hallucinogens bring psychotic dissociation, and stimulants can cause violent antisocial behavior. In their action, they affect different parts of the brain."

Drugs that affect the brain stem, like opiates, GHB, and methanqualone, can result in stupor or coma. Drugs that affect the cerebellum result in stumbling, jerky eyes, and shaking tremor. "Almost all drugs effect the midbrain, an area that deals with emotion (limbic system), regulation of temperature and pain (thalamus), hormone production (hypothalamus), memory (hippocampus) and the sense of reward (nucleus accumbens)." It has been determined that addiction is associated with the midbrain.

The part of the brain that is important in cognitive functions, the cortex, is also influenced by psychoactive drugs. Different regions of the cortex, for example, the frontal cortex and visual cortex, react differently to certain drugs and different types of drugs.

Substance Abuse and Child Development

Drug use and abuse is also threatening to an unborn child (Chiriboga, 2003). Mothers on drugs may not even know they are pregnant or may not care. To the unborn individual, however, drugs can be very toxic, resulting in a great risk of miscarriage and development of sudden infant death syndrome, as well as infant deformity and addiction to the drugs being used by the mother.

In a report in Teachnology (*The Effects of Substance Abuse on the Development of Children: Educational Implications*), Colleen Meade (1999) reviewed current investigations about substance abuse (based on research by numerous scientists) and pointed out that "the birth of the Thalidomide babies in the early 1960s awakened the world to the fact that drugs ingested by the mother can severely impact the development of a growing baby." Thalidomide was widely prescribed to relieve morning sickness in pregnant women. It was the use of Thalidomide that affected nearly 12,000 infants in 46 countries and caused defective limbs, eyes, ears, genitals and internal organs. The birth of these children soon became known as history's greatest medical disaster.

"To study the effects of substance abuse or misuse during pregnancy," she says, "one must determine exactly what substances will be included. While much uproar has [been] formulated around the use of illegal drugs during pregnancy, one cannot eliminate the effects of legal drugs such as, nicotine, alcohol, caffeine and over the counter and prescription medications":

> Additionally, in order to determine the long-term effects of substance abuse on children, one must include the study of other environmental issues. The development of children impacted by maternal substance use does not stem from prenatal exposure alone, but rather is the result of a constellation of biological, psychosocial, and environmental circumstances.

Meade also pointed out that, "While the specific effects of substances taken during pregnancy are unclear, much recent research has shed light on a number of physical and cognitive abnormalities mainly found in connection with exposure to substances *in utero*. The most commonly researched substances are: alcohol, cigarettes, cocaine (in various forms) and heroin." While more information is consistently available about the effects of these substances, which does not mean that they are the only substances that will cause damage to a fetus. These and other chemicals that are known to result in embryological problems are as follows:

Alcohol: "Alcohol," Meade says, "is a teratogenic substance (relating to substances or agents that can interfere with normal embryonic development) which, in connection with pregnancy, causes a fetus to be born with a condition termed fetal alcohol syndrome (FAS). Due to the commonalities between signs and symptoms of children affected by FAS, this has received a tremendous amount of attention by researchers because it is more easily detectable than the effects of other drugs on the fetus. Currently, alcohol abuse during pregnancy hails as the third leading cause of birth defects and one of the leading causes of mental retardation."

Diagnosis of FAS in children is made after three major clinical manifestations are seen in the patient prenatal and/or postnatal growth deficiency, a specific pattern of facial anomalies and indications of central nervous system dysfunction. The physical signs of FAS are as follows: short palpebral fissures, flat midface, indistinct philtrum, thin upper lip, epicanthal folds, low nasal bridge, minor ear anomalies, micrognathia strabismus, ptosis of the upper eyelid, narrow receding forehead, and a short upturned nose.

These physical deficiencies make it difficult to overlook the damage from alcohol on children displaying these characteristics. While the physical effects of FAS are quite recognizable, it is more difficult to ascertain whether a child has sustained any of the cognitive effects. It is believed that many children with FAS might also display hyperactivity, fine and gross motor developmental delays or incoordination, impaired language development, impulsivity, problems with memory, poor judgment, learning problems, distractibility, seizures, and structural abnormalities of the brain which are indications of deficient brain growth.

Smoking: Because there are over 2500 chemicals found in a single cigarette, it is difficult to analyze the effects of each specific chemical and easier to analyze the effects of cigarettes in general on a growing fetus (Meade, 1999). However, researchers have established that the majority of adverse effects are attributable to two main ingredients: nicotine and carbon monoxide.

One reason for concern regarding the effects of cigarette smoking on pregnancy is that it is estimated that over 22 million women in this country smoke, and among women of reproductive age, almost one-third are smokers. Smoking during pregnancy causes a variety of problems, including low birthweight, fetal, neonatal, and perinatal mortality, sudden infant death syndrome and spontaneous abortion. It is also believed that the problems associated with smoking during pregnancy might affect children later in life.

Cocaine and crack: Cocaine and crack (the freebase form of cocaine) use in the United States has increased in recent years (Meade, 1999). In 1997, the National Household Survey on Drug Abuse estimated that 1.5 million Americans were using cocaine and 2.6 million were considered occasional users. While the long-term effects of cocaine exposure *in utero* have not been conclusively established, many researchers agree that children who were exposed to cocaine *in utero* most likely sustained some degree of damage.

The difficulty researchers find with labeling specific problems associated with cocaine use during pregnancy is the fact that many women abuse more than one drug (polydrug use), thereby making it difficult to isolate the effects of cocaine. However, brain, kidney and urogenital system malformations in neonates are strongly associated with maternal cocaine intake. Gastroschisis (a birth defect that causes a baby's intestines to extend outside of the body through a hole in the abdomen) is one other fetal anomaly that may be related to cocaine use.

"Many infants whose mothers used cocaine while pregnant," Meade states, "are born addicted to cocaine. The withdrawal symptoms exhibited by cocaine-addicted infants include: irritability, tremulousness, course tremors, feeding difficulties, tachypnea (an elevated respiratory rate, or more simply, breathing that is more rapid than normal), diarrhea, vomiting, high-pitched crying and seizures":

> One reason that the unborn child easily becomes affected by cocaine use by the mother is because cocaine easily and rapidly crosses the placenta and is not significantly metabolized during maternal–fetal transfer. While evidence exposing long-term effects of cocaine exposure in utero remains inconclusive, some research does suggest that these children will have difficulty with self-regulation and impulse-control.

Heroin: Heroin is classified as an opiate drug (Meade, 1999). When discussing heroin use during pregnancy, one must also study methadone use during pregnancy. Methadone is a drug given to heroin-addicted individuals to alleviate the severe withdrawal symptoms experienced when these individuals stop using heroin. Infants born addicted to heroin and methodone typically exhibit the presence of neonatal opiate abstinence syndrome (NOAS). This syndrome is characterized by dysfunction of the CNS, autonomic nervous system, gastrointestinal tract, and respiratory system.

The specific symptoms of NOAS include: irritability, tremulousness, hypertonia, excessive crying, voracious appetite, exaggerated sucking drive, abnormal coordination between sucking and swallowing, regurgitation, pulmonary aspiration, and abstinence-associated seizures. The quantity and severity of the symptoms varies. Therefore, the treatment of NOAS must be tailored to specific situations. Some treatments include providing the drug for the child to wean him [or her] off the drug without causing severe health problems:

As is the case with cocaine, conclusive data relating to long-term effects of prenatal exposure to heroin and methadone are inconclusive. As many as 40% of heroin-exposed children required special educational classes, and 25% needed to repeat one or more grades. However, it has not been determined whether this low performance is a result of early exposure to heroine or environmental factors.

Other drugs: It is known that many other drugs may affect the development of a fetus (Meade, 1999). "Since concrete evidence may be difficult to find, many frustrated doctors feel they cannot give patients definitive information. Additionally, there are many drugs that are ingested regularly, which are so ingrained in society that they are viewed as harmless. For example, caffeine is a drug that is widely used and socially promoted in contemporary society, and yet, problems can occur as a result of excessive caffeine intake."

ENVIRONMENTAL FACTORS AND DRUGS

It is virtually impossible to discuss the long-term effects of substance abuse in gestation on child development without also discussing environmental factors. These factors include "nutrition, familial conditions (substance abuse, child abuse, and so on), socioeconomic status, and issues related to general health care (Meade, 1999)." It seems that the higher the socioeconomic status, the better opportunity a family has for good healthcare and nutrition. Additionally, empirical studies and clinical experience show that addiction or substance abuse interferes with parenting and contributes to developmental, behavioral and health problems.

Meade mentions that "while exposure to drugs and alcohol in the womb might cause damage to the developing fetus, environmental factors might further damage the development of a child, leading to secondary disabilities." Secondary disabilities include mental health problems, inappropriate sexual behavior, disrupted school experience, trouble with the law, confinement through incarceration for a crime or impatient treatment for mental health or alcohol and drug abuse problems.

"The chaotic lifestyle of the addicted mother tends to lend itself to a home environment containing neglect and poor parental influences. Often times, a woman who abuses drugs during pregnancy will abuse drugs after the birth of the child." Drug and alcohol abuse by any member of the family can lead to problematic behavioral conditions, for example, chronic instability, disharmony, and possible violence such that a child's psychosocial, developmental, behavioral, and learning competencies can become seriously compromised. Additionally, substance-using mothers have been found to be more likely to be hospitalized as a result of violence:

Other common factors in the homes of drug-abusing parents are a lack of adequate health care and good nutrition. Studies have shown that women who use cocaine during pregnancy were more likely to use other drugs, consume alcohol or smoke, they typically had a lower socioeconomic standard and were more likely to be malnourished. The poor parental care found in many drug-abusive homes leads to neglect of the children, which retards their natural development. The multiple risk factors in the lifestyle of drug-abusing pregnant women appear to be major factors in the poor growth reported both prenatally and postnatally.

NEURAL TISSUE DAMAGE AND MODIFICATION

Damage to neural circuitry may also lead to rule-breaking behavior, which is common in people with antisocial, violent, and psychopathic tendencies (Meade, 1999). Damage to areas, such as the prefrontal cortex, for instance, can result in violent behavior. Antisocial individuals with damage to certain parts of their brain, especially the dorsal and ventral prefrontal cortex, the amygdala and angular gyrus, have difficulty in making moral judgments.

Hyperactive Amygdala

Teenage boys who express inappropriate levels of aggression to perceived threats may do so because of a hyperactive amygdala, an area of the brain that processes information regarding threats and fear (Yang et al., 2008) (Fig. 5.3). There also may be a lessening of activity in the frontal lobe, a brain region linked to decision making and impulse control.

Misinterpretation of Surroundings

Adolescent boys who react aggressively to situations sometimes misinterpret their surroundings, feel threatened, and act inappropriately. They have a tendency to strike back during episodes of being teased, they blame others when getting into fights, and they overreact to accidents.

Reactive–Affective–Defensive–Impulsive Disorder and Impulsive Aggression

Adolescents with reactive–affective–defensive–impulsive (RADI) disorder are at an increased risk for experiencing a lifetime of problems associated with impulsive aggression (Karnik et al., 2008). Their brains exhibit greater activity in the amygdala and less activity in the frontal lobes when responding to fear-inducing images than the brains of adolescents without the disorder.

Unchecked Aggressive Behavior

It has been found that unchecked aggressive behavior, as seen in certain younger people, can eventually change the brain in ways that result in a lowering of serotonin levels and an increase in violent behavior (Anderson and Bushman, 2002; Buss, 1961).

Social Stressors and Relationships

Sources of social stressors and negative social relationships are believed to contribute to stress-related disorders (Hales and Zatazick, 1997; Lawrence and Lawrence, 2015). Stress has a profound influence on neuroendocrine and other neurochemical systems which, in turn, result in chemical changes in many areas of the brain. Some of these areas are involved in emotion, for example, the prefrontal cortex, the hippocampus, and the amygdala.

Media and the Acquisition of Aggressive Tendencies

Some correlations between the media and aggression in youths have shown that 8 year olds who watch a lot of violent television appear to be more likely to be arrested by the time they are 30 years old (Huesmann et al., 2003). The reader may relate to what was said in Chapter 2 when types of aggression were defined and discussed. However, such studies have been highly criticized because they apparently do not show significant differences.

Others have found that the effects of video game violence exposure largely depend on whether an individual has an aggressive personality (Huesmann et al., 2003). Highly aggressive personalities and high exposure to video game violence seems to form a complex association that may lead to an increase in aggressive behavior while individuals with a lower aggressive personality and high exposure to video game violence appears to be relatively nonaggressive.

Likewise, exposure to weapons seems to have a correlation with increased aggression (Frodi, 1975). The presence of aggression is associated with a lack of rational thinking and a likelihood of aggressive–impulse reactions during provocation.

Rejection in Children

Feelings of being rejected have been found to be a cause of aggression (Brainerd et al., 2003).

Behavioral Variability

Due to its complexity, understanding expressions of dominance and aggression in humans is quite difficult. It is obvious that the human population has tremendous behavioral, physiological, and anatomical variability, and each individual acts differently to situations. In addition, contemporary society is complex and stressful. Therefore, to understand aggression, it must be approached from a variety of different perspectives, taking into account an individual's genetics, learning in the formative years and all of the body's systems that may affect rising levels of aggression.

Mazur (2005) points out the importance of properly interpreting the influence of chemicals on dominance and aggression because (as we have seen here) they differ in their causes and effects. A person is said to "act aggressively if his or her apparent intent is to inflict physical injury on a member of his or her species." On the other hand, a person is said to "act dominantly if [their] apparent intent is to achieve or maintain high status" (as in the acquisition of power, influence, or valued prerogatives) over a member of the same species.

It is plain to see that combining these two features (acting aggressively and dominantly), especially in conjunction with a genetic predisposition for violence, early behavioral trauma, developmental anomalies and poor nutrition, can lead to an explosive personality (McElroy, 1999). It is also clear why it is so difficult to analyze emotional problems when there are so many variables to complicate the issue.

According to Mazur, interpersonal behavior is often overtly or subtly involved with managing dominance and subordinateness without causing physical harm. "Sports, spelling bees, elections, critics, competitions for promotion and academic jousting all involve domination without aggression." While it is more difficult to identify instances of aggression devoid of a dominating motive, aggression beyond the need to establish a position in the dominance hierarchy is most often avoided by animals. As pointed out a number of times in these pages, organisms generally (but not always) tend to establish their differences in more subtle or less threatening ways.

Life is complex in a modern world, and problems associated with our minds can further complicate our lives. Chemicals of any sort can alter the way our brains develop in prenatal, postnatal, adolescent, and later periods. Understanding that our personalities are determined early in life should give us a heads-up about how we should consume food and other chemicals and how we should act toward young people so that they may end up reasonably stable citizens. In short, it will determine how they function in a modern society, under the influence of stress and environmental threats.[7]

Note

Superscript numerals appearing in this chapter refer to additional text/explanation given in the appendix.

11

Dominance and Aggression in the Workplace

The fundamental concept in social science is power, in the same way that energy is the fundamental concept in physics. Power and dominance submission are two key concepts in relationships, especially close relationships, where individuals rely on one another to achieve their goals, and as such, it is important to be able to identify indicators of dominance. *Dunbar & Burgoon, 2005* Perceptions of Power and Interactional Dominance in Interpersonal Relationships. Jour. Soc. Pers. Rel.

Statements in the last few chapters point to behavioral differences in the lives of humans due to a complex suite of factors, stemming from both genetic and learned origins. Behavioral expressions are further compromised by nutritional inadequacies and the intake of toxic substances either prenatally or after birth. These genetic, learned, toxic, and nutritional circumstances infiltrate our society and provide a wide assortment of human behaviors, many of which affect the way we get along in our jobs. Behavioral variation in the workplace may be influenced by any of these parameters.

CONFLICT IN THE WORKPLACE

Dominance interactions and aggression are commonly found expressed between individuals in the workplace (Argyris, 1957; Johnson, 1976). How one goes about dealing with this often hinges on the personality of individuals in high positions, those climbing the corporate ladder, and those doing their job without aspirations for climbing the ladder (see discussion on personalities in the current chapter and in Chapter 8, *Human Nature*). The relationships between individuals can often be determined through logical approaches, but they become complex when asocial and antisocial behaviors play a role in how people dominate their subordinates and how subordinates react to the way they are treated.

In order to accomplish their work and arrive at their desired position in the working world, most individuals may logically assume that doing their job properly and paying attention to rules and regulations will pave the way to their success. Under ideal circumstances, this may be true. However, given the complex personalities of people in general and the complications that can arise from them, agonistic behavior is often expressed in all phases of the working world and for many reasons, and it is not uncommon for an individual to experience conflicts that result in disharmony.

In *Understanding Conflict in the Workplace*, J. Gatlin et al. (2007) have reviewed the works of several authors in an attempt to understand and resolve the origins of conflicts. The following summary from their book offers some reasons why people do not find the happiness they seek in their jobs:

- *Conflicting needs* result when workers compete for scarce resources, recognition, and power in order to establish themselves in the company's hierarchy but fail to achieve these goals because of favoritism or differential treatment. We must remember that people in high places are generally dominants. How they express their dominance and degree of tolerance may result in stress in their subordinate worker force. For example, people who get differential treatment from their bosses may feel intolerant and complain.
- *Conflicting working styles* result when different individuals approach people and problems in different ways. For example, workers with different personalities or environmental requirements may face a problem with strikingly different approaches.

Dominance and Aggression in Humans and Other Animals
http://dx.doi.org/10.1016/B978-0-12-805372-0.00011-0

- *Conflicting perceptions* result when workers perceive the same incidents in dramatically different ways. As an example, a newly hired worker may be looked at by some individuals as a form of assistance in getting a job done (teamwork), while others may perceive them as a message that they are not performing their job properly. How a mentor presents instruction, delegates responsibility, and perceives the work ethics of a worker (involving tolerance, care, and understanding) are important in avoiding conflicts.
- *Conflicting goals* result if workers approach a goal in different ways. As an example, a worker may be slower at his or her job but more thorough, while another may be faster and less concerned with details. Thus, their goals and methods of reaching them may vary.
- *Conflicting pressures* result when two or more workers are assigned different actions with the same deadlines. How workers decide to resolve the situation may represent a form of conflict.
- *Conflicting roles* result when a worker is required to perform a function that is outside his or her job requirements. Such a situation may contribute to power struggles.
- *Different personal values* may result in segregation in the workplace and contribute to a lack of team work.[1]
- *Unpredictable policies* can result in misunderstandings.

THE SERIAL BULLY

Much of what has been said up to this point deals with logical approaches to internal problems and conflicts. Most businesses have such variation in their workforce at one time or another. However, all situations in the workplace do not express themselves in a logical manner. There are certain personalities in the working world, just as there are in the nonworking world, which parasitize coworkers and the system to which they belong. As we have seen, discrepancies in human behavior may have many sources, making human expressions the complex result of almost unlimited malfunctions.

The United Kingdom National Workplace Bullying Advice Organization, for instance, discusses the serial bully in the workplace and points out his or her dominating psychopathic behavior (Rayner and Keashley, 2005). "Most organizations have a serial bully," says one *bullyonline* commenter (Fig. 11.1). "It never ceases to amaze me how one person's divisive, disordered, dysfunctional behavior can permeate the entire organization like a cancer." According to the organization, a "common objective of these offenders is power, control, domination and subjugation."

Bullies are commonly encountered, and it is not only in the workplace that these individuals are found. Evidence indicates that the serial bully expresses themselves in the workplace, in the community, and at home (Brank et al., 2012).

True bullies possess the characteristics of a psychopath. What varies between the two is the way in which their violent nature is expressed. Most offenders with psychopathic behavior, for instance, commit criminal or arrestable offenses, while the serial bully commits mostly nonarrestable offenses. As we may expect here, the degree of bulliness expressed by humans shows a gradation effect.

The following exhausting list presents characteristics found in a variety of serial bullies: negligence, incompetence, maladministration, neglect of duty, dereliction of duty, misappropriation of budgets, financial irregularities and fiddling with the books, fiddling with expenses, pilfering, stealing, diverting, skimming, or "losing" clients'

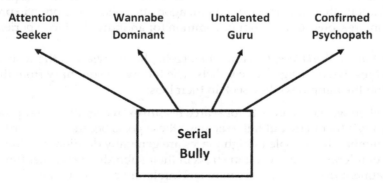

FIGURE 11.1   Four types of serial bully, as pointed out by the United Kingdom National Workplace Bullying Advice Organization (Rayner and Keashley, 2005). Embittered by an abusive upbringing, seething with resentment, irritated by others' failure to fulfill his or her superior sense of entitlement and fueled by anger resulting from rejection, the serial bully displays an obsessive, compulsive and self-gratifying urge to displace their uncontrolled aggression onto others whilst exhibiting an apparent lack of insight into their behavior and its effect on people around them.

money and investments, embezzlement, fraud, deception, malpractice, misrepresentation, conspiracy (such as obstructing or perverting the course of justice), using the employer's resources to run their own business on the side, moonlighting for employer's clients or competitors, leaking information to people who should not be in possession of that information, awarding contracts to family and friends, failure to fulfill obligations, breaches of health and safety regulations, breaches of rules and regulations, breaches of codes of conduct, improper use of fraternal allegiances, indiscretions, impropriety, inappropriate sexual conduct, being the target of previous grievance and disciplinary action, being the target of previous legal action (unfair dismissal, harassment, personal injury, etc.), fraudulent qualifications and misleading or bogus claims of professional affiliation, collusion, corruption, being sacked or asked to leave their previous job(s), recruitment through nepotism or favoritism rather than ability, and extramarital affairs; at home, they have a poor credit rating, and they exercise verbal and domestic abuse, neglect, and abandonment.

The serial bully is consistent in doing what they do repeatedly, and they can be either employer or employee. In the workplace, they are "often found in a job which is a position of power, they have a high administrative or procedural content but little or no creative requirement which provides opportunities for demonstrating a 'caring' or 'leadership' nature." This is such an important characterization to watch for in the workplace, that I find it important to quote further from The United Kingdom National Bullying Advice Organization:

Embittered by an abusive upbringing, seething with resentment, irritated by others' failure to fulfill his or her superior sense of entitlement, and fueled by anger resulting from rejection, the serial bully displays an obsessive, compulsive, and self-gratifying urge to displace their uncontrolled aggression onto others while exhibiting an apparent lack of insight into their behavior and its effect on people around them. Jealousy and envy motivate the bully to identify a competent and popular individual who is then controlled and subjugated through projection of the bully's own inadequacy and incompetence. When the target asserts their right not to be bullied, a paranoid fear of exposure compels the bully to perceive that person as a threat and hence neutralize and dispose of them as quickly as possible. Once a person has been eliminated, there is an interval of between 2 days and 2 weeks before the bully chooses another target, and the cycle starts again.

Dario Maestripieri (2012) points out in *Games Primates Play* that Machiavellian strategy is commonly demonstrated in human and other animal societies. He uses rhesus macaque monkeys to show that different strategies are used to gain status, a process utilized by serial bullies.

The United Kingdom National Workplace Bullying Advice Organization points to four primary types of serial bully:

1. the attention seeker, in which they are a control freak, manipulative, and narcissistic;
2. the wannabe person in power who craves respect for being competent and professional, in spite of lacking in these qualities, and who is deceptive;
3. the guru who appears task focused but in truth is confused and does not understand how others think and feel;
4. the psychopath whose motivation is power, gratification, personal gain, and survival, with a mindset of manipulation, deception, and evil.

In Maestripieri's example, a woman entered a major business and immediately began acquiring information on the other workers and people in more dominant positions. In phase 2 of her strategy, she began to disqualify certain individuals who surrounded her, and in stage 3, she began spreading rumors. Once she acquired followers and caused her boss to lose his job, she began working on the next step to move to the top of her department. Maestripieri found similar behavior in societies of macaques.

The United Kingdom National Workplace Bullying Advice Organization points out that "the serial bully is able to exert a hold over people for a variety of reasons." Their targets are "disempowered such that they become dependent on the bully to allow them to get through each day without their life being made hell." The bully "is often able to bewitch an emotionally needy colleague into supporting them," and this person "becomes the bully's spokesperson and advocate."

The use of other people in power struggles is actually a coalition in which two or more individuals cooperate against a target. Maestripieri recognizes three basic types of coalitions in macaques, and all three are found in the human workplace:

1. those in which two or more males are higher ranking than the target (referred to as conservative coalitions);
2. those in which one coalitionary male is high ranking and the other is lower ranking than the target (called bridging coalitions);
3. those in which the two coalitionary males are both lower ranking than the target (revolutionary coalitions).

People with these qualities strategize to get to the top and are not concerned with the psychological damage they inflict on their targets. In the process, they may use different strategies, depending on costs and benefits. Maestripieri states that "in a highly despotic group, particularly if the group is small and the alpha male does not have a great deal of support behind him or her, it pays to directly challenge the leader. In more democratic human groups or societies in which power and resources are more evenly distributed among individuals on different steps of the status hierarchy, it pays to wait and rise in rank through seniority, particularly in large groups with a lot of social inertia."

When called to account for their behavior, the bully instinctively exhibits the following recognizable behavioral responses:

1. they deny everything, trivializing the problem, talking about a fresh start, abdicating responsibility, and diverting attention by using false conciliation;
2. they counterattack, lying and deceiving to avoid answering questions;
3. they feign victimhood by manipulating people through their emotions, including crying.

Serial bullies harbor a particular hatred of anyone who can articulate their behavior profile, either verbally or in writing. While an organization should carry out an in-depth study of this type of personality in their workforce, they are sometimes difficult to recognize. Superiors should be familiar with this personality type. The bully organization concludes with a statement from Stanton E. Samenow's book, *Straight Talk about Criminals* (1998):

Certain nonarrestable criminals behave criminally toward others, but they are sufficiently fearful so that they do not commit major crimes. We all know them: individuals who shamelessly use others to gain advantage for themselves. Having little empathy, they single-mindedly pursue their objectives and have little remorse for the injuries they inflict. If others take them to task, they become indignant and self-righteous and blame circumstances. Such people share much in common with the person who makes crime a way of life. Although they may not have broken the law, they nonetheless victimize others.

TOXIC LEADERS

Toxic leaders (often referred to as boardroom hitlers, little hitlers, managers from hell, or bosses from hell) (Whicker, 1996) are bullies who have responsibility over a group of people or an organization and abuse the leader/follower relationship by leaving the group or organization in a worse-off condition than when they first found them. This is similar to the behavior of certain despotic national and international leaders who sometimes commence their leadership roles with what appears to be good intentions but subsequently create chaos and cause the death of large numbers of people (current chapter and Chapter 13).

The basic traits of a toxic leader are seen in the following list. The reader may recognize them as features found in the *Hare Psychopathy Checklist* of antisocial personalities (Lykken, 1995):

- Oppositional behavior
- Corporate power politics
- An over-competitive attitude toward other employees
- Perfectionistic attitudes
- Abuse of the disciplinary system, for example, removing a workplace rival
- A condescending/glib attitude
- Poor self-control and restraint
- Physical and/or psychological bullying
- Procedural inflexibility
- Discriminatory attitudes, for example, sexism and others
- Causes workplace division instead of harmony
- Uses "divide and rule" tactics on their employees

Many toxic leaders (as reviewed in a recent Wikipedia article) are also "authoritarian and/or control freaks to varying degrees, who tend to use both micromanagement, over-management and management-by-fear to keep a grip of their authority in the organizational group. Micromanagers usually dislike a subordinate making decisions without consulting them, regardless of the level of authority or factual correctness."

A toxic leader can be both hypocritical and hypercritical of others, seeking the illusion of corporate and moral virtue to hide their own workplace vices. Hypocrisy involves the deception of others and is thus a form of lying. Such people can also be both frightening and psychologically stressful to work with.

The US Army defines toxic leaders as commanders who put their own needs first, micromanage subordinates, behave in a mean-spirited manner, or display poor decision making. A study for the Center for Army Leadership found that toxic leaders work to promote themselves at the expense of their subordinates and usually do so without considering long-term ramifications to their subordinates, their unit, and the Army profession.

According to the Army Center, toxic leaders operate in different ways to discredit individuals around them:

- *Workload*: The "Setting up to fail" procedure is in particular a well-established workplace-bullying tactic that a toxic leader can use against his or her rivals and subordinates.
- *Corporate control systems*: They could use the processes in place to monitor what is going on. Disciplinary systems could be abused to aid their power culture.
- *Organizational structures*: They could abuse the hierarchies, personal relationships, and the way that work flows through the business.
- *Corporate power structures*: The toxic leader controls who, if anyone, makes the decisions and how widely spread power is.
- *Symbols of personal authority*: These may include the right to parking spaces and executive washrooms or access to supplies and uniforms.
- *Workplace rituals and routines*: Management meetings, board reports, disciplinary hearing, performance assays, and so on may become more habitual than necessary.

Inevitably, the victim's workplace performance, self-esteem, and self-confidence will decline as employee(s)' stress inclines. Heavy running costs and a high staff turnover/overtime rate are often also associated with employee-related results of a toxic leader.

Jean Lipman-Blumen (The Allure of Toxic Leaders: Why We Follow Destructive Bosses and Corrupt Politicians—and How We Can Survive Them) (2005) explains that there was and still is a tendency among contemporary society to seek authoritative, even dominating characteristics among our corporate and political leaders because of the public's own personal psychosocial needs and emotional weaknesses.

She noticed that "toxic leadership" was not about run-of-the-mill mismanagement. Rather, it refers to leaders, who, by virtue of their dysfunctional personal characteristics and destructive behaviors, "inflict reasonably serious and enduring harm" not only on their own followers and organizations, but on others outside of their immediate circle of victims and subordinates as well. A noted rule of thumb suggests that toxic leaders leave their followers and others who come within their sphere of influence worse off than they found them either on a personal and/or corporate basis. Relate this to despotic leadership, as discussed in political circles.

Lipman-Blumens' core focus was on investigating why people will continue to follow and remain loyal to toxic leaders. She also explored why followers often vigorously resist change and challenges to leaders who have clearly violated the leader/follower relationship and abused their power as leaders to the direct detriment of the people they are leading. She suggests that there is something of a deeply psychological nature going on. She argues that the need to feel safe, special, and belongingness in a social community all help explain this psychological phenomenon.

In *Bad Leadership: What It Is, How It Happens, Why It Matters*, Barbara Kellerman (2004) suggests that toxicity in leadership (or simply, "bad leadership") may be categorized into seven different types:

- *Incompetent*: The leader and at least some followers lack the will or skill (or both) to sustain effective action. With regard to at least one important leadership challenge, they do not create positive change (http://en.wikipedia.org/wiki/Toxic_leader - cite_note-Spot-1).
- *Rigid*: The leader and at least some followers are stiff and unyielding. Although they may be competent, they are unable or unwilling to adapt to new ideas, new information, or changing times (http://en.wikipedia.org/wiki/Toxic_leader - cite_note-Spot-1).
- *Intemperate*: The leader lacks self-control and is aided and abetted by followers who are unwilling or unable to effectively intervene (http://en.wikipedia.org/wiki/Toxic_leader - cite_note-Spot-1).
- *Callous*: The leader and at least some followers are uncaring or unkind. The needs, wants, and wishes of most members of the group or organization are ignored and discounted, especially subordinates (http://en.wikipedia.org/wiki/Toxic_leader - cite_note-Spot-1).
- *Corrupt*: The leader and at least some followers lie, cheat, or steal. To a degree that exceeds the norm, they put self-interest ahead of public interest (http://en.wikipedia.org/wiki/Toxic_leader - cite_note-Spot-1).
- *Insular*: The leader and at least some followers minimize or disregard the health and welfare of those outside the group or organization for which they are directly responsible (http://en.wikipedia.org/wiki/Toxic_leader - cite_note-Spot-1).
- *Evil*: The leader and at least some followers commit atrocities. They use pain as an instrument of power. The harm can be physical, psychological, or both.

In *Understanding Ethical Failures in Leaders*, Terry Price (2005) argues that the volitional account of moral failures in leaders does not provide a complete account of this phenomenon. Some have suggested that the reason leaders behave unethically is because they willingly behave in ways that they know to be wrong. Price, however, suggests that leaders can know that a certain kind of behavior is generally required by morality but are mistaken as to whether the relevant moral requirement applies to them in a particular situation and whether others are protected by this requirement. He demonstrates how leaders make exceptions of themselves, explains how the justificatory force of leadership gives rise to such exception making, and develops normative protocols that leaders should adopt.

Corporate management analyst Gillian Flynn (1999) describes a toxic manager as the manager who bullies, threatens, and yells: whose mood swings determine the climate of the office on any given workday and who forces employees to whisper in sympathy in cubicles and hallways about the backbiting, belittling boss from hell.

The reader should recognize that these characteristic features of a bully and toxic leader are the same characteristics that personify many dominant (often psychopathic) individuals in all sorts of human activities, including those mentioned in the following chapters. No human activity is without them.

Note

Superscript numerals appearing in this chapter refer to additional text/explanation given in the appendix.

12

Dominance in Religion

Propitiation and sacrifice, which are near-universals of religious practice, are acts of submission to a dominant being. They are one kind of a dominance hierarchy, which is a general trait of organized mammalian societies. **E.O. Wilson,** *Consilience, Ethics and Religion*

THE BEGINNINGS OF RELIGION

Religion is often defined as an organized collection of belief and cultural systems that relate humanity to spirituality and, sometimes, to moral issues (Davis, 1985). Religion most likely was initiated in early human societies simply as an attempt to explain the unknown, and it got its strength when humans came together with the commencement of civilizations (King, 2007). Strangely, this approach is reminiscent of the beginnings of science, which arose out of Greek mythology due to the curious nature of humans.

Religion's roots in human society are found in the many humans who had an uncontrollable desire to have a spiritual dominant (a God) in charge of their life who offers them spiritual comfort, guidance, and an eternal existence. Thus, individuals who abide by the rules of their religion generally believe they will be judged by God and sent to heaven where they will live an eternal, sublime life.

In most religions, there is an opposing force, such as a Devil (the personification of evil and adversary of God) accepted by its followers, a mythological essence that humans can blame their hardships on. Unless forgiven by God, individuals who express bad traits during their life on Earth are believed to go to hell, a place of fire and brimstone, torment, and destruction.

There are many forms of religion on Earth (Fox, 2008). The establishment and building of a religion goes something like this: From initial curiosity and the desires mentioned earlier, an idea arises about how humans came to be on Earth, how they should act toward their fellow humans, and who they should relate to for a source of wisdom, and if the idea gains the interest of groups of people, a cult is formed. Religions stem from cults that attract a strong following.

The initial concepts of a religion generally stem from an individual (prophet) who is said to be inspired directly from God (or a similar fictitious spiritual leader). The details of how each prophet became inspired is generally unknown, but as indicated in Chapter 11, many early religions sprang from the thoughts of individuals who were possibly under chemical influence, which allowed them to connect to what they felt was the divine maker of all things.

Initially, formative religious concepts appear to have commenced prior to or during the Middle Paleolithic era (100,000–70,000 years ago) (Smith, 1992). Organized religion traces its roots to the Neolithic Revolution, which began 11,000 years ago in the Near East, accompanying the rise of civilizations. At the time, there was little known about the world and its physical, chemical, and biological nature. Without a scientific base, it was natural to wonder about the unknown and imagine a God or Gods who directed events that were little understood. Long before and during this period, humans had experimented with the use of plants, along with their nutritious, medicinal, and recreational values.

Paul Gahlinger (*Illegal Drugs*) (2004) points out that the first evidence of drug use (the opiates) "was in Mesopotamia, roughly at the site of the biblical Garden of Eden, and dates back to 4000 BC—about the time, coincidentally, that fundamentalist Christians believe the world was created." Very little was known about drugs at the time, but humans understood their chief feature: "to cause altered perceptions or sense of increased insight or awareness."

The earliest writings of the Bible were prepared in clay about 3500 years ago, around 1400 BC. The writings of the 30 or so other contributors to the Old Testament span a 1000 years.

Dominance and Aggression in Humans and Other Animals
http://dx.doi.org/10.1016/B978-0-12-805372-0.00012-2

Drugs and their effects were accepted by dominants and subordinates alike in these early times, and their use became important to citizens and the developing medical field. Opium preparations, for instance, are described in the Thebes papyrus of 1552 BC, which recommends its use in over 700 remedies. Humans determined early in their relationship with this drug that there were both benefits to its use and serious consequences to its overuse. Romans and their enemies, for instance, viewed opium not only as a painkiller but also as a poison—especially as a pleasant means of suicide.

"The first written record of marijuana can be found in the Vedas, a collection of holy books set down in India from 2000 to 1400 BC. According to the Atharva Veda, Lord Shiva took pity on humans and brought cannabis from the Himalayas to give health and pleasure. These sacred psalms revered marijuana as freeing the soul from distress and anxiety. By the 10th century AD, a liquid cannabis preparation called bhang was described as the food of the Gods":

> The ancient Vedas of India also extolled psychoactive drugs, for example, a bright red mushroom (*Amanita muscaris*), which the northern tribes called Soma. It was revered as a God. In the Rig-Veda, over 100 holy hymns are devoted to the use of Soma.

Drugs that we proclaim to be dangerous to modern society had many seemingly harmless or even beneficial uses that, at times, promised to be the answer to many societal problems during the 1800s and 1900s. Medical drugs such as morphine (derived from opium), codeine (methylmorphine), and heroine (diacetylmorphine) were treated as wonder drugs. Laudanum, a tincture of opium, was praised for its calming effects and healing qualities against gastrointestinal illnesses, food poisoning, and parasites. Before societal problems were recognized, many drugs were accepted and used by many people in all walks of life. Scrooge drank it in Charles Dickens' *Christmas Carol*, and Ben Franklin died addicted to opium, which he had started taking for gout.

Cocaine (a tropane alkaloid that is obtained from the leaves of the coca plant) was particularly useful in surgery on the face, eyes, nose, and throat, and thus it was adopted by physicians. Coca plants are represented by a number of species: *Erythroxylum coca coca*, *Erythroxylum coca ipadu*, *Erythroxylum novogranatense novogranatense*, and *Erythroxylum novogranatense truxillense*. It was praised by kings, queens, and people of renown from Thomas A. Edison to Ulysses S. Grant, who depended on it for inspiration to write their memoirs (Gahlinger, 2004).

Coca's origin is believed to be somewhere in the eastern Andes. The coca plant was initially viewed as having a divine origin, and its cultivation became subject to a state monopoly. Its use was restricted to nobles and favored classes by the rule of the Topa Inca (1471–1493). Subsequently, as the Incan empire declined, the leaf became more widely available. Following the invasion of the New World by Europeans, Philip II of Spain recognized the drug's religious importance to the Andean Indians, but urged missionaries to end its religious use.

Traces of coca have been found in mummies dating back to 3000 years. Other evidence dates the communal chewing of coca with lime 8000 years ago (Dillehay et al., 2010; Rivera et al., 2005). Evidence for the chewing of coca leaves extends from at least the 6th century, based on mummies found with a supply of coca leaves, pottery depicting the characteristic cheek bulge of a coca chewer, spatulas for extracting alkali and figured bags for coca leaves and lime made from precious metals, and gold representations of coca in special gardens of the Inca in Cuzco.

While those with a religious need would argue that considering the sacred scriptures as products of prophets who were under the influence of drugs is preposterous, there is strong evidence that this is precisely the environment in which many prophets lived to receive inspiration for their writings. Drugs were commonly used by many cultures to make the connection between earthbound activities and God, and experiments have shown that people exposed to certain drugs do, in fact, claim to have had a God-like connection.

Initial Development of Religions

Religious prophets have been many, and while most contemporary human societies follow religions that developed within the past 2000 years, initial forms of religion had their beginning in aboriginal societies (Joyce, 1913). Even Neanderthals appear to have followed some sort of a religious ritual in burying their dead.

An article in the Proceedings of the National Academy of Science by William Rendu et al. (*Evidence Supporting an Intentional Neandertal Burial at La Chapelle-aux-Saints*) (2014) revealed recent findings that pertain to a 50,000-year-old Neanderthal skeleton discovered in a cave in La Chapelle-aux-Saints, France, in 1908. In 1999, French researchers began reexamining the site and concluded in 2012 that the depression where the skeleton was found was at least partially modified to create a grave. Moreover, unlike an array of reindeer and bison bones that were also present in the cave, "the Neanderthal remains contained few cracks and showed no signs of weathering-related smoothing or disturbance by animals." They concluded that the grave was intentionally made.

The idea that Neanderthals buried their dead fits with recent findings that they were capable of symbolic thought and of developing rich cultures. For example, findings show they likely decorated themselves using pigments, and wore jewelry made of feathers and colored shells.

Thus, most primal belief systems arose within small groups of people who had similar thoughts and interpretations of the events of life. Some of these belief systems, however, were increasingly accepted over time due to dominant personalities of their time. The Indian emperor Ashoka, for instance, became a Buddhist in the 3rd century, subsequently renouncing war and establishing rules for his society that were based on respect for all life forms (Smith, 1901).

Even though a countless number of humans believe in and follow a particular religion, the basis of religious organization has been poorly represented in historic documents often because they were unwritten or lost scrolls, deeply hidden (sometimes in secret societies), or poorly understood. Known religions generally commenced along with the earliest civilizations, which have frequently been associated with the Ancient Near Eastern Chalcolithic (Ubaid period, Naqada culture), especially in the Fertile Crescent (Mesopotamia and Levant) and Egypt, but also extended to sites in Asia Minor (Anatolia), Armenia, and the Iranian Plateau (Elam).

Official religions appear to have commenced somewhere in the Middle East (centered in western Asia and Egypt) as prehistoric cults, most of which are little known because of a lack of records (Olson, 2006; Stark and Brainbridge, 1996). Before written records, teachings of these religions were handed down through generations as parables (short allegorical stories designed to illustrate or teach some religious principle or moral lesson) and were later recorded in the beginning of the historic period.

There was certainly a long line of Gods in the earliest religions, even prior to the development of civilizations. We know very little of these Gods because there are no written accounts of them. According to E.A. Wallis Budge, in *The Gods of the Egyptians* (1969), Osiris was one of the greatest Gods, his presence being known in religious and mythological systems of the Egyptians as far back as the 1st Dynasty (over 5000 years ago). "He was an indigenous God of Northeast Africa," his home and origin were possibly Libyan. While he was initially seen as a river God, he, in due course, became the king of Egypt and was believed to have "taught men husbandry, established a code of laws, and made men worship the Gods."

When Egypt had become peaceful and prosperous, it is believed that Osiris "set out to instruct the other nations of the world." Osiris "came to be identified with the Greek God Dionysus" and was "identified with the sun and moon, and with the great creative and regenerative powers of Nature, and he was at once the symbol of rejuvenescence, resurrection, and of life of every sort and kind, which has the power of renewing itself." He eventually became associated with Åmen-Ra, who became a prominent God in Upper Egypt around the time of the 12th Dynasty (1971 BC to 1802 BC). Stories of these Gods and the events in their lives came to be the basis of prominent religions in the world, including those of today.

Many of the records of early religions have been lost or tampered with. In *The Secret Teaching of All Ages*, Manly Hall (2011) states that "the early Christians used every means possible to conceal the pagan origin of their symbols, doctrines, and rituals. They either destroyed the sacred books of other peoples, among whom they settled or made them inaccessible to students of comparative philosophy, apparently believing that in this way they could stamp out all records of pre-Christian origin." Speaking of the faith of Islam, Hall mentions that "if you will take a brief view of the Alcoran, you shall find it a hodgepodge made up of. . . four ingredients: (1) contradictions, (2) blasphemy, (3) ridiculous fables, and (4) lies."

Cults, from which religions developed, are based on questionable stories that had been handed down through generations and documents, some of which are treated as the word of God himself. Further, in their original written form and in translations, they have been modified and used by people (e.g., prophets and other dominant societal leaders) to lead others in what is presumed to be a moral life.

Once cults and new religions are established, followers sometimes grow tired of them. On occasion, new religions may spring from old religions that for one reason or another do not precisely fit the needs of some of the people. Under certain circumstances, people who have followed a religious path may abandon that life and follow a more secular one.

It is clear that not all contemporary people accept religious concepts in any form, and there have been disbelievers throughout history. Today, many skeptics are found in people of all walks of life, especially in certain of the sciences.

This is because individuals in the various fields of science have investigated past and present cosmological, physical, chemical, and biological phenomena and strongly feel that there is a more rational explanation for the origin and progression of life through the process of evolution, which is supported by a massive amount of evidence from the various fields of science.

For some who have accepted a religious explanation for the origin and progression of life, their religion sometimes becomes the most important aspect of their life. For others, religion may function as little more than a social convention, and still others feel religion is an obstruction of the truth.

For many people, a reliance upon a supreme being has not changed. Over time, people from around the world who require religious direction have developed a variety of religious and other primal beliefs, all of which rely on faith and scriptures claimed to be sacred. Some such beliefs have expanded in human populations to become major religions.[1]

THEISTIC SYSTEMS

In religions that accept a God as their supreme deity, concepts developed by a dominant individual or multiple codominants are utilized as the basis for theistic (God-based) beliefs (Appleby, 2006) (Fig. 12.1). In its broadest sense, theism (the belief that at least one deity exists) and the concepts that define it may be divided into subcategories, for example, deism, pantheism, polytheism, and monotheism.

Deism is the belief that reason and observation of the natural world are sufficient to determine the existence of God, a dominant individual who has made the world and its biota and guides the populace in the right direction. As stated by Rana Williamson in *American History* (2005), "deism portrayed God as the master clockmaker, who had planned and built the universe, set it into motion, and left it to its own fate."

Pantheism is the belief that everything composes an all-encompassing, immanent God and that nature and the universe are identical with divinity (Picton, 1905). Thus, the essence of pantheism is a profound reverence for nature and the much larger universe and awed recognition of their power, beauty, and mystery. Some pantheists use the word "God" to describe these feelings, while others prefer not to, so as to avoid ambiguity.

Polytheism is the worship or belief in multiple deities (portrayed as Gods and Goddesses), along with their religions and rituals. This approach to the unknown was the typical form of religion during the Bronze and Iron Ages and up to the Axial Age (800–200 BC) and the gradual development of monotheistic, pantheistic, and atheistic concepts. It is well documented in historical religions of classical antiquity, especially Greek and Roman polytheism, and after the decline of Greco-Roman polytheism in tribal religions, such as Germanic paganism or Slavic mythology. There are various polytheistic religions practiced today, for example, Chinese folk religion, Candomble, Druidry, Hinduism, Odinism, Shintoism, Taoism, Thelema, and Wicca.

Monotheism is defined as a belief in the existence of a single God or in the oneness of God (McGrath, 2003). The various forms of monotheism, all of which have significant parallels in polytheistic religions, are generally classified into three major religions: Judaism, Christianity, and Islam.

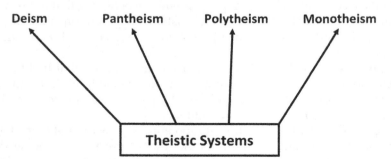

FIGURE 12.1 Theistic systems and its subcategories. In its broadest sense, theism is defined as the belief that at least one deity exists. Deism is the belief that reason and observation of the natural world are sufficient to determine the existence of God, a dominant individual who has made the world and its biota and guides the populace in the right direction. Pantheism is the belief that everything composes an all-encompassing, immanent God and that nature and the universe are identical with divinity. Polytheism is the worship or belief in multiple deities (portrayed as Gods and Goddesses), along with their religions and rituals. Monotheism is defined as a belief in the existence of a single God or in the oneness of God.

The Formative Stage of Religions

According to anthropologists John Monaghan and Peter Just (*Social and Cultural Anthropology*) (2000), many major world religions appear to have begun as revitalization movements of some sort, stemming from visions of dominant and charismatic prophets who were able to fire the imaginations of people seeking a more comprehensive answer to problems that they felt were provided by customary beliefs. These authors suggest that even though some such charismatic individuals have emerged at many times and places in the world, it seems that the key to long-term success for a religion has a lot to do with not only the prophet who brings a new concept to people but also the group of followers who are able to institutionalize the movement.

A good example is seen in the rise of Åmen as a prominent deity in Egypt. Budge (1969) states that "up to the time of the 12th Dynasty, Åmen was a God of no more than local importance, but as soon as the princes of Thebes had conquered their rival claimants to the sovereignty of Egypt and had succeeded in making their city a new capital of the country, their God Åmen became a prominent God in Upper Egypt." Over time, "all the attributes of all the great Gods of Memphis were contained in Åmen also. Thus, by these means, the priests of Åmen succeeded in making their God, both theologically and politically, the greatest of the Gods in the country."

Many stories that have arisen in aboriginal societies throughout the world to explain the unknown represent the material with which different groups of aboriginal people have established their individual cultures, their belief systems, and moral codes. Interestingly, many of the mythological stories that spring from the human mind in many early religions explain the beginning of the world in which they live in similar ways.

All formal religions appear to have had common roots, stemming from stories and beliefs that existed in the earliest civilizations (Budge, 1969). The major religions of the world show many parallels with the belief systems of early civilizations. But over time, religions change. Old religions sometimes bend their thoughts to apply to more modern populations, and new religions make new rules and interpret earlier doctrines to suit their needs.

Creation Stories

Most mythologies incorporate a creation account, in which one or more deities bring forth some form of what appears to be reality, often in a way that reveals the close intimacy between the creator and the creation (Hamilton, 1988). Many mythologies include a savior of sorts whose main desire is to liberate mankind from its own violent and ignorant (animalistic) nature. In the process, this savior makes rules to follow and requires absolute obedience to them; if the rules are not followed, there are consequences that may extend into an afterlife.

It appears that human behavior is expressed toward these rules in different ways. Many humans believe they require their religion and attempt to follow the way of their God, using his or her (God's) strength to carry out their earthbound activities on a daily basis and thanking their God for this strength. While moral issues are part of religious philosophy, whether individuals in this group adopt a moral code depends on their particular human nature.

Other humans utilize religious concepts when it is advantageous for them to do so. In these individuals, moral issues may or may not be important in their lives. Still others practice moral, immoral, and amoral ways of life without considering an intervening God-like force that guides them. As noted in these chapters, morality, immorality, and amorality may or may not have to do with the possession of a religion.

The Basic Nature of Indigenous Religions

Religious systems have led to approaches for understanding the mythological roots of humans and other organisms that have led to more than 730 established religions in the world, which form more than 3200 different subsects. Christianity, for example, is one of the major religions but has more than 200 subsects, each with its own unique traditions and interpretations of the Bible.

The three religions most often brought up in conversations (all considered monotheistic) are Judaism, Islam, and Christianity. Christianity is the largest religion, with two billion followers, Islam is 2nd largest, with 1.3 billion followers, and Judaism is 12th largest with 14 million followers. The founder of Judaism is believed to be Moses of Abraham; the founder of Islam is believed to be Muhammad; and the founder of Christianity is believed to be Jesus. The sacred books for each is the Bible (Jewish Bible and New Testament) for Christianity, the Bible for Judaism, and the Koran (Qur'an) for Islam.

Exactly when Judaism (the oldest religion) was founded is not known, although around the 1st century BC, there were already several small Jewish sects; Islam was founded in 622 BC, and Christianity began as a Jewish sect in the Levant of the middle east in the mid-1st century.

Human nature, as defined by each of these religions, is as follows: Judaism—there are two equal impulses, one good and one bad; Islam—there is an equal ability to do good or evil; and Christianity—the original sin was inherited from Adam, and there is a tendency toward evil. The means of salvation for each of these religions is: Judaism—belief in God and good deeds; Islam—correct belief and good deeds; and Christianity—correct belief, faith, good deeds, and sacraments (some Protestants emphasize faith alone).

It is interesting that the people who belong to these religions and subsects believe that their form of religion is the right one and that other religions are less acceptable. Their choice to follow a particular religion, of course, depends on how they were raised, what supervision they had during the remainder of their young life, and on conclusions they have derived through their individual thinking process. At times, feelings about their religions are strong enough to cause them to reject other religions and even fight one another over their differences.

Strange, New Religions

The Internet reveals many religions (or cults) that may seem strange to most humans, but they show how people choose to ignore rational thinking and are able to abscond from well-established religions and accept the unusual. The following few examples are from a 2014 article by Josh Parylak (2014) (*8 Incredibly Strange Religions From Around the World*) at whatculture.com:

- *The John Frum Cult*: This cult is based on Vanuatu, an island nation in the South Pacific. The actual person, "John Frum," is something of an enigma, and his exact ethnicity is also something of a mystery. He is depicted as an American serviceman, or sometimes a native man dressed as a Westerner, who visited the island and promised the natives that if they followed him, he would bring them wealth and happiness. For this to happen, they had to reject Christianity and other western cultural ideas.
- *The Prince Philip Movement*: Coming from a nearby island, people of the village Yaohnanen worship Prince Philip of Edinburgh as a God. The origins of the cult are a bit foggy, but the cult itself dates back to before Prince Philip visited the island in 1974. This trip only strengthened the cult as they had an opportunity to observe the Prince for the first time.

  The origins are theorized to be due to interactions with the British colonists on the island in the years prior to the Prince's visit. The villagers were exposed to the Christian belief that one day a messiah would return to the world, and also the great respect that the British had for Prince Philip. At some point, the two ideas became combined.
- *Jediism*: The origin of this cult is not in question. Though having no clear founder (unless you count George Lucas), the tenants of the faith are based on the fictional order depicted in the Star Wars films and media.

  The religion is not a centralized one, with various Jedi churches throughout the world, like the Temple of the Jedi Order, which is registered in Texas. The various Jedi churches acknowledge the Star Wars origin, but are also clear that they are not using it as an excuse to be an organized group of super-fanboys. Followers of Jediism do not run around learning how to fight with light sabers or move things with their mind. Rather, they follow the basic tenants of the Order as shown in the movies. In short, they avoid obsessive attachments, they try to remain mindful of their feelings, and try to foster peace and harmony.
  In the past, there have been various grassroots movements to get citizens to list "Jedi" or "Jedi Knight" as their religion on the national census. In 2001, 390,127 people in England and Wales listed Jedi on their census forms, placing it before Sikhism, Judaism, and Buddhism and making it the fourth largest listed religion.
- *The Church of Euthanasia*: This church, on its own website lists its purpose as follows: "a nonprofit educational foundation devoted to restoring balance between Humans and the remaining species on Earth." Their one commandment is "Thou shalt not procreate," and membership in the church is seen as a vow not to do so.

  "The Human population is increasing by one million every four days. This is a net increase of 95 million per year. Even major wars or epidemics hardly dent this rate of growth, and modern wars also have tremendous environmental consequences. It is for these practical reasons, as well as moral ones, that we support only voluntary forms of population reduction":

  > The Church has only one commandment, and it is 'Thou Shalt Not Procreate.' In addition, we have four 'pillars' or principles, which are Suicide, Abortion, Cannibalism, and Sodomy. Note that cannibalism is only required for those who insist on eating flesh, and is strictly limited to consumption of the already dead. Also note that sodomy is defined as any sexual act not intended for procreation.

- *Nuwaubianism*: The Nuwaubian Nation was founded by Dwight York. The group began as a Black Muslim group during the 1970s New York. York went through a few pseudo-religious organizations before officially founding Nuwaubianism. His teachings started with heavily modified Islamic themes and has since formed a formal document of cult teachings, borrowing from Ancient Egypt, UFO culture, Cryptozoology, and conspiracy theorists.

  In the 1990s, York built Tama-Re in Georgia. It was a heavily Egyptian-themed compound for his followers that was demolished in 2005 after it was sold. York himself was arrested in 2002 and convicted on charges of child molestation. He now claims that he should be immune from prosecution because he is a Consul to Liberia.

  Five randomly selected beliefs of Nuwaubianism are shown below, taken from the writings of York, many of which were sent out as letters from prison (collected by Wikipedia from his many letters and writings):
  1. It is important to bury the afterbirth so that Satan does not use it to make a duplicate of the recently born child.
  2. The physical body, when one dies, recycles from the ground in which it is planted; it then penetrates into the atmosphere as film or dust particles, and are breathed into the nostrils of human beings where they are recycled by triggering nerves in the sinus cavity. This sinus cavity works like a kidney to sift out the unneeded particles and relates the chemical composure of that which is needed into the mother's bloodstream and the human flesh is reanimated, and this is the process by which a child is given form or flesh in the mother's womb.
  3. Women existed for many generations before they invented men through genetic manipulation.
  4. The Koran called the Holy Qur'aan or the glorious Qur'aan as held in the hands of Muslims today is a product of Jewish scholars and the Catholic Church's branch of the Jesuit priest under Pope Augustine."
  5. The Earth is hollow and contains cities populated by different species of people, such as the Deros, the Teros, the Flugelrods, the Duwanis, the Dunakial, and the Anunnaqi. The legends of the Sumerians, Ancient Egyptians, Aztecs, Olmecs, Mayans, Hopi, and Hindus speak of these things (for instance, the story of the river Styx). Many of the chambers of the Egyptian pyramids lead to this subterranean world.
- *Apatheism*: Apatheism is not an organized viewpoint or church but rather a type of belief, or lack thereof. Although not technically a religion, it is presented here to explore the mental meanderings of humans:
  1. Theists believe in some sort of deity.
  2. Atheists believe that there is no deity of any kind.
  3. An Agnostic is generally the term given to someone who is unsure of a belief in a deity.

  Apatheists present a view in which they do not care about whether there is a deity. They feel that believing in a deity is irrelevant to their everyday lives. The idea of "Is there a God or is not there?" cannot be proven or disproven by any amount of arguing.
- *Raëlism*: Raëlism is a UFO (unidentified flying objects) religion. It was founded by Claud Vorilhon, now known as Raël. Based on the number of followers, it is the largest UFO religion in the world (unless you count Scientology as one).

  According to Raëlism, life on Earth was created by the Elohim, a group of extraterrestrials who posed as angels and deities in their dealings with early humans. Raël is the most recent (and last) human contacted by the Elohim to prepare us for them for his return.

  The Raëlists believe that if humanity could free itself from war, we too could go on to create life on other planets. They also believe that humans can achieve eternal life through cloning. One would clone his or herself, age that clone to adulthood, then transfer their mind/consciousness into that new body. They believe that most of the major religions on Earth are the doings of the Elohim and have alternate explanations for most major religious stories, such as the Noah's Ark (a spaceship to preserve DNA and recreate animals following the Great Flood, which was caused by the Elohim sending a nuclear missile to Earth) and the Garden of Eden (a research laboratory of the Elohim located on an artificial continent).
- *Happy Science*: Founded in Japan by Ryuho Okawa in 1986, this religion concerns itself with making people happy. A large number of their teachings reflect Buddhism and the concept of "Right Mind."

  Okawa claims to be the incarnate form of the "Supreme Being," the deity that Happy Science calls "El Cantare," which they state is the true hidden name of God in the Old Testament. Okawa also regularly interviews the "Guardian Spirits" of other political figures and publishes them in the group's newsletter and in books.

They believe that China and North Korea are plotting to use nuclear weapons on Japan so that they can invade and colonize it.

While some of these religions may have been developed in gest, some, possibly all, have followers. It is often the case that they were established by crazed, or psychopathic dominants who found followers who wanted to believe in something that would make their life more important.

Hierarchical Nature of Religions

Every religion seems to hold to some chain of command (dominance hierarchy), an organized line of increasing authority up to the God they worship. One of the best examples is the extremely well-defined, complex hierarchical nature within the Catholic church, which is a structural organization considered by those who belong to it to be of divine institution.

Its hierarchical nature refers to the ordering of ministry in the church, under God and Christ, into the threefold order of episcopate, presbyterate, and diaconate, which is considered to be divinely instituted and essential to the Church itself. In some cases, it refers only to the magisterium, the official teaching body of the church, the bishops, excluding deacons and presbyters (priests). At the bottom of the list are its followers, the common, subordinate people who follow the religion.

There is, in addition, an order of precedence of the various offices and ministries, which indicates the precedence, rank, or position of various ministers and offices in the Church for use during liturgies or other ceremonies where such protocol is helpful.

As of 2008, the Catholic Church was comprised of 2795 dioceses, each overseen by a bishop. Dioceses are divided into individual communities called parishes, each staffed by one or more priests. Priests may be assisted by deacons. All clergy, including deacons, priests, and bishops, may preach, teach, baptize, witness marriages, and conduct funeral liturgies. Only priests and bishops are allowed to administer the sacraments of the Eucharist, Reconciliation (Penance), Confirmation (deacons may administer these sacraments with prior ecclesiastic approval), and Anointing of the Sick. Only bishops can administer the sacrament of Holy Orders, which ordains someone into the clergy.

Thus, it appears clear that there is a strong desire by most humans to believe in a supernatural force, a force that gives them strength and hope to get through life with as few malfunctions as possible. Yet, it is sometimes the malfunctions that stimulate some humans to seek assistance from a deity. When times get rough, people often turn to their God to make things better, and they are willing to take a place in a hierarchy, the different positions depending on what is most comfortable for them. And in the process, they must believe (or pretend to believe) in the religion, the God, and the stipulations put forth by sacred scriptures.

Mesoamerican Religious Hierarchy

Even in ancient cultures, hierarchies are quite clear. In Mesoamerica, for instance, one of the basic features of the traditional Indian peasant communities is the civil-religious hierarchy, which combines most of the civil and ceremonial offices of the town's organization in a single scale of yearly offices. According to Pedro Carrasco at the University of California in Los Angeles (*The Civil Religious Hierarchy in Mesoamerican Communities: PreSpanish Background and Colonial Development*) (2009), "The higher offices are those of town councilmen (*regidores*) and judges or mayors (*alcaldes*) in the civil government and several ceremonial stewardships (*mayordomias*) in the cult organization."

When a town is subdivided into wards, each ward most often participates equally in the higher levels of the hierarchy; there are parallel offices of the same rank, one for each ward, or a single position rotates year after year among the different wards. Generally, a man alternates between civil and religious positions and, after filling an office, takes a period of rest during which he does not actively participate in the town's civil or ceremonial organization until the time comes again for him to occupy a higher office. As a citizen of the community, he has the obligation to serve, and social pressure to that effect is always strong. Citizens are also driven to apply for offices in order to raise their social status. In the ceremonial organization, office-holding can also be the outcome of a religious vow by which an individual attempts to obtain supernatural help through participation in or sponsorship of a public ceremony.

Participation in the lower ranks of the ladder (the subordinates) simply involves the performance of menial tasks, such as sweeping, carrying messages, or policing the town. Higher offices (those who are higher in the dominance hierarchy), carry higher responsibilities in the political and ceremonial organizations and usually demand a number of expenditures in the form of sponsorship of festivals and the banqueting connected with the transmission of office:

> These celebrations are, in effect, feasts of merit in which the consumption of an individual's wealth results in his enhanced social status, and a number of reciprocal exchanges of goods and services center around the organization of festivals. Thus, the operation of the hierarchy's ladder also implies that all share, in turn, in the financing of the town's government and ceremonials.

The Value of Contemporary Religions

As can be imagined, contemporary religions are complex and include social institutions and morality. However, they also include superstitions or make use of magical thinking. While adherents of one religion often think of their religion as real and true, they sometimes think of other religions as superstition. Atheists, some deists, and skeptics regard all religious belief as superstition.

According to a recent article in the Washington Post, "more than eight in ten people [worldwide] identify with a religious group," based on a new comprehensive demographic study of more than 230 countries and territories conducted by [a 2010] Pew Research Center's Forum on Religion and Public Life. There are 5.8 billion religiously affiliated adults and children around the globe, representing 84% of the 2010 world population of 6.9 billion.

While atheists disregard higher God-like authorities, many people from all walks of life depend on one to get them through life. The following quote is from a Luminosity article entitled, *Types of Religion* (2015):

> Religion adds meaning and purpose to the lives of followers, granting them an appreciation of the past, an understanding of the present and hope for the future.

This is a natural reaction by many people to the unknown, especially in right-brained individuals who want and need to believe in an all-powerful deity and choose to disregard a more logical explanation for the presence of humankind and other life forms. Such beliefs are often in direct contrast with the findings of science, a field that has delved into factual information about the origin of Earth and life and the progression of events that have taken place on Earth since its formation and the evolution of biotic populations.

Controversy Between Religion and Science

Science and religion in contemporary times are far from compatible, although these approaches to understanding life do not currently share the competitive edge that they had in earlier times.[4] There are scientists today who follow one religious sect or another and accept the concept of evolution as well. How they rationalize the two and live a life that has realism and fantasy as opposite components is difficult to understand.

Religion in general has cast a deceptive shadow over the scientific world since the commencement of philosophical thought, not relenting until philosophers of the Renaissance and the rise of Humanism forced religion to lose its dominant grip on world thinking (Bauman, 1999). It was through the release of religious prejudice that science began to survey the world like no one had done before, and it has been through this release that humans have come to examine and accept the evolving nature of their species and other life forms.

Strangely, the field of science actually had its beginnings in philosophical thought, which initially stemmed from Greek mythology and a religious base (Cartwright, 2013). Therefore, a scientific approach, like that of religion, was fed by the curious nature of humans and the desire to understand the unknown.

Greek philosophy is usually divided into thoughts that existed before Socrates (pre-Socratic), during the period when Socrates was alive (Socratic), and after Socrates (post-Socratic) (Kahn, 1998). In pre-Socratic philosophy, religion was highly influential, and thoughts of a scientific nature opposed the thinking of the day. Thus, although freethinkers attempted to explain the unknown, they were considered by religious followers to have a skewed view of the world, and science was destined to be ridiculed from its inception. Religious dominants and freethinking dominants were intolerant of one another.

One of the most influential of the pre-Socratic philosophers was Parmenides who stated that change cannot occur (May, 2000). This concept of a changeless world presided in the minds of early philosophers, including Socrates, Plato, and Aristotle. As we now realize, the early concept of a changeless world has been dramatically revised. Scientific research into the physical Earth and every form of life and its origin have shown that change is present everywhere and has occurred throughout the 4.5 billion years since Earth's formation.

Such expressions about a lack of change had been influenced by religious thought that everything was put on Earth in the form it now held. It is important in contemporary biological thinking that we realize that change is inevitable, that it has always occurred, that it is occurring at all times, and that it will continue to change until the Earth itself dies. Its remains will somehow, at some point, be used again, and thus change will continue in a more universal form through shifting forms of cosmological energy. We may not always recognize change in a mere human lifetime, but over long periods, it has been occurring ever since our world began and will continue until the end of time.

A modern religious approach to change is grossly different. Since God is responsible for putting humans on Earth in their present form, the evolutionary steps referred to are not accepted. Thus, fossil evidence of early humans is not acceptable to these people. Neither is the amount of time that has passed since the beginning of simple, unicellular life forms and the chemical, anatomical, genetic, and embryological evidence that supports change.

Science accepts and religions generally reject the idea that change is found in every aspect of life on Earth, as well as in association with abiotic constituents. The universe is in constant motion. Our world is directly influenced by events in our solar system and in our galaxy, possibly in the universe or multiple universes. Weather constantly changes, the position of planets that orbit our sun changes, there are polar shifts, seasonal weather shifts, global warming, global cooling, and many other planetary fluctuations that provide a diverse assortment of stresses for populations, all of which show variation due to mutations.

Mutagenic changes within populations give a species its variability (Sinha and Nussinov. 2001). Environmental changes interact with population variation, putting pressure on populations and resulting in natural selection. To say that the world was static was an indication of how little was known about science (even in the greatest of minds) during its formative period and how influential were concepts guided by religious thought. Many people still cling to these ideas, often because of what has been written in religious scriptures or in how they are interpreted.

Let us examine the thinking of some of these early philosophers and their concepts of the world. While Socrates (469 BC) was an early freethinker, his knowledge of the world was limited. In addition, he was biased because of religious influence and yet, he felt there was no higher virtue than that of wisdom. He freely admitted that he knew nothing but in doing so he claimed he was much the wiser for knowing that he knew nothing. At the age of 70, he was sentenced to death for refusing to acknowledge the Gods of Athens, but it has been said that his real crime was his unrelenting passion for philosophy or thinking of the unknown, attempting to understand it and openly expressing his thoughts:

> Men of Athens, I honor and love you, but I shall obey God rather than you, and while I have life and strength, I shall never cease from the practice and teaching of philosophy.

With no original body of his work to refer to, we must rely on secondary sources, the most reliable of which was the work of Plato. Plato (427–347 BC) believed that the world was made up of objects, which he called forms (such as birds and trees) and ideas (such as virtue and equality). He, like Socrates, said that forms are eternal, changeless, and predetermined, another indication of how powerful religious thought was at that time.

Aristotle (384–322 BC) was the most methodical and systematic of the three great ancient Greek philosophers, and yet, he was influenced by contemporary religious thought and by those who came before him. His metaphysics (i.e., the philosophical understanding of reality) is essentially a modification of Plato's theory of ideas.

Following these philosophers and after the Athenian city state began to decline, two schools of philosophical pessimism emerged: cynicism and skepticism. In many ways, one can think of skepticism as Greek philosophy returning, full circle, back to the philosopher who originally gave it its energy, namely Socrates. Nevertheless, in spite of this more humanistic approach, religious dominance continued to influence the thoughts of philosophers for many years to come.

The intellectual energy of the Renaissance (14th and 15th centuries in Europe) sparked an adventurous exploratory spirit in Europe (Stark, 2005). Based on the information he had at the time, Copernicus claimed to have discovered that the sun rather than the Earth was the center of the Universe (Rabin, 2007). The movement that is most commonly associated with the Renaissance is Humanism, a movement dedicated to the study of the cultural artifacts of human beings as opposed to the divine creations of God.

In the late 15th century, a resentment of the corruption, indulgences, and moral and political dominance of the Catholic church had developed (Blockmans and Hoppenbrouwers, 2014). Francis Bacon (1561–1626) was a champion of a movement opposed to the dominance of theology over scientific matters (Gaukroger, 2001). He argued that it is only through reason that humans can hope to understand and control the laws of nature. Therein originates the famous maxim, "Knowledge is Power."

At the time, a control of the laws of nature seemed like a plausible and obvious way to improve life, but it may have partially stemmed from the religious connotation that we were given the right to control nature by God. Unfortunately, our control of nature has since become one of our most daunting attributes.

It was not until the 7th century that philosophers began to think of themselves as distinct from theologians, but the basis for change began to occur much earlier. For example, unlike his predecessors, Jon Du Scot (AD 800–877) was prepared to indulge in classical scholarship rather than accept concepts already established merely for the benefit of the church. For Scott, reason presided over religion.

Nevertheless, the Enlightenment actually established a greater realization of the conflict between science and religion. Scientific discoveries of the 17th and 18th centuries questioned the wisdom of religious orthodoxy by establishing laws of nature that did not originate in divinity. Galileo (1564–1642), for instance, strongly believed that the laws of the natural world were firmly within the grasp of human rationality and not hidden in the hand of God.

John Locke (1632–1704) regarded himself as a practical philosopher and believed philosophy should be built upon reason and common sense rather than metaphysical speculation. Immanuel Kant (1724–1778) followed Locke's reasoning and has been regarded as the most acute incarnation of the Age of Reason (Wellbery, 1984). He was one of the first thinkers to question concepts that were not scientifically based. He thus entered the Age of Reason without defending the religious authority, which the age had threatened (Geisler and Turek, 2004).

Philosophers of the time, however, still thought of humans as distinct from other animals. Indeed, many modern humans fail to see the resemblances between humans and other animals, in spite of the many similarities. The Enlightenment view held that the human attribute that most sharply distinguished human beings from animals was reason.

New rational thought was not accepted by everyone. For the Romantic, the intellect can never be separated from the passions of life that embody it. For example, truth and justice, they believe, cannot be accounted for at the expense of emotion and desire. Rationality was thus seen by some as a threat to human individuality and creativity. Friedrich Nietzsche (1844–1890) believed that truth (i.e., scientific thought) is merely a cultural necessity and that there is no moral truth and no scientific natural truth that governs our existence.

Based on a continuing philosophical approach, times and approaches to science and determining a factual basis for human existence grew more favorably in the minds of science-based individuals. This approach was never more energetic than during the latter half of the 19th century and beginning of the 20th century.

It has been within the last couple of centuries that science has developed into its individual fields. Sincere changes in the human approach to biology came about when Charles Darwin published his most famous book, *Origin of Species*, in 1859.

The Vienna School was made up of a group of philosophers, scientists, and mathematicians who met during the 1920s and 1930s to discuss the foundations of science and philosophy. They wanted to rid philosophy completely of what they saw as meaningless metaphysics. Bertrand Russell (1872–1970) also hoped to purify philosophy by eliminating the speculative content and allowing only pure unmediated fact to serve as its foundation (Ryle, 1970). Like other freethinkers of his time, it was his opinion that "truth can be reduced to a series of logically independent facts, which have a logical coherence."

Karl Popper (1902–1995) wondered why academics should pursue philosophy at all if science can provide answers to the meaning of existence with more vigor and certainly more facts (Miller, 1997). However, part of scientific investigation is precursory (philosophical) speculation about a particular subject. Needless to say, scientific discovery is usually based on the knowledge and ability to intelligently speculate when approaching a research project.

Early theories about the scientific method involved the process of induction in which facts were gathered and examined for patterns and connections that may suggest a certain hypothesis about a certain phenomenon. Popper vigorously rejected this theory, arguing that science proceeds in precisely the reverse direction (deductive reasoning), beginning with the premise of a theory and then accumulating facts. In contemporary science, both approaches are valid.

According to Popper, no theory should ever be closed to scrutiny and therefore falsification. In scientific circles, this is a wise approach. Many "discoveries" have been made over time that were eventually determined to be invalid. Science, irrespective of the greatness of an individual and his theories, often begins with speculation, follows a path through hypotheses and testing, and leads to the formation of a conclusion. Nevertheless, conclusions for one reason or another are sometimes incorrect, but the scientific approach remains an endeavor to determine the truth, and with repeated testing, the truth will ultimately be determined.

Since Darwin, numerous philosophers and scientists have repeatedly speculated about the details of evolution, but even then, many individuals, including some scientists, were not ready to accept humans as full-fledged animals. The controversy that resulted from Ed Wilson's *Sociobiology* points out the reluctance of humans to accept their place in the kingdom Animalia.

Summary of Concepts on Change

Although early philosophers felt that organisms were static, we have since learned that:

- Change is evident all around us. It is seen in the weather, ice ages, global warming and cooling, plate tectonics, earthquakes, volcanic activity, mountain formation, erosion, continental drift, and polar and vegetation shifts; change is even recognized throughout the universe. Change is inevitable and continuous. Nothing is really static in our world. Everything is constantly changing. In terms of organisms, mutations are constantly occurring, and genes are variably selected during the meiotic divisions that lead to the production of spermatozoa and ova, and the determination of spermatozoa to fertilize, causing offspring to be different and variable.
- All organisms vary tremendously in their anatomical makeup, behavior, chemistry, physiology, and many other ways. Because of these variations, organisms have a variable toleration for changes in their environment. This can cause them to migrate, emigrate, disperse, survive, develop new species, die, or become extinct.
- Environmental changes put stress on populations to change. While changes within an individual's somatic (body) cells may alter the individual's behavior, such changes are not inherited by the offspring unless they are based on changes within the germinal cells. When they occur in the germinal tissues, they affect subsequent generations, and thus acceptance of change in germinal tissues is a population issue.
- It is the selection of features already inherent in individuals within populations by nature, a process referred to as natural selection, that is behind the process we call evolution. Survival of the fittest is simply the selection of individuals that are most suited to their changing environments and interactions with other organisms. The continuous selection of individuals that best suit their environmental needs, generation after generation (maximization of fitness), leads to population changes.

And yet, quite a bit of our DNA remains in our cells which is somehow lost in expressions of our phenotype (physical appearance), DNA that has been referred to as junk DNA, selfish DNA, and pseudogenes. There are developmental stages that we go through during our embryological period that are turned on and off, which reflects on evolutionary stages our predecessors have passed through millions of years ago. We carry genetically related behavioral features in these genes as well.

Religions With the Term Science in Their Name

Religions such as Scientology, Christian Scientists, and others with the term "science" in them have nothing to do with science, but they use the image of science to gather followers who like to believe they are capable of believing in logic. It is here that I feel we should take a closer look at what the word "science" means.

Based on a Wikipedia review (http://en.wikipedia.org/wiki/Science) (2014), science is recognized as a systematic enterprise that builds and organizes knowledge in the form of testable explanations and predictions about the universe.

- Since classical antiquity, science as a type of knowledge has been closely linked to philosophy. In medieval times, foundation for scientific method was laid which emphasized on experimental data and reproducibility of its results. In the early modern period, the words "science" and "philosophy of nature" were sometimes used interchangeably. By the 17th century, natural philosophy (which is today called "natural science") was considered a separate branch of philosophy (see the forgoing discussion in this chapter on science and religion).
- In modern usage, "science" most often refers to a way of pursuing knowledge, not only the knowledge itself. It is also often restricted to those branches of study that seek to explain the phenomena of the material universe. During the 17th and 18th centuries, scientists increasingly sought to formulate knowledge in terms of laws of nature such as Newton's laws of motion. And over the course of the 19th century, the word "science" became increasingly associated with the scientific method itself, as a disciplined way to study the natural world, including physics, chemistry, geology, and biology. It is in the 19th century also that the term scientist was

created by the naturalist–theologian William Whewell to distinguish those who sought knowledge on nature from those who sought other types of knowledge.

- However, the term "science" has also continued to be used in a broad sense to denote reliable and teachable knowledge about a topic, as reflected in modern terms such as library science or computer science. This is also reflected in the names of some areas of academic study, such as "social science" or "political science."

Fields of science belong mostly to four groups: natural sciences, social sciences, formal sciences, and applied sciences. The natural sciences may be subdivided into subgroups, such as physical, Earth, and life sciences. Physical science includes such fields as physics and chemistry. Earth science includes such fields as ecology, oceanography, geology, and meteorology. Life science includes biology, which can be further subdivided into zoology, human biology (the animal we know most about), and botany. The formal sciences include such fields as decision theory, logic, mathematics, statistics, systems theory, and theoretical computer science. We will not include the social sciences and applied sciences here because they do not reflect upon the concepts we wish to bring forth.

While all of the natural sciences and formal sciences collide with religion to some degree, it is mainly the field of biology that religious concepts are antagonistic to, because it is primarily biology that holds to the concept that all living forms evolve, that they were not put on Earth in the form in which they currently appear. All life forms represent the simplest of prokaryotes (unicellular organisms that do not have a nucleus) or forms that arose from them through natural selection. Evolution is a progression of change within populations to environmental pressures. The chapters in this book discuss the similarities and differences between organisms in detail, showing that all living organisms share many of their biotic features, and each species is the result of isolation mechanisms of some sort.

Seventh-day Adventist writer, Ted Wilson, summarizes the way most dyed-in-the-wool religions look at evolution. "We believe that the biblical creation account in Genesis 1 and 2 was a literal event that took place in six literal, consecutive days recently, as opposed to deep time. It was accomplished by God's authoritative voice and happened when He spoke the world into existence." Wilson also points to Psalm 33:6–9, which proclaims that "By the word of the Lord, the heavens were made, and all the host of them by the breath of His mouth. . . for He spoke, and it was done; He commanded, and it stood fast."

Evangelist Garner Ted Armstrong often chose scientific papers to discuss on his radio programs. He commenced his analysis of a scientific paper, often showing a reasonable understanding of the work done, and subsequently he proceeded to discredit the paper based on scriptures, thus first appealing to his reader's intelligence and logic and subsequently converting his thoughts that he felt showed the paper to be wrong. Evangelists like him and religions that portray themselves as individuals or groups based on logical thinking evidently feel threatened and choose to constantly eat away at the scientific findings that relate to evolution and the origin of life.

Scientology is a "religion" that appears to appeal to certain influential individuals in contemporary times. The following information was taken from a 2014 Wikipedia article of that name (http://en.wikipedia.org/wiki/Scientology):

- Scientology is a body of beliefs and related practices created by science fiction writer and religious dominant, L. Ron Hubbard (1911–1986), beginning in 1952 as a successor to his earlier self-help system, Dianetics. Hubbard characterized Scientology as a religion, and in 1953 incorporated the Church of Scientology in Camden, New Jersey.
- Scientology teaches that people are immortal beings who have forgotten their true nature. Its method of spiritual rehabilitation is a type of counseling known as auditing, in which practitioners aim to consciously reexperience painful or traumatic events in their past in order to free themselves of their limiting effects. Study materials and auditing sessions are made available to members on a fee-for-service basis, which the church describes as a "fixed donation." Scientology is legally recognized as a tax-exempt religion in the United States, Italy, South Africa, Australia, Sweden, the Netherlands, New Zealand, Portugal, and Spain; the Church of Scientology emphasizes this as proof that it is a bona fide religion. In contrast, the organization is considered a commercial enterprise in Switzerland, a cult (secte) in France and Chile, and a nonprofit [organization] in Norway, and its legal classification is often a point of contention.
- A large number of organizations overseeing the application of Scientology have been established, the most notable of these being the Church of Scientology. Scientology sponsors a variety of social-service programs. These include the Narconon antidrug program, the Criminon prison rehabilitation program, the Study Tech education methodology, the Volunteer Ministers, the World Institute of Scientology Enterprises, and a set of moral guidelines expressed in a booklet called *The Way to Happiness*.
- Scientology is one of the most controversial new religious movements to have arisen in the 20th century. The church is often characterized as a cult, and it has faced harsh scrutiny for many of its practices that, critics contend, include brainwashing and routinely defrauding its members, as well as attacking its critics and perceived enemies with psychological abuse, character assassination, and costly lawsuits. In response,

Scientologists have argued that theirs is a genuine religious movement that has been misrepresented, maligned, and persecuted. The Church of Scientology has consistently used litigation against its critics, and its aggressiveness in pursuing its foes has been condemned as harassment. Further controversy has focused on Scientology's belief that souls ("thetans") reincarnate and have lived on other planets before living on Earth, and that some of the related teachings are not revealed to practitioners until they have paid thousands of dollars to the Church of Scientology. Another controversial belief held by Scientologists is that the practice of psychiatry is destructive and abusive and must be abolished.

Dominance Conflicts Between Religions With War-Like Tendencies

Certain individuals in all cultures will continue to follow one religion or another. In many free societies, they retain the right to believe whatever they want. Fortunately, we live at a time and place in which nonbelievers of deities most often have the same rights. However, some religions are based on intolerance, mostly because of dominant prophets who have been recognized as leaders of a particular religion and their intolerant followers, and this has led to cultural segregation. In extreme cases, cultural segregation and intolerance have led to religious war and genocide.

People with different religions have been concerned with domination over one another throughout time. As pointed out by Richard Dawkins (2006a) in *The God Delusion*, "Christianity was founded by Paul of Tarsus as a less ruthless monotheistic sect of Judaism." Their guide book and interpretation of the human spirit is the Bible.

"Several centuries later, Muhammad and his followers reverted to the uncompromising monotheism of Jewish origin and founded Islam upon a new holy book, the Koran or Qur'an, adding a powerful ideology of military conquest to spread the faith. Christianity, too, was often spread by the sword, wielded first by Roman hands after the Emperor Constantine raised it from the level of an eccentric cult to an official religion, then by the Crusaders and later by the conquistadores and other European invaders and colonists, with missionary accompaniment." The following statement attests to the war-like qualities of Islam:

> The Qur'an tells us: "not to make friendship with Jews and Christians" (5:51), "kill the disbelievers wherever we find them" (2:191). "Murder them and treat them harshly" (9:123). "Fight and slay the Pagans, seize them, beleaguer them, and lie in wait for them in every stratagem" (9:5). The Qur'an demands that we fight the unbelievers, and promises "If there are twenty amongst you, you will vanquish two hundred. If a hundred, you will vanquish a thousand of them" (8:65). *Institute for the Secularization of Islamic Society*

The God that people pray to in contemporary times is a powerful God, a dominant, often intolerant (despotic) God who strives to maintain dominance over all of mankind. This appears to be true in any religion, but God's position of dominance and the way that humans interpret it is questioned by some philosophers. Dawkins, who vehemently expresses an atheistic view, opens his third chapter in *The God Delusion* with the following statement, indicating his version of God's dominant position in the world:

> The God of the Old Testament is arguably the most unpleasant character in all fiction: jealous and proud of it; a petty, unjust, unforgiving control-freak; a vindictive, bloodthirsty ethnic cleanser, a misogynistic, homophobic, racist, infanticidal, genocidal, filicidal, pestilential, megalomaniacal, capriciously malevolent bully.

Regardless of who the worshippers are and what God they worship, God-loving people look at him (or her) in a positive manner and think of their deity as a good God. Believers and nonbelievers may ask: Do people who worship God actually believe their God has these characteristics or do they understand that it is simply their interpretation? What is the difference?

God and the expectations of God have been designed by prophets and their followers in the first place. Is God looked at as the same God, no matter what religion it is, or is God a different God in each culture or interpreted in different ways by different cultures?

It appears that the majority of people in the world (those following the major religions) must think they believe in different Gods because they are willing to fight for their own God. Prophets took it upon themselves to write down their interpretations of the God they believed in, making it a different God for each religion.

The God of Islam is an intolerant God. The God of Christianity, also mostly intolerant, has many personalities, depending on the subsects one follows (Dawkins, 2006a). The God of other religions may be a kind or despotic God. The God of yet others may be energy or nature itself.

As in any social group of humans, religious leaders who express an undying obedience to their God form a dominant (beta) position over the members of their submissive congregations, to which they broadcast what they believe to be the traits of a kind and understanding God. The degree of dominance expressed by religious leaders varies with different religions.

According to Dawkins, the predisposition of humans to be easily seduced by confident, charismatic leaders may be best demonstrated in religious organizations. Through religious leaders that claim special access to a deity, the image of a supreme being is reinforced by broadcasting myth and eucharistic rites.

It is understandable that many nonscience-related, mythologically-dependent individuals throughout the world turn to some form of religion. As pointed out by E.O. Wilson in *Consilience*, "the vast majority of our political and community leaders are trained exclusively in the social sciences and humanities and have little or no knowledge of the natural sciences." The same can be said about people who follow them. Thus, most are not familiar with or even interested in the biological concepts of variability, isolation, natural selection, changing populations, and coexistence among all living things. To them, as in other people who indulge in religious practices, there is beauty in religious affirmations where one can find the process of life and death beautiful and eternally hopeful in the scriptures. It is more appealing to go to heaven and live a sublime afterlife than to simply return to nature.

Moral Codes, Intolerance, and War

Most religions project some form of moral code on their followers. Yet, our moral obligations as citizens of the world have fallen by the wayside when religions guide the various people in the world to intolerance of populations and religions other than their own. It is yet another form of dominance: my religion is better and truer than yours, and if you do not follow mine, you are infidels. What is the beauty in intolerance? It undermines togetherness and compatibility.

It is also an ecological tragedy that emphasis is taken away from Earth and all of its living forms and put on the salvation of the human soul. The soul is a human concept, but preservation of the soul is an extension of the preservation of one's self, which is animalistic. Through our population's ignorance and abuse of the biotic world, we may all expect to test the authenticity of the human soul at the premature termination of the Anthropocene and the Cenozoic Era.

In the *Encyclopedia of Wars*, authors Charles Phillips and Alan Axelrod (Lincove, 2005) present a comprehensive listing of wars in history, documenting 1763 wars overall, 123 (7%) of which are classified as involving religious conflicts. William T. Cavanaugh (2009) argues that "what is termed 'religious wars' is actually largely a Western dichotomy of different power configurations which serve a Western consumer audience."

To be honest, many wars that are considered religious wars have a number of economic and political ramifications (e.g., land acquisition, control of trade routes, and dynasty changes) that we could use to question the apparent reasons behind such conflicts (Kippenberg, 2011; Ring, 2000). Differences in religion can readily inflame the participants of a war being fought for other reasons. Historically, places of worship have been destroyed to weaken the morale of the opponent, even when the war itself is not being waged over religious ideals.

Religious designations are sometimes used as a form of shorthand for cultural and historical differences between combatants, often giving the misleading impression that the conflict is primarily about religious differences. For example, there is a common perception of the troubles in Northern Ireland as a religious conflict. One side (Nationalists) is predominantly composed of native Catholics and the other (Unionists) stems from British-sponsored immigrant Protestants (Leonard, 2006). However, upon analysis, it is found that the more fundamental cause is the attachment of Northern Ireland to either the Republic of Ireland or the United Kingdom, and while religion plays a role as a cultural marker, the conflict is in fact ethnic or nationalistic rather than religious in nature.

Religious Harmony and Disharmony

As pointed out earlier, much of the world is Christian, especially in the United States, and thus it is appropriate to use Christianity as an example of a religion that can provide harmony in the lives of its followers, and at the same time, there are drawbacks as well. It, like other religions, is based on mythology, and factual information that contradicts its beliefs is rejected. Nevertheless, for many people, Christianity is a positive force in their lives. Many Christians depend on their relationship with God and Christ to get them through their days of strife. They thank their deities (or deity) for their blessings and pray for help when they pass through dark periods in their life. The concepts and their deities comfort them.

To most believers, God is a kind God, a tolerant God, and yet, strong believers are often intolerant of ideas and people outside of their religion and will force their beliefs upon anyone who believes otherwise, including their children. At times, they will go to any length to achieve their goals.

Dawkins speaks of Pastor Keenan Roberts and a method of indoctrinating children into the religious world with what he refers to as Hell Houses. "A Hell House is a place where children are brought, by their parents or their Christian schools, to be scared witless over what might happen to them after they die. Actors play out fearsome tableaux of particular 'sins,' such as abortion and homosexuality, with a scarlet-clad devil in gloating attendance. These are a prelude to the *piece de resistance*, Hell Itself, complete with sulfurous smell of burning brimstone and the agonized screams of the forever damned."

On a planetary level, Christianity and some other religions, as briefly mentioned earlier in this chapter, may adversely affect the human population by suggesting that humans reproduce and subdue the Earth, as well as expressing a relationship of dominance over all creatures, allowing us to decentralize the importance of living in harmony with our fellow creatures. Examples are found in Genesis in the Bible:

> God created man in his own image, in the image of God created he him; male and female created he them. And God blessed them, and God said unto them, *Be fruitful and multiply and replenish the earth and subdue it.*
> God said, let them have complete authority over the fish of the sea, the birds of the air, the beasts, and over all of the earth, and over everything that creeps upon the earth.

While such statements may have seemed reasonable when they were written, our breeding has now led us to an overpopulated world, and our inconsiderate domination (=authority) over our environment and its biota has led to a planetary environmental disaster. The writer of such a passage (a prophet) would have best included the thought of cohabitation between humans and other planetary life forms rather than dominance over everything.[5]

Consider overfishing alone as a faulty use of our dominant nature. According to an article published by Overfishing. org, "the world's fish stocks are either overexploited or depleted. Another 52% is fully exploited; these are in imminent danger of overexploitation (maximum sustainable production level) and collapse. Thus, a total of almost 80% of the world's fisheries are fully-to-over-exploited, depleted, or in a state of collapse. Worldwide, about 90% of the large predatory fish stocks are already gone. In the real world," they point out, "all this comes down to two serious problems":

- We are losing species as well as entire ecosystems. As a result, the overall ecological unity of our oceans is under stress and at risk of collapse.
- We are at risk of losing a valuable food source many depend upon for social, economic, and dietary reasons.

Also, read what is said in Chapter 16 about the pollution of Earth's oceans and lack of respect for Earth's biota in general. It is not simply in the ocean that we are losing biota. It is throughout all ecosystems. The world has become a garbage heap.

The authors of this article (Pauly and Palomares, 2005) point out that the single best example of the ecological and economical dangers of overfishing is found in Newfoundland, Canada. In 1992, the once-thriving cod-fishing industry came to a sudden and full stop when no cod appeared at the start of the fishing season. Overfishing allowed by decades of fisheries mismanagement was the main cause for this disaster that resulted in almost 40,000 people losing their livelihood and an ecosystem in [a] complete state of decay. Now, 15 years after the collapse, many fishermen are still waiting for the cod to return, and communities still have not recovered from the sudden removal of the region's single-most important economical driver. The only people thriving in this region are the ones fishing for crabs, a species once considered a nuisance by the Newfoundland fishermen. And [with the human population steadily increasing], this shall decline as well:

> It is not only the fish that are affected by fishing. As we are fishing down the food web, the increasing effort needed to catch something of commercial value, for example, marine mammals, sharks, sea birds, and noncommercially viable fish species in the web of marine biodiversity are overexploited, killed as by-catch and discarded (up to 80% of the catch for certain fisheries), and threatened by industrialized fisheries. Scientists agree that at current exploitation rates, many important fish stocks will be removed from the system within 25 years.

In a *Science* article in 1998 (*Fishing Down Marine Food Webs*) and another in *Marine Science Bulletin* (*Fishing Down Marine Food Webs: It is Far More Pervasive than We Thought*), Daniel Pauly and other environmental critics state that "the big fish, the bill fish, the groupers, the big things will be gone (Pauly et al., 1998; Pauly and Palomares, 2005). It is happening now. If things go unchecked, we'll have a sea full of little horrible things that nobody wants to eat. We might end up with a marine junkyard dominated by plankton." Pauly's later articles point out that earlier accounts of overfishing and its damage to ecosystems have been grossly underestimated.

Religious dominants have argued that God most certainly would not mean for us to abuse nature and that the original writing may be misinterpreted to mean extreme dominance when it should mean domination within limits. Yet, these are the words that many people believe and follow, and they have taken it to an extreme. Why could not it or an additional commandment (added to the top of the list) have said "Love nature as you would love yourself [or better than you would love yourself] and treat it kindly, and with respect" rather than "dominate it with complete authority." Another commandment may be added as well, "Do not overpopulate the world." Maybe Mel Brooks was right. Maybe there were 15 commandments.

Relationship Between Religions and Human Behavior

In summary, it is true that there have been numerous religious wars that have resulted in numerous deaths through time. Over the years, I have listened to people ridicule religious doctrine as the reason behind many wars, and I have entertained similar thoughts for years as well, but I sincerely believe it is not precisely the religions that are at fault, at least much of the time. In the first place, inspired or not, humans wrote the text upon which religions are based. It also is humans who interpret the text. Whatever doctrine that suggests or supports war originated in the human mind. Also, humans obviously have varying degrees of intolerance toward one another and have a warring nature and will use anything to war on groups of their own species that are unlike themselves (see Chapter 15).

Because of the association between religion and war, as well as the pseudo-religious approach people have toward life itself, many antisocial individuals use religions and other socially acceptable groups to disguise their true nature. Many nonreligious writers suggest that religions sometimes lead to disharmony in human populations. In truth, the concept of religion (in spite of its mythological basis) is often good for those who need it to cope with daily life, but the actions of people who initially instill a war-like quality in their religion and people who later employ this aggression against other people who differ in their religious beliefs are what leads to incompatibility and religious war.

Religious and Nonreligious Morals

This brings us to wonder if people who follow religions are associated with any more moral fiber than those who do not have a religion. The answer appears to be no.

There are many types of religious morals. Modern monotheistic religions, such as Islam, Judaism, Christianity, and certain others, define right and wrong by the laws and rules set forth by their respective Gods (or the human prophets who wrote about them) and as interpreted (or misinterpreted) by religious leaders (dominants) within the respective faith. Polytheistic religious traditions tend to be less absolute. For example, within Buddhism, the intention of the individual and the circumstances should be accounted for to determine if an action is right or wrong (Harvey, 2000; Keown, 1992).

In certain other religions, for example, Hinduism, right and wrong are decided according to the categories of social rank, kinship, and stages of life (Flood, 1996). Also, there is no absolute prohibition on killing in this religion. It recognizes that it may be necessary in certain circumstances.

It has been found that religion is not always positively associated with morality and that nonreligious individuals not only have a moral code but also are generally more tolerant toward individuals unlike themselves and actually practice their moral code more so than those with religious backgrounds (Rachels and Rachels, 2011). In fact, major crimes have been found, in many instances, to be compatible with a superstitious piety and devotion.

Sexual abuse of minors by Catholic priests was found to be rampant in the United States and apparently around the world, based on articles printed in the Boston Globe in 2002 (Investigative Staff, 2015; Miles, 2012; Moore, 1995; Podles, 2007). Coverage of an initial few priests and their sexual urges encouraged other victims to come forward with allegations of abuse, resulting in a number of lawsuits. What initially appeared to be a few isolated cases of abuse subsequently exploded into a national scandal involving thousands of victims over several decades and encouraged other nations to bring forth allegations of abuse as well. Allegations, of course, caused the Catholic church to reassign the accused to other parishes, apparently allowing them to continue their crimes.

According to Nicolai Senneles (2014), sexual abuse is also widespread among Muslims. Thus, it is justly regarded as unsafe to draw any inference in favor of a man's or woman's morals from the fervor or strictness of his or her religious or nonreligious exercises.

Religious intolerance is an outstanding feature of many religions. An example can be seen in the invasion of North America by Europeans who in the late 1400s were escaping from their homeland to find gold, better health, and a freedom of religion. One of their first chores in the New World was to dominate "the savages," convert what they referred to as "heathens" to their way of belief, and kill those who would not convert.

Religions and Crime

The overall relationship between religious faith and crime is unclear. Some researchers argue for a positive correlation between the degree of public religiosity in a society and certain measures of dysfunction (Canasova, 1980). Others have concluded that a complex relationship exists between religiosity and homicide, with some dimensions of religiosity encouraging homicide and other dimensions discouraging it. Still other studies seem to show positive links in the relationship between religiosity and moral behavior, even altruism.

Criminological research also acknowledges an inverse relationship between religion and crime. Analyses on religion and crime have shown that religious behaviors and beliefs exert a moderate deterrent effect on an individual's criminal behavior. It appears that determining positive and negative correlations between religion and criminal behavior depend on the individual features of each religion and not on religions as a whole. Also, understanding the nature of humans, whether religious or otherwise, is complicated by a diverse assortment of personalities and deceptive behavior.

Note

Superscript numerals appearing in this chapter refer to additional text/explanation given in the appendix.

13

Dominance in Politics

The vast majority of our political leaders are trained exclusively in the social sciences and humanities and have little or no knowledge of the natural sciences. *E.O. Wilson, Consilience, The Great Branches of Learning*

Individuals who enter political office in any country are among the world's most dominant individuals in our society in terms of the influence they have on making the rules we follow, and therefore, they are in a position to greatly influence the populace and the world around them (Suttner, 2006). These are generally individuals who, in a democratic society, we choose as our political leaders, based on their platform and reputation of former policies prior to their election to higher offices of leadership.

Unfortunately, while the direction in government that a political dominant takes may initially appear to be related to a desire to improve how people are governed and the rights of the populace, it all-too-often becomes a personal endeavor by the individual to improve their own status and immediate environment (Williamson, 2013), sometimes at the expense of the people being governed by them. After all, we have become familiar with the personalities of dominant people and know that they are dominant because they seek the benefits of being dominant, for example, status, power, recognition, choice of their surroundings, wealth, and material resources.

While many laws have been passed to protect the environment, the environment is most often not a prime issue when we select a leader. Many laws about the environment are passed by pressure from ecologically minded individuals who are more atuned to the importance of environment and leaning on political figures to work it all out (Brulle, 2010a,b). Otherwise, most politicians do not usually understand environmental issues and do not seek changes without outside pressure, although they may profess to take an interest in it.[1]

Thus it is as much the fault of the people who choose a leader who does not hold the environment in the number one position of importance, as it is the fault of the leader. Most people make their political decisions on what they understand, and most people do not understand or care about environmental degradation if it does not apply directly to what they see as their well-being (Stamm et al., 2000).

In addition, knowing human behavior as we do, we may not actually have the ability to choose a leader who has a good moral character. Politicians train themselves in speaking, creating an environment in which they dominate and say the things their electors want to hear (Edgar, 2012).[2]

As humans, we generally want to employ trust in our decisions to choose a leader, but when we apply what we know about the various personalities found in human populations, their dominant and subordinate traits and their potential for following a moral code, we begin to realize that what we hear and what is on the mind of a dominant (and/or the party to which they belong) may not be the same. As discussed in Chapter 9, personality type is a perplexing feature of humans, and individuals who have serious personality disorders are capable of covering up what they do not want others to see.

According to BusinessDictionary.com, leadership is characterized by a clear line of authority that gives a leader the power of delegation to control a subordinate's level of participation in the decision-making process. Thus a dominant personality is associated with the most predominant forms of leadership in our societies. We choose dominant people for political office.

In a publication called Why Do Dominant Personalities Attain Influence in Face-to-Face Groups? The Competence-Signaling Effects of Trait Dominance, Cameron Anderson and Gavin J. Kilduff (2009) point out that dominance is demonstrated in ways that show up in expressions of personality. The concept of trait dominance deals with a person's characteristic feelings of control and influence over everyday situations, events and relationships, versus feelings of being controlled and influenced by circumstances, for example, changes in life and work situations.

Based on studies of assertive, forceful, and self-assured behavior, it has been found that individuals higher in personality trait dominance tend to attain more influence in face-to-face groups than others. They speak more, gain more control over group processes, and hold disproportionate sway over group decisions.

"At first glance," Anderson and Killduff say, "the reason why trait dominance leads to influence seems obvious: Individuals high in trait dominance are assertive and motivated to lead and thus take control through the force of their personality. However, prior research has shown that [under normal circumstances], individuals cannot take charge of groups simply through force; rather, they must [at least] seem to possess superior task and social competence. Because trait dominance is unrelated to many of the tasks and social competencies required in leaders, it is unclear why dominance relates to influence in groups so strongly and consistently."

Selection of such individuals, of course, depends on who is making the selection. Under normal circumstances in a democratic society, it is the populace that is choosing their leader. However, the populace can make mistakes by not understanding the political motivation and true personality of the politician being judged. Based on information that was covered in Chapter 9 on aberrant behavior, it can be very difficult to identify the true personality of an individual. In addition, in countries in which a dictatorship is in force, it is the government and the dominant people who compose the government which are making the selection.

In their research, these authors proposed that "dominant individuals achieve influence because they tend to appear competent to others, even when they actually lack competence. This brings fourth thoughts of toxic leadership and deception, as discussed in earlier chapters. Specifically, dominant individuals behave in ways that make them seem both expert at the task and socially skilled, which leads groups to afford them influence and control," that is, they are acting out their strategies rather than having the roots of knowledge they should be relying upon.

They tested this hypothesis in two studies of task groups, using a social relations model analysis of peer ratings of competence and influence. They also used outside observers to rate individuals' competence and used frequency counts of discrete behaviors to measure the display of competence-signaling behaviors during group discussions. Influence in face-to-face groups is a process in which individuals modify others' behaviors, thoughts, and feelings.

They determined that their assumptions were true, that "dominant individuals [did] attain influence because they behave in ways that make them appear more competent along both task and social dimensions—even when they actually lack competence." These authors based this argument on two sets of findings.

First, because individual member's abilities are typically hidden from each other, groups can only allocate influence on the basis of what they believe each group member's competence to be. These beliefs are often based on superficial cues such as nonverbal behaviors. For example, individuals are perceived to be more competent when they use more certain and factual language, speak more often and in a fluid and assertive manner, use more direct eye contact and use a relaxed and expansive posture.

"Second, individuals higher in [personality] trait dominance tend to display more of these superficial 'competence cues.' In group settings, individuals higher in [personality] trait dominance make more suggestions and express their opinions more frequently, speak in more assertive tones" and follow through with eye contact and postural expertise.

THE POWER OF POLITICS

The fate of people, destinies of nations, and history itself are determined to a substantial extent by the strategies of political leaders (Bratsis, 2014; Garifullin, 2012). When electing a politician, voters frequently base their decisions on the image created by mass media and professional image makers. These, in turn, are based on the appearance of dominance portrayed by their selection of a leader, as discussed earlier.

Voters also tend to take into account only current events and tasks. However, this method, along with a poor understanding of the politician's true personality, can result in bringing to power a politician who eventually damages the country's position or even causes national or international cataclysms.

The performance of a political government can be measured by the increase of the quality of life in the respective country during the time of ruling. Personality traits of many political leaders of different countries in various

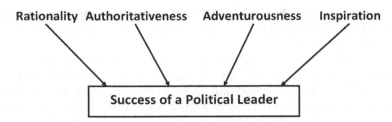

FIGURE 13.1 Four personal characteristics on which success of a political leader crucially depends. These personality traits determine what underlies the politics of the leader on the basis of what factors he or she uses to understand the situation and make decisions. Unfortunately, decisions made by voters often do not take into consideration the hidden personality variations in dominant leaders and their influence on people who seek trust in the individual who leads them.

historical periods were analyzed and juxtaposed with the performance of their governments (Cowley, 1931; Derue et al., 2011). This approach revealed four personal characteristics on which success of a political leader crucially depends (Fig. 13.1):

- *Rationality*: an exercise of reason.
- *Authoritativeness*: the quality of possessing authority; the quality of trustworthiness and reliability.
- *Adventurousness*: inclined to undertake new and daring enterprises; hazardous; risky.
- *Inspiration*: an inspiring or animating action or influence; something inspired, as an idea; a result of inspired activity; a thing or person that inspires.

TOXIC LEADERSHIP

We discussed toxic leadership in Chapter 12, but the highlights of that concept are worth repeating because of its implications in understanding politics and political dominants. Recall that Colonial George E. Reed (*Toxic Leadership*, 2004) stated that the terms toxic leader, toxic manager, toxic culture, and toxic organization appear with an increasing frequency in business, leadership, and management literature. This concept can just as well be used for a political leader, although there are some differences between the two.

A toxic leader in business is generally a manager who bullies, threatens, and yells (see Chapter 11); they have mood swings that determine the climate of the office on any given work day; they force employees to whisper in sympathy in cubicles and hallways; they are the backbiting, belittling boss from hell. In "some people," Reed states, "the sheer shameful force of their personalities makes working for them rotten." This same concept can be applied to many political leaders in the contemporary world and in earlier historical times.

The world has had its share of overly dominant (despotic) and toxic leaders, many of them with extremely serious personality disorders (Ceumern-Lindenstjerna et al., 2002). Political leaders are like other alpha individuals who may each demonstrate a level of dominance which suits their position in life. Once in power, some utilize their position to obtain their personal goals, sometimes at a catastrophic disadvantage to their citizens.

There are other factors to consider in determining their degree of dominance. Whether they become despots, for instance, depends not only on their behavior but also on the type of government they head. The most psychotic and intolerant individuals have been associated with governments that condone absolute rule.

B. Brunner's article on *The World's Most Notorious Despots* (2015) gives us some idea of the types of individuals who take charge of such governments. Each leader, he said, "had a penchant for sadism. And each would no doubt claim that the murder and mayhem that took place under his rule was necessary for law and order. The genuine reason," he says, "was an overwhelming egotism at the expense of everything and everyone else."

Tamerlane, descendant of Genghis Khan and ruler of the Timurid Empire from 1370 to 1405 spent his entire life conquering the inhabitants of Asia (Marozzi, 2004). When 24-year old, Tamerlane became the head of his tribe, and in a few years, he established himself as the leader of the entire Mongolian race. He was legendary for his ruthless savagery and lack of mercy, massacring entire populations, including 90,000 residents of Delhi. He "razed whole cities, leaving behind nothing but rubble."

Ivan the Terrible, Tsar of Muscovy, was the son of Vasily III Ivanovich, grand duke of Muscovy. As the first czar of Russia, he ruled until 1954. He is best known for his brutal ruling, centralized administration of Russia and expansion of the boundaries of the Russian Empire.

According to Brunner, "his early reign was primarily spent in battle in an effort to expand Russian land." He was considered a man of great natural ability, and his political foresight was claimed to be extraordinary. It has been said that he anticipated the ideals of Peter the Great, and only failed in realizing them because his material resources were inadequate. Brunner states that admiration of his talents should not blind us to his moral worthlessness, nor is it right to cast the blame for his excesses on the brutal and vicious society in which he lived. The same society that produced his infamous favorites also produced St. Philip of Moscow, and by refusing to listen to St. Philip, Ivan sank below even the amoral standard of his own age.

Brunner claims that he certainly left the Muscovite society worse than he found it, and thus prepared the way for the horrors of "the Great Anarchy." Later in life, he became paranoid and vindictive, at which time he exhibited tyrannical cruelty. Violence became prominently exhibited in his behavior in general. For instance, in addition to unleashing trained killers on the landed gentry, he was also guilty of domestic violence.

Maximillien Robespiere, a French lawyer, politician and one of the best known and most influential (dominant) figures of the French Revolution, was elected to the position of Estates General in 1789 by Artois (Carr, 1972). He joined the extreme left wing and immediately commanded attention. His influence in governmental matters grew daily, his followers admiring his earnest cant and incorruptible boasting about the left-wing bourgeoisie. He was described as being physically unimposing yet immaculate in attire and personal manners. Apparently not familiar with his true personality, his supporters called him "The Incorruptible," while his adversaries called him dictateur sanguinaire (inhuman dictator).

By many, he has been referred to as a mastermind of the Reign of Terror between 1793 and 1794, the dark underside of the French Revolution which perverted its lofty ideals of democracy with fanaticism. As a sign of his despotic and intolerant nature, he turned France into a police state, sending enemies of the nation to the guillotine without benefit of a public trial or legal representation. In the process of maintaining his dominance, about 40,000 French men and women were executed or died in prison, and another 300,000 were imprisoned. Only Robespiere's own beheading ended the slaughter.

Joseph Stalin attended a local church school as a young boy (Hingley, 1974; Montefiore, 2005). Despite health problems, he made good progress at school and eventually won a free scholarship to the Tiflis Theological Seminary. While studying at the seminary, he joined a secret organization called Messame Dassy. Members were supporters of Georgian independence from Russia. Some were also socialist revolutionaries, and it was through the people he met in this organization that Stalin first came into contact with the ideas of Karl Marx (revolutionary, sociologist, historian, and economist who collaborated with Friedrich Engels) to publish *Manifest der Kommunistischen Partei* (1848), commonly known as the Communist Manifesto, the most celebrated pamphlet in the history of the socialist movement.

He subsequently was among the Bolshevik revolutionaries who brought about a revolution in Russia in 1917 and later held the position of General Secretary of the Communist Party of the Soviet Union's Central Committee from 1922 until his death in 1953. He was a totalitarian leader of the USSR from 1929 to 1953.

In spite of his religious and seemingly moral upbringing, he crushed the Soviet people with his megalomania and repressive version of communism, his name becoming associated with cruel, draconian socialism. Millions of Soviets were forced to enter labor camps, and a vast number of ethnic groups were persecuted, especially Jews and Ukranians. In the process, about 20 million people died from starvation, executions, forced collectivization, and life in labor camps. Another 20 million survived imprisonment and deportation.

As a young boy, Adolph Hitler (born on April 20, 1889, in the small Austrian town of Braunau near the German border) did extremely well at primary school, and it appeared at that time he had a bright academic future to look forward to (Aigner, 1985). He was popular with other pupils and was much admired for his leadership qualities. He was also considered a deeply religious child and for a while considered the possibility of becoming a monk.

He later became the leader of the National Socialist German Workers Party (German: Nationalsozialistische Deutsche Arbeiterpartei, commonly referred to as the Nazi Party). He was chancellor of Germany from 1933 to 1945 and dictator of Nazi Germany (as Führer and Reichskanzler) from 1934 to 1945.

Hitler has been considered by many one of history's most intolerant and chilling tyrants. During his control of Germany from 1933 to 1945, he performed facist maneuvers in order to achieve world domination; he dreamed of a Teutonic master race and subjugated all non-Germanic peoples. As an intolerant despot, he was responsible for the

FIGURE 13.2   While Hitler believed he had his reasons (as told in Mein Kempf), a combination of hatred, extreme prejudice, and a psychopathic personality led him to accomplish despicable things during his political reign. He demonstrated intolerant despotic behavior that is so characteristic of psychopaths (as implied in Figs. 8.9 and 9.1). While Hitler has been considered by many one of history's most intolerant and chilling tyrants, there is a long list of similar leaders, as depicted in this chapter, who eliminate certain groups of humans that they have learned to hate, without feelings of compassion or empathy. At least 60 million men, women, and children have been victims of genocide and mass killing in the past century alone (Waller, 2007). Differences among people become ugly and unacceptable in an intolerant society, sometimes leading to violence and genocide. *Source: Anton_Ivanov/Shutterstock.com*

genocide of 6 million Jews, as well as the slaughter of Gypsies, Slaves, homosexuals, communists, and other people he considered undesirables and decadents (Fig. 13.2).

According to Edwin Black in *War Against the Weak* (2012), Hitler's means of dealing with population subordinates actually was an extension of eugenic thoughts that stemmed from England and the United States, under the control of wealthy dominants. Racial prejudice was and still is the force behind the elimination of what leaders consider undesirables in the human population.

Mao Zedong (Mao Tse-tung), commonly referred to as Chairman Mao (1893–1976), was a Chinese Communist revolutionary, guerrilla warfare strategist, Marxist political philosopher and leader of the Chinese Revolution (Carter, 1976; Chang and Halliday, 2005). He was the architect and founding father of the People's Republic of China from its establishment in 1949 and held control over the nation until his death in 1976. To many, he was an intolerant despotic ideologue. During his reign, he subjected the Chinese people to massive social experiments, ruling with absolute intolerance of dissent and opposition. His economic plans resulted in the worst famine of the century which claimed between 23 and 30 million lives. Other citizens in his realm were imprisoned, tortured, and murdered as suspected class enemies.

Proclaiming himself President-for-Life, Fransois (Papa Doc) Duvalier (1907–1971) was the President of Haiti from 1957 until his death in 1971 (Juang and Morrissette, 2008). Duvalier first won acclaim in fighting diseases, earning him the nickname "Papa Doc." He opposed a military coup d'état in 1950 and was elected president in 1957 on a populist and black nationalist platform. Subsequently, he and his henchmen terrorized and murdered potential political foes and ordinary citizens in Haiti. Up to 60,000 Haitians died under his reign between 1957 and 1971. His son, Jean-Claude Duvalier, nicknamed "Bébé Doc" or "Baby Doc" (born in 1951), succeeded his father as the ruler of Haiti upon his father's death in 1971. After assuming power, he introduced additional cosmetic changes, during which thousands of Haitians were killed or tortured, and hundreds of thousands fled the country.

Nicolae Ceausecu was a Romanian Communist politician. He was General Secretary of the Romanian Communist Party from 1965 to 1989, and as such was the country's last Communist leader. He was also the country's head of state from 1967 to 1989. His rule was marked in the first decade by an open policy toward Western Europe and the United States, which deviated from that of the other Warsaw Pact states during the Cold War. He continued a trend first established by his predecessor, Gheorghe Gheorghiu-Dej, who had tactfully coaxed the Soviet Union into withdrawing its troops from Romania in 1958. Ceaușescu's second decade was characterized by an increasingly brutal and repressive regime by some accounts, the most Stalinist regime in the Soviet bloc. He thus ran a neo-Stalinist police state in Romania between 1967 and 1989. His economic criminality turned a moderately prosperous country into a region with desperate shortages of food, fuel, and other essentials. His corruption and nepotism are considered legion, leaving Romania with the worst human rights contradiction in the Eastern bloc. He and his wife were executed following a televised and hastily organized 2-h court session.

Claiming to have a direct connection with God, Idi Amin was the military dictator and third President of Uganda from 1971 to 1979. Amin joined the British colonial regiment, the King's African Rifles, in 1946. Eventually, he held the rank of Major General in the postcolonial Ugandan Army and became its Commander before seizing power in a military coup in January, 1971, deposing Milton Obote. He later promoted himself to Field Marshal, while he was the head of state. He was a brutally authoritarian president, overseeing the torture and murder of an estimated 300,000 of his own people. He also orchestrated the persecution of the Lango and Acholi tribes and expelled all 60,000 ethnic Asians from the country, destroying the economy in the process. He gained an international reputation as a parody of an African despot. His savagery and repression have deeply scarred Uganda.

Pol Pot was a Cambodian Maoist (a view that the industrial-rural segments of a population represent a major division exploited by capitalism) revolutionary who led the Khmer Rouge from 1963 until his death in 1998 (Kiernan, 1985). From 1963 to 1981, he served as the General Secretary of the Communist Party of Kampuchea. From 1976 to 1979, he also served as the prime minister of Democratic Kampuchea. Pol Pot became leader of Cambodia in 1975. As a radical Marxist leader from 1975 to 1976, he led the state-sponsored extermination of citizens, during which one to two million people were massacred or worked to death through forced labor. His radical vision of transforming the country into a Marxist agrarian society led to the virtual extermination of the country's professional and technical class. He died in 1998.

These despots are just a few of the world leaders who have expressed themselves in an overtly dictatorial and intolerant way. They represent some of the extreme alphas, individuals comparable in dominant traits to despotic wasps on a nest that they rule with an iron thumb, or tarsus. And as was the case with extremely despotic wasps, some of the human despots suffered the consequences of being overly dominant and intolerant by being killed or ousted from their native society.[3]

In the meantime, they created havoc in their population, murdering, and torturing individuals they did not want in their society. At times, their desires to eliminate extended into other societies and world domination, influencing the formation, and use of armies for war.

Like most biological concepts, dominance behavior is not a black and white phenomenon; there are gradations, just like there are gradations in innocent and psychopathic behavior. There are many dominant individuals in the world who do not go to the extremes that these intolerants reach, but they possess the traits of dominants nevertheless and often live an egotistical life, generally hurting people in their path to gaining dominance. We may wonder, though, whether possessing dominant traits alone is enough for them to reach their goals. Are they so powerful that they can forge ahead without help from others?

While the finger for despostic dominance always points at one individual in the alpha position, we must remember that such individuals cannot do what they do alone. They, as in the prominent religions of the world, must have a following of individuals who will be loyal to their commands and a political system that punishes those individuals who are not obedient to their leader. In *Tyrants, The World's 20 Worst Living Dictators*, David Wallechinsky (2006) states the following:

> We like to think of dictators, such as Saddam Hussein, Kim Jong-il and Musmmae al-Qaddafi, as inhuman personifications of evil. But what is more disturbing is that each of these tyrants is a human being. Each of them had parents and each of them made a series of choices to gain power and to use that power in a cruel manner. What is more disturbing is that none of these dictators can maintain his grip on his victims without the active support of accomplices. And what is most distressing is that many of today's worst tyrants are financed, aided and abetted by the United States government and by U.S.-based corporations. These unholy alliances have continued whether the president of the United States is a Democrat or a Republican and whether Congress is controlled by liberals or by conservatives. *David Wallechinsky Tyrants, The World's 20 Worst Living Dictators*

We have heard such comments as those in Wallechinsky's statement above from others and may wonder how humans have become so corrupt and deceptive. Such is the case with regard to subordinates who were involved with the elimination of millions of individuals under the rule of Adolf Hitler. But is this dominance behavior any different than the behavior of lower animals where agonistic behavior, intolerance, deception and corruption had its beginnings? With them, we call it survival strategy.

While other world leaders are all dominant individuals, they demonstrate a wide array of degrees of tolerance and moral obligation, sometimes even assisting their societies to live a relatively happy, rewarding life (Piot et al., 2007).

However, even in these circumstances, the motives for carrying out a particular movement are unclear and often relate to benefits that can be reaped by those in a dominant position.

As in any population, dominant status and intraspecific tolerance by leading personalities appears to hold the key to keeping the human population under control but relatively happy. On the other hand, despotic dominance and intolerance leads to a psychopathic condition in which the populace of home and foreign countries suffer from their leader's amoral personality.

Note

Superscript numerals appearing in this chapter refer to additional text/explanation given in the appendix.

14

Human Aggression: Killing and Abuse

Modern man inherits all the innate pugnacity and all the love of glory of his ancestors. *William James, The Moral Equivalent of War*

While working on this book, I had the opportunity to watch the 2012 Memorial Day ceremonies on television. Story after heartbreaking story were told by spouses who lost their loved ones to recent wars, and dignitaries spoke of the tragedies of war. Similar stories accompanied Memorial Day ceremonies in subsequent years. It was a depressing and tear-jerking day, and for me, it supported thoughts about how animalistic we really are and how little we understand the roots of our actions.

In Chapter 8, we had a chance to look at human nature. We subsequently had an opportunity to see that most humans are quite willing to live a moral life as long as it does not interfere with personal pleasures and that humans in a dominant position often are capable of destroying a complete society by way of their antisocial behavior, selfishness, and intolerance. We also examined aberrant behavior and realized that humans are extremely complex animals, driven by animalistic, dominance-related interactions and a desire to be recognized. It is this human animal who can sometimes live a peaceful life and at other times commit violent crimes toward other animals, members of its own species, and Earth, often without remorse.

KILLING

As a group, many writers claim that we are a killing and abusive species, while others claim that we are not (Zuckerman, 1995). Some claim that we are basically moral, while others claim that we are corrupt. If we learn anything from the examples of human behavior presented in this book, we should realize that as a species, we are a mixture of good and bad traits (an antagonistic duality, expressed in historic, religious, and philosophical literature), occasionally expressing altruistic behavior, selflessness, remorse, and sympathy, and at other times we are often prejudiced, egotistical tyrants, selfish to the core, and unable to understand the feelings of other humans (Simon, 2008).

It is no secret that human behavior is diverse, and attitudes toward killing vary considerably. In addition, certain events in the lives of people often change them in one way or another, bringing out pleasant features at times and unpleasant features at others. War and the killing that accompanies it, for instance, are especially detrimental to the mental homeostasis of a person and his or her family who experience it (Livshits and Kobyliansky, 1991).

Lacking predators, we have modified the forces that influence our environmental carrying capacity (e.g., predators, parasites, and inadequate supplies of food); we have also overpopulated the Earth, and with our dominant and aggressive dispositions, we have put undue pressure on ourselves, providing an incentive for increased dominance struggles, including warring on one another (Russell, 2001).

Unpleasantly, we can relate the human population and the Earth to a cage of rats. The scenario goes something like this: Purchase a pair of rats, confine them to a cage, feed them, and let them reproduce. At first, the rats are generally a harmonious group, but with an estrus cycle of 4 days and an ability to have litters almost every 21 days, the population will soon grow out of bounds, and they will become more aggressive toward one another. Certain individuals will become overbearingly dominant, despotic, if you will. Eventually, they will kill one another and pollute their environment. We can allow them to continue increasing their population size and suddenly make their food scarce and watch closely at how their degree of aggressiveness rises.

Dominance and Aggression in Humans and Other Animals
http://dx.doi.org/10.1016/B978-0-12-805372-0.00014-6

209

In many respects, we are those rats, and our confinement is within the bounds of Earth. If we could, we might provide them with a cognizant brain, technological skills, and other human attributes and record the resultant increase in behavioral complexity and deception and intolerance.

We, like them and all other animal species, ignore the signs of overpopulation and impending doom. Under normal environmental circumstances, the numbers of organisms are brought back into bounds because of predation, parasitism, and the abundance of food. Because we ignore these signs and lack predators to keep our population down, we have additional problems, all contributing to a stressful existence, for example, several forms of environmental pollution, global environmental changes, dwindling resources, and all of the consequent environmental abuse pointed out in Chapter 16. As our populations grow out of bounds, we will find that certain resources will be in more demand, and we will fight other subpopulations of our species with technologically advanced weapons to get what we want. And in the process, we will destroy the only cage we have.

We should once again recall the outstanding differences between human personalities. There is a gradation of behavior from subordinates to tolerant dominants to towering intolerant, despotic dominants in groups of people around the world. Certain individuals in these hierarchies are more willing to kill than others. The perpetrators of killing sometimes go to extremes in their killing, while some get others to do it for them.

Some people say that there are legitimate reasons for war, but most wars occur because of differences between dominant, aggressive, and intolerant world leaders and the coalitions or covenants to which they belong.

Origin of a Killing Trait

As a species, we are certainly complex. We have pointed out that we sometimes demonstrate altruistic behavior, going out of our way to help others, and at other times, we seem to have very little respect for life. As members of what we may refer to as "normal" society, we routinely kill living things, often without ill feelings, and those who kill nonhuman animals often attempt to distinguish between which organisms have the right to live and those which do not. Humans who commit genocide even make this distinction between different groups of people (Andreopoulos, 1994).

Killing appears to be well engrained in our species, and it is found in most other predaceous animal species as well. It is a trait that actually assists all species in their survival and reproduction. Kill or be killed, express yourself in ways that tell your foe that you are not to be dealt with or hide to avoid killing. At times, however, we (psychopaths and "normal" people alike) kill for pleasure.

At the other end of the behavioral scale, there are people who abhor killing of any sort. Many individuals in our population even refrain from eating meat because of what they feel are atrocities committed to domestic animals from which the meat is taken (Andreopoulos, 1994). Yet, as a species, we most often feel no guilt for killing another animal: we kill agricultural animals; we fish certain species even to the brink of extinction; we hunt wild species as a sport, killing multitudes of feral individuals simply for our pleasure or by taking land for building upon or growing crops to feed our masses; and we sometimes kill one another, and often without feelings of empathy for the people we kill or their distraught families. But where does this killing trait originate? It is worth looking into this phenomenon to understand more about who we are and why we do these things.

In *Becoming Evil, How Ordinary People Commit Genocide and Mass Killing*, J. Waller (2007) discusses not only why humans commit genocide but the type of person who carries it out. While genocide and other forms of mass killing are complicated, prejudicial affairs, Waller's conclusion was that the people who commit atrocities against humanity are people just like us.

Although this may be a bitter pill to swallow for most of us, we may take solace in knowing that killing is not always the nature of most individuals who take part in genocide. The idea, brought out in the last chapter and in the earlier statement , is generally initiated by dominant psychopaths and followed by group action.[1]

For a moment, let us entertain the thought that certain types of killing are accepted in human society (although some individuals within the society may feel otherwise). Other forms of killing may be determined to be expressions of amoral, selfish, or psychopathic behavior.

Interspecific Killing

Predator–prey (interspecific) interactions are absolutely necessary in all natural communities since predators are heterotrophs (they eat other animals) and depend on prey to survive (Ballard, 2011). In the process, they function in keeping the populations of prey within certain limits, maintaining adequate food chains and ecosystem homeostasis (Trojan, 1984). Thus, such killing has both instinctive and learned components, and for feral populations of other animals, they are matters of survival and reproduction. Since humans do not have predators and the process of dying at the population level is slow compared to their reproductive rate, the population continues to grow and will reach a point of total collapse unless we somehow lower our biotic potential (Dailykos, 2014).

There are other ways of lessening the pressure on human reproductivity. Humans may choose not to represent themselves as predators or carnivores, for instance, by eating lower on the ecological trophic scale (e.g., eating plants). But most humans like to eat meat (as of February, 2012, McDonald's restaurants alone were selling an average of 4500 hamburgers every minute of the day). The consummation of beef and other meats by humans has required the production of increased numbers of cattle and other animals, putting an excessive drain on and a loss of ecosystems.

In addition to purchasing meat, humans have retained a killing nature and use it on occasion to obtain the meat by acquiring wild animals from natural environments. Whether through fishing or hunting, certain individuals take pleasure in obtaining their high-protein diets, and an entire industry has been built upon the human need to kill. Such killing is so engrained in our behavior that it is not looked upon as killing in the eyes of most people, but rather a form of a sport (Peta, 2015). But let us investigate the seemingly paradoxical phenomenon further.

Upon analyzing the process of fishing, most would agree that using a steel, barbed device to snare an organism, allowing the device to become lodged in the organism's throat, under its tongue, or even further into its digestive system where it cannot be retrieved or that removing it, possibly tearing some of its vital tissues in the process, is not only an aggressive act but one that is premeditated. Yet, most people do not think of fishing in this way. Rather, it is perceived as an enjoyable pastime (Quora, 2015). Likewise, using a fire-powered device to shoot metal into the body of a nonarmed animal, causing it pain and eventual death, is also premeditated aggression. Most people do not think of these events as aggressive because they are carried out on a subhuman organism and in a "sporting" fashion. If we did them to another human, even to domesticated animals, we most likely would go to jail. Of course, this depends on the society to which we belong.

In considering these forms of human behavior, several questions about them come to mind. The following is not meant to be a critique of human behavior but merely an attempt to describe who we are:

- Why do people enjoy the sports of fishing and hunting, particularly if they do not need the flesh of an animal for survival?
- What is it about killing animals that people feel is normal human behavior?
- Do humans fail to realize they are taking lives when hunting and fishing?

Growing and Killing Domestic Animals

In human populations, there are various reasons for killing and not killing. According to James A. Serpell (*One Man's Meat. Further Thoughts on the Evolution of Animal Food Taboos*) (2012), there are people throughout the world who display strong aversions to killing and consuming particular animals, and thus the choices people make for killing and preserving life often vary unpredictably from culture to culture, place to place, and between individuals. There may be ethical issues involved, resulting in sparing the lives of animals. However, when pressed, humans usually elect to ignore ethics and morality, opting instead for pleasure (Brandt, 1979; Schneewind, 1977, 2002; Smith, 1954). Recall that moral issues in human society are viable as long as they do not interfere with anthropocentric reasoning.

Most contemporary humans have arisen from more ancient societies that were made up of hunter–gatherer consumers of meat (Barnard, 2004; Bettinger, 1991; Bowles and Gintis, 2011; Brody, 2001; Lee and DeVore, 1968; Meltzer, 2009; Morrison and Junker, 2002; Panter-Brick et al., 2001; Turnbull, 1987). However, in the modern world, there are concerns by certain individuals and groups about the ethics of killing for food.

Peter Singer (*Animal Liberation*) (2001) and his followers believe that if alternative means of survival exist for humans, one should choose the option that does not cause unnecessary harm to animals. People who are considered ethical vegetarians argue that the same reasons should exist against killing animals to eat that exist against killing humans to eat. Killing animals, they claim, can only be justified in extreme circumstances and that consuming a living creature for its enjoyable taste, convenience, or nutritional value is not sufficient cause.

While morality is considered important in most societies, moral values change, depending on the society we are talking about (Bowles and Gintis, 2011). Also, there is variation within each population as to how people within their society interpret morality and feel about killing. Societies, especially dictatorial societies, function around their dominant leaders who may not stick to moral codes. Atrocities committed by such leaders depend on how immoral they can be (De Vries, 2003).

Some opponents of ethical vegetarianism within human society describe the thought of comparing the eating of livestock with the killing of people to be fallacious (Donaldson and Kymlicka, 2011). According to this view, humans are cultural organisms, innovative and capable of acting in an ethical manner. It is common to think that they also have a soul. Animals, they say, are not morally equivalent to humans and thus do not possess the rights that a human has. For example, shooting a moose or other large mammal, killing a mouse, deliberately running over a snake or turtle in the road, or stepping on an insect are not moral equivalents to committing homicide. Such acts point out that many humans do not mind killing. They, in fact, may even enjoy it, depending on the context and the worthiness of the species being killed (Bourke, 1999).

As it turns out, when humans want to hunt and eat meat, they will hunt and eat meat, making arguments about whether killing animals for food is ethical a pointless endeavor. Thus, people who want something are willing to ignore taboos on the killing of domestic animals to satisfy themselves. Is the lack of concern for the lives of animals any different in cases of feral species? Not really. Most people are tuned in to conveniences, and when wildlife interferes with convenience, it must go.

Another example of how people rationalize about animals is found in their domestic pets (Fraser, 1998). People who adore their dogs or cats still put some distance between them and feral animal populations. However, the distance is shortened under certain circumstances. When witnessing the killing of a young feral animal on television (especially if it is cute), people become upset. Yet, no one becomes alarmed when vast amounts of habitat are destroyed, resulting in the killing of an untold number of animals, in order to provide pastureland, a space upon which buildings will be constructed, or simply for the beautification of an area (Cincotta and Engelman, 2000; Geist and Lambin, 2002).

Also, all animals are not created equally in the eyes of humans. Domestic pets are high on the list of animals to protect. They are worthy because they are our pets and companions. In a way, we can consider this a selfish act because we are actually thinking about our pleasures when we accept a pet. When a person sees a dog that has been killed on the road, they often express some form of grief. Seeing a reptile or any lower-than-domestic organism dead on the road brings less grief or no grief at all.

When it comes to ethics, humans have a skewed idea of what is and is not important, depending on what is most personal to them (Heter, 2006). Most people do not consider the results of taking land from nature. Nevertheless, the growing and killing of animals for food represents an ecological problem because we need land to grow them on, which in reality results in an unethical treatment of our natural world.

As an example, growing crops for farm animals (including the 95 million cattle and 60 million hogs in the United States alone) requires nearly half of the water supply in the United States and 80% of its agricultural land (Nickerson and Borchers, 2012). Animals raised for food in the United States consume 70% of the country's grain, 80% of the corn, and 90% of the soy crop. It is clear that producing animal feeds in what is referred to as factory farms is typically much less efficient than the harvesting of grains, vegetables, legumes, seeds, and fruits for direct human consumption (Baumann, 2000). In essence, if humans need food, they will get it, killing the world's natural populations in the process. Consequently, an enormous amount of natural environment is taken from nature, no matter how the land is used. With our human population running rampant, we are far from maintaining a homeostatic balance with the natural world (Esmay, 1969).

Thus, anthropocentric reasoning supports the growing of animals for slaughter, case closed. Most of us may wonder why there is any argument about it. Killing has its place in modern society. But how far will the human animal go to claim humane treatment of animals, domestic or wild?

FISHING

Fishing is a sport that is loved by many people. However, as explained, it is a form of killing or torture in the eyes of the animal being caught, as well as in the eyes of some humans. People who are avid fisherpersons are generally not concerned with using a barbed hook or a net to catch, subdue (dominate), and otherwise traumatize a fish, whether it is for sport, food, or some mystical ingredient to promote sexual dominance in those who eat it. When it jumps out of the water, we get excited and comment about the pleasure of the fight. As it gets close to the boat, we may remark about its natural beauty. If the fish being brought in is large enough and strong enough, we often do not hesitate to comment about how much of a fighter it was, and then, for our safety, we sometimes clobber it mercilessly with a blunt instrument before bringing it into the boat. If it is a trophy fish, it may end up on the wall for all to see.

In a Florida Fish and Wildlife Conservation article about why people fish, Rich Abrams (2012) says, "Some people find solace in a fishing experience." What is better than anticipating the trip, rising early in the morning, watching the sun rise in the east, and traveling over intoxicating blue water to an ultra-serene location far away from modern society (the part of life that gives us stress)? "The smell of the salt air, [the quietness,] the sun on your shoulders while casting a lure across a calm water surface." Our pleasure in fishing, he says has to do with "the anticipation of a fish striking the lure as the water erupts with unseen energy, the tug of the rod and line stripping off the reel, the pleasure in releasing the fish (by some), and watching it glide through the water back to its spot. By releasing it, we have done a good thing. We have given its life back, and it makes us feel good. Its torn mouth, gills, and digestive system often do not enter the equation." If it does bother us, we may comment that we hope it will be alright, and without further thought, we return to fishing:

> "Others fish and develop a relationship with their adversary. [Seasoned] anglers, for instance, study the life history of the fish they seek, speculate about where fish will be during a certain time of year or day, what lures or baits will work best under different conditions and the best method to present [bait] to a fish. They study their foe as if preparing for a battle."

In all honesty, it is simply a game of dominance to many a fisherperson, a game to conquer the unknown, the pleasure of subduing a wild animal and pacifying it, the bigger the better. Subduing their small-brained fish brings untold happiness to a fisherperson and carries with it stories about the adventure and the one that got away. But, so is war a game to certain people, a game in which members of their own species are subdued. Is there a difference?.

Food is the goal of others who fish. They spend the day on the water catching fish that will be cooked and enjoyed that night (sometimes in spite of warnings of polluted waters), and they save some for other meals. They know which fish taste better over the grill, battered, and deep fried or blackened with lemon squeezed over it.

To remember their fishing trips, many people take pictures. Some humans who have their prizes mounted snuff out its life without taking a bite. Again, the bigger they are, the better the prize, dominance over nature. We want to remember the story and tell others about it.

OVERFISHING

If we take a several-day cruise to anywhere, we are awed with the vast amount of water Earth has. Looking out in the deep blue, our first reaction may be that it is an untouched underwater wilderness, full of exotic fish that we can depend on indefinitely. Not so. Oceanic waters have undergone dramatic changes within recent years, and fish populations have seriously dwindled because of overfishing (Coleman et al., 2004). While important examples are expansive oceanic waters in the North Sea of Europe, the Grand Banks of North America, and the East China Sea of Asia, there are many other examples from around the world.

Until the middle of the 20th century, the blue walleye was a commercially valuable fish, with about a half-million tonnes being landed between 1880 and the late 1950s (Hartman, 1972). The populations subsequently collapsed, apparently through a combination of overfishing, anthropogenic eutrophication, and competition with introduced rainbow smelt. The blue walleye in the Great Lakes became extinct in those waters in the 1980s.

While anchovies were a major natural resource in Peru, yielding 10.2 million metric tons of fish in 1971 alone, the Peruvian coastal anchovy fisheries subsequently crashed due to overfishing and an El Niño season (Fimrite, 2015). This was a major loss to Peru's economy.

According to the United Kingdom's official Biodiversity Action Plan, "the sole fisheries in the Irish Sea, the West English Channel, and other locations have become overfished to the point of virtual collapse. Oil spills and other forms of pollution also may very well have had a hand in dwindling populations. While the United Kingdom has devoted considerable thought to ways of restoring this fishery, the expanding global human population and the never-ending demand for fish has reached a point where demand for food threatens the stability of these and other fisheries, if not the species' survival. Could it be that we are experiencing signs of human overpopulation?

The collapse of the cod fishery off Newfoundland is also a dramatic example of the consequences of overfishing (Hutchings, 1996). The 2006 Northwest Atlantic cod quota was 23,000 tons, representing half the available stocks, while the Northeast Atlantic quota was 473,000 tons. Pacific cod are currently receiving strong global demand. The 2006 total allowable catch (TAC) for the Gulf of Alaska and Aleutian Islands was 260,000 tons.

According to a CBS news report, the numbers of predators such as tuna, cod, and swordfish have dropped by two-thirds over the past century, while small prey fish have sometimes doubled due to the lack of predation (Carey, 2006). Fifty-four percent of the predatory fish population has declined in the last 40 years alone. Is it possible that there are too many people to feed with the size of available fish populations?

It thus is clear that the culprit in dwindling fish populations in global oceans is worldwide overfishing, especially of tuna, groupers, and other big, tasty fish that are slow to reproduce. According to a Monteray Bay report, 85% of the world's fisheries are either fully exploited, overexploited, or have collapsed.

When we see such signs within natural populations between predator and prey, we can predict what will happen. When the predator population is too large, the prey population dwindles. When the prey population dwindles, there is not enough food for the predators and they compensate by producing fewer offspring, or there is a complete population crash. If fewer numbers of predators are produced, there may be a chance for the populations of prey to reestablish themselves in community homeostasis. Allow the predators to continue to overkill, and it could lead to extinction.

HUNTING

There are people who hunt and fish and people who strongly criticize those who do them. The US Fish and Wildlife Service found that almost 75% of Americans believe in hunting. However, they may not agree with the methods used by people who fail to teach themselves proper hunting procedures. For instance, 62% of them believe that many hunters violate laws and use dangerous or unethical hunting practices (Posewitz, 1994). In addition, poachers, people who take plants and animals illegally, are commonly found in all countries, often resulting in the dwindling of populations to unstable numbers, in spite of laws to protect them.

According to a group called People for the Ethical Treatment of Animals (PETA), one of the biggest arguments against hunting with respect to life is actually an argument against doing it as a sport. "Humans," they acknowledge, "used to hunt out of necessity to feed and clothe their families." Today, we can buy our food from stores (due to the raising and killing of domestic animals) and purchase clothes whenever we need them. So the question becomes: Why do we still hunt?

PETA says that "arguments against hunting with respect to life and hunting in general include:

- Sportsmen who hunt for the thrill of the kill, which brings out large numbers of men, women, and youths who have little or no experience with hunting. This makes wild lands dangerous to nonhunters and hunters alike. Every year, deer season brings with it news reports of accidental shootings and deaths. Thus, the accidental killing of people is bad, but the purposeful killing of mammals (other than humans) and birds is apparently good.
- Killing animals can damage the social structure of an animal society." One may ask, "Why is the social structure of an animal society so important? Is it because it will hurt the society and the individuals in it or destroy hunting privileges for humans?
- Hunters can use unethical means to hunt—baiting, hounding, and predator-calling—to easily lure animals to them. Humans in PETA want the killing of wildlife to be fair.
- Studies show that bow hunting is often an unreliable means of hunting. One out of two animals shot with a bow will die painfully and slowly from the wounds. So it appears that there is some degree of caring or recognition of moral obligations for which humans feel they are responsible, although hunters that use primitive hunting techniques may not see it that way. However, it appears that there are acceptable ways for humans to kill in the context of morality.
- Hunting can increase disease by killing healthy animals and those with natural disease immunities. Actually, hunting results in the killing of the best animals, for example, the biggest deer with the largest rack of antlers, the deer that would most probably have established an alpha role in the process of reproduction, whereas nature generally works in a reverse, more natural-selective way to weed out the unfit and preserve the fit. To the hunter, the more powerful the animal is, the more they feel the pride of the conquest. The concept is similar to bringing in the largest and most challenging fish.
- Hunters use telescopic, high-powered weaponry to hunt, which negates any opportunity of a fair fight. Thus, hunting is a kind of mental and lethal fight between dominance-seeking humans with sophisticated killing devices and animals with cunning survival instincts, an interspecific competition. The dominant individual wins.

PETA does not acknowledge hunting with respect to life specifically, but it deems hunting as a cruel and unnecessary sport. As a leading organization for the protection of animal rights, PETA holds strong convictions mostly about animals being injured, not killed, during the hunt, resulting in suffering for an extended period.

PETA also feels that hunting disturbs migratory, familial, and hibernation patterns. They also claim that hunting creates abnormal stress that, in turn, disturbs feeding patterns. "This causes an inability to store energy or food properly in preparation for winter."

Since hunting disrupts ecological balance due to the killing of the healthiest individuals of a species (as stated earlier), not the weakest as natural predators do, PETA prefers the natural means of population control, such as disease and starvation, letting nature ensure that the healthiest animals survive to strengthen the rest of the herd.

They acknowledge hunting accidents and the deaths or injuries that are propagated by hunters being untrained in hunter safety. They offer alternatives to maintain healthy populations of animals, such as sterilization and birth control. They consider private, for-profit hunting reserves and game ranches unjust. Animals in such areas are often put there or grown there for the purpose of hunting, and they become accustomed to humans or have a disadvantage by not being able to flee from enclosed spaces. Also, they are often lured to hunting hot spots via feed lots or salt blocks.

Reasons for Hunting

When asked, "Why do you want to hunt?," many hunters say, "I do not know—I just like it."

In an article published by the International Hunter Education Association, it was simply said that "hunters enjoy hunting." They have some very deep feelings about hunting, but [they] often find it difficult to express those feelings in ways [that] others can understand.

Almost every hunter will tell you they love animals. Yet, hunters kill animals. How do you explain such a paradox? "It is a little like farming or gardening," they explain. "People protect and care for their chickens and their vegetable plants, only to end up using them for food. Hunters support conservation of wild places and laws that protect wildlife populations, but they use some of the wildlife for hunting—not just killing, and not just eating, but to experience hunting." Is hunting then a selfish behavior for human pleasure, or is there any consideration whatsoever for the hunted species?

It is a lot easier, and often less expensive, to buy food and clothing at the store than to make them or grow them yourself. But people are "do-it-yourselfers" because they enjoy the activity, not just the product. Hunting is a lot like that.

"Some people say 'hunting is in my blood' because our ancestors were hunters. It feels good to know we can still survive on our own in Nature the way our ancestors did." Actually, it is a lot easier for us to kill than it was for our ancestors because we have higher-powered guns and telescopes that were not available to humans in the past. "Hunting, growing food, and making things by hand are all things that people love to do. Even though those things may not seem to be absolutely necessary, they are very important to people who enjoy them."

In an article published by the International Council for Game and Wildlife Conservation (*Why Do We Hunt?*), based on the question of hunting at the North American Wildlife and Natural Resources Conference, J.J. Jackson said, "We know that the public perception of hunting and acceptance of hunting varies dramatically with what they perceive the motivations for hunting to be (2012)":

"The deep, complex, philosophical, and personal motivation for hunting relates more to the value of hunting in human terms and more to what we are and what is important to our essence than with our normal preoccupation with the animals we hunt. Though both nonhunters and hunters themselves do not fully appreciate that hunting is the greatest generator of wildlife conservation, there is even less understanding of its importance and worth to humans for itself and what it uniquely provides to man himself."

It is currently "the morality of hunting that is most under attack," Jackson said. "At Conservation Force, we believe that attack on hunters is immoral! We believe that there is a moral right to hunt within sustainable limits and that it is so important to man in human terms that it is deserving of protection on moral grounds. It is antisocial, offensive, and immoral for antihunters to attack what is so unique and fundamentally valuable in human terms to the significant minority who hunt."

Speakers also described some of the human virtues nurtured by hunting. "It furthers character virtues, such as self-reliance, responsibility, competence, discipline, and resolve. It employs and awakens our senses and our physical condition. As beings, we are programmed or designed to be hunters. It is our essence. Hunting made us human. It has shaped our evolution and development. It is our 'authenticity'":

"Hunting uniquely provides self-actualization, completeness, and expression that are complex, higher-order needs deserving of protection. These are human needs [that are] higher on the needs scale than food and security. It puts us in touch with our past and with ourselves. It is recreational only in that it is not commercial, but it is much more than just a recreational pursuit. If we were deprived of it, we would lose more than recreation. It is more than our heritage and culture, it is our essence. Hunting and our prey made us what we are today. We are wired to hunt. It helped define who and what we are."

A recent study demonstrates three of the values of hunting in human terms (values that hunters themselves use to hunt): (1) it is a form of historical reenactment; (2) it embodies an honest relationship between humans and nature; and (3) it keeps alive the ideal of self-reliance that is considered by some scholars to be the only distinctive American contribution to political philosophy.

"The historical reenactment aspect is why hunting gives a hunter a deeper appreciation of the past." For example, since many hunters feel a kinship to and respect for American Indians, "hunting is often like entering a time capsule, putting them in touch with a life they dream about or a time they imagine would have been simpler and somehow more virtuous." It includes a relationship with nature that is uniquely honest, bringing them in contact with the natural world in the most natural way. It has a sacredness to those who hunt that will never end.

That relationship teaches us basic truths of life and death while almost everything else in life masks reality and how life works, not just where our food comes from. While those who have illusions about nature evade responsibility and even an acknowledgment of the life and death process, hunters take personal responsibility for the conservation use they make of nature. Witness the enormity of our conservation system. Those who kill and eat what they harvest know a sense of self-sufficiency and more truthful worldview than those who evade the truth and allude themselves about their consumption and responsibility:

> "Hunters embrace and celebrate the eating of the game they have shot. Nonhunters never experience this. It is symbolic of self-sufficiency and self-reliance while it is an expression of a particular sense of self."

With the risk of presenting information about hunting in an even more exhaustive manner, I would like to present one last explanation. In *On Second Thought*, a publication of the North Dakota Humanities Council, Lawrence Cahoone (2012) discusses the virtues of hunting in an article called *Wild Business, a Philosopher Goes Hunting*. While I will not repeat many of the points he made (since most of them are covered in the previous paragraphs), he had some interesting final comments about conservation that should be brought out.

"Farming kills animals by taking away habitat; by poisoning ground water with pesticide and fertilizer; by pollution from fossil fuel–using machinery, refrigeration, and transportation of produce; by the intentional killing of local herbivores to protect crops; and by machinery passes that kill ground nesting animals. Farming is by definition more environmentally intrusive than the selective, in-person culling of one or a few wild animals." Cahoone has made an important point.

"Data on the number of animals killed by farming [or the taking of land for any purpose] are hard to come by, but some sources indicate that farming's animal deaths per nutritional unit is comparable to or greater than the animal deaths per nutritional unit for some kinds of hunting, for example, large animals, such as deer. Stated in another way, it is probably true that a unit of protein from hunted venison kills fewer animals than the same unit of protein from farmed soybeans when you add farming's habitat destruction, the environmental harms to the deer or groundhogs the farmer shoots to save his crops, and all the rabbits, birds, snakes, and frogs killed by tilling, planting, and harvesting machinery.

"The philosophical, or moral, point is clear," Cahoone says: "Farming and hunting both kill animals. The question is which type of farming or hunting, in which environment, regarding which crop or animal species, kills more or fewer animals for the same meal. The choice between hunting and vegetable farming is not a black-and-white choice between killing and not killing, but a choice between grey shades of more and less killing."

Cahoone pointed out that the second session of the conference he attended dealt with the conservation benefits of the hunting system. "Hunting is a conservation generator," he says, "that is both enormous and indispensable." We have been the caretakers of America's wildlife as long as man has accepted that responsibility.

"Hunting serves both people and animals." Our system insures it will do both perpetually. Hunters' responsibility for habitat protection was particularly enlightening. Hunters are largely responsible for the creation of the National Wildlife Refuge System and have funded 69% of its 93 million acres. The same is true for 86.3 million acres of state wildlife management areas funded with license fees and the Pittman–Robertson Act firearm and ammunition revenue. The sportsmen-sponsored Conservation Reserve Program (CRP) has added another 35 million acres and the Wetland Reserve Program (WRP) has enrolled more than 912,000 acres of private lands, mostly in permanent easements. These alone total more than 1.1 billion acres!

"National Wildlife Refuge System lands represent twice the amount provided by the Bureau of Land Management (BLM), US Forest Service and National Park lands combined, and they do not include most private hunting habitat, which has not been quantified. US hunters spent $923 million in 1996 on hunting leases alone. Organizations such as those mentioned earlier also enhance habitat improvement, for example, Ducks Unlimited's restoration and enhancement of more than 10 million acres of wetlands, National Wild Turkey Federation's 2.22 million acres, the Rocky Mountain Elk Foundation's 3 million acres, Quail Unlimited's 400,000 acres, Pheasant Forever's 2 million acres, and Ruffed Grousse Society's 450,000 acres.

"The success of our wildlife conservation system," Cahoone says, "is unprecedented. The very formation, structure, and evolution of that system and modern wildlife agencies have arisen from hunting. Of course, hunting provides the largest source of revenue. Hunting and fishing license fees and excise taxes on firearms provide \$2 billion a year in funding that is indispensable. Recent surveys show that influence has not diminished."

According to Cahoone, "the Government receives another \$1.4 billion (1996) in state income tax revenue and \$1.7 billion in federal income tax. In 1996," he says, "the federal income tax revenue alone was nearly twice the 1996 budgets of the US Fish and Wildlife Service, Bureau of Land Management, National Biological Service, and National Park Service combined. The combination of hunters to wildlife agencies' conservation mission are long, wide, and deep":

> "No one has ever restored more wildlife. Hunters were and continue to be the force behind 'the greatest environmental success story of the 20th century,' according to Valerious Geist and many scholars. Finally, Aldo Leopold's 'land ethic' in *A Sand County Almanac* summed up the modern relevance of hunting. A thing is right when it tends to preserve the integrity, stability, and beauty of the biotic community. It is wrong when it tends to do otherwise."

Although we have gotten off the subject of why people kill, these statements are important in revealing the amount of protection these organizations contribute to national biota. It is difficult to argue against what they are doing when many other human endeavors are cutting into wildlife protective lands, illustrating that most humans have little to do with wildlife conservation.

WILDLIFE EXPLOITATION

Other perspectives of killing may not present such a positive approach to the subject, and it may add support to the arguments already presented here, pointing out that people do things that create disharmony to ecosystems. In an online article in Articlesbase.com, called *How are Humans Killing Wildlife: A Level Environmental Studies*, it was pointed out that "humans pose many threats to wildlife and ecosystems through deliberate exploitation, accidental harm from human activities, introduction of species into habitats, [and] habitat change and destruction. These activities result in species becoming vulnerable (under threat) or endangered (low in numbers and in danger of extinction)."

Examples of exploitation presented in the earlier article were related to: food, fashion, furniture and ornaments, traditional medicines, entertainment, and a variety of other products, such as whale oils and cosmetics. Also, numerous animals are utilized in research, there are numerous cases of accidental harm from human activities, a deliberate eradication of certain species and threats from the introduction of numerous exotic species.

While many people oppose the eating of fish or the flesh of any other animal, disliking the thought of killing for any reason, there are many other people who could not care less about the lives of fish or any other animal, sometimes harvesting animals for body parts because of some mythological or pseudomedical reason.

Traditional Chinese medicine, for instance, often incorporates ingredients from all parts of plants and animals (Britannica, 2007). The sale of parts of endangered species (e.g., seahorses, rhinoceros horns, and tiger parts) has resulted in a black market for poachers who hunt restricted animals. Examples of popular anatomical items are tiger penis, culturally believed to improve virility, and tiger eyes. Deep-seated cultural beliefs in the potency of tiger parts are so prevalent across China and other East Asian countries that laws protecting even critically endangered species fail to halt the display and sale of these items in open markets.

Exploitation of Whales—Whales are marine mammals in the order Cetacea. Other cetaceans include porpoises and dolphins. Whales range in size from the blue whale (30 m, 98 ft) to various pygmy species, such as the pygmy sperm whale (3.5 m, 11 ft). Whales inhabit all the world's oceans and number in the millions. They have a history of being hunted for their meat and other products used by humans (Christensen et al., 1992).

According to a 2013 Wikipedia article on whaling, the hunting of whales primarily for meat and oil, dates back to at least 3000 BC. "Various coastal communities have long histories of sustenance whaling and harvesting beached whales. Industrial whaling emerged with organized fleets in the 17th century; competitive national whaling industries [emerged] in the 18th and 19th centuries; and the introduction of factory ships along with the concept of whale harvesting [emerged] in the first half of the 20th century.

As technology and demand for [whale] resources increased, catches began to far exceed the sustainable limit for whale stocks. "In the late 1930s, more than 50,000 whales were [harvested] annually, and by the middle of the century, whale stocks were not being replenished. In 1986, the International Whaling Commission (IWC) banned commercial whaling so that stocks might recover." With this attempt to ban commercial whaling, human deceit began to take hold of populations who depended on the large mammals for sustenance.

This article points out that "while the moratorium has been successful in averting the extinction of whale species due to overhunting, contemporary whaling [has been] subject to intense debate. Pro-whaling countries, notably Japan, wish to lift the ban on stocks that they claim have recovered sufficiently to sustain limited hunting. Anti-whaling countries and environmental groups say whale species remain vulnerable and that whaling is immoral, unsustainable, and should remain banned permanently."

Since whale oil is little used today, modern commercial whaling is carried out for food. "The primary species hunted are the common minke whale and Antarctic minke whale, two of the smallest species of baleen whales. Recent scientific surveys estimate a population of 103,000 in the northeast Atlantic. With respect to the populations of Antarctic minke whales, as of January, 2010, the IWC states that it is 'unable to provide reliable estimates at the present time' and that a 'major review is underway by the Scientific Committee.'"

According to this article, international cooperation on whaling regulation began in 1931 and culminated in the signing of the International Convention for the Regulation of Whaling (ICRW) in 1946. "Its aim is to provide for the proper conservation of whale stocks and thus make possible the orderly development of the whaling industry."

The International Whaling Commission (IWC) was set up under the ICRW to decide hunting quotas and other relevant matters based on the findings of its scientific committee. Nonmember countries are not bound by its regulations and conduct their own management programs. The IWC voted on July 23, 1982, to establish a moratorium on commercial whaling beginning in the 1985–86 season. Since 1992, the IWC's Scientific Committee has requested that it be allowed to give quota proposals for some whale stocks, but this has so far been refused by the plenary committee.

Canadian whaling is carried out in small numbers by various Inuit groups around the country and is managed by Fisheries and Oceans Canada. Harvested meat is sold through shops and supermarkets in northern communities where whale meat is a component of the traditional diet, but typically not in southern, more urban cities, such as Vancouver, Toronto, or Montreal. The Whale and Dolphin Conservation Society says that "Canada has pursued a policy of marine mammal management which appears to [have] more to do with political expediency rather than conservation."

While Canada left the IWC in 1982, the only species currently harvested by the Canadian Inuit that is covered by the IWC is the bowhead whale. As of 2004, the limit on bowhead whale hunting allows for the hunt of one whale every two years from the Hudson Bay–Foxe Basin population, and one whale every 13 years from the Baffin Bay–Davis Strait population. This is roughly one-fiftieth of the bowhead whale harvest limits in Alaska.

Around 950 long-finned pilot whales (*Globicephala melaena*, actually a species of dolphin) are caught each year around the Faroe Islands, mainly during the summer. Other species are not hunted, though, occasionally, Atlantic white-sided dolphins can be found among the pilot whales. Faroese whaling is regulated by Faroese authorities but not by the IWC, which does not regulate the catching of small cetaceans.

Most Faroese consider the hunt an important part of their culture and history, and arguments about the topic raise strong emotions. Animal-rights groups criticize the hunt as being cruel and unnecessary and economically insignificant. Hunters claim that most journalists lack knowledge of the catch methods used to capture and kill the whales.

Greenlandic Inuit whalers catch around 175 whales per year, making them the third largest hunt in the world, after Japan and Norway, though their take is small compared to those nations, who annually averaged around 730 and 590 whales, respectively, in 1998–2007. The IWC treats the west and east coasts of Greenland as two separate population areas and sets separate quotas for each coast. The far more densely populated west coast accounts for over 90% of the catch. In a typical year, around 150 minke and 10 fin whales are taken from west coast waters, and around 10 minkes are from east coast waters. In April, 2009, Greenland landed its first bowhead whale in nearly 40 years after being given a quota by the IWC in 2008 for two whales a year until 2012.

The Inuit [had] already caught whales around Greenland since the years 1200–1300. Vikings on Greenland also ate whale meat, but archaeologists believe they never hunted them on sea.

Iceland did not object to the 1986 IWC moratorium. Between 1986 and 1989, around 60 animals per year were taken under a scientific permit. However, under strong pressure from anti-whaling countries, who viewed scientific whaling as a circumvention of the moratorium, Iceland ceased whaling in 1989. Following the IWC's 1991 refusal to accept its Scientific Committee's recommendation to allow sustainable commercial whaling, Iceland left the IWC in 1992.

The idea of "scientific whaling," a way of circumventing the rules and regulations set up by the IWC, nevertheless was catching on and would subsequently be used by a number of countries to harvest whales in the interest of collecting scientific data. It is yet another deceitful maneuver by humans to get what they want, regardless of what it will do to the animal populations they were harvesting.

Iceland rejoined the IWC in 2002 with a reservation to the moratorium. Iceland presented a feasibility study to the 2003 IWC meeting for catches in 2003 and 2004. The primary aim of the study was to deepen the understanding of fish–whale interactions. Amid disagreement within the IWC scientific committee about the value of the research and its relevance to IWC objectives, no decision on the proposal was reached. However, under the terms of the convention, the Icelandic government issued permits for a scientific catch. In 2003, Iceland resumed scientific whaling which continued in 2004 and 2005.

Iceland resumed commercial whaling in 2006. Its annual quota was 30 minke whales (out of an estimated 174,000 animals in the central and northeastern North Atlantic) and 9 fin whales (out of an estimated 30,000 animals in the central and northeastern North Atlantic). For the 2012 commercial whaling season, starting in April and lasting 6 months, the quota was set to 216 minke whales.

Lamalera, on the south coast of the island of Lembata in Indonesia, and Lamakera on the neighboring island of Solor are two remaining Indonesian whaling communities. The hunters obey religious taboos that ensure that they use every part of the animal. About half of the catch is kept in the village; the rest is bartered in local markets. In 1973, the UN's Food and Agriculture Organization (FAO) sent a whaling ship and a Norwegian whaler to modernize their hunt. This effort lasted three years, and was not successful. According to the FAO report, the Lamalerans "have evolved a method of whaling that suits their natural resources, cultural tenets, and style."

"When the commercial whaling moratorium was introduced by the IWC in 1982, Japan lodged an official objection. However, in response to US threats to cut Japan's fishing quota in US territorial waters, under the terms of the Packwood–Magnuson Amendment, Japan withdrew its objection in 1987." According to the BBC, "America went back on this promise, effectively destroying the deal." Since Japan could not resume commercial whaling, it began whaling on a scientific-research basis. Australia, Greenpeace, the Sea Shepherd Conservation Society, and other groups dispute the Japanese claim of research "as a disguise for commercial whaling, which is banned." The Sea Shepherd Conservation Society has attempted to disrupt Japanese whaling in the Antarctic since 2003.

The stated purpose of the research program is to establish the size and dynamics of whale populations. The Japanese government wishes to resume whaling in a sustainable manner under the oversight of the IWC, both for whale products (meat, etc.) and to help preserve fishing resources by culling whales. Anti-whaling organizations claim that the research program is a front for commercial whaling, that the sample size is needlessly large, and that equivalent information can be obtained by nonlethal means, for example, by studying samples of whale tissue (such as skin) or feces. The Japanese government–sponsored Institute of Cetacean Research (ICR), which conducts the research, disagrees, stating that the information obtainable from tissue and/or feces samples is insufficient and that the sample size is necessary in order to be representative.

It is easy to see that "Japan's scientific whaling program is controversial in anti-whaling countries (Normile, 2013)." Countries opposed to whaling have passed nonbinding resolutions in the IWC, urging Japan to stop the program. Japan claims that whale stocks for some species are sufficiently large to sustain commercial hunting and blame filibustering by the anti-whaling side for the continuation of scientific whaling. Deputy whaling commissioner, Joji Morishita, told BBC News that the reason for the moratorium [on commercial whaling] was…"for the sake of collecting data, and that is why we started scientific whaling. We were asked to collect more data."

Norway registered an objection to the International Whaling Commission moratorium and is thus not bound by it. Commercial whaling ceased for a five-year period to allow a small scientific catch for gauging the stock's sustainability and resumed in 1993. Minke whales are the only legally hunted species. Catches have fluctuated between 487 animals in 2000 to 592 in 2007. For the year 2011 the quota is set at 1286 Minke whales. The catch is made solely from the Northeast Atlantic minke whale population, which is estimated at 102,000.

Russia had a significant whaling hunt of orcas and dolphins along with Iceland and Japan. In 1970, a study that included photographic recognition of orcas found a significant difference in the suspected ages of whale populations and their actual ages. Following this evidence, the Russians continued a scientific whale hunt, though the verisimilitude of the intentions of the hunt over the last 40 years are questioned. Currently, Russians in Chukotka Autonomous Okrug in the Russian Far East are permitted under IWC regulation to take up to 140 gray whales from the Northeast Pacific population each year.

Natives of Saint Vincent and the Grenadines on the island of Bequia have a quota from the International Whaling Commission of up to four humpback whales per year using traditional hunting methods and equipment.

"In early July, 2012, during IWC discussions in Panama, South Korea said it would undertake scientific whaling as allowed despite the global moratorium on whaling." South Korea's envoy to the summit, Kang Joon-Suk, said that consumption of whale meat "dates back to historical times" and that there had been an increase in the minke whale population since the ban took place in 1986. "Legal whaling has been strictly banned and subject to strong punishments, though the 26 years have been painful and frustrating for the people who have been traditionally taking whales for food." Joon-Suk said that South Korea would undertake whaling in its own waters:

> "New Zealand's Commissioner, Gerard van Bohemen, accused South Korea of putting the whale population at risk. He also cited Japan as having not contributed to science for several years despite undertaking scientific whaling. New Zealand's stated position may be seen by its media as less solid than Australia's on the matter given that its indigenous people are pushing forward with plans, unopposed by the government, to recommence whaling there. The people of Ulsan have also traditionally and contemporarily eaten whale meat. South Korea's representative at the IWC said that "this is not a forum for moral debate. This is a forum for legal debate. As a responsible member of the commission we do not accept any such categorical, absolute proposition that whales should not be killed or caught.""

Whaling is carried out by nine different indigenous Alaskan communities in the United States. The whaling program is managed by the Alaska Eskimo Whaling Commission that reports to the National Oceanic and Atmospheric Administration. The hunt takes around 50 bowhead whales a year from a population of about 10,500 in Alaskan waters. Conservationists fear this hunt is not sustainable, though the IWC Scientific Committee, the same group that provided the previously mentioned population estimate, projects a population growth of 3.2% per year. The hunt also took an average of one or two gray whales each year until 1996. The quota was reduced to zero in that year due to sustainability concerns. A future review may result in the gray whale hunt being resumed. Bowhead whales weigh about 5–10 times as much as minke whales.

The Makah tribe in Washington State also reinstated whaling in 1999, despite protests from animal-rights groups. They are currently seeking to resume whaling of the gray whale, a right recognized in the Treaty of Neah Bay.

The final statements in this article point to the corrupt and deceitful mind of humans and a distinct lack of moral responsibility for life, and while the cited organizations (backed by actual scientific findings) are attempting to protect the species being hunted, a percentage of humans appear to have little concern about the reduction and even extinction of certain magnificent species. It may be true that certain segments of the human population require food that is accessible, but does that tell us something about the human population size. If we have to exploit such animal populations to their extinction in order to feed our masses, does that not say something about our reproductive rates? The human population is concerned about managing the populations of other animals but not its own.

The World Wide Fund for Nature says that 90% of all northern right whales killed by human activities are from ship collision, calling for restrictions on the movement of shipping in certain areas. Bycatch (organisms caught unintentionally by fishing gear) also kills more animals than hunting. Some scientists believe pollution to be a factor. Moreover, since the IWC moratorium, there have been several instances of illegal whale hunting by IWC nations. In 1994, the IWC reported evidence from genetic testing of whale meat and blubber for sale on the open market in Japan in 1993. In addition to the legally permitted minke whale, the analyses showed that 10–25% of tissue sample came from non-minke, baleen whales, neither of which were then allowed under IWC rules. Further research in 1995 and 1996 shows a significant drop of non-minke baleen whale samples to 2.5%.

It was revealed in 1994 that the Soviet Union had been systematically (grossly) undercounting its catch. For example, from 1948 to 1973, the Soviet Union caught 48,477 humpback whales rather than the 2710 it officially reported to the IWC. On the basis of this new information, the IWC stated that it would have to rewrite its catch figures for the last 40 years. According to Ray Gambell, then Secretary of the IWC, the organization had raised its suspicions with the former Soviet Union, but it did not take further action because it could not interfere with national sovereignty.

It appears clear that the key elements of the debate over whaling include "sustainability, ownership, national sovereignty, cetacean intelligence, suffering during hunting, health risks, the value of 'lethal sampling' to establish catch quotas, the value of controlling whales' impact on fish stocks, and the rapidly approaching extinction of a few whale species."

"At the 2010 meeting of the International Whaling Commission in Morocco, representatives of the 88 member states discussed whether or not to lift the 24-year ban on commercial whaling. Japan, Norway, and Iceland have urged the organization to lift the ban. A coalition of anti-whaling nations has offered a compromise plan that would allow these countries to continue whaling, but with smaller catches and under close supervision. Their plan would also completely ban whaling in the southern Ocean. More than 200 scientists and experts have opposed the compromise proposal for lifting the ban, and have also opposed allowing whaling in the Southern Ocean, which was declared a whale sanctuary in 1994. Opponents of the compromise plan want to see an end to all commercial whaling, but are willing to allow subsistence-level catches by indigenous peoples."

In an article on sustainability, Nalen Uk at the University of California, Irvine, stated that "modern commercial whaling is continuously being overexploited as whaling companies strive to maximize their profits, despite international protest and a ban by the International Commission."According to Uk, many species of whales have become extinct, and the relatively few that remain are extremely vulnerable.

After reviewing the true nature of the whaling industry, Uk states that "the factors that [have] led to their overexploitation in the past have not changed. Norway, one of the several countries involved in illegal whaling, has again refused to accept the international moratorium on whaling and has announced that its catch quotas for whales have risen. In 1995, Norwegian whalers caught 217 whales out of a quota of 232. In 1996 the quota was 425. At the beginning of the 1997 whaling season, the year's quota was increased to some 700–800 whales." Uk points out that the rise is highly questionable, especially due to the fact that 50 tons of whale meat from one of the earlier years had to be frozen and put into storage due to the lack of demand. "How many whales," he says, "will be massacred this year just to be frozen?"

Within recent years, a Norwegian smuggler was caught in Japan trying to smuggle 6 tons of whale meat disguised as mackerel. Later, it was discovered that the consignment was part of a conspiracy to smuggle a total of 60 tons of meat to Japan. The sale of whale meat can be very lucrative, considering that whaling companies make $10,000 for each whale captured.

UK repeats that "Japan is another country that exploits whaling for what they call "scientific analysis." However, he mentions that carcasses are distributed to wholesalers and restaurants after research is carried out. "Japan remains the world's largest consumer of whale meat, supplied in part by black market trade in species of minke, humpback, and fin whales.

Countries who refuse to follow moral conduct on harvesting whales offer what they feel are legitimate reasons for continued whaling. Several justifications that Norway and Japan have given for the endless massacre of innocent whales, for instance, are: (1) "it is a long established Norwegian 'tradition' of whaling"; (2) Japan hunts whales under "scientific research"; and (3) both countries argue that minke whales are not endangered and that an uncontrolled population of the whales would threaten valuable fish stocks.

"However, none of these arguments use scientific evidence to justify the killings of whales. In fact, the cloak of 'tradition' in Norway is being used to disguise the fact that the price of whale meat is $290 a kilogram, which is a very lucrative profit. Japan's excuse for 'scientific research' is a poor one, considering money is made when the whale meat is sold to restaurants and wholesalers after the so-called 'scientific research' is completed. In addition, the relationship between fish stocks and whales need not be questioned. The fishery is a commercial industry, not a subsistence way of life. Minke whales are not participating in a commercial industry."

UK concludes by saying that "the unreasonable justifications for murder cannot be accepted by any decent human being because whales are mammals, not fish, but they have been historically treated as fish by the commercial whaling industry. A great number of fish species reproduce by laying an enormous quantity of eggs into the sea for fertilization by males. Even though only a small percentage of these eggs will survive and develop into mature adult fish under normal circumstances, they have the ability to reproduce more eggs."

In short, their biotic potential is high, much higher than that of mammals, such as whales, which produce a single calf at a time and take 7 to 10 years to become reproductive. "Whales have a long gestation period and usually give birth every one to two years to a single calf that requires more than one year of maternal care before it can survive independently. To make matters worse, whaler's love catching pregnant females because they yield more meat." For these reasons, it is difficult for whales to recover from the exploitation by commercial whaling.

In addition, after several years of research, the growth rate of the whale populations is unknown due to the fact that whales are highly migratory, long-lived, slow-reproducing mammals that make studying them extremely difficult. Thus, fishing industries that fear a ban on whaling will lead to a ban on fishing should not regard the vastly different characteristics of two entirely different groups of species in the same manner.

There is controversy that whales have killed fish stocks in many parts of the world. Norway, Japan, and Iceland argue that this is a means for killing whales to protect the fishing industry. However, whales are not the only consumers of fish. Large populations of fish are consumed by other fish and seabirds. Uk points out that "killing whales does not protect the fish that would eventually be consumed by other species in the sea." He claims that "there has been no study that shows that killing whales will increase the fish stocks in the world." Quite to the contrary, whales have inhabited the Earth's oceans for millions of years and play an important part in maintaining homeostasis in the ocean ecosystems.

"Despite the International Whaling Commission's ban on whaling, various countries still whale illegally, which has led to the extinction of several whale species. The Atlantic population of gray whales went extinct in the late 17th century. In the different oceanic environments, gray whales are critically endangered. Blue whales are critically endangered or vulnerable, while fin whales, North Pacific right whales, North Atlantic right whales, and sei whales are endangered. Sperm whales are vulnerable, while other species may be at lower risk of extinction. In other reports, whaling and other threats have led to at least five of the 13 great whales being listed as endangered.

After the decline or extinction of larger whales, commercial whalers have resorted to the smallest species of the great whales, for example, the minke whales (*Balaenoptera acustorstrata*). The minkes, although larger than elephants were too small for whalers to make an advantageous profit until the early 1970s when most of the larger whale populations were extinct or endangered. Still, Norway, Japan, Iceland, and perhaps other unknown countries disregard all laws that protect whales from what Uk calls a "worldwide massacre":

> "Fortunately, Greenpeace has been influential in creating a whale sanctuary in Antarctica where The International Whaling Commission has achieved a worldwide moratorium on commercial whaling. However, we need to enforce these international laws more strictly, to prevent illegal killing and smuggling of whale meat for…profit. Even though strict reinforcement may not cease the massacre completely, at least some whales will be saved, and smugglers will be caught."

THE PET TRADE

Based on reports by E. Chivian and A. Bernstein (*Sustaining Life: How Human Health Depends on Biodiversity*) (2008) and B.A. Stein et al. (*Heritage: The Status of Biodiversity in the United States*) (2000), a National Wildlife Federation article called *Overexploitation* points out that people have always depended on wildlife and plants for food, clothing, medicine, shelter, and many other needs. However, it was also pointed out that we are taking more than the natural world can supply, a phenomenon referred to as overexploitation. The danger is that if we take too many individuals of a species from their natural environment, the species may no longer be able to survive. In addition, the loss of one species can affect many other species in an ecosystem.

The hunting, trapping, collecting, and fishing of wildlife at unsustainable levels is not something new. The passenger pigeon was hunted to extinction early in the last century, and overhunting nearly caused the extinction of the American bison and several species of whales:

> "Today, the Endangered Species Act protects some US species that were in danger from overexploitation, and the Convention on International Trade in Endangered Species of Fauna and Flora (CITES) works to prevent the global trade of wildlife. But there are many species that are not protected from being illegally traded or overharvested."

Plants are vital to our survival and are the foundation of most of the Earth's ecosystems. People harvest plants for food, medicine, building materials, and as raw materials for making numerous other products. But we are taking too many plants from the wild. Some plants, such as orchids, are so prized by collectors that they are now endangered and legally protected from poaching by international law. Some medicinal plants, such as American ginseng, have also been so enthusiastically collected that it is now very hard to find them in the wild. A number of tree species that are prized for their wood, such as mahogany, are under threat because of overharvesting.

Invertebrates make up at least 75% of all known animal species. Insects, oysters, octopi, crayfish, sea stars, scorpions, crabs, worms, and sponges are all kinds of invertebrates. Today, many invertebrates—particularly marine invertebrates—are at risk from overharvesting. Chesapeake Bay oysters, once an important part of the Bay economy, are now in decline. Horseshoe crabs, whose eggs provide food for migratory birds, fish, and other organisms, are being harvested as bait for eel and whelk fishing. Octopi are suffering declines worldwide due to heavy fishing pressure. Shells and corals are collected for ornaments and jewelry.

As fishing gear and boats have improved, the fishing industry has become very efficient at harvesting fish and shellfish. The industrialization of the fishing industry and the increasing world demand for seafood have people taking more fish from oceans, lakes, and rivers than is sustainable. Prized fish, such as swordfish, cod, and tuna, have undergone dramatic declines. In the Great Lakes, overfishing has caused whitefish, walleye, and sturgeon populations to decline. Beyond their role in the food supply, freshwater and marine fish are also trapped for the aquarium trade and fished for sport.

Amphibians are collected and shipped all over the world for the pet trade, medicine, education (frogs are dissected in many biology classes), scientific research, and for food (frog legs are a delicacy in many parts of the world). The California red-legged frog, now a federally protected endangered species, was overhunted for food, and its numbers seriously depleted during the Gold Rush in the area around San Francisco.

Reptiles are harvested and traded around the world for their skins or shells, their eggs, their meat, and for the pet trade (Fig. 14.1A). Overharvesting of the Kemp's ridley sea turtle's eggs nearly led to its extinction, and today it is still

FIGURE 14.1 Exotic animals. Because of the pet trade, many animals are removed from their natural environments and sold to people who want unusual animals as pets. (A) Chameleon. (B) Red-eyed tree frog. (C) Poison arrow frog. (D) Giant (cane, marine) toad. *Arrow* in D points to the toad's parotid gland, a venomous gland that sometimes kills animals that eat them. When people grow tired of the animals they bought as pets, they often release them in local environments. Introducing exotic animals to new habitats often results in a loss of ecological homeostasis. When exotic species negatively affect the populations of indigenous species, we say the exotic species is invasive.

an endangered species. In the United States, box turtles are being collected at unsustainable levels for the overseas pet trade. Some reptile skins—such as crocodile, python, and monitor lizard—are highly prized as exotic leathers.

"Birds are collected or hunted for sport, food, and the cage-bird pet trade (parrots and songbirds are prized as pets). Millions of birds are traded internationally each year. Close to 30% of globally threatened birds are threatened by overexploitation, particularly parrots, pigeons, and pheasants. The indigenous Carolina parakeet was once the only species of parrot in the United States, but it was hunted to extinction early in the last century for food, to protect crops, and for its feathers (which adorned ladies' hats)." At this time, there are many established populations of exotic parrots and other birds that have been brought into the United States for various reasons, but they are exotic species that will likely affect the lives of indigenous species.

"People have always hunted mammal species—for fur, food, sport, and for their horns or antlers. Mammals are also trapped for the pet trade, zoos, and biomedical research. Today, illegal hunting still threatens many species, especially large mammals, such as tigers, rhinoceros, bears, and even primates, whose body parts are highly valued in some parts of the world for traditional medicine."

Most pets in the United States are dogs and cats. Dogs number more than 78 million, and cats number close to 80 million. Fish top the list at 172 million, many of which are released into native waters and become established, often upsetting indigenous ecosystems. Other pets fit the following categories: small mammals (16 million), birds (15 million), reptiles and amphibians (14 million), and equines (about 13 million).

Domesticated pets (any animals that have been tamed and made fit for a human environment) are the most common types of pet. Domesticated animals have consistently been kept in captivity over a long-enough period of time and thus exhibit marked differences in behavior and appearance from their wild relatives. These may include alpacas, asses, cats, cows, dogs, ferrets, hedgehogs, horses, goats, pygmy goats, llamas, pot-bellied pigs, rabbits, a number of different rodents, sugar gliders, and sheep. Birds include such parrots as budgerigars, lovebirds, monk parakeets, cockatoos, African gray parrots, macaws, and rose-ringed parakeets, as well as toucans, domestic canaries and other finches, chickens, domestic turkeys, domestic ducks and geese, peafowl, ravens, and crows. Popular fish are koi, goldfish, and Siamese fighting fish.

Wild animals (any species of animal that has not undergone a fundamental change in behavior to facilitate a close coexistence with humans) are often kept as pets as well (Kendall, 2012). The term wild in this context specifically applies to animals that are normally in their natural environment, free from human contact. Some species listed here may have been bred in captivity for a considerable length of time, but are still not recognized as domesticated. Many of these pets, such as insects and fish, are kept as a hobby, rather than for companionship.

The pet trade has also been involved in the collection and dispersal (exploitation) of exotic species to just about every country in the world. There are certain ways in which people legally bring animals to the pet trade, and there are number of people who choose to bring in animals illegally.

One of the dangers of removing species from their natural habitat is that if we take too many individuals, the species may no longer be able to survive. In addition, the loss of one such species can result in a loss of population and ecosystem stability. Other dangers come from the introduction of exotic species to habitats to which they do not naturally belong, resulting in the establishment of a species that will possibly cause a disruption in the populations of indigenous species. Thus, an overexploitation of wildlife and plant species by people for a wide variety of reasons often leads to a loss of population and ecosystem homeostasis and possibly extinction of species in those areas.

The hunting, trapping, collecting, and fishing of wildlife at unsustainable levels is not something new. The passenger pigeon was hunted to extinction early in the last century (Avery, 2014). The Carolina Parakeet mentioned earlier is believed to have died out because of a number of different threats: (1) to make space for more agricultural land, large areas of forest were cut down, taking away its habitat; (2) the bird's colorful feathers (green body, yellow head, and red around the bill) were in demand as decorations in ladies' hats (Tebeau, 1963).

Other bird species suffered population decline for similar reasons. By the late 19th century, plume hunters had nearly wiped out the Snowy Egret population of the United States. Flamingoes, Roseate Spoonbills, Great Egrets, peafowl, and bird of paradise were also targeted by plume hunters (Grunwald, 2006). Victorian era fashion included large hats with wide brims decorated in elaborate creations of silk flowers, ribbons, and exotic plumes. Hats sometimes included entire exotic birds that had been stuffed. Plumage often came from birds in the Florida everglades, some of which were nearly extinguished by overhunting. By 1899, early environmentalists were engaged in efforts to curtail the hunting for plumes. By 1900, more than 5 million birds were being killed every year, including 95% of Florida's shore birds.

We have been exposed to the overhunting of American bison which nearly caused their extinction, along with several species of whales (Clark, 1973). Today, the Endangered Species Act protects some US species that were in danger from overexploitation, and the Convention on International Trade in Endangered Species of Fauna and Flora (CITES) works to prevent the global trade of wildlife. But there are many species that are not protected from being illegally traded or overharvested.

What species are currently being overexploited? As fishing gear and boats have improved, the fishing industry has become very efficient at harvesting fish and shellfish. The industrialization of the fishing industry and the increasing world demand for seafood have people taking more fish from oceans, lakes, and rivers than is sustainable. Prized fish, such as swordfish, cod, and tuna, have undergone dramatic declines (Olden et al., 2007). In the Great Lakes, overfishing has caused whitefish, walleye, and sturgeon populations to decline. Beyond their role in the food supply, freshwater and marine fish are also trapped for the aquarium trade and fished for sport.

Birds are collected or hunted for sport, food, and the cage-bird pet trade (parrots and songbirds are prized as pets). Millions of birds are traded internationally each year. Close to 30% of globally threatened birds are threatened by overexploitation, particularly parrots, pigeons, and pheasants.

As pointed out earlier in this chapter, people have always hunted mammal species—for fur, food, sport, and for their horns or antlers. Mammals are also trapped for the pet trade, zoos, and biomedical research (BornFree, 2003). Today, illegal hunting still threatens many species, especially large mammals, such as tigers, rhinoceros, bears, and even primates, whose body parts are highly valued in some parts of the world for traditional medicine.

Invertebrates make up at least 75% of all known animal species. Insects, oysters, octopus, crayfish, sea stars, scorpions, crabs, and sponges are all kinds of invertebrates. Today, many invertebrates—particularly marine invertebrates—are at risk from overharvesting (Jamieson, 1993). Chesapeake Bay oysters, once an important part of the Bay economy, are now in decline. Horseshoe crabs, whose eggs provide food for migratory birds, fish, and other organisms, are being harvested as bait for eel and whelk fishing (Shuster et al., 2003). Octopi are suffering declines worldwide due to heavy fishing pressure. Shells and corals are collected for ornaments and jewelry.

Plants are vital to our survival and are the foundation of most of the Earth's ecosystems. People harvest plants for food, medicine, building materials, and as raw materials for making other products. But we are taking too many plants from the wild. Some plants, such as orchids, are so prized by collectors that they are now endangered and legally protected from poaching by international law (Doyle, 1995). Some medicinal plants, such as American ginseng, have also been so enthusiastically collected that it is now very hard to find them in the wild. A number of tree species that are prized for their wood, such as mahogany, are under threat because of overharvesting.

The legal wildlife trade includes specimens of species that are not listed in any of the three CITES Appendices and specimens of species that are listed by CITES and that are traded internationally with the appropriate documents. These documents include permits and certificates for the import, export, reexport, and introduction from the sea which are issued by the CITES Management Authorities of the respective countries. The UNEP–WCMC manages a trade database on behalf of the CITES Secretariat, where records of trade in wildlife and scientific names of taxa listed by CITES are reported annually.

Note that wildlife trade that is legal is not necessarily sustainable. Hence, much legal trade in wildlife may still be a significant conservation concern. Unsustainable wildlife trade may be addressed in a number of ways, including listing a species of concern on the appropriate Appendix of CITES.

Hunting for the illegal wildlife trade has the greatest potential to do maximum harm in minimal time, and is a serious threat to a number of endangered and vulnerable species. Illegal wildlife trade and contraband includes live pets, hunting trophies, fashion accessories, cultural artifacts, ingredients for traditional medicines, and wild meat for human consumption. Bushmeat trade is considered illegal when imports occur in contravention of the Washington Convention on International Trade in Endangered Species of Fauna and Flora (CITES), national quarantine laws and other laws that ban the trade of specific animals.

Illegal wildlife trade is broadly defined as an environmental crime, which directly harms the environment. Wildlife trafficking is driven by organized groups that exploit natural resources and endanger threatened species and ecosystems in contravention of CITES. Environmental crimes by their very nature are trans-boundary, using porous borders, and involve cross-border criminal syndicates characterized by irregular migration, money laundering, corruption, and the exploitation of disadvantaged communities.

The links between wealth, poverty, and engagement in the wildlife trade are complex: people involved in the trade are not necessarily poor, and the poor who are involved do not capture the majority of the trade's monetary value. In 2002, the illegal wildlife trade was estimated to be the second-largest illegal trade, second only to the drugs trade, with a value of at least £10 billion. In 2008, it was estimated that it is worth at least US$5 billion, and may potentially total in excess of $20 billion annually. This ranks the illegal wildlife trade as among the most lucrative illicit economies in the world, behind illegal drugs and possibly human trafficking and arms trafficking. Due to its clandestine nature, the illegal trade is difficult to quantify with any accuracy. Potential areas of market growth include the Internet, where traders use chat rooms and auction websites to engage in illicit wildlife sales.

In Europe, the revenues generated by the trafficking of endangered species are estimated at 18 to 26 billion euros (about 32 billion US dollars) per year, with the EU being the foremost destination market in the world (Warburton, 2014). The trade is principally coordinated by well-organized, loose networks based in the EU and in the source regions. It is now common for perpetrators to use couriers and air mail-orders. Animals from several destinations are concentrated in one place, from which it is possible to organize transit into the EU.

A number of highly sophisticated Colombian groups manage the supply chains for a wide variety of species. Chinese-organized crime groups, based mainly in Hong Kong, have specialized in the supply of traditional Chinese medicine products containing derivatives of endangered species to several companies across the EU, particularly in Northwest Europe (Nuwer, 2014). Within the EU, dedicated organized crime groups often exploit legitimate business structures to facilitate the importation and retail of specimens. Groups in Northwest Europe, for instance, cooperate with breeders in other member states to launder "wild-caught" animals, using false documents to trade them as captive bred on the legitimate market.

Difficulties in ascertaining the authenticity of foreign certificates frustrate enforcement efforts. Of note, there is evidence that trafficking in endangered species is of increasing interest to polycriminal-organized crime groups. Groups involved in high-level drug trafficking in Brazil, Colombia, and Mexico have established a notable role in the illegal supply of endangered species to the EU and US markets. As a result, some of the concealment methods developed for drug trafficking are now used to traffic endangered species.

In the Middle East, there is a large illegal trade in free-living houbara bustards, trapped in Pakistan, Iran, and Afghanistan (Bailey et al., 2000), which are exported to the Middle East where they are used by some falconers to train their falcons. By the end of the 1990s, 4000–7000 houbaras were traded in this way from Pakistan each year. In 1998, initiatives were taken in Abu Dhabi and Sharjah to confiscate illegally imported houbara bustards when they entered the United Arab Emirates, while government agencies in Pakistan actively confiscate birds as they are smuggled out of the country through air and sea ports.

After the Haj of 2010, skins of pythons and other reptiles, tigers, and of Arabian leopards poached in Yemen were offered among the products being sold in the tent city of Mina, Saudi Arabia (Saudi Trade, 2014). Endangered animal parts are smuggled into the Kingdom, often with little effort in disguising or hiding the items.

In May 2011, a United Arab Emirates citizen was arrested as he was preparing to fly first class from Bangkok to Dubai with various rare and endangered animals in his suitcases, which included four leopard cubs, one Asiatic black bear cub, and two macaque monkeys.

A substantial portion of the global illegal wildlife trade—possibly the largest in the world—takes place in Asia, where demand is driven by the need for specific animal parts to practice traditional Asian medicine, for human consumption, and as symbols of wealth (Illegal Trade, 2013). Demand for illegal wildlife is reportedly increasing in Southeast Asia due in part to the region's economic boom and resulting affluence. Southeast Asia is also a key supplier of wildlife products to the world (Felbab–Brown, 2013).

Live animals, such as tortoises, freshwater turtles, snakes, sharks, pangolins, and monitor lizards, and animal parts are sold in open-air markets and end up as pets, trophies, or in specialty restaurants that feature wildlife as gourmet dining (Jenkins et al., 2011).

China is the world's largest importer of wildlife products, including an insatiable demand for turtles, ivory, tigers, pangolins, and many other species used for food or medicine (Actman, 2015). http://en.wikipedia.org/wiki/Wildlife_trade-cite_note-hance2009-13 India and Nepal feature as source and transit for the trade in body parts of tigers, rhinos, leopards, snow leopards, otters, and musk deer for usages in traditional Chinese medicine, and for decorative use by the neo-rich. Traders use land routes via Sikkim, Ladakh, and Tibet as borders are porous and customs lax. Skins and body parts of 783 tigers, 2766 leopards, and 777 otters were seized between 1994 to August 2006 in India alone, probably representing a tiny fraction of the actual trade bound for Tibet and China. Among the many seizures of live and dead pangolins in Southeast Asian countries, these were the largest ready for export to China: in spring 2008, two shipments containing dead pangolins and scales were discovered in Vietnam but originating in Indonesia; in July, 2008, frozen Malayan pangolins and their scales were seized in Sumatra.

In Thailand, the Chatuchak weekend market in Bangkok is an important hub for the sale of freshwater turtles and tortoises for pets (Drews, 2002). People from Japan, Malaysia, and Singapore are known to purchase large numbers of turtles from the dealers for retail in their respective countries. The majority of the illegally sourced species observed during surveys carried out in 2006 and 2007, or confiscated in recent years were not native to Thailand, but originated in Indonesia, Indochina, Madagascar, Congo, Uganda, Kazakhstan, Lebanon, Barbados, and Venezuela.

The illegal trade in live elephants and ivory still flourishes (Kramer, 2008; Lavers, 2009). Seizures in Asia and Africa in recent years appear to have severely reduced the availability of African raw ivory, and prices increased on average over 300% since 2001. However, Thailand still has one of the largest and most active ivory industries seen anywhere in the world. Every year, many elephants are illegally imported from Myanmar for use in the tourism industry; elephant calves are slated for begging on the streets.

The sale of lizards, primates, cats, and other endangered species has been widely documented. The Suvarnabhumi International Airport offers smugglers direct jet service to Europe, the Middle East, North America, and Africa. Trade routes connecting in Southeast Asia link to the United States for the sale of turtles, lemurs, and other primates, Cambodia to Japan for the sale of slow lorises as pets, and the sale of many species to China.

In Vietnam, 14,758 cases involving wildlife hunting and trade violations were identified and prosecuted from 1996 to March 2007, and about 635 tons of wildlife with a total of 181,670 individual animals were confiscated. The data showed an increasing trend in the number of wildlife violations, from 1469 cases in 2000 to 1880 cases in 2002.

Expansion of markets and price acceleration have contributed an important boost for the development of illegal wildlife trade that was identified as the most important factor contributing to the significant depletion of populations of some species, such as cats, bears, pangolins, amphibians, reptiles, orchids, agarwood, and some other endemic plants. The quantity of wildlife provided for the Vietnam markets is estimated at about 3400 tons and over 1 million heads per year.

Japan appears frequently on the top three list of importing countries in endangered species with official permission under the regulation of CITES. Japan is a major importer of live reptiles, mostly tortoises and freshwater turtles, but also American alligator, reticulated python and Nile crocodile, although export of these species is restricted in the countries of origin, or species are listed on the IUCN Red List of Threatened Species.

In August, 2010, a notorious Malaysian wildlife trader was arrested after having tried to smuggle about 100 live snakes to Indonesia. Since the early 1980s, he legally wholesaled tens of thousands of wild reptiles annually, many of which were on sale in American pet stores. But he allegedly commanded one of the world's largest wildlife trafficking syndicates, and using a private zoo as a cover, also offered a large array of contraband, including snow leopard pelts, panda bear skins, rhino horns, rare birds and Komodo dragons, chinchillas, elephants, gorillas, and tigers, and critically endangered wildlife was smuggled from Australia, China, Madagascar, New Zealand, and South America to markets largely in Europe, Japan, and the United States.

In North and South America, the United States is the second-largest importer of wildlife products and a large destination for the illegal pet trade (Broad et al., 2003). Every month, for instance, many tons of bushmeat arrives from Africa. During the period between 2001 and 2005, over 11,000 specimens (live animals and wildlife products of birds, reptiles, marine turtles, corals, and mammals) were seized in shipments from countries such as Costa Rica, Dominican Republic, El Salvador, Guatemala, Honduras, and Nicaragua. Amazon rainforest animals are smuggled across borders the same way illegal drugs are—in the trunks of cars, in suitcases, in crates disguised as something else. In August, 2011, a couple was arrested when they tried to smuggle jaguar pelts into the United States from Mexico. The two had made repeated trips to Florida and had offered to sell the skins to customers in Texas and Florida or through Internet sites. Both face up to 5 years in prison and criminal fines of up to $250,000 each.

Animal trading in Latin America is widespread as well (Ringwald, 2015). In open-air Amazon markets in Iquitos and Manaus, a variety of rainforest wildlife is sold openly as meat, such as agoutis, peccaries, turtles, turtle eggs, walking catfish, and others. In addition, many species are sold as pets. The keeping of parrots and monkeys as pets by villagers along the Amazon is commonplace. But the sale of these "companion" animals in open markets is rampant. Capturing baby tamarins, marmosets, spider monkeys, saki monkeys, and other species in order to sell them, often requires shooting the primate mother out of a treetop with her clinging offspring; the youngster may or may not survive the fall. With the human population increasing, such practices have a serious impact on the future prospects for many threatened species.

In addition to deceptive means of obtaining illegal animals for profit, humans cause catastrophic mass killings of wildlife because (1) much land is required for agricultural or other use and (2) numerous types of wildlife are lost through oil spills and other environmental catastrophes.

We must admit that many humans think very little of wild organisms and their protection. Their uncontrollable dominance of the environment and its biota has run amuck. Where do they draw the line on killing and using inhumane techniques in their acquisition and sale of nonhuman animals, either domesticated or wild? What about others who kill members of their own species, often without remorse? Is there a correlation between killing animals in any form or phyletic position?

Note

Superscript numerals appearing in this chapter refer to additional text/explanation given in the appendix.

15

Killing Humans

To kill another human being is always murder and should never be called anything else. In a war, the members of two rival groups try to murder the opponents. If there is such a thing as a just war, then murder can be necessary for the sake of justice. **W.H. Auden,** _Poem about Pieter Brueghel's The Fall of Icarus (c.1565), Musée des Beaux Arts'_

ABUSE TO HUMANS AND OTHER ANIMALS

According to a report entitled _Animal Abuse and Human Abuse: Partners in Crime_ that was released by PETA, "Acts of cruelty to animals are not mere indications of a minor personality flaw in the abuser; they are symptomatic of a deep mental disturbance. Research in psychology and criminology shows that people who commit acts of cruelty to animals don't stop there—many of them move on to their fellow humans. Murderers… very often start out by killing and torturing animals as kids."

Studies have shown that violent and aggressive criminals are more likely to have abused animals as children than criminals who are considered nonaggressive (Nockleby, 2008). A survey of psychiatric patients who had repeatedly tortured dogs and cats when they were young found that all of them had high levels of aggression toward people as well. According to a New South Wales newspaper, a police study in Australia revealed that 100% of sexual homicide offenders examined had a history of animal cruelty.

To researchers, a fascination with cruelty to animals is a red flag in the backgrounds of serial killers and rapists. These are the kids who never learned it is wrong to poke out a puppy's eyes.

Pet Abuse by Notorious Killers

History is replete with serial killers, some of which have been mentioned in Chapter 9, whose violent tendencies were first directed at animals. "Albert DeSalvo (the 'Boston Strangler'), who killed 13 women, trapped dogs and cats and shot arrows at them through boxes in his youth. Serial killer Jeffrey Dahmer impaled frogs', cats', and dogs' heads on sticks. Dennis Rader (the BTK killer), who terrorized people in Kansas, wrote in a chronological account of his childhood that he hanged a dog and a cat. During the trial of convicted sniper Lee Boyd Malvo, a psychology professor testified that the teenager, who killed 10 people with a rifle, had pelted—and probably killed—numerous cats with marbles from a slingshot when he was about 14."

"The deadly violence that has shattered schools in recent years has, in most cases, begun with cruelty to animals (Benzinger and Console, 2014). High-school killers, such as Kip Kinkel in Springfield, Oregon, and Luke Woodham, in Pearl, Mississippi, tortured animals before starting their shooting sprees. Columbine High School students Eric Harris and Dylan Klebold, who shot and killed 12 classmates before turning their guns on themselves, spoke to their classmates about mutilating animals."

There appears to be a common theme in the behavior displayed by individuals who were involved in all of the shootings of recent years. Children who grow up to be murderers have symptoms of aggression toward their peers, a more-than-casual interest in fire, a fixation toward cruelty to animals, a propensity to enjoy social isolation, and many warning signs that schools have ignored. Sadly, childhood violence in many of these criminals went unexamined until their killing was directed toward humans.

Dominance and Aggression in Humans and Other Animals
http://dx.doi.org/10.1016/B978-0-12-805372-0.00015-8

FIGURE 15.1   History is replete with serial killers, whose violent tendencies were first directed at animals. Children who are animal abusers may be repeating lessons that they learned or practiced at home; often abused themselves, they subsequently react to anger or frustration with violence. According to PETA, "60% of more than 50 New Jersey families that had received treatment as a result of incidents of child abuse also had animals in the home that had been abused."

Connecting Animal Cruelty and Family Violence

Because abusers target individuals who are less dominant than themselves, crimes against animals, spouses, children, and the elderly often go hand in hand (Herrenkohl, 2005). Children who are animal abusers may be repeating lessons that they learned or practiced at home; often abused themselves, they subsequently react to anger or frustration with violence. Their violence is often directed at animals because they are the only individuals in the family who are more vulnerable than themselves. There are definite connections between animal abuse, domestic violence, and child abuse (Fig. 15.1).

PETA states that "60% of more than 50 New Jersey families that had received treatment as a result of incidents of child abuse also had animals in the home that had been abused." In three separate studies, more than half of the battered women surveyed reported that their abuser also threatened or injured their animal companions. In one of those studies, one in four women said that she stayed with the batterer because she feared leaving the animal behind.

"Stephen Williams was charged with cruelty to animals, child cruelty, and aggravated assault in Georgia after allegedly hacking his wife's puppy to death with an ax and threatening to decapitate her with the same weapon, all in front of three horrified children." Scott Maust of Pennsylvania was charged with the corruption of minors, making terroristic threats and committing cruelty to animals by allegedly shooting his family's dog with a 0.22-caliber firearm, ordering his four children to clean up the bloody scene and threatening to kill them if they told anyone.

Stopping the Cycle of Abuse

"Schools, parents, communities, and courts are beginning to realize that shrugging off cruelty to animals as a 'minor' crime is like ignoring a ticking time bomb. Some courts now aggressively penalize animal abusers, examine families for other signs of violence, and order perpetrators to undergo psychological evaluations and counseling." Whether such counseling will facilitate a turnaround in the behavior of a perpetrator has not been shown.

In March 2006, Maine Gov. John Baldacci signed a law (the first of its kind in the United States) that permits judges to include animal companions in court-issued protection orders against domestic abusers. Other states, including Vermont, New York, California, and Colorado, followed suit. People who harm animals in violation of a court order can face fines and jail time.

A handful of states require animal control officers and spousal/child abuse investigators to share information when animal abuse or domestic abuse is found in a home. Professor Ascione, who also advises law enforcement officials in abuse cases, told The New York Times that cross-reporting requirements have helped foster early intervention.

PETA offers advice about what can be done as precautionary measures to avoid abuse and stress that abusive children represent a possible sign of forthcoming abusive behavior in adults.

Killing Individuals of One's Own Species

Humans, like certain other primates, often go beyond the boundary of simple forms of hunting, even beyond the killing of pets, and kill conspecific (human) individuals for one reason or another. Some humans even appear to enjoy or, for some reason, require the process of killing members of their own species.

We may ask, as D. Peterson and R. Wrangham have done in *Demonic Males, Apes and the Origins of Human Violence* (1997), "Is it natural for animals [in general] to try to kill their own species?" It appears that in certain societies, in the minds of certain individuals, and in certain cases it is. The following information shows that the killing of individuals who belong to one's own species, even one's own family, is apparently as common to some animal species as procreation, appearing to be a broadly based biological phenomenon throughout the natural world.

Infanticide

According to Wrangham and Peterson (1996) in *Infanticide: An Anthropological Analysis*, the killing of offspring by conspecific adult feral animals is routine. Male lions, for instance, are well-known participants in infanticide, eliminating offspring that were produced by other males. A wide assortment of species from microscopic rotifers and insects to fish, amphibians, birds, and mammals (including certain nonhuman primates) do it as well.

There may be other reasons for eliminating offspring. Besides maximizing one's own fitness by eliminating the genes of former dominants, it may sometimes be a part of natural selection to cull runts and young with anatomical abnormalities in order to maximize the survivability of the strongest offspring (Maestripieri, 2000). In certain cases, it may even be a common practice to feed upon the young (oophagy, cannibalism).

Oophagy (feeding upon eggs) has been reported to commonly occur in colonies of social insects. In paper wasps, for instance, females sometimes feed upon eggs that were deposited by other females (Hermann and Dirks, 1975). I have found it primarily on nests with multiple cofoundresses in which a dominance hierarchy was poorly developed. On nests in which the queen is clearly dominant, oophagy occurs less often. Other social insects apparently feed on eggs for their nutritive content.

Strangely, two forms of intrauterine cannibalism are known in certain sharks (Fox, 1975; Polis, 1981). In the most extreme form, embryophagy (=adelphophagy), the largest and strongest embryo actually consumes its womb-mates. In the most common form, oophagy (=oviphagy), developing embryos feed on unfertilized eggs. Both forms of intrauterine cannibalism continue throughout embryonic and fetal development.

Infanticide is not restricted to nonhumans (Marks, 2009). It, in fact, has been quite common in human populations around the world at earlier times, and it is found in contemporary populations as well, although to a lesser degree. Infant cannibalism is even found on occasion. Reasons for human cannibalism, including the eating of infants, have been: (1) sanctioned by a cultural norm (e.g., in endocannibalism and exocannibalism by certain cultures); (2) necessity in extreme situations of famine; and (3) mental illness, insanity, or social deviancy.

J.B. Birdsell (1986) and other anthropologists have calculated that infanticide rates in human populations during Paleolithic and Neolithic times were between 15% and 50%, the rates persisting until the development of agriculture, which was associated with the onset of civilizations, more abundant local foods, and a more sessile existence during a time referred to as the Neolithic Revolution.

Early New World infanticide appears to have been associated with religious practices (e.g., sacrificial infanticide) by humans in Mesoamerican and Inca empires. Sacrificial infanticides in the Old World have been found in such places as Sardinia, Babylonia, Pelasgia, Syria, Gezer, Egypt, Carthage, Tophet, Phoenicia, Canaan, Moaba, and Sepharva.

Although historical Greeks considered sacrificial infanticide barbarous, it nevertheless went on for certain reasons. If husbands rejected a child, for instance, it was put to death, the preferred method of disposal being exposure to the elements. Other forms of disposal of infants were drowning, suffocation, and throwing infants to the animals. Such practices were prevalent in ancient Rome as well.

Christianity rejected infanticide, and while pre-Islamic Arabia practiced it, infanticide is explicitly prohibited by the Qur'an. Sacrificial infanticide was practiced in Russia during the early 900s, and babies in Karnchatka were thrown to the dogs. Infanticide was still common in northeastern Siberia in the 19th century. China and South Asia, Kutch, Kehtri, Bengal, Mazed, and Kalouries are known to have practiced sex-selective infanticide in earlier times.

African children were killed if it was suspected that they were evil or unlucky. There also were ritual killings of twins. In addition, newborns were killed if the mother died in childbirth.

In Australia, 30% of aboriginal infants at one time were killed at birth. Inuits killed children by throwing them into the sea. Infanticide and cannibalism were practiced by Dane and other Indians in Canada, and female children were sometimes killed by the Eastern Shoshone, Maidu, and Mariame Indians. Infanticide was practiced by pre-Columbian tribes in Mexico and parts of South America.

In modern times, infanticide has become less common in the Western world, but it continues in West Africa, China, India, Pakistan, and Oceana. England and Wales had 30-50/million child homicides between 1982 and 1996. Although the reasons for child death are often not reported, the United States ranked eleventh in the world for infant deaths under 1 year old, and fourth for children killed between the ages of 1 and 14 during 1983. Canada had 114 cases of child murder between 1964 and 1968.

Traditional neonaticide is apparently mostly due to an inability to properly care for an infant, while older infants are more typically killed because of child abuse, domestic violence, or mental illness. A family's history of violence, violence during current relationships, child neglect, personality disorder, and depression in adults who generally care for younger people may be associated with some infanticidal behavior. In a modern world, there may be economic hardships associated with the root of such problems. It is for this reason that abortions are sometimes carried out.

Abortions

In an eMedicineHealth article, Suzanne R. Trupin (2015) states that, "Abortion is one of the most common medical procedures performed in the United States each year. More than 40% of all women will end a pregnancy by abortion at some time in their reproductive lives. While women of every social class seek terminations, the typical woman who ends her pregnancy is either young, white, unmarried, poor, or over the age of 40." In the United States and worldwide, abortion (known also as elective termination of pregnancy) remains common.

Wrangham and Peterson (1996) state that "the US Supreme Court legalized abortion in 1973; currently, about 1.2 million abortions are performed each year in the United States. Worldwide, some 20–30 million legal abortions are performed each year, with another 10–20 million abortions performed illegally." "Abortion in China, a country with an extreme overpopulation problem, has become a government service available on request for women. In addition to the virtually universal access to contraception, abortion is a way for China to contain its population in accordance with a one-child policy." Its policy has recently undergone some changes.

It is quite clear that illegal abortions are unsafe, accounting for 13% of all deaths of women because of serious complications. Death from legal abortion is almost unknown in the United States or in other countries where abortion is legally available.

In spite of the introduction of newer, more effective, and more widely available birth control methods, more than half of the 6 million pregnancies occurring each year in the United States are considered unplanned by women who are pregnant. Of these unplanned pregnancies, about half end in abortion.

From 1900 until the 1960s, abortions were prohibited by law. However, the Kinsey report noted that premarital pregnancies were nevertheless electively aborted, and public and physician opinion began to be shaped by the alarming reports of increased numbers of unsafe illegal abortions.

In 1965, 265 deaths occurred [in the United States] due to illegal abortions. Of all pregnancy-related complications in New York and California, 20% were due to abortions. A series of [United States] Supreme Court decisions granted increased rights to women and ensured their right to choice in this process. No decision was more important than *Griswold v Connecticut*, which, in 1965, recognized a constitutional right to privacy and ruled that a married couple had a constitutional right to obtain birth control from their health care provider.

People against abortion argue that parents need to be informed about and approve an abortion for a daughter younger than 18 years. Those supporting the rights of a woman to choose abortion say parental consent is not required for a woman to carry a pregnancy to term (the birth of a baby), nor do parents need to give permission for a woman seeking birth control methods, such as pills or an intrauterine device (IUD). Parents are also not consulted when a woman seeks treatment for a sexually transmitted disease.

Research shows that many young women younger than 18 years [of age] do involve their parents in their decision to abort (45%). Laws requiring parental consent are forcing minors to obtain abortions much later in their pregnancies. Some minors must travel great distances to states with no such law.

Most women seeking abortion are white (53%); 36% are black, 8% are of other race, and 3% are of unknown race. Abortion rates are highest among 20–24-year-old women. Rates are lowest among women younger than 20 or older than 40 years, but these women are far more likely to have an abortion if they become pregnant.

"Abortion causes at least 13% of all deaths among pregnant women. New estimates are that 50 million abortions are performed worldwide each year, with 30 million of them in developing countries. Approximately 20 million of these are performed unsafely because of conditions or lack of provider training."

Killing of Adults

It is the killing of adults that appears to be relatively rare among nonhuman animals. Even in expressions of agonistic behavior in which two individuals are vying for the alpha position, death is unusual unless both possess alpha qualities. So why do animals kill infants more readily than adults? Researchers on the topic state that it is simply because "infants are easier to kill." While this may be true, there may be underlying reasons for infanticide, as mentioned earlier (Hausfater and Hrdy. 1984).

Altercations between human adults that lead to murder are sometimes expressed by young adults toward their caregivers as an act of violence against authority or adults of equal age over domestic differences (Sinha and Milligan, 2010). Murders outside the family are generally the result of antisocial personalities. Killing by governments generally is the result of political differences, dominance establishment (especially with relationship to dictatorial governments), or a need for particular resources. Such governments are generally led by dictatorial individuals (the subject of Chapter 13).

It makes no difference what the sex or species is. Even in the social wasp colonies that were described in an earlier chapter, dominant females did what was necessary to establish and maintain themselves in a dominance hierarchy. They generally did not fight to the death, but if either of two females that expressed dominant behavior would not step down, death occasionally occurred.

We must recognize that agonistic behavior is variable, just like any other behavior in a population. There are different individuals who express dominance from one extreme to the other. Natural selection works on such behavioral variation just as it does on any other feature that is important to a species for survival and reproduction. Extremely dominant and extremely subordinate individuals are not as successful at spreading their genes as wasps that are dominant but tolerant.

This phenomenon of tolerance appears to work well in most vertebrate and invertebrate species. However, when we apply this to a code of ethics in human populations, it may not be the tolerant individuals who obtain the most benefits. Alphas, even dictatorial alphas, in human society sometimes may be selected because of their apparent leadership capabilities and ability to deceive those who surround or follow them.

Other than in predatory/prey situations, the behavior of not killing is more of a "norm" in most nonhuman species than that of killing. Wrangham and Peterson (1996) point out that conspecific killing is what "makes the lethal raids of chimpanzees and humans (and perhaps macaques) an interesting puzzle." "Killing by humans and chimpanzees is dominated by males and characterized by raiding, a phenomenon which is also demonstrated in certain aboriginal human groups; for animals such as lions and certain other mammals, killing is demonstrated by males participating in takeovers; in animals such as hyenas, killing is initiated by female-dominated gangs."

People Who Kill

While there seem to be reasons for the killing of humans and warfare, the nature of killing lies in the human mind, and the desire to kill varies tremendously among people. Often, these variations are expressions of personality differences.

As expressed in Chapter 8, the development of a personality is dependent upon a variety of both genetic and environmental causes. While most people refrain from killing, some enjoy it, and there are varying levels of these two extremes found in many people in the human population.

In an attempt to understand the roots of killing, we may ask the following questions and attempt to address them during the remainder of the chapter:

- Why do people respect life so little that they kill their spouses, their friends, their children, and themselves?
- Why is there so little respect for other humans that rape and abuse are so common?
- Why do we feel the need to war on others?
- Why do we repeatedly witness ethnic cleansing in the world?
- Why are there murderers and serial murderers who kill for the sport of killing?
- Are we all potential killers?
- Why are there gangs who choose to isolate themselves from others and kill those who are different?
- Why are there terrorists?

Family Violence and Abuse

Family violence and abuse are among the most prevalent forms of interpersonal violence that are carried out against women and children (both young boys and girls; Widom and Maxfield, 2001). As stated by the US Department of Justice, the sexual abuse of a child should never be considered "just a family matter." It is not only a crime against an individual, but an individual who is an innocent victim, and who has little defense against the instigator. It often traumatizes victims who survive or witness such actions, and it has a long-lasting effect on the lives of those individuals. With this said, our first question may be: Why do some people commit such a crime?

According to an article by Roxanne Dryden–Edwards in MedicineNet.com (2016), sexual assault is defined as "any illegal sexual contact that involves forcing a person without their consent or inflicting such contact on someone who is unable to give consent due to their age or to physical or mental incapabilities. Sexual assault may also involve sexual contact that is inflicted by someone who is trusted by or has authority over the victim. Incest is one example of childhood sexual abuse that is perpetrated by a trusted person who often has authority over the victim, as when a parent or sibling engages in such a violation on another family member."

The causes of sexual violence are complicated, often stemming from one or a combination of the following factors: evolutionary pressures, attitudes toward victims, anger, power, sadism, sexual pleasure, socioeconomics, psychopathy, ethical standards, and laws (Holmes et al., 1996; WHO, 2002; Widom and Maxfield, 2001). As indicated in earlier chapters, such behavior may stem from genetic predisposition (including chemical malfunction), some form of trauma that upsets homeostatic mechanisms, or learned behavior. It is often the case that those who are abused become abusers themselves.

Many cases of child abuse are unknown since children are often afraid to report an incident to the police because the abusers are too often a family friend or relative. Abouty one-third of all juvenile victims of sexual abuse cases are children younger than 6 years of age.

Lack of Respect for Other Humans

According to a US Department of Justice report, 354,670 women were the victims of a rape or sexual assault in 1995 alone. A woman was raped every 2 min in America at that time (Rand, 2005). The FBI in this year estimated that only 37% of all rapes are reported to the police. US Justice Department statistics were lower (26%).

"The FBI estimates that 72 of every 100,000 females in the United States were raped in that year. Approximately, 28% of victims were raped by husbands or boyfriends, 35% by acquaintances, and 5% by other relatives. A 2010 report shows that rape cases were lower in that year than in several of the previous years, 27.3 rapes occurring per 100,000 citizens," and the number of cases appear to continue declining over the years.

According to a July, 2014, Justice Department report (http://www.statisticbrain.com/rape-statistics/), the average number of rape cases reported in the United States annually is 89,000. The percent of women who experienced an attempted or completed rape is 16%. The percent of men who experienced an attempted or completed rape is 3%. The decline in rape rate since 1993 is 60%. The percent of rapes that are never reported to authorities is 60%. The percent of college rapes that are never reported to authorities is 95%. The percent of rapes where both victim and perpetrator had been drinking is 47%.

As mentioned earlier, "one of the most startling aspects of sex crimes is how many go unreported" (Langston et al., 2010). The most common reasons given by women for not reporting these crimes are that it is a private or personal matter and they are driven by fear that there may be a reprisal from the assailant.

"Both physical and emotional trauma are commonly expressed in rape victims. According to US Department of Justice statistics, nongenital physical injuries occur in approximately 40% of completed rape cases. As many as 3% of all rape cases have nongenital injuries requiring overnight hospitalization."

How does rape affect a victim? "Victims of rape often manifest long-term symptoms of chronic headaches, sleep disturbance, recurrent nausea, decreased appetite, eating disorders, menstrual pain, sexual dysfunction, and suicide attempts" (Ward et al., 1994). In a longitudinal study, sexual assault was found to increase the odds of substance abuse by a factor of 2.5.

"Victims of marital or date rape are 11 times more likely to be clinically depressed and 6 times more likely to experience social phobia than are nonvictims (Ward et al., 1994). Psychological problems are still evident in cases as long as 15 years after the assault. Fatalities occur in about 0.1% of all rape cases."

Ethnic Cleansing: Genocide

In spite of all humans belonging to a single species, many people throughout the world (social intolerants) have imagined problems with a particular ethnic group other than their own. In their particular society, they often claim a right to eliminate what they feel is a scourge on human society (Grosby, 1994; Horowitz, 1985). Such emotional feelings are generally learned, with respect to either their parent's or society's perspective on the matter or the religious groups to which they belong. This is a topic that is more comprehensively discussed later in this chapter.

Eugenics

The concept of eugenics (a field that deals with the improvement of hereditary qualities of a race or breed) is, in a way, an extension of ethnic cleansing. According to Edwin Black in *War Against the Weak, Eugenics and America's Campaign to Create a Master Race* (2012), the eugenics society looks at the human species the way others would look at domestic animals: Preserve the best and eliminate the rest.

Commencing in England during the late 1800s and early 1900s, subsequently spreading to the United States and eventually becoming a worldwide movement, "the victims of eugenics were typically poor urban dwellers and rural 'white trash' from New England to California, immigrants from across Europe, Blacks, Jews, Mexicans, Native Americans, epileptics, alcoholics, petty criminals, the mentally ill, and anyone else who did not resemble the blond and blue-eyed Nordic ideal the eugenics movement glorified."

It was and still is a game of dominance. Dominant individuals in the world feel a necessity to improve the human species, and they apparently will go to extremes to have it happen.

"The [initial] goal in the early 20th century was to immediately sterilize 14 million people in the United States and millions more worldwide—the 'lower tenth'—and then continuously eradicate the remaining lowest tenth until only a pure Nordic super race remained. Ultimately, some 60,000 Americans were coercively sterilized and" according to Black, "the total is probably much higher."

Over time, the eugenics society promoted their ideas for improving the human animal, often drawing on dominant, wealthy supporters who could back research on how to reach their goals. Prior to and during the Second World War, eugenic philosophy was running wild, a segment of the concept spreading to world leaders, for example, Hitler. According to Engdahl and other authors, Hitler was being financed by certain individuals in the United States until he attacked Poland, and support was terminated. Hitler's eugenic approach was similar to that of many dominant people in the world. His behavior, on the other hand, was more manifest.

A contemporary approach to their problems resulted from a statement made by Henry Kissinger: Control the oil and you control nations; control the food, and you control people. Largely because of world overpopulation, the concept of eugenics rages on today in what has been called the "Green Revolution," as expressed by F. William Engdahl in *Seeds of Destruction, The Hidden Agenda of Genetic Manipulation* (2007). The more people that enter the world, the more food must be produced. While genetic engineering appears to have helped this ever-growing population feed itself, toxins that sterilize and kill humans have been threatening side effects.

Murderers and Serial Killers

According to Federal Bureau of Investigation statistics, "the United States' national crime rate in the mid-1990s was 3466 crimes per 100,000 residents, which was actually down from 3680 crimes per 100,000 residents over thirty years earlier in 1969 (−9.4%). Two-fifths of the victims were young people between the ages of 10 and 29 who were killed by other young people. While the crime rate apparently fluctuates, the estimated volume of violent crimes in 2010 dropped 6 percent compared to the 2009 figure, the fourth consecutive year it has declined. For the eighth consecutive year, the volume of property crimes went down as well—2.7 percent."

Statistics Canada says the rate of crimes reported to police in Canada "reached its lowest level in 40 years in 2011." Total violent crimes in the United States for 2012 show an increase of 0.30% over 2011, while total property crimes show a decrease of − 0.91% over 2011.

"An estimated 520,000 people were murdered in 2000 around the globe. The likelihood of committing and falling victim to crime," they say, "depends on several demographic characteristics, as well as the population's location. Overall, men, minorities, the young, and those in financially less-favorable positions are more likely to be not only crime victims but also the instigators of crimes. Crime in the United States and other countries is also higher in certain areas."

As of 2010, statistics reveal that of 16,277 murders committed in the United States, "10,568 were committed by males, 1176 were by females, and 4533 were by offenders in which their sex was unknown. Further, 5334 murders were committed by white offenders, 5943 were by black or black and Hispanic offenders, 273 were by offenders of other races, and 4727 murders were committed by offenders whose race was not known."

Murder rates vary greatly among countries and societies around the world. "In the Western world, murder rates in most countries have declined significantly during the 20th century and are now between 1 and 4 cases per 100,000 people per year. Murder rates in Japan, Ireland, and Iceland are among the lowest in the world (around 0.5 cases per 100,000 people per year); the rate in the United States is among the highest of developed countries (around 5.5 in 2004, with rates in larger cities sometimes over 40 per 100,000). In the United States, 666,160 people were killed between 1960 and 1996."

Honor Killings

An honor killing is defined as "a homicide of a member of a family or social group by other members, commonly as the result of a belief by the perpetrators that the victim has brought dishonor upon the family or community

(Burke, 2010). Dishonors are normally the result or suspicion of certain behaviors, for example, dressing in a manner unacceptable to the family or community, wanting to terminate or prevent an arranged marriage or desiring to marry by one's own choice (especially if to a member of a social group deemed inappropriate), engaging in heterosexual acts outside of marriage, and engaging in homosexual acts."

It is suspected that more than 20,000 women lose their lives in honor killings in the Middle East and Southwest Asia each year. At times, women other than family or community members are attacked. For example, women who come to the United States or other country that has different customs may desire to accept their new society, resulting in their bridging social divides, publicly engaging other communities, or adopting some of the customs or the religion of an outside group are especially vulnerable to attack. It has also been shown that low-status immigrant men and boys who have immigrated to countries with different customs have asserted their dominant patriarchal status by inflicting honor killings on female family members who have participated in what they consider dishonorable behavior, for example, feminist and integration politics.

"At the time of this writing, honor killings had been reported in Bangladesh, Brazil, Ecuador, Egypt, India, Israel, Italy, Jordan, Lebanon, Morocco, Pakistan, the Syrian Arab Republic, Sweden, Turkey, Uganda, Yemen, and other Mediterranean and Persian Gulf countries, and they had also taken place in western countries, such as France, Germany, and the United Kingdom, within migrant communities."

According to Widney Brown, advocacy director for Human Rights Watch, the practice of honor killing "goes across cultures and across religions." In a 2009 article in *Middle East Quarterly*, Phyllis Chesler (2010) states that although there are not many cases of honor killings within the United States, 90% of those known to have taken place in Europe and the United States from 1998 to 2008 were perpetrated by Muslims against Muslims.

Are We All Potential Killers?

In a BBC Magazine article called *Are We all Capable of Violence?* Diene Petterle (2009) points out that this is one of the least-understood topics of the 20th century, and it remains a conundrum today. The question that stimulated her to write her article was, "Do all people who we would label as 'ordinary' potentially possess a violent nature?"

Petterle is right in saying that many humans have two opposing inner forces that make them react in different ways to their surroundings. Upon examining other animals, we would find these opposing forces in them as well. However, with increased population complexity in humans, especially due to differences in religions and homeland customs, human reasons for killing are quite different than for most nonhuman animals. At the same time, there are some common features in both. As is often the case, how population members are treated in life generally brings out one side or the other of human personality.

In *Becoming Evil, How Ordinary People Commit Genocide and Mass Killing*, James Waller (2007) points out that at least 60 million men, women, and children have been victims of genocide and mass killing in the past century alone. Genocide has occurred throughout the world, including the United States (Lawrence, 1991). The country was involved in genocide from the time foreigners landed on its shores in the late 1400s and early 1500s. At the time of invasion, there may have been as many as 100 million indigenous people who inhabited the western Hemisphere. The population diminished by some 90% two centuries later. "Over the past 500 years, indigenous peoples have had to cope with destruction of their habitats, epidemic disease, hunger, warfare, despair, and violations of their human rights."

The discrimination of people by other people has spanned the history of the human population around the world, involving race, gender and gender identity, sexual orientation, national origin, color, ethnicity, religion, language, disability, age, socioeconomic status, and other characteristics. It appears that people often think of themselves as belonging to an elite group, and they often attempt to exclude others who are not like them.

In every walk of life, discrimination raises its ugly head. M. Rubin and M. Hewstone (2004) have distinguished between three types of discrimination:

1. *Realistic competition* is driven by self-interest and is aimed at obtaining material resources (e.g., food, territory, and customers) for the in-group (e.g., favoring an in-group in order to obtain more resources for its members, including one's self).
2. *Social competition* is driven by the need for self-esteem and is aimed at achieving a positive social status for the in-group relative to comparable out-groups (e.g., favoring an in-group in order to make it better than an out-group).

3. *Consensual discrimination* is driven by the need for accuracy and reflects stable and legitimate intergroup status hierarchies (e.g., favoring a high-status in-group).

While numerous forms of discrimination continue to go on in the world, the United Nations' stance on discrimination includes the following statement: "Discriminatory behaviors take many forms, but they all involve some form of exclusion or rejection." The United Nations Human Rights Council works toward the elimination of discrimination between people.

We all belong to one species. The differences between different people in the world are mainly cultural, differences that can be accompanied by fascination and beauty in a curious and tolerant society. Differences among people become ugly and unacceptable in an intolerant society, sometimes leading to violence and genocide.

Attractive Nature of Violence

Even in individuals who we suspect are nonviolent, there seems to be an attraction by many people in our population to learn about violent events. A survey of current television documentaries and programs about people with a violent nature (*On The Case with Paula Zahn*; True Crime with Aphrodite Jones; *Fatal Encounters*; *Deadly Sins*; Dangerous Persuasions; Frenemies; ID Addicts; Breaking Point Moments; Who the (Bleep) Did I Marry?; Scorned: Love Kills; Dateline: Five Minute Mysteries; Deadly Women; Bizarre Ends, Desperate Measures; Deadly Affairs; Disappeared; Homicide Hunter; The Will; Very Bad Men; Pretty Bad Girls; Final Cut; Wives with Knives; Fatal Vows; Sins and Secrets; Unusual Suspects; Scorned; Wicked Attraction; Mansion Murders; Swamp Murders; and *I (almost) Got Away With It*) should give us some idea of the current popularity of observing violence on television.

"The human race is both appalled and fascinated by violence," Petterle says. "Human aggression" spans the globe "from terrorist attacks to guerrilla wars to gang-related crime." Being everywhere, it is shared by people of all nations and all races. But what are its roots? Is it instinctive or learned?

"Most of us think of ourselves as calm and peaceful people. We are most often brought up to try and resolve conflicts peaceably and tend to think that violence is something that 'other' people commit. But is it?" Is it possible that we, our mothers, fathers, daughters, or sons, could ever be driven to commit a violent crime? Is there a potential for violence in all of us? Based on the TV shows mentioned earlier, violence is all around us, often involving people that look to us to be "regular" people.

Under the right circumstances, it appears that it is possible for many of us to commit a violent crime. It may depend on an unfortunate circumstance or a series of unfortunate circumstances. While Petterle feels that people are born violent, philosophers have long debated this concept and generally agree that violence is built into some of us and has both genetic and learned roots.

It is true, as pointed out in Chapter 10, that we can receive genetic input that will function in the formation of our brains. We have also seen that learned behavior and even the food we eat may alter who we are and sometimes lead to violent behavior. During our formative years, up until the age of about three to five, our personality is developed, but it can be altered at any point (as indicated in cases of personal and emotional trauma), even in later years. Once our personality is formed, the urges that come from the emotional center in our brain can direct how we react to a current and changing world.

As we mature, we all develop that part of the brain that allows us to control aggression— the prefrontal cortex (seeChapter 5). In truth, how well this control mechanism works depends on our predisposition for violence (genetic), the food we eat (chemical, which can be acquired or genetic), and experiences, especially those in our formative and adolescent years (learned).

As this author points out, positive learning and nonviolent early years in the lives of children goes a long way toward building a proper foundation to progress naturally through life without resolving conflicts in a violent fashion. Learning in a positive environment is powerfully influential toward building a nonaggressive approach to life's events. Proper education in the early stages of brain development can actually change the physical structure of the brain, resulting in a less-aggressive approach to most of life's problems.

However, trying to resolve conflicts peaceably is not something all individuals and cultures are capable of or willing to do. In the Bolivian Andes, for instance, one tribe settles disputes that arise throughout the year in an annual festival of violence, known as the Tinku (Valeriano T'ula, 2003). Traditionally, men, women, and even children learn to fight, sometimes to the point of killing one another.

They have programmed themselves their entire life to resolve conflicts in their own violent fashion. While we know nothing about their predisposition to violence, their behavior is more-than-likely either partially or wholly learned as a societal norm. The reader is directed to Chapter 10 where genetic and chemical variations in the human brain, along with learned behavior, are discussed in more detail.

Gangs and Hate Groups

Gangs and hate groups have already been discussed in Chapter 9. Both groups are composed of people who are discontent either with their position in life or with people around them. While many individuals in such groups may express their discontent verbally, others are explicitly intolerant and express their emotions in violent or criminal ways, often including murder.

WARFARE

Like gangs, warfare combines the human affinity to kill with territorial needs, the acquisition of resources, and defense, but on an even bigger scale. Contemporary warfare is carried out with sophisticated tools, but early human populations utilized instruments that were initially designed to be used in predatory forays:

> War… is not only a universal factor, but also… a biologically determined factor arising out of human nature. *F.M. Keesing, Cultural Anthropology, the Science of Custom*

Territoriality

In Chapter 2, territory was defined as an area (usually fixed) that is defended by a member or members of a species to keep other members of the same species or other species out. Territories are typically familiar to the species occupying them and used for feeding, mating, and rearing the young. The size of a territory varies, depending on the species that occupies it.

As an example, song sparrows may require $3000\,m^2$ of space to carry out activities during their breeding season (Chamberlain et al., 2003; Feldhamer et al., 2004). Male bobcats require an average of 4900 acres and females require about 2900 acres to survive and reproduce. Although of large size, sea lion males defend relatively small areas that are used for mating. Red squirrels have relatively large territories that are necessary for feeding purposes.

Territorial displays in populations of animals other than humans are most often displays of tool-less defense. They utilize whatever defenses they naturally have or whatever they can construct based on the capabilities of their brain. While many birds and mammals may use tools, they are most often utilized to obtain food or mates. It is primarily in the higher primates where both offensive and defensive tools become important as defensive weapons.

According to an article by P.R. Ehrlich et al. (1988, 1994), "territory size [in populations of birds] varies enormously from species to species, and even within species, from individual to individual. Golden eagles have territories of some 35 square miles; least flycatchers' territories are about 700 square yards; and sea gulls have territories of only a few square feet in the immediate vicinity of the nest. Territory size often varies in the same species from habitat to habitat."

"Some birds," they say, "defend their entire home range. Others defend only their food supply, a place to mate, or the site of their nest. Some tropical hummingbirds chase most other hummingbirds and other nectar-feeding birds (and some butterflies) away from favorite patches of nectar-bearing flowers. On their leks (patches of ground traditionally used for communal mating displays), grouse, some sandpipers, and a few other birds defend small territories. Most colonial-nesting seabirds simply defend the immediate vicinity of their nests—presumably to protect their eggs and, at least in the case of some penguins, the pebbles from which the nest is constructed."

"Territoriality tends to space some species of camouflaged (cryptic) birds and their nests rather evenly throughout their habitat; it prevents them from occurring in flocks or clusters while breeding. This, in turn, may reduce danger from predation, since many predators will concentrate on one kind of prey after one or a few individuals of that prey type are discovered (i.e., the predator forms a 'search image'). Clustering can promote the formation of a search image by predators and thus reduce the security of each individual prey (birds that are not cryptic, however, may gain protection in clustering)."

Territoriality has been shown to offer benefits that outweigh energy costs necessary to defend the territory, and this translates into an increase in species fitness.

Territorial displays in social groups may directly affect the individual displaying them or indirectly affect them through the use of altruistic or reciprocity behaviors in their subordinates. Such behavioral strategies provide a compilation of benefits, expressed as an inclusive fitness that benefits group members by allowing them to produce adequate numbers of offspring.

Altruistic behavior is most often expressed as a coefficient of relatedness. The closer an individual is related to the altruist, the more likely it will receive the benefits of an altruistic act. In terms of biological success, the target of an altruistic act will most likely be in a better position to pass on genes, which are similar to those of the altruist. Close relatives of the altruist possess a genetic makeup that is similar to that of the close relative they are performing an altruistic act for, and therefore the altruist is allowing some of the genes that he or she has to be protected and passed on to succeeding generations.

Ehrlich et al. (1988) point out that "to minimize the need for actual physical contact in order to defend territories, animals have evolved 'keep-out' signals to warn away potential intruders. In birds, of course, the most prominent expressions of territoriality are the songs of males. Far from being beautiful bits of music intended to enliven the human environment (as was long assumed), bird songs are, in large part, announcements of ownership and threats of possible violent defense of an area. If, of course, the aural warning is ineffective, the territory owner will often escalate its activities to include visual displays, chases, and even combat. This territorial behavior is typically quite stereotyped, and can usually be elicited experimentally with the use of recorded songs or with stuffed taxidermy mounts."

Similar behavior is seen in colonies of social wasps, although it is expressed in nonviolent demonstrations rather than in songs (Hermann et al., 2017). When a colony of paper wasps is approached, for instance, they wave their antennae and forelegs, raise their body, flap and buzz their wings, and jump forward, behaviors that function in warning an intruder. If these warnings are not recognized and the intruder chooses to continue approaching the nest, stinging generally ensues. Thus, territories are protected by these species because of their investment in a nest, protection of the queen and offspring and the resources associated with them, for example, food and mates.

Human Territories

The difference for having territories between humans, as compared to other animals, is that social behavior in humans requires more or different resources, for example, fossil fuels for a growing number of vehicles and other energy needs, land for agriculture, and the building of housing for an ever-growing population (Reinhardt, 1998). To get these things, humans may initially choose to negotiate with other humans (equivalent to the warning behavior expressed in nonhuman animals) who can provide them. If negotiation (warnings or reciprocity) fails, humans may choose to go to war.

In an interesting article called *Territoriality*, Serge Kahili King (Florian, 2015) defines human territories as patterns "of boundaries imposed on something by individual decision or group agreement." He further points out that in human populations, "an individual or group can decide or agree to change the pattern."

"The pattern itself is established by a set of rules that define the pattern, so changing the pattern is done by changing the rules. The rules are changed by changing the symbols we use to form the rules, i.e. words, images, feelings, and actions. When a territory is threatened, then, we can either defend it, attack the territory of the attacker, suffer the trauma of invasion, or change the rules of the territory."

In considering types of territories, King brings to our attention the concepts of physical territory (the physical body, real, and personal property); social territory (family, intimates, clan, tribe, peer group, club, associations, and other social, cultural, or religious groupings); and psychic territory (thoughts, opinions, theories, philosophies, plans, purposes, dreams, memories, and time).

"It is possible," he says, "to think of humans as basically territorial animals, to think of evil as based on a violation of territory or territorial rights, and good as based on an opening up or giving up of territory or territorial rights. Among the things we consider evil and worthy of punishment are the destruction, damaging, or appropriation of life and property; the breaking up of social ties and reputations; and the theft of ideas, the breaking of rules, the attacking of ideas, the frustration of plans, the invasion of dreams, the changing of memories, and the wasting of time."

"Interestingly, any of these are [often thought of by humans as] less than evil and perhaps not evil at all when done to a being not of one's own social territory. To the warrior-oriented mind, the invasion, destruction, and/or appropriation of territory belonging to another social group may even be deemed as good, since it expands the territory or territorial influence of one's own group." This, of course, is an animalistic view, not a moral one.

Among the things we consider good, loving, and worthy of reward are the giving up or allowing the use of life and property; the strengthening and extension of social connections; respect or tolerance for psychic territory; and the healing of territorial violations.

"A curious aspect of human territoriality," King says, "is how we sometimes assign human territorial concepts to certain nonhuman entities and not others. For instance, many humans will acknowledge the spirit of a mountain, a glen, or an ancient structure and show respect for that spirit by asking permission to cross into what is perceived as its territory. Yet, those same humans will blithely trespass the territorial boundaries of…birds, [other mammals], insects, and plants which live in the area. The logic is something like asking the spirit of a city for permission to enter and then freely roaming through anyone's home that you care to."

Thus, the point is made that while human territories may entertain the same basic concept as those of other animals, human territories may be changed according to the needs of dominant human factions. While this concept may relate to other animals, as well, humans add other factors based on cognitive reasons. Examples are found in the making and breaking of territorial treaties by the United States government and Native Americans, leading to such situations as *The Trail of Broken Treaties*, a cross-country protest in the United States by Native Americans and First Nations organizations that took place in the autumn of 1972 (Deloria, 1974). It was designed to bring attention to Native American issues, such as treaty rights, living standards, and inadequate housing.

Treaty-breaking has not been restricted to the United States (Andelman, 2008). The making and breaking of treaties have been carried out by all major governments around the world throughout historic times, making humans one of the most deceptive political animals on Earth.

Comparing Humans With Other Simians

Upon viewing human characteristics, it is commonplace to compare them with features in populations of lower animals, particularly other simians. A passage comparing rhesus macaque behavior with that of humans has already been presented in Chapter 7. Macaque monkeys have succeeded in establishing themselves in a wide variety of habitats and have obtained the greatest range of any other simians on Earth other than humans (Groves, 2005).

The simian most often compared to humans is the chimpanzee, a great ape that is currently endangered, mostly because of: (1) massive destruction of its natural habitat by humans; (2) humans capturing it for the purpose of trading; and (3) humans hunting it down for the consumption of its meat (Cowlishaw et al., 2005). Recent increases of victims of Ebola virus in that area are especially threatening to chimpanzee and bonobo populations because their numbers are already down. However, let us briefly look at each of the three points mentioned.

According to an article on Buzzle.com, there was a time when more than 1 million chimpanzees lived in the forests of at least 25 countries in Africa. At present the figure has gone down to as low as about 200,000. The dramatic decrease in the number of chimpanzees began in earnest in 1960. Currently, there are hardly six countries in Africa that have conditions suitable for the breeding of wild animal species in general. The remaining countries have destroyed their forestland to convert them into farmlands. Other cutting of forest trees has taken place to supply logs to various industries and to build new roads.

Chimpanzees are caught illegally and sold as pets. People who live in and around forests also hunt them down to obtain bushmeat. However, they kill them not only for their own pleasure and consumption but also for the purpose of trading, because there are customers who are ready to pay huge sums of money to have chimpanzees as pets or to eat their meat, and sellers make large profits at the cost of endangering the survival of an entire species. It is estimated that at least 4000 chimpanzees are killed annually. Such mass killing represents the biggest threat to chimpanzees, which may jeopardize their existence on Earth.

The population of bonobos is even more threatened (Dupain et al., 2000). Estimates of the number of bonobos range from 10,000 to 20,000 left in the wild. In addition to their low numbers, their populations are fragmented and decreasing. Some of the reasons for their threat of extinction are: civil unrest, habitat degradation, and a lack of information about the species because of its remote territory. As already mentioned, disease transmission also poses a silent threat to them (along with gorillas and chimpanzees), especially in recent years when great apes occupying West Africa have succumbed to the Ebola virus.

Conservationists are trying to spread awareness among people in order to save these endangered animals from extinction. As pointed out repeatedly in these pages, there are striking resemblances in the behavior of chimps and humans. Besides saving these animals because of ethical, moral, and behavioral reasons, studying their biology can provide us with a lot of valuable information to not only save them from extinction but also provide data that may lead to solving the mysteries of our own social development:

> If chimpanzees are any indication, warfare is even more ancient than mankind. **James Case**, *Competition, The Birth of a New Science*

As pointed out in works such as *Demonic Males*, by Richard Wrangham and Dale Peterson (1996); *The Most Dangerous Animal*, by D.L. Smith (2007); and *Competition, The Birth of a New Science*, by James Case (2007), the warlike qualities of chimpanzees, animals that use an array of tools, including some crude nature-made implements of aggression, may (as most of these researchers have stipulated) provide us with some insight into where the origin of warlike behavior in humans may have arisen. We may also look at the war patterns that are demonstrated by rhesus macaques.

However, primate behavior is quite diverse, and as stated earlier, we should be very careful about how we make comparisons. As S. MacKinnon and S. Silverman point out in *Complexities Beyond Nature and Nurture* (2005), chimpanzees and bonobos are equally related to the most recent common ancestor from which human beings have arisen. As with most other primates, chimpanzee and bonobo behavior varies tremendously, especially in terms of dominance establishment and aggressiveness, often between individuals and within different groups. There are yet other patterns of behavior seen in gorillas, gibbons, and orangutans, more distant relatives of humans but great apes nevertheless.

Forming isolated groups, these primates have continued to evolve since splitting from the lines that led to hominines (6–12 million years ago). As pointed out in Chapters 7 and 8, any contemporary behavior probably evolved independently and cannot be assumed to be a model for common ancestral behavior without some question about the stability of behavior over millions of years. There is no one generalized "primate pattern" found in common ancestors but rather a variety of patterns with some common themes.

In spite of the independent rise of social development in separate groups of organisms, it is nevertheless inviting to approach the subject of human sociality by delving into what appears to be a commonality between them (as many investigators have already done). The following information from Wrangham and Peterson's *Demonic Males* (1996) provides a synoptic glimpse into the lives of chimpanzees and indicates how close their patterns of agonistic behavior are to those of humans.

These and most other authors feel that the social world of chimpanzees and humans is extraordinary when comparisons are made. Very few animals live in patrilineal, male-bonded communities wherein females routinely reduce the risks of inbreeding by moving to neighboring groups to mate. Only two animal species are known to do so with a system of intense, male-initiated territorial aggression, including lethal raiding into neighboring communities in search of vulnerable enemies to attack and kill. Of 4000 types of mammals and several million other animal species, this suite of behaviors is known only among chimpanzees and humans.

However, our species has added other facets to its territorial displays to dominate its enemies, as seen in the evolution of their tools of aggression and defense. Thus, while humans have acquired the animalistic territorial trait from predecessors, they have carried it to an extreme, mostly due to their superior brain. In addition to increasing the variety and complexity of their territorial implements, they have increased the numbers of individuals who partake in territorial displays to the point of having major global wars.

HUMAN WARFARE

Ultimately, most wars are irrational. The forces that impel men to throw away their lives and fortunes are compounded primarily of greed, power drives, and dark passions left over from the ages before civilization. **Jeremy Barnes**, *The Pictorial History of the Civil War*

It is doubtful that isolated groups of people ever got along. They had to compete with one another for territorial rights and the resources that would keep them alive, lacking trust in any other hominine forms or in conspecific members of groups other than their own. In the process, they sometimes entered altercations and killed one another for a variety of what appeared to be legitimate reasons. As the groups got larger, skirmishes turned to war, and more people were killed in the process.

Warring involves more than killing. It incorporates increases in group size, territoriality, defensive and premeditated offensive group behavior, and intelligence (Mullen, 2011).

Warring behavior by humans is an old business. In *The Social Conquest of Earth*, E.O. Wilson (2012) points out that, "Burial sites of foraging people of the Upper Paleolithic and Mesolithic of the Nile Valley and Bavaria include mass interments of what appear to be entire clans. Many had died violently by bludgeon, spear, or arrow. From the Upper Paleolithic, from 40,000 to about 12,000 years ago, scattered remains often bear evidence of death by blows to the head and cut marks on bones. This was the period of the famous Lascaux and other cave paintings, some of which include drawings of people being speared or lying about already dead or dying."

Over time, with an increase in the size of the human population and an increase in social dependency on one another, human battles increased in complexity, and isolated battles evolved into larger and more serious wars (states of open, armed, and often prolonged conflict that is carried on between organized groups, for example, nations, states, or parties). According to the authors of Barnes and Noble's *Battles of the Ancient World* (B&N, 2005), warfare had already become an established practice by the time the Mesopotamian and Egyptian civilizations were founded.

Based on early evidence of weapons, spears and arrows were among the first weapons of war, representing significant improvements over stones and other poorly defined instruments generally found in more primitive societies. There is no doubt that some significant improvement in battles had already been made by the time genuine wars had come about. Even chimpanzees have been found to occasionally make rustic spears to use in subduing their prey.

Once the initial period of weapon innovation had commenced, the human mind continued the movement toward more complex weaponry. There was a tremendous advancement in military technology in the latter part of the stone age (4.000–10,000 BC), during which bows, daggers, slings, maces, hammers, and armor were present.

War tactics changed considerably when copper was alloyed with other metals around 4000 BC to form bronze which was made into tools and weapons. Thus, ancient warfare got a boost when the Bronze Age issued in, resulting in the domination of the Middle East by Egypt and Mesopotamia.

Lands of the Upper and Lower Nile were united by Egypt's first pharaoh, Narmer or Menes, their concerns and the concerns of other early pharaohs being mostly defensive. Egypt's Old Kingdom collapsed around 2050 BC, due to long periods of warring between its smaller entities.

Short-range weaponry, such as daggers, spears, and axes, remained in vogue into the Bronze Age, although some longer-range weapons, such as slings, bows, and javelins were also evident. Protective armor was developed as well.

By 145 BC, Mycenaean green armies were intricately organized. Sailing vessels came into use early (Chatterton, 1915). Mycenaean armies maintained extensive fleets for conquest and piracy, and light chariots were in use on land. Subsequently, Greece continued a civil war that ushered in the Dark Ages.

What followed were assorted Greek wars. Horses had been utilized in pulling wagons and chariots prior to 800 BC. The Hittites may have introduced the horse as a cavalry mount (Crouwel, 2013). Chariots and individual combats were replaced by other uses of the horse and formations known as the phalanx. The phalanx was an arrangement of soldiers who stood in a tightly formed square. Simple as it was, it proved to be a formidable form of organization.

With new developments, defending groups were forced to construct almost impenetrable fortresses. By the 4th century BC, an enormous siege ladder, called a sambuca, was developed to gain access to the tops of walls. The sambuca could be manipulated to place it across moats and ditches (Campbell, 2005); it was covered to protect the soldiers from fire, and soldiers were not required to know the height of enemy walls.

By 305 BC, a huge device, called a heliopolis or city taker, had been invented for warfare (Campbell, 2003). Two hundred men could inhabit it and operate weapons, called lithoboloi, that could launch huge stones. The outer walls were iron plated.

Through the years, increasing organization of forces led to the formations of legionaries that improved on the phalanx concept and developed such battle tactics as the tortoise formation to overcome their adversaries (Cowan, 2007; Rance, 2004). Legionary discipline depended on loud and proper signaling through use of a high-frequency horn, called a cornu.

Defensive maneuvers, weapons, and tactics continued to improve over time. Some early battling devices were siege towers that soldiers could use to traverse castle walls, trebuchets that hurled large stones, ballistas that were large crossbows, and rams.

A device called green fire was invented by a Byzantine chemist in AD 670 to spray or launch petroleum products at the enemy (Dennis, 1984). Early Chinese philosophers described how the shock and awe of fire, the most frightening weapon in the ancient arsenal, could impress an enemy.

The Chinese had explored the military potential of gunpowder by at least the 11th century, and by the end of the 13th century, they had primitive guns that fired stone or iron balls (Kingsbury, 1849). By 1326, the Italian city of Florence was ordering metal cannons for defensive purposes. Canons played a marginal part in the battle of Crécy between French and English forces in 1346.

The growing importance of gunpowder-driven weapons was a vital factor in land warfare from the 16th century onward. Small-arm technology developed rapidly in the early modern period. The muzzle-loaded, smoothbore cannon—made of bronze, brass, or iron—was the basic field-artillery weapon from the late 15th century through to the 19th century.

A portable device called the coehoorn mortar was invented in 1674 to fire grenades at a fixed angle of 45 degrees. By the second half of the 19th century, improvements in engineering had made it possible to manufacture reliable multiple-fire weapons, such as the Gatling gun.

Orville and Wilbur Wright have been credited with inventing and building the world's first successful airplane and making the first controlled, powered, and sustained heavier-than-air human flight, on December 17, 1903 (Anderson, 2004). While tethered observation balloons had already been employed in several wars, aircraft were first used in the First World War (1911). Tanks made their appearance at the same time. Poison gas, including chlorine, phosgene, cyanide, and mustard gas, was used extensively on both sides of the war.

Air power increased in the Second World War with aircraft such as the Boeing B-17 Flying Fortress bomber and the Vought F4U Corsair fighter (nuclear-powered submarine in 1954; Bowman, 2000).

Contemporary wars have shown that the technological expansion of military capability is indeed complicated. The world has now entered a chemical and electronic age where weapons are not only powerful but also extremely accurate. Explosives and other devices, including atomic weapons, are available to a vast array of peoples around the world, and all for what reason? People, demonstrating their animalistic traits of dominance, territoriality, greed, and selfishness, cannot get along with one another.

Old and new secret weapons, for example, dirty bombs and anthropological weaponry are finding a continuing and sometimes more intimate relationship with warring techniques. David Price's *Weaponizing Anthropology* (2011) presents political and ethical critiques of a new generation of counterinsurgency programs, such as Human Terrain Systems, and a broad range of new academic funding programs, such as the Minerva Consortium, the Pat Roberts Intelligence Scholars Program, and the Intelligence Community Centers of Academic Excellence, that now bring the CIA and Pentagon onto university campuses.

Aboriginal Humans as Warring Animals

During the 1960s, I had the opportunity of carrying out research in a tropical rain forest about 120 miles east of Quito, Ecuador. The village I stayed at was called Limoncocha because of abundant lemon trees that grew around a two-mile-long lake. The area was inhabited by Quechua people, a relatively docile group of humans who lived off the land. Limoncocha was also a missionary base that was visited by individuals of other tribes to learn about the modern world. It was called the "Summer Institute."

During one summer there, Limoncocha was visited by Rachael Saint, a missionary whose brother (Nate) and some friends had been killed by spears from a tribe she then referred to as Aucas (now referred to as the Huaorani, Waorani, or Waodani).

They were a small tribe occupying the jungle of eastern Ecuador between the Napo and Curaray Rivers, an area of about 20,000 km^2 (7700 mi^2). They numbered about 600 people, and were split into three groups, all mutually hostile—the Geketaidi, the Baïidi, and the Wepeidi. They lived on the gathering and cultivation of plant foods (e.g., manioc and plantains), as well as fishing and hunting with spear and blowgun. Family units consisted of a man and his wife or wives, their unmarried sons, their married daughters, sons-in-law, and their grandchildren.

Prior to Rachael's visit to the tribe, the Waodani (Aucas) once believed that the entire world was a forest. The Oriente's rainforest remains to this day the essential basis of their physical and cultural survival. For them, the forest is and always has been home, while the outside world is considered unsafe. Living in the forest offered protection from the witchcraft and attacks of neighboring people. Because they lived in fear of being killed by others of their or closely related tribes, they were fierce because they had formed a way of life in which they would kill before being killed.

As told to me by Rachael, their decision to kill had developed from a story passed down over time from one generation to another. The story pointed out that it was common for individuals to be caught by some unknown group of people and used initially as slaves. According to their beliefs, they would be kept and used until a time came when they would be eaten. At times they were kept in a cage that was fabricated by vegetation, and at other times, they were given some freedom in the village, depending on the chores to be done and trust that they had earned.

One of their activities was to collect fireflies to use as lights in the evening. When I was there, I remember most of the bioluminescent light in the forests as coming from click beetles (family Elateridae), not the beetles *Photinus* and *Photuris* (family Lampyridae) mentioned by Thomas Eisner in *For Love of Insects* (2005). Their chore was to go out into the jungle every evening and collect bioluminescent insects and return to the village. With time, individuals who were reliable were trusted to return, and they gained even more freedom.

With some planning, the enslaved people who had the privilege of brief jaunts into the surrounding jungle kept going further and further from the camp each day to get their specimens. On one particular occasion, they went out to collect and continued through the jungle in an escape that would take them back to their own village. They decided that from that time onward, they would never allow their enslavement to happen again, that they would kill anyone who came into their village and even raid other villages to war upon and kill its occupants before they themselves were killed. They had learned to trust no one.

It was an unlucky day for Nate Saint when he landed his plane on a sandbar on one side of the Napo River, because in spite of his friendly approach, he and four other missionaries were killed by Auca spears the following morning. It was by meeting an Auca woman who was escaping certain death that Rachael Saint learned about their tribal ways, the story of which had eventually been made into a book and movie called *End of the Spear*. It showed how aggressive humans can be, even in modern times, and how one group of people could develop warlike behavior.

Whether this story is the true reason that a violent approach to life was developed by these people, it does indicate that their behavior had been acquired as a human endeavor, with assistance from some thought and superstition, and just like many forms of primate behavior, it may have had nothing to do with the warring behavior of chimpanzees. However, what it does not show is the temperament of the people who created fear in this tribe or the true nature of the Auca people before they became violent.

It also points out that the nature of their defense was a cooperative behavior, a behavior that could not have been developed and carried out without the presence of a social system. Members of their tribe decided what their lifestyle would be, and they carried out warring behavior and murder in order to survive and reproduce.

Wrangham and Peterson (*Demonic Males, Apes and the Origins of Human Violence*, 1996) discuss the rise of bipedal humanoids from tropical rain forest to woodland ecosystems, feeding on roots in addition to their usual foods of meat, fruits, mushrooms, and honey. Our ancestors from this line began shaping stone tools and relying much more consistently on meat around 2 million years ago (Wilford, 2015). Their brains began expanding toward human size about the same time. They may have tamed fire a million and a half years ago (James, 1989), perhaps developed human language by 150,000 years ago (Stam, 1976; Tallerman and Gibson, 2012), and some cultures developed agriculture as early as 10,000 years ago.

These authors point to aboriginal societies, such as the Yanomamö, "a cultural group of some 20,000 people living in southern Venezuela and northern Brazil in the lowland forest of the Amazon basin, as a base for comparing warlike behaviors with chimpanzees." These people represent "the largest tribe on Earth that has not yet been pacified, acculturated, destroyed, or integrated into the rest of the world."

They are well known for their intense warfare. They are an agricultural society, with men remaining in the village of their birth while women emigrate from the village either before or at the time of marriage.

Intervillage war is declared over such things as suspicion of sorcery or trivial arguments, but most often over women. Men often have chest-slapping and side-slapping duels that sometimes result in serious injury. They may continue with clobbering each other over the head with building poles or club fights. If peace does not come about, the villages move on to war.

One form of battle may involve the use of deceit, for example, hospitality that is followed by an attack or raids when their enemies are most relaxed. In raids, they may kill one or two men and escape, sometimes abducting a woman from the enemy's village. The woman is raped by all raiders, raped again by men in their village, and given to one of their men as a wife.

Some of the most prominent men in Yanomamö societies have killed often. One man was known to have killed 21 times. As much as 30% of all Yanomamö men die from violence.

Wrangler and Peterson ask the question: Do people [such] as the Yanomamö suggest to us that early human and chimpanzee violence is linked to human war? "Clearly," they say, "they do." As indicated earlier, there are opposing views on the legitimacy of this comparison.

The appetite for engagement, the excited assembly of a war party, the stealthy raid, the discovery of an enemy and the quick estimation of odds, the gang-kill, and escape are common elements that make intercommunity violence possible for both.

Lethal raiding is not displayed solely by the Yanomamö. Wrangler and Peterson state that "it has also been one of the commonest styles of primitive war throughout the world."

In *Primitive War*, Harry Turney-High (1971) stated that primitive people have killed each other with ganglike tactics as eagerly and ordinarily as anyone else. North American Indians employed surprise as their central tactic. DeVries used the term "treachery" to describe the Native American approach to war. According to Wrangler and Peterson, "surprise—the ambush, the trap, the sneak attack, the midnight raid—has been a fundamental fighting tactic of tribal groups from North America to Africa, Europe to Melanesia."

According to S.C. Gwynne (*Empire of the Summer Moon*, 2010), "American Indians were warlike by nature, and they were warlike for centuries before Columbus stumbled upon them," Comanches were especially brutal defenders and aggressors toward both white intruders and other groups of Native Americans.

Gwynne quotes Comanches as saying that "The white man comes and cuts down the trees, building houses and fences, and the buffalos get frightened and leave and never come back, and the Indians are left to starve, or if we follow the game, we trespass on the hunting ground of other tribes and war ensues… If the white man would draw a line defining their claims and keep on their side of it, the redmen would not molest them."

These authors link human aggression to a more sedentary lifestyle, those associated with agriculture rather than a foraging culture. There are exceptions, however. The Comanches, one of the most aggressive groups of Native Americans, represent a foraging culture of people who used their horses in warfare. "A global assessment of the ethnographies for 31 hunter–gatherer societies found that 64% fought wars less often; 10% were considered to fight wars rarely or never."

Besides the Yanomamö (in which a calculated 60% die in confrontations), Wrangler and Peterson point out that "violence accounted for the deaths of about 19.5% of the Huli, between 25% and 28.5% of the Mae Enga and Dugum Dani of highland New Guinea, and 28% for the Murngin of Australia."

They compare male-bonded, patrilineal kin groups of humans with those of chimpanzees, pointing out that bonded males of both groups form aggressive coalitions with each other in mutual support against others. Examples, they say, are seen in such groups as the Hatfields and the McCoys, Montagues and Capulets, Palestinians and Israelis, Americans and Vietcong, and Tutsis and Hutus.

It is always the males who instigate a fight or fight in defense of the group. "This is true even in villages labeled by anthropologists as 'matrilineal' and 'matrilocal,' where inheritance (from male to male) is determined according to the mother's line, and where women stay in their native villages to have children—such villages operate socially as subunits of a larger patrilineal whole." In short, "the system of communities defended by related men is a human universal that crosses space and time."

In chimpanzee society, patriarchy also rules. "Communities persist through a line of father–son relationships. Males are inheritors of territory. Males conduct the raids and the killing. Males are dominant. Males gain the spoils":

"A lost territory means death for adult males. Conquered land may include a larger foraging area, as well as new females. Such is not true for females. Females have more options, more freedom, even if they suffer to take advantage of it."

The same patterns described for chimpanzees have been recognized among humans, and yet, war for humans is quite different. With a more highly developed brain and an ever-increasing size of their society, humans have carried aggression to new heights, and the sizes of human battles have expanded.

The Atrocities of War

Our governments and news agencies never reveal the true nature of war by humans. Men and women who fight these wars often are not aware of what the war is all about. Many fight these wars out of sheer patriotism, and many of them die in the process or suffer the loss of limbs and a chance to ever lead a life they could have had if they had not partaken in battle, and our dominant leaders know this. And for their sacrifice, do they get rewarded? No, they do not. When they return home, they are expected to carry on just as they had done before suffering great emotional and physical trauma.

In being one of these people who devote their life to patriotism or talking with them (if they care to talk at all about it), we come to understand something about the true atrocities of war. The glorification of war, as seen in our movies and on TV, is a case of major deception. Many humans may possibly consider refraining from war if the truth was revealed. In *The Most Dangerous Animal* (a book that I sincerely believe everyone should read), D.L. Smith (2007) briefly describes the naked truth behind war.

"War is mangled bodies and shattered minds. It is the stomach-churning reek of decaying corpses, of burning flesh and feces. It is rape, disease and displacement. It is terrible beyond comprehension, but," he says, "it is not senseless. Wars are purposeful. They are fought for resources, lebensraum, oil, gold, food and water or peculiarly abstract and imaginary goods like God, honor, race, democracy and destiny."

Smith goes on to point out that the reality of war is quite different from what we understand from its coverage by news agencies. He further expresses the true nature of war, "bodies which had been cut in half at the waist, legs and arms, heads bearing only necks," tanks that "grind human flesh into the dirt to make a bloody paste, their massive treads plastered with pieces of human bodies. War," he says, "is grotesque, real and... terrible. War is a bad-taste business."

"Like it or not," he states, "war is distinctively human. There is nothing in nature that comes anywhere near approximating it," and millions of people have died in battles throughout time for purposes that they often did not understand. Smith's appendix, a partial list of what he refers to as democides committed during the past 100 years, summarizes the number of people and their ethnic groups who have died in various wars:

- Eight million residents of the Congo Free State killed by Belgians between 1877 and 1908.
- Sixty-five thousand Namibian Herero killed by the Germans between 1904 and 1907.
- One and a half million Armenian Christians killed by Muslim Turks in 1915–16.
- Five million Ukranians killed in 1931–32 by the Soviet Union's perpetration of famine.
- Over 4 million Soviet citizens killed by their own government in the Great Terror of 1937–38.
- Over 300,000 Chinese residents of the city of Nanking killed by the Japanese in 1937.
- Eleven million Jews, Romans, Poles, homosexuals, and others killed by the Germans during the 1940s.
- Over 250,000 Muslims, Serbian Orthodox Christians, Romans, and others killed in death camps run by the Roman Catholic Ustashi regime in Croatia between 1941 and 1945.

- More than 200,000 Muslims killed by the French in the 1954–62 war for Algerian independence.
- Around 1 million Indonesians killed by their own government in 1965–66.
- One million, 700,000 Cambodians killed by the Khmer Rouge during the 1970s.
- Roughly 2.5 million people, many of whom were Hindus, killed by the Muslim Pakistani army in East Bengal in 1971.
- Around 150,000 Hutus killed by Tutsis in Burundi in 1972.
- Around 200,000 Maya killed by the government of Guatemala between 1970 and 1996.
- Two hundred thousand Muslims killed by Serbian Orthodox Christians in Bosnia–Herzegovina during the 1990s.
- Close to 1 million Tutsi killed by the Hutu majority in Rwanda in 1994.
- Two hundred thousand Roman Catholics in East Timor killed by the Muslim Indonesian occupation force between 1975 and 1999.
- An as-yet undetermined number of Muslims killed by Serbian Orthodox Christians during the 1990s.
- Around 2 million black Sudanese killed in Darfur by the government of Sudan, which was ongoing at the time of Smith's writing.
- An undetermined number of Anuak killed by the government of Ethiopia, ongoing at the time of Smith's writing.

Smith further points out that "other victims of 20th-century genocides include the Bubi of Equatorial Guinea, the Dinka, Nuba and Nuer of Sudan, the Isaak of Somalia, the Karimojong of Uganda, the San of Angola and Nambia, the Tuareg of Mali and Niger, the Tyua of Zimbabwe, the Atta of the Philippeans, the Auyu of West Papua and Indonesia, the Dani of Papua, New Guinea, the Hmong of Laos, the Kurds of Iraq, the Nasioi of Papua, New Guinea, the Tamil of Sri Lanka, the tribal peoples of Bangladesh, the Ache of Paraguay, the Arara, Ticuna, Nambiquara,and Yanomani of Brazil, the Cuiva, Nunak and Paez of Colombia, the Mapuche of Chile, the Maya of Guatemala and the Miskito of Nicaragua."

Add to this the casualties of the invasion of North America in the 15th century by foreign factions (unknown number of deaths), First World War, referred to by David Price (2011) as the Chemist's War (19 million), Second World War, referred to by Price as the Physicist's War (40 to 71 million), Korean War (2.5 to 3.5 million), Vietnam War (4.95 to 5.2 million), and other of the numerous wars that have taken place through time. As listed in Wikipedia's *List of Wars and Disasters by Death Toll*, we have a faint idea of some of the deaths that have occurred among humans at their own hand (humans killing humans).

Of the 87,500,000 or so deaths in 20th-century wars, around, 54,000,000 or just over 60%, were civilians. During the 1990s, 75 to 90% of the world's fatalities were civilians. Noncombatants are routinely shot, bombed, raped, starved, and driven from their homes.

In 2001, 40 million people were displaced from their homes because of armed conflict or human rights violations. Five million Europeans were uprooted from 1919 to 1939. Second World War displaced 40 million non-Germans in Europe, and 13 million Germans were expelled from countries in eastern Europe. About, 2.5 million of the almost 4.5 million people in Bosnia and Herzegovina were driven from their homes during that region's war in the early 1990s. More than 2 million Rwandans left their country in 1994. In 2001, 200,000 people were driven from Afghanistan to Pakistan.

As Smith points out, many of these fatalities have been children. More than 2 million children were killed in wars during the 1990s. Three times that number were disabled or seriously injured. Twenty million children were displaced from their homes in 2001. Many were forced into prostitution. A large percentage of those will contract or have contracted AIDS.

Based on these statistics, is there any question that humans represent the most dangerous animal, in terms of killing its own species, that ever existed? And what have these humans fought for? As Smith points out, "today's genocides and ethnocides often take place at the behest of multinational corporations eager to acquire resources, typically by dispossession and environmental degradation." These include widely divergent resources, for example, diamonds in Africa, oil interests in Ecuador, Burma and Nigeria, copper and gold mining in West Papua, farming in Tanzania, logging in Malaysia, and uranium mining in Australia. There are numerous other corporate benefits of war, all directed by dominants who are willing to sacrifice lives for their benefit.

The People Who Fight in Wars

And what about the people who engage in battle, those heroic individuals who are acting altruistically for their government? Are they the killers that war makes them out to be? Smith presents some interesting statistics that make us realize that the desire for war originates not with them but with their leaders or the dominants who are above the leaders. Is it our governments who depend on its citizens for loyalty or the dominant people who run it? Is there a difference between the two?

FIGURE 15.2 Soldier suffering from battle fatigue. Most people, including those who go to war, abhor killing and also fear being killed. Statistics show that as much as 75% to 80% of American soldiers have refrained from firing their weapons in battle. Under battle conditions, "only 2% of enlisted men do not break down, no matter how long they have been subjected to the stress of war. After 60 days of continuous fighting in Normandy, 98% of the survivors suffered psychiatric damage."

While "84% of women and 91% of men admit to daydreaming about killing people whom they dislike," suggesting once again that "there may be a homicidal streak in human nature, only a small percentage of people with such daydreams actually get to the point of committing themselves to it." We may wonder how much killing there would be if laws did not exist against it.

On a personal level, most people, including those who go to war, abhor killing and also fear being killed (Fig. 15.2). There are statistics to show that as much as "75% to 80% of American soldiers have refrained from firing their weapons in battle." Smith states that "during World War II, less than 5% of United States fighter pilots were responsible for 30 to 40% of the kills, and most pilots did not shoot down any planes at all. Their governments, of course, select those individuals who do kill and reward them with medals, other honors, and promotions, and they show no compassion for those who choose not to kill."

Under battle conditions, "only 2% of enlisted men do not break down, no matter how long they have been subjected to the stress of war (Fig. 15.2). After 60 days of continuous fighting in Normandy, 98% of the survivors suffered psychiatric damage." Smith quotes Dave Grossman as saying that the other 2% "were crazy before they got there."

But how can humans abhor battling other members of our species and yet proceed to kill them? Smith suggests that we tend to picture our enemies as less than human. In contemporary times, the word infidels, people who have subhuman beliefs about who God may be, has come into vogue.

"Dehumanizing the enemy in warfare draws on ancient biological dispositions to overcome the problem posed by the taboo of killing members of one's own species. Although dehumanizing one's enemies goes a long way toward freeing a soldier from the oppressive burden of conscience, it helps more if he perceives his enemies as some kind of nonhuman beings who must be met with violence. Killing people is easiest if there is something about them that makes you want to kill them—something that arouses deep aggressive passions." Opposing governments are quick to point out these characteristics to make their fighters do what they feel they must do.

He further suggests that we also perceive our enemies as predators and parasites. "We equate them with disease or with vectors of disease, such as lice, flies, and rats." Once this mode of thinking is activated, the stage is set for genocide.

This process of treating our enemies as something other than human is a form of deception, commencing with governments that oppose them; humans are good at deceiving themselves, as pointed out by Robert Trivers in *The Folley of Fools* (2011), because self-deception, "lubricates the psychological machinery of slaughter, providing balm for an aching conscience."

In following this direction of thought, it is believed that aggressors in wars are often inspired by what appears to be moral feelings. As a matter of good and evil, aggressors generally feel a need to purify others. "The Khmer Rouge purified Cambodia of nearly 2 million human beings during the 1970s. Puritan settlers of New England branded native Pequot Indians [as] agents of Satan and rejoiced in their death."

George W. Bush proclaimed in his 2003 state of the Union address that "God told me to strike at Al Qaeda, and I struck them. And then he instructed me to strike on Saddam, which I did. With the might of God on our side, we will triumph."

The target of his aggression was simultaneously speaking to his God. "On the eve of the American invasion of Iraq, Saddam Hussein entreated, 'O Arabs, O believers across the world, O enemies of evil, God is on your side. Rely on God and the soldiers of the merciful on our land will be granted victory'."

ORIGIN OF AGGRESSIVE BEHAVIOR IN HUMANS

Where does our aggressive behavior come from? We have treated this to some degree in Chapter 9. Smith points to our primate ancestors, the chimpanzees as the origin, but the roots of our aggressive nature actually have more distant biological roots and a much older connection with other forms of life.

Two of the most outstanding qualities of all organisms are a desire (actually the necessity) to survive and reproduce. These two qualities are absolutely required for a species to continue existing. How organisms go about accomplishing them is extremely diverse among species. Often, they choose deception as a means to protect themselves from environmental threats.

Manifestations of survival techniques in natural populations of animals take such forms as akinesis (pretending to be dead), crypsis (pretending to be something else and blending in with the environment), aposematic coloration and patterns (pretending to be dangerous, advertising the promise of a bad experience), and threatening behavior (advertising behavior that indicates an organism is bigger or more ferocious than it actually is).

Biologists agree that these survival strategies represent methods of defense that actually keep species from fighting or keep predators at bay. Likewise, some people look at war as a means of saving lives if the threat they are warring about promises the loss of a greater number of people. Did dropping atomic bombs on the Japanese during Second World War save lives or destroy them?

While the formation of societies (banding together as a group and cooperating in defense) is not a form of deception, it does represent a highly effective means of defense. In the context of sociality (as expressed in the first few chapters), members of a species may become deceptive in order to dispel any indication of inferiority on their part. Displays of social behavior and the animal species that express them are discussed in earlier chapters as well.

As we have seen, family groups go into the making of local societies, and they are naturally the defenders of their society. A question that remains unanswered is: What constitutes a family? Biologically, small groups of related individuals are distinctly a family group. What about a school or professional team, a company, a city, or a nation? Are they thought of as families as well? If so, who is the instigator of aggressive behavior? Is it everyone who belongs to the group, or is it a dominant individual (a coach, a CEO, a mayor, senator, congressman, a president, or the party to which they belong) or some other type of coalition of dominant individuals (corporate dominants) who make the rules and insist that we follow them?

Final Note

What has been presented here represents a few examples of dominant and aggressive organisms and what they do to obtain resources and defend themselves. They show that the seeds of aggression and colony defense are well established in all animals, especially in social species, invertebrates, and vertebrates alike. Such behavior has been important in their survival and reproductivity.

In a social group, interactions between colony members require both collaboration (to defend and carry out various colony functions) and the application of agonistic behavior to establish a dominance hierarchy. In such a setting, aggression is often turned toward other colony members, the consequence of which establishes who the alpha individuals are. Ideally, with dominant status comes certain responsibilities that are generally carried out, resulting in social homeostasis. But with human cognizance and antisocial personalities most often prevalent on the side of dominant personalities, is the ideal ever reached and is homeostasis the issue?

In a social insect colony, dominance establishment determines who the principal egg layer is and who the initial colony defenders will be. Obtaining certain rights by dominant individuals and the following of the alpha by more subordinate society members is just as true in a social group of vertebrates. Primary defenders may work alone (as the alpha individuals of a group) or form coalitions with other dominant members of the group to defend the colony against nongroup members of the same species. In the case of apes (including humans), group defense, especially under the deceptive guidance by dominant members of our society, may turn to extremely strong aggression, including war.

In spite of what has been said about being careful when comparing aggressive and other behaviors in primates, choosing the chimpanzee to compare with agonistic behavior in humans seems to be a natural approach, since we share a similar genetics, as long as we understand that agonistic behavior is more deeply seated and actually developed long before primates ever arose.

Chimpanzees are extremely status conscious (Wrangham, 1996). They live in communities of related individuals in which an alpha male dominates their social group. Dominant male chimpanzees, like dominant humans, live a stressful life. They must exert their dominance on a continuous basis (either as overt aggression toward another group member or as cues that are understood by subordinates), and they (like the dominants in our governments) are primarily responsible for their group's protection (or destruction). Do they, like humans, deceive other group members into believing they must form coalitions to war upon other chimpanzees? It seems apparent that they do.

While they generally tolerate members of their own social group, they often viciously attack and kill members of other groups that they encounter. They form coalitions and attack others to secure their territories, acquire additional territories, acquire females, and protect resources for them and their group. As pointed out by Smith, intimidating or exterminating rivals, a group of chimpanzees can gain access to new territory and simultaneously absorb fertile females into their group.

Humans strive to attain the same things for the same reasons. As a group, they protect territories, strive to acquire new territory, and seek resources, for example, fossil fuels and religious and ideological converts. At the head of their coalitions is generally a dominant and his in-group, making the rules and carrying them out with the remainder of the population's individuals (the subordinates).

The difference between human and chimpanzee may simply be in the types of resources and the way they are obtained. As an example, Spanish conquistadors ransacked Native American cultures in an attempt to find the gold of the mythical El Dorado. In short, humans fight for resources and even go so far as to seek and fight other groups to obtain potential resources. They do these things for themselves or for a higher power, a religious figure, a dominant individual who was the alpha male or female and leader of their country, or for the country itself.

At the same time, they ravage an enemy's females, spreading their genes to individuals who are out of their territory. Men, like other animals, appear to unconsciously desire to spread their genes, and while we profess a monogamic nature (a single female for every male), only 15% of the world's cultures are strictly monogamous, and even those that profess to be monogamous consist of individuals who demonstrate promiscuity (Lehrman, 2002).

The heads of human governments often express their dominance behavior in a much safer environment than the military personnel they send to their death, generally in debates. The dominant who wins the debates acquires adequate numbers of votes to become the alpha of the office they desire, and they get to rule in a dominant position until society sees fit to choose another dominant. Whether the position they acquired was through deception (which it usually is), it was acquired because voters felt they were the right choice. Do we ever get enough correct information that we can make a truly unbiased choice? For most of us, in spite of what we feel we know about the dominants who run our governments, we actually know very little.

Status is a resource commonly sought by animals. Most males who have a number of females, including humans, enjoy high status. Muhammad is said to have had 16 wives and 6 concubines. It also has been said that King Solomon had 700 wives and 300 concubines. Moulay Ishmael the Bloody, the last Sharifian emperor of Morocco, apparently had 1000 wives and 888 children.

Men who do not enjoy the opportunity to experience such marital bliss, especially under warlike conditions, often forcibly copulate with women. During war, rape is very common, even for men considered to be virtuous.

According to Smith, "rape occurs in virtually all armed conflicts. An orgy of sexual violence erupted when the Red Army advanced into Germany during the closing days of World War II." "Japan imprisoned up to 2000 (mainly Korean) 'comfort women' in brothels for the pleasure of the Imperial forces. Somewhere between 20,000 and 70,000 Bosnian women were raped during the war in Kosovo. At least 5000 Kuwaiti women were raped in the short-lived Iraqui occupation of 1990. Rape was also encouraged in the Rwandan genocide."

Smith summarized the comparison between humans and chimpanzees by stating that "we are the products of a long process of biological evolution," going back to the beginning of animal evolution. We are also social animals and share a xenophobic existence (hostile to individuals living outside the boundaries of their social group) with chimpanzees.

However, we must consider other, more personal and recent, characteristics of humans that are not shared with our closest primate relatives. While there are many similarities between chimpanzees and humans, there are also "a great deal of differences between the chimpanzee mind and the human mind, and a vast gulf separating the small-scale brutalities of chimpanzee 'war' from the colossal horrors of human carnage."

Note

Superscript numerals appearing in this chapter refer to additional text/explanation given in the appendix.

CHAPTER

16

Are We Our Own Worst Enemy?

From prehistory to the present time, the mindless horsemen of the environmental apocalypse have been overkill, habitat destruction, introduction of animals such as rats and goats, and disease carried by these exotic animals. In prehistory, the paramount agents were over-kill and exotic animals. In recent centuries, and to an accelerating degree during our generation, habitat destruction is foremost among the lethal forces. **E.O. Wilson,** *The Diversity of Life, Environmental Impact by Humans on Earth*

CHANGES IN HUMANS AND THEIR ENVIRONMENT

Commencing a long, torturous journey as *Homo sapiens* about 200,000 years ago (Table 5.1), the human animal was initially, like its predecessors, just another social animal species among numerous others, its existence largely influenced by mutations, variability, environmental pressures, survival strategies, and biological inheritance. While human populations were experiencing constant mutations, resulting in genetic, anatomical, behavioral, and physiological variation, the human nervous system was undergoing a unique brain explosion in an already-enlarged brain around the time when *H. sapiens* was evolving from a predecessor hominine (an earlier ancestral human form).

In the process of human evolution, human bodies and minds were undergoing a transformation into bipedal technologists with a variety of unique features. In the early human animal (as with more contemporary forms), all of these features were subjected to natural selective forces, culling the unfit and maximizing the human species' fitness.

Early humans were fine in the world as animals in terms of their impact on Earth. However, as with any other population in nature, it was a wicked existence, and they periodically had a tough go of it.

It is certain that human populations fluctuated, depending on environmental changes. In addition, it appears that humanity has had more serious, near-extinction bouts with nature on several occasions.

Researchers can detect near-extinctions in several ways (MacPhee and Sues, 1999). By investigating DNA samples, recording mutations, and measuring genetic diversity (e.g., Alu sequences and mitochondrial DNA being especially important), they can detect population bottlenecks. A bottleneck is a time in the past that a species had significantly low numbers. A reduction in the number of individuals reduces genetic diversity, since every member of the species in future generations is descended from this low number of individuals. Environmental catastrophes, for example, extremely divergent temperature swings, the eruption of supervolcanoes, or asteroid impacts, may result in population bottlenecks.

With genetic investigations, scientists can pinpoint the dates of eruptions. During eruptions, massive amounts of CO_2 and toxic chemicals may be released into the atmosphere, leading to environmental warming (due to a greenhouse effect) and areas that are uninhabitable. Huge eruptions may be followed by a period of prolonged environmental cooling (a volcanic winter).

Based on studies of mitochondrial DNA, modern humans may have had a shaky beginning when they arose from a predecessor hominine, with the first individuals possibly arising from what scientists refer to as a single mitochondrial "Eve." Later in human evolution, it has been estimated that the number of hominines may have shrunk to a population size of 2000 before it began to expand in the early Stone Age (over 2 million years ago) (Sahlins, 1972).

Dominance and Aggression in Humans and Other Animals
http://dx.doi.org/10.1016/B978-0-12-805372-0.00016-X

Situated in the African Pleistocene, the human animal was provided with some great opportunities in life, but as pointed out by C. Boehm, it also passed through some potentially very dangerous climatic instabilities, frequent hunger, and hardships. Between 135,000 and 90,000 years ago (before early humans left Africa), for instance, East Africa experienced a severe drought, possibly contributing to severe human population changes that resulted in numerous deaths and group isolation (Campbell, 2009). As we know from studying nonhuman populations, isolation (allopatry) over time results in anatomical, physiological, and behavioral differences between population segments and eventually (if the periods of isolation are long enough) speciation.

Subsequently, humans have had a very close brush with extinction 70,000 years ago (Extinction, 2008). As advanced primates, humans continued their role as a social foraging species, initially prone to function in familial groups to obtain food and protect themselves from predation and aggressive members of their own species.

To remain as a species, they had to survive threatening circumstances and dramatically changing environments while simultaneously producing viable offspring to continue their line. Being social, they undoubtedly had to concern themselves with extended child care, feeding strategies, territorial complications, and choosing a leader (a dominant) who would strive to protect them and their group members.

As we will see, their leader most likely had a dominant personality and was assertive but somewhat tolerant to maintain an almost egalitarian societal homeostasis and, yet, possess what appeared to be conflicting qualities, such as selfishness and some degree of altruism, cooperation and agonistic behavior, and reciprocity and deception. Any advantageous feature common to a social group was important to their survival and reproductivity.

As animals, early humans had very little readily apparent sophistication to distinguish them from other members of the animal kingdom, except for their social behavior, unusual bipedal gait, and wit. These primitive humans, along with other animal forms, were coinhabitors of Earth. They took what they needed from the environment, generally thanking nature or a mythological figure for their bounty, and left the rest for nature's natural balance. They blended with and often respected nature.

Their bond with nature was evident in their every activity and often vividly expressed in their music (Hermann, 2011). There is some evidence that flutes, for instance, may have been used by Neanderthals 45,000 years ago. There is direct evidence of twenty-four and thirty-thousand-year-old flutes found in Les Roches and La Rogue, Dordogne, France, and Asian flutes, as Z. Juzhong and L. Y. Kuen stated in the *Magic Flutes*, that range from 7000 to 9000 years ago. Even with some sophistication in their societies, as long as their populations were low and they lacked urbane technological skills, they did not represent a particular threat to a relatively homeostatic Earth.

Along the path to developing a more sophisticated brain and language, humans diverged somewhat in their appearance, and they commenced to form separate social groups of myth-seeking creatures, basing their good and bad times on the unknown or some sort of a mythological premonition. They sometimes isolated themselves from other human groups and continued to achieve superficial anatomical and behavioral differences (e.g., different skin colors, body styles, and facial features).

LANGUAGE

By utilizing language, each subpopulation began to preserve its heritage by passing stories from one generation to the next, sometimes influencing the lives of one another in the process (Ruhlen, 1994). Their stories had multiple functions. In addition to conveying important information to its recipients about life in general, any particular advice passed along by elders was contained in the stories rather than in the form of criticism, and it brought people together. With no scientific studies under their belt, they had no way of understanding the complexity of their world except by reacting animalistically to environmental circumstances and inventing imaginary beings who represented both good and evil.

ACQUIRING INNOVATIVE SKILLS

Needless to say, humans have undergone vast changes in most of these categories as a distinct species, and within the past several thousand years, human influence upon the Earth and its biota has steadily increased with the acquisition of additional innovative skills, an ever-expanding ability to inflate their numbers, advances in communication skills, and an overwhelming hunger for a wide variety of Earth's resources.

As human skills and populations grew, their societies became bigger and more complex. They sometimes formed egalitarian groups which had members that were dependent upon one another for their survival. Over time, humans established a sedentary lifestyle (about 3000 to 12,000 years ago) with increasing agricultural expertise, and as populations grew, certain groups formed complex civilizations and lost their egalitarian features due to the rise of despotic dominants. Within the last few thousand years, many individuals and their societies have transformed from an egalitarian status to become more independent, evolving from a more-or-less eusocial existence to a more-or-less asocial (solitary or private) one, even though we live and work closer together in larger populations.

Many humans have progressively become so preoccupied with modern life, their jobs, the political establishment, and individual desires that they have lost their connection with Mother Nature, the biological system that bore them. Except for a small but very valuable, sometimes heroic few, many contemporary humans could 'not care less about other species that they coinhabit Earth with or the ecosystems that support them. Many humans are even losing touch with each other.

CONTEMPORARY HUMANS

Within recent years, humans have raced almost unconsciously toward their biotic potential as an ultramodern, overpopulated cosmopolitan species. The question we must now ask ourselves is this: With all of the evolutionary changes the human species and its predecessors have undergone, have they managed to develop features that keep them from being overbearingly threatening to other species, themselves, and the world, or have they remained more animalistic and increased their high degree of arrogant dominance, aggression, and technological capabilities to the point of putting everything they know in harm's way? I think the answers to these questions are obvious, and one of the points of this book has been to reveal our diverse personalities and attitudes toward and influence upon the nature of our environment and Earth:

> And a Man sat alone, drenched deep in sadness. And all the animals drew near to him and said, "We do not like to see you so sad. Ask us for whatever you wish and you shall have it." The Man said, "I want to have good sight." The vulture replied, "You shall have mine." The Man said, "I want to be strong." The jaguar said, "You shall be strong like me." Then the Man said, "I long to know the secrets of the Earth." The serpent replied, "I will show them to you." And so it went with all the animals. And when the Man had all the gifts that they could give, he left. Then the owl said to the other animals, "Now the Man knows much, he'll be able to do many things. Suddenly I am afraid." The deer said, "The Man has all that he needs. Now his sadness will stop." But the owl replied, "No. I saw a hole in the Man, deep like a hunger he will never fill. It is what makes him sad and what makes him want. He will go on taking and taking, until one day the World will say, 'I am no more and I have nothing left to give.'" **Old Story Teller, *Apocalypto*, Written by Mel Gibson and Farhad Safinia**

OPENING THE CAN OF WORMS

In the eyes of many scientists, the human species, to be blunt but realistic, is in serious trouble. By saying this so bluntly, I may lose a few readers who regard such a negative statement as an unsupported overstatement, but let us follow through with this thought and see where it takes us.

Throughout human evolution and even immediately prior to the 1960s, humans continued to have a somewhat animalistic approach to the environment and its resources. The Earth appeared to have plenty to give, and we needed Earth-giving resources to survive. Commencing in 1962, Rachael Carson and others began to point out a genuine need to constantly monitor our impact on the environment and the world. The human species has responded to these warnings by establishing numerous rules to follow to avoid ecological disaster. But because our populations have increased at such a rapid rate and many of our political leaders seem to have a blatant disregard for environmental deterioration or scientific information in general, we have fallen short of the required goals to maintain a homeostatic Earth. Although many of us have come to realize the nature of rapidly growing overpopulation problems, we often have ignored or underestimated our impact and responsibilities to keep abreast of where we stand among all the other biota we share Earth with and the harm we are doing to them and their environments.

The human population has grown technologically primarily since the commencement of Mesopotamian civilizations in Sumer and the Akkadian, Babylonian, and Assyrian empires, all native to the territory of modern-day Iraq. Without predators to keep our numbers down and in spite of our wars, disease epidemics, flare-ups of genocide, and secret societies that have been adamant about eliminating what they feel represents the weaker segment of the human species, our cosmopolitan population has grown all too rapidly.

As with any animal population, when the population and competition for resources increase, we choose an animalistic approach and kill one another. We have not learned to keep our population within a homeostatic balance with other biota and within itself. We also have not learned that not keeping our population within limits will result in a tremendous number of deaths even if we did not have wars because dense animal populations often face increased threats from food shortages, fierce competition for a number of other resources, and pathogenic organisms.

Laurie Garrett, *The Coming Plague, Newly Emerging Diseases in a World Out of Balance* (1994), brings this point out and has other things to say about humans and disease: "The world has rapidly become much more vulnerable to the eruption and, most critically, to the widespread and even global spread of both new and old infectious diseases. The dramatic increases in worldwide movements of people, goods and ideas is the driving force behind the globalization of disease."

Based on these statements, it appears that we have arrived at a point in our evolution where we must take a serious stand and reconsider our responsibilities to maintain ecological harmony throughout the world or suffer the consequences. We must step back and take a close look at who we are, what we have done as humans, where we are headed, and what will be waiting for us when we get there.

But the big question is: Are we capable of doing this? Are we, with our egotistical, dominant, and competitive nature, in control of our destiny or are we in some sort of a Catch-22 situation in which there is no way out? Are we capable of reacting to our ecological and behavioral problems only in an animalistic way, or are we a new sort of animal that can analyze our problems and with a logical mind overcome the stresses we put upon ourselves?

In *The Great Game of Life*, we have examined humans and other animals in a comparative way and have gradually brought the human animal to the forefront in an attempt to understand its most outstanding attributes, dominance and aggression. It has been repeatedly stated by writers of human behavior and strongly implied in the foregoing chapters that humans, the most dominant forms of life on Earth, represent their own worst enemy and that they are the most dangerous animals to ever have lived on the planet.

With life commencing on Earth about 4 billion years ago and the many millions upon millions of diverse, sometimes formidable, species that have existed since the primordial ooze and its anaerobic atmosphere generated enough complex chemicals to give rise to simple protobiont life forms, some may think this negative statement about our importance to the world is an unfair assessment. What about all the good we have done? First things first.

DOMINANT LIFE FORMS ON EARTH THROUGH GEOLOGICAL TIME

Even with considering the abundance of and dominant nature expressed by such life forms as the extant and extinct marine and terrestrial megafauna of the Paleozoic, Mesozoic, and Cenozoic eras (encompassing about 500 million years of evolution) and the diabolical, self-perpetuating approach to life that got them there, biologists would strongly agree that human life has changed Earth and its inhabitants like no other.

Most animals, regardless of size, have treated our planet in a much kinder fashion than we have. Dinosaurs, for instance, have been some of the most formidable terrestrial vertebrates prior to and during the rise of mammals and birds, and they have had a much longer reign on the planet (about 165 million years, possibly more if you consider birds as avian dinosaurs) than humanity can ever expect to have unless we make some very significant changes in human population growth and establish a more nonaggressive attitude toward each other. As the most intelligent animal on Earth, we are racing almost aimlessly toward our biotic potential with ever-increasing speed and an apparent lack of concern for the consequences of surpassing our usual environmental carrying capacity.

In most cases, we are not intentionally destroying habitats, polluting the environment, and causing stress to populations of other organisms. We are simply living our lives, facing our daily problems, doing our jobs, raising our families, paying our taxes, and making the best with what we have. Most of us generally do not think of other species' populations and their rights to live on Earth or even how or why they are dying. We do not believe or want to believe that the human population is large enough to be doing the damage scientists are claiming is being done. We do not consider the importance of releasing exotic animals and plants into areas occupied by indigenous species. We do not understand the atrocities that occur just from using pesticides and artificial fertilizers on land to grow crops or the reasons for developing bioengineered produce for our consumption. We are most often not educated to understand these things, so we do not worry about them.

A brief overview of some of the things we have done to the planet may give us some indication of what this means and why we hold this elite status. These final chapters will imply that our destructive nature is expressed in a variety of ways and that we are largely neglecting the signs of impending adverse conditions.

As one recent example of the negative influence we are capable of leashing upon the environment, let us look at the disaster we are responsible for in the pollution of marine waters and adjacent terrestrial ecosystems before, during, and after one of the most important oil spills in the Gulf of Mexico. It was an act that came about because of our need for increasing amounts of energy (mostly due to an overabundance and ever-increasing number of people in the world), a dominant attitude, and technological skills.

Why should we consider the health of an oceanic environment when we live a terrestrial existence? Part of the answer is in the fact that terrestrial and oceanic environments are inextricably linked, and what happens to one influences changes in the other. In addition, when we combine the effects of oil spills with other ways that we are affecting oceanic environments, we see emerging difficulties of immense proportions.

Ehret et al. (2000) lists 10 anthropocentric reasons why oceans are vital to planet Earth and its most dominant inhabitants. After all, it has been for anthropocentric reasons that Earth is in the shape it is in. We look at the value of the ocean in a more biological light later in this and other chapters:

1. Oceanic life forms generate a good deal of the oxygen we breathe. As a matter of fact, it is estimated that oceanic chlorophyll-harboring organisms (many of which are unicellular) produce between 70% and 80% of the oxygen in the atmosphere. The ocean is where the first photosynthetic organisms arose 3.5 billion years ago, and it has been a major supplier of this gas ever since.
2. The ocean occupies the vast majority of the surface of the planet (about 71%) and provides a vast array of resources.
3. Oceans drive climate and weather through the transfer of water (water cycle) and heat.
4. Most US and international commerce travels through the nation's ports.
5. Oceans account for a $20 billion recreational fishing industry…, not to mention, a $60 billion annual seafood industry.
6. Oceans represent an estimated $8 trillion in oil and gas reserves.
7. Oceans support nearly 50% of all species on Earth.
8. Over 50% of our nation's human population live in coastal areas.
9. Oceans mitigate the effects of CO_2 in the atmosphere.
10. Marine animals and plants produce a myriad of compounds that prevent and treat human disease.

While these points indicate the importance of marine environments to the human population, it also implies that much human activity is in and around marine environments, and those are areas that are particularly open for environmental abuse. It is the abuse of such areas that I want to direct our attention to.

The Power of Petroleum

Petroleum (consisting mostly of crude oil with a wide array of hydrocarbons) is the product of large quantities of decomposing organisms, usually zooplankton, algae, and possibly land plants from the Paleozoic era, that became buried in bogs, swamps, bays, and beneath sedimentary rock millions of years ago. The process of petroleum synthesis was dependent upon anaerobic (anoxic) decomposition under intense heat and pressure over long periods of time.

Petroleum products have been in demand since ancient times, although its demand was initially small, and it did not appear to be an environmental concern when it was first used (Owen, 1975). The first petroleum products used were mostly those that seeped to the surface from underground sources. In ancient times, for instance, Egyptians are known to have coated mummies with pitch made from natural asphalt. In 1000 BC, Chinese salt miners found natural gas while drilling, subsequently using the new gas for fuel.

Asphalt (a mixture of dark bituminous pitch with sand or gravel, currently used for surfacing roads, flooring, roofing, and other uses) was employed by King Nebuchadnezzar in Babylonia about 600 BC to build walls and pave streets. Assyrians and Persians also used asphalt in building. Because of its hydrophobic properties, boatmen on the Euphrates River made vessels of woven reeds that were smeared with asphalt.

American Indians used petroleum products hundreds of years before Europeans invaded the New World. Bitumen (a product of petroleum) was used by the Toltec Indians of Mexico to set mosaic tiles. Bitumen was used to seal flutes, instruments of cultural importance (Hermann, 2011). It was also one of the first pesticides used in attempts to control house-invading insects.

The remains of ancient oil wells have been discovered in the oil regions of Pennsylvania, Kentucky, and Ohio which are currently covered by trees that are hundreds of years old. Jesuits reported that Native Americans collected oil from surface pools in the early 1600s, subsequently using it for both fuel and medicine.

Many oil seepages had been found in New York, Pennsylvania, and West Virginia by 1750. In addition, wells drilled for salt often produced oil.

Samuel Kier began bottling petroleum for medicine about 1847 in Pittsburgh. Around the same time, frontiersman Kit Carson is said to have collected oil from a seepage in Wyoming which he sold to pioneers as axel grease for wagons. Kerosene (a combustible hydrocarbon liquid that is currently widely used as a fuel in industry and in households) was discovered in 1852 by Abraham Gesner, a Canadian geologist.

Romania may have had the first oil industry, producing about 2000 barrels in 1857. The oil was brought up from hand-dug wells by workmen, using bags and buckets.

After digging an oil well in 1857, James Miller Williams of Canada established a refinery near present-day Oil Springs, Ontario. He subsequently distilled and sold oil for lamps.

The large-scale beginning of the oil industry has been attributed to the drilling of a well near Titusville, Pennsylvania, by Edwin Drake. Striking oil at a depth of about 70 ft in 1857, he commenced producing 10 to 35 barrels a day. By the 1860s, thousands of wells had been drilled in the Pennsylvania hills. Initially, wagons and river barges carried oil from the fields to refineries. Subsequently, railroads built branch lines to the fields. The first oil pipeline was constructed in 1865 by Samuel van Syckel.

According to James Case in *Competition, the Birth of a New Science* (2007), the first offshore oil well was drilled near Summerland, California, in 1887. It was only 300 ft from the water's edge and connected to the land by a long wharf. The first well in marine waters that was not connected to the land was drilled at Caddo Lake, Louisiana, in 1911:

> During the 1920s and 1930s, wells were drilled in the lakes, swamps, and bayous of southern Louisiana. A Frenchman named Etienne Lenoir patented the first practical gas engine in Paris in 1860 and drove a car based on the design from Paris to Joinville in 1862 (Berger, 2001).

The first gasoline-powered car in America was built by Charles and Frank Kuryea in 1893. Henry Ford had constructed an engine by then and had made great strides on experiments with gasoline engines after 1896. The Model T was introduced in 1908. By 1916, half of the cars in the United States were Model Ts (Berger, 2001; Williams et al., 1992).

After passing through many years of turmoil, Henry Ford became one of the richest and best known people in the world until his death in 1947. By that time, petroleum products were in greater demand. Eventually, the automobile changed the face of small town America. Midway through the century, cars had become a central feature of life for young people.

Oil sales had been sluggish during the Depression era, but demand picked up following World War II when wartime technologies proved adaptable to new demands. By the time Congress passed the US Submerged Lands Act in 1953, establishing that the federal government would control mineral rights located three or more miles from shore, and the state nearest the well would control rights within that limit, Henry Ford II had taken over the company's vehicle production, and the demand for production was on the upswing.

A lot has happened since, putting a demand on petroleum production that could have never been predicted. By 2012, there were 240.5 million cars and light trucks being driven in the United States alone, the largest vehicle population in the world. According to a report from Ward's Auto, the global number of cars exceeded 1.015 billion in 2010.

As the need for petroleum products increased, oil wells took their place on land and sea. In a 2014 anonymous online article presented by Energy4Me, the United States was listed 14th in the top 20 petroleum-producing countries. The 13 countries that outproduce the United States are the following in their order of production: Saudi Arabia, Canada, Iran, Iraq, Kuwait, Venezuela, United Arab Emirates, Russia, Libya, Nigeria, Kazakhstan, Qatar, and China. The frontier of oil exploration within recent years has been and continues to be offshore, over 10,000 ft/3048 m below sea level.

Buying and selling petroleum products around the world has been big business, and getting petroleum from one country to another requires movement over water. That oil was being harvested for some time under marine waters and transported to other locations has been a disaster in the making. By the end of the 20th century and the beginning of the 21st century, there had been oil spills all over the world, caused either by naval accidents or during major wars (Oil Spills, 2015).

- *Liberian Amoco Cadiz*: On March 16, 1978, the Liberian super tanker Amoco Cadiz [was] stranded on Portsall Rocks off the coast of Brittany, France, because of failure of the steering mechanism at 9:15. Although Captain Pasquale Bandari hoisted the international signal for "Not Under Command" almost immediately, he did not request assistance until 11:20, when his engineer determined that the damage was irreparable. The Amoco Cadiz started drifting to shore where touching the bottom ripped open the hull and storage tanks.

  Although the crew of the tanker was rescued by helicopter, the ship broke in two, releasing 230,000 tons of crude oil, which subsequently spread through the English Channel. The oil spill polluted about 300 km of coastline, destroying fisheries, oysters, and seaweed beds. Beaches of 76 Briton communities were polluted by oil.

As a result of the isolated location of the grounding and rough seas, cleanup efforts were restricted for 2 weeks following the accident. Severe weather eventually caused the complete breakup of the ship before any additional oil could be removed from the wreck.

At the time, this was one of the largest environmental disasters known to have ever happened. It took 10 years before the resulting lawsuits were complete. In 1988, a United States federal judge ordered Amoco Oil Corporation to pay 85.2 million dollars in fines, consisting of 45 million dollars for the costs of the spill and an additional 39 million dollars interest.

- *Occidental Petroleum and Texaco's Piper Alpha*: On July 6, 1988, an explosion occurred on the oil and gas production platform of Piper Alpha of Occidental Petroleum Ltd. and Texaco in the North Sea. Piper Alpha was located on the Piper Oilfield, about 190 km from Aberdeen in 144 m of water. There were about 240 people working on the platform. The explosion and resulting fire killed 167 of them. It was established that evacuation plans were inadequate and thus failed to prevent any of the deaths. By the time rescue helicopters arrived, flames over 100 m in height prevented a safe approach, and only 62 workers were pulled from the sea alive.

  A nearby platform called Tartan continued to pump gas into the upstream pipelines of Piper Alpha after the explosion because workers did not have the authority to shut down production, even when the Piper Alpha caught fire. The released gas caused a second explosion and the fire increased, covering the entire platform.

  The personnel who had the authority to order evacuation of the Piper Alpha had been killed during the first explosion, which destroyed the control room. It led to people still trying to leave the platform hours after the fire had started.

  The Cullen Enquiry was set up in November, 1988, to establish the cause of the disaster. It was determined that the initial explosion was caused by a leakage of natural gas condensate that built up beneath the platform due to maintenance work on a pump and a faulty safety valve. It resulted in the ignition of secondary oil fires and the melting of upstream gas pipelines. Piper Alpha's operator, Occidental, was found guilty of having inadequate maintenance procedures.

- *Exxon Valdez*: In 1989, the American oil tanker Exxon Valdez clashed with the Bligh Reef, causing a major oil leakage. The tanker had left the Valdez terminal in Alaska, navigating through Prince William Sound. Captain Joseph Hazelwood informed the coast guard they would change course to avoid collision with some small icebergs present in the region. The coast guard instructed the captain to sail north.

  After passing Busby Island, the tanker was supposed to turn back south, but it did not turn rapidly enough, causing the collision with the reef. This caused an oil spill of between 41,000 and 132,000 m^2 of area, polluting 1900 km of coastline. The oil spill killed about 250,000 sea birds, 2800 sea otters, 250 bald eagles, and possibly 22 killer whales.

  Exxon Mobil, the owner of the Exxon Valdez, has paid about 3.5 billion dollars in connection with the accident, of which 2.1 billion was meant for the cleanup operation. Both Exxon and the government have ordered investigations for the disaster, because of the large sums of money involved.

  Ironically, official NOAA (National Oceanic and Atmospheric Administration) investigations have shown that most of the damage from the oil spill was caused by the cleaning operation that followed the disaster. It is claimed that pressure-washing was responsible for killing most of the marine life. On stretches of beach that were uncleaned, life seemed to recover somewhat after 18 months, whereas on the cleaned parts of the beach it did not recover for the next 3–4 years. Oil spill cleanup continues to be performed because the public believes this is still the way to save most animals.

  The Exxon Valdez oil spill had, and still has, a large deal of media attention. Many people still remember the spill today. However, Exxon Valdez did not cause the largest oil spill in human history. According to Bjorn Lomborg, it is not even in the top 10. A much larger disaster was caused, for example, during the Gulf War in 1991.

- *The Gulf War*: In August, 1990, Iraqi forces invaded Kuwait, starting the Gulf War in which an allegiance of 34 nations worldwide was involved (Salinger and Laurent, 1991). In January, 1991 (during the Gulf War), Iraqi forces committed [acts of violence, resulting in] two environmental disasters. The first was a major oil spill 16 km off the shore of Kuwait by dumping oil from several tankers and opening the valves of an offshore terminal. The second was setting fire to 650 oil wells in Kuwait. These incidents point out how tenaciously irresponsible and destructive humans can be.

  The apparent strategic goal of the action was to prevent a potential landing by US Marines. American air strikes on January 26 destroyed pipelines to prevent further spillage into the Gulf. This, however, seemed to make little difference. About 1 million tons of crude oil were already lost to the environment, making this the largest oil spill of human history.

In the spring of 1991, as many as 500 oil wells were still burning, and the last oil well was not extinguished until a few months later (November). They did considerable damage to life forms in the Persian Gulf. Several months after the spill, [it was determined that] the poisoned waters [had] killed 20,000 seabirds and had caused severe damage to local marine flora and fauna. The fires in the oil wells caused immense amounts of soot and toxic fumes to enter the atmosphere. This had great effects on the health of the local population and biota for several years. The pollution also had a possible impact on local weather patterns.

- *Bahaman Container Ship Kariba*: In the early hours of December 14, 2002, the Norwegian ship Tricolor collided with the Bahaman container ship, Kariba, in the French Channel. The accident was caused by fog and human errors. The Kariba was heavily damaged, but managed to reach the Antwerp Harbor. The crew of Tricolor was rescued by emergency teams, which experienced low visibility that made the rescue operation very difficult to carry out.

  Despite warning signals on the location of the Tricolor, the Nicola collided with the wreck on December 16. The Nicola could be safely removed from the scene, but the Tricolor was now much more severely damaged. The ship was declared a total loss, and Berger Smit started to pump 2200 tons of oil from the wreck.

- *Oil Tanker Vicky*: In January 2003, the oil tanker Vicky collided with the Tricolor, causing some oil from the Vicky to flow into the sea which subsequently reached French and Belgium shores. Fortunately damages were limited, and the Tricolor did not leak any oil.

  By the end of January even more extreme weather caused Berger Smit to collide with the Tricolor wreck, which then started leaking oil. It became apparent that at least 1000 tons of oil had leaked into the Channel. When the oil reached France and Belgium, it resulted in thousands of dead seabirds washing ashore.

After this third collision, a confederation was ordered by the French government to remove the wreck to prevent further environmental damage. Eventually, the ship was broken into nine small pieces, heaved from the water, and carried away:

> At this moment the sea soil still contains hundreds of car wreckages that were once transported by the Tricolor. The freight value was approximately 49 million euros. The ship has an additional value of 40 million euros.

Today, crude oil production is used to make many products, including fuel oil, gasoline, asphalt, tar, paraffin wax, lubricating oils, and a large assortment of chemicals that are used in the production of numerous other products. The demand for petroleum products has never been higher, and they are transported by ships throughout the world. Consequently, the accidents in waiting have happened on numerous occasions and will continue to happen as long as there is a need for them.

Other Oil-Related Accidents

The following list shows examples of the major oil spills throughout the years, most of which did not receive as much media attention as the ones mentioned earlier. We will return to discuss BP's Transocean Deepwater Horizon after noticing the long list of oil spills around the world and the amount of petroleum products which was released into the environment before the last spill:

- 1967: Liberian tanker Torrey Canyon spilled 120,000 tons of oil near Cornwall.
- 1968: Witwater tanker spilled 14,000 barrels of oil near the Panama coast.
- 1969: Tanker Hamilton trader spilled 4000 barrels of oil in Liverpool Bay, England.
- 1970: Tanker Arrow spilled 77,000 barrels of oil near Nova Scotia, Canada.
- 1971: Tanker Wafra spilled 20,000 barrels of oil near Cape Agulhas, Africa.
- 1972: Tanker Sea Star caught fire after collision in Gulf of Mexico.
- 1974: Dutch tanker Metulla spilled 53,000 tons of crude oil near South Chili.
- 1976: Liberian tanker Argo Merchant spilled 29,000 m^2 of oil near the Massachusetts coast.
- 1976: Spanish tanker Urquillo spilled more than 100,000 tons of oil near Spain.
- 1977: Tanker Al Rawdatain spilled 7350 barrels of oil near Genoa, Italy.
- 1977: Tanker Borug spilled 213,692 barrels of oil near the coast of Taiwan.
- 1978: Brazilian Marina spilled 73,600 barrels of oil near Sao Sebastiao, Brazil.
- 1979: Betegeuse spilled 14,720 barrels of oil near Bantry Bay, Ireland
- 1979: Ixtoc I exploratory well in Mexico blew out and spilled 600,000 tons of oil.
- 1984: Alvenus tanker grounded southeast of Cameron, Louisiana, and spilled 65,000 barrels of oil.
- 1985: ARCO Anchorage spilled 5690 barrels of oil near the coast of Washington.

- 1986: An unknown oil spill reached the coast of Georgia and was later appointed to the Amazon Vulture tanker.
- 1989: Aragon tanker spilled 175,000 barrels of oil near Madeira, Portugal.
- 1990: Tanker American Trader grounded near Huntington Beach, California, and spilled 9458 barrels of oil.
- 1990: Cibro Savannah tanker caught fire and spilled 481 m² of oil.
- 1990: Jupiter tanker caught fire in Bay City, Mexico, and caused an oil spill.
- 1990: Mega Borg tanker caught fire and spilled 19,000 m² of oil near Galveston, Texas.
- 1991: Tanker Bahia Paraiso spilled 3774 barrels of oil near Palmer Station, Antarctica.
- 1992: Greek tanker Aegean Sea spilled 70,000 tons of oil near Galicia.
- 1993: Bouchard B155 tanker spilled 1270 m² of fuel oil after collision with two ships.
- 1996: Liberian tanker Sea Empress spilled 147,000 tons of oil near Wales.
- 1999: Maltese tanker Erika spilled 30,000 tons of oil near Brittany.
- 2001: Tanker Jessica spilled 900 tons of oil near the Galapagos Isles.
- 2002: Bahamese Prestige spilled oil near Galicia.
- As of this writing, the most disastrous petroleum accident in history was one of the most recent events in the Gulf of Mexico: BP's Transocean Deepwater Horizon. Over time, it will also be [known as one of] the most recorded petroleum-associated events, including the environmental damage it has been responsible for.
- *BP's Transocean Deepwater Horizon*: The 2010 [accident] in the Gulf of Mexico [was not an oil spill (Harlow et al., 2011). It] resulted from a leak in a well that was drilled by the Transocean Deepwater Horizon, a giant British Petroleum offshore drilling rig that exploded and sank. The subsequent oil spill was reported to be leaking about 42,000 or more gallons a day from a well nearly a mile deep. By the time the oil leak was alleviated, about 100 million gallons of oil had reportedly entered the marine waters in the Gulf of Mexico.

According to an August 2010 report by the Center for Biodiversity, "the BP oil spill" threatened some of the most productive and fragile marine ecosystems in the United States. About 25% of the nation's wetlands lie in the Mississippi River Delta, providing habitat for nesting seabirds and resting migratory birds. The Gulf itself is home to dozens of threatened and endangered species, as well as commercially important fish, crabs, and shrimp that provide much of the basis of the Gulf Coast economy.

While the response to the oil spill largely focused on stopping oil from reaching shore, the offshore ecosystem, from plankton to dolphins, were expected to suffer devastating impacts. Endangered sperm whales and dolphins were spotted passing through the oil slick—which has spoiled critical habitat for federally protected piping plovers on the Chandeleur Islands. Oiled gannets and brown pelicans were the first victims discovered by response teams; the goo permeated mangroves and soaked birds and their eggs. Heavy oil also moved into the Queen Bess Island pelican rookery, a nesting site that has been essential to the recovery of the brown pelican population. Experts worry that the spill will set back the Louisiana state bird's recovery from near-extinction.

The Center for Biodiversity stated that "the timing of the spill could not have been worse." Imperiled species including the Atlantic bluefin tuna, Kemp's ridley sea turtles, loggerhead sea turtles, piping plovers, and sperm whales flock to the Gulf to spawn, migrate, and feed. For many of them, there is nowhere else to go. And in a disturbing development, large numbers of sharks, other fish, and marine animals gathered in shallow inshore waters, presumably seeking areas where oxygen has not been depleted by oil and the microbes that eat it.

Marine animals can die when oxygen levels in the water drop below two parts per million, which had happened early, even in inshore areas. Moreover, creatures congregating near the shore risked getting trapped between shore and the oil and experiencing depleting oxygen levels, even in refuge areas.

During oil response efforts, concern arose that sea turtles were in oiled sargassum mats that were lit afire to burn off the oil. During the week of June 28, the Center initiated litigation on multiple fronts to prevent the injury and death of sea turtles during controlled burn operations, which resulted in requirements for observers to rescue sea turtles from this unnecessary and brutal threat.

THE INDUSTRIAL REVOLUTION AND TECHNOLOGICAL EXPERTISE

Such disasters, of course, could not have happened until humans allowed their populations to grow so large and developed their technological expertise to a point of being able to allow these things to happen.[3] As hunter-gatherers in bands of 50 or fewer individuals, early humans survived by attempting to understand their natural environment, and populations grew very slowly. Even with an increasing population growth, tools, and fire, human exploitation of the environment by aboriginal societies was usually readily repaired. Yet, there were early signs that the human population was beginning to have an adverse effect on Earth's ecosystems (Wackernagel and Rees, 1998).

As an example, American Indians in the United States, as in most early societies, lived with nature, showing great respect for the land and its biota, but even their growing populations required food and may have had a hand in the extinction of certain animals that were preyed upon (including 33 of 45 genera of large mammals and about 19 genera of birds). Some argue that it was not spears that killed them but pathogens brought to this continent by the dispersal of people (Paleoindians) from Siberia.[7]

Besides North America, the Quaternary period (a period between 2.6 million years ago and the present, in which glaciations took place repeatedly) was a time of mass extinction of predominantly large megafauna (over 100 kg), mostly in South America and Australia (Johnson, 2009). About 10,000 to 12,000 years ago (about the same time that many North American animal extinctions occurred), the agricultural revolution (due to increasingly large populations of people with new skills directed at planting and growing crops) led to increased sedentary behavior, a greater depletion of local resources, and new tensions between settled subpopulations.

Over time, the human population continued to grow, and the industrial revolution, commencing in the 1700s and 1800s (Vitousek et al., 1997), led to greater productivity and many accompanying behavioral shifts in forms of human existence. These changes and the trend to globalization, especially since 1970, resulted in a more integrated world and more rapid human population increases (Cohen, 2003).

With a continuing population increase, humans began to investigate and utilize more resources and pollute the world at a more rapid rate than ever before. Through industrialization, technological innovation, and population increase, we have become modernized to the point of disregarding Earth's biodiversity and our impact on world resources.

According to authors such as G. Tyler Miller in *Environmental Science* (Miller, 1993), our tendency to dominate the world has progressed to a point where many influential world leaders and developers (some of the most dominant individuals in our populations) continue to disregard the importance of preserving the planet's biodiversity. Often with their eyes trained on economic growth and potential voters, political leaders have a tendency to disregard education in the sciences, refrain from talking about biological problems, and underplay the importance of paying attention to environmental troubles. However, in spite of what world leaders say, there is a significant amount of data to show that we are destroying Earth and its biota, sometimes at record speed. The most recent oil spill in the Gulf of Mexico is actually one small example of how serious our destructive nature can be.

We must stop and admit that we understand the world's human population requires energy to function. We must seek this energy to maintain planetary homeostasis. However, does not the speed of world movement, our dependence on constantly finding new ways to feed our growing masses, and mining more petroleum to enable the world to continue functioning— even at the expense of world pollution, resource depletion, and mass extinction of nonhuman organisms—indicate that our cosmopolitan population is too large?

According to the US Environmental Protection Agency, an estimated 10–25 million gallons of oil are spilled annually. These oil spills and our most immediate observations of their effects are just the beginning of horrendous ramifications of its consequent influence on wildlife. Oil spills have major, irreversible effects on wildlife and humans, as well as air quality, whether the spill happens on land or in water (Johnson, 2009).

A report by the Australian Maritime Safety Authority gives a brief explanation of the most immediate damages an oil spill can cause (Authority, 2005). They point out that not all oils are alike. There are many different types of oil, making each oil spill different. Thus, each oil spill has a different impact on wildlife and the surrounding environment, depending on the type of oil spilled, the location of the spill, the species of wildlife in the area, the timing of breeding cycles, seasonal migrations, and the weather at sea during the oil spill.

Vertebrate wildlife is affected by the coating of their bodies with a thick layer of oil. Many oils also become stickier over time (a process called weathering, due to the release of volatiles), resulting in an even stronger adherence of the oil to their coats. Since most oil floats on the surface of the water (it is lighter than water), it can affect many marine animals because they do not understand what an oil spill is and may not avoid it. Some fish are even attracted to oil because it looks like floating food. This, in turn, endangers sea birds, which are attracted to schools of fish and may dive through oil slicks to get to them. Some marine mammals, such as seals and dolphins, are known to swim and feed in or near oil spills (Fig. 16.1).

Unrefined petroleum products that stick to fur or feathers can cause many problems, some of which are:

- Hypothermia in birds by reducing or destroying the insulation and waterproofing properties of their feathers.
- Hypothermia in mammals by reducing or destroying the insulation of their fur. Adult mammals with blubber may not suffer from hypothermia if oiled, although they may be affected in other ways. Dolphins and whales are examples of mammals that do not have fur, so oil will not easily stick to them, although they also may nevertheless suffer from oil spills.
- Birds become easy prey, as their feathers become matted by oil, making them less able or unable to fly.

FIGURE 16.1 Oil spill. Oil spills have occurred around the world, resulting in the destruction of habitat and loss of life in millions of organisms (A–C). Some of the world's most valuable ecosystems have been destroyed by the human need for petroleum products. In addition, waste materials, for example, plastic bottles, and cups. (D), end up in large garbage dumps in certain of the planet's oceans. Not only do these materials pollute the world's ecosystems, they often are fed upon by many animals or otherwise threaten wildlife safety. *Source: (C) Joseph Sohm/ShutterStock.com (D) FashionStock.com/ShutterStock.com.*

- Marine mammals become easy prey if oil sticks their flippers to their bodies, making it difficult for them to escape predators.
- Birds sink in water or drown because oiled feathers cannot trap enough air between them to keep them buoyant.
- Marine mammals lose body weight when they cannot feed due to contamination of their environment by oil.
- Birds become dehydrated and can starve as they give up or reduce drinking, diving, and swimming to look for food.
- Inflammation or infection occurs in certain mammals, for example, dugongs and manatees, and difficulty eating due to oil sticking to the sensory hairs around their mouths.
- There is a reduction in the disguise of scent that some adult and immature mammals rely on to identify each other, leading to rejection, abandonment, and starvation.
- There is damage to internal organs of animals, for example, causing ulcers or bleeding in their stomachs, if they ingest oil.
- While some of the following effects on sea birds, marine mammals, and turtles can be caused by crude oil or bunker fuel, they are more commonly caused by refined oil products. Oil in the environment or oil that is ingested can cause:
- Poisoning of wildlife higher up the food chain if they eat large amounts of other organisms that have taken oil into their tissues.
- Interference with breeding by making the animal too ill to breed, interfering with breeding behavior, for example, birds sitting on their eggs or by reducing the number of eggs a bird will lay.
- Damage to the airways and lungs of marine mammals and turtles, congestion, pneumonia, emphysema, and even death by breathing in droplets of oil, or oil fumes or gas.
- Damage to a marine mammal's or turtle's eyes, which can cause ulcers, conjunctivitis, and blindness, making it difficult for them to find food and sometimes causing starvation.
- Irritation or ulceration of skin, mouth, or nasal passages.
- Damage to and suppression of a marine mammal's immune system, sometimes causing secondary bacterial or fungal infections.
- Damage to red blood cells.
- Organ damage and failure, for example, in bird or marine mammal livers.
- Damage to a bird's adrenal tissue, which interferes with a bird's ability to maintain blood pressure, and concentration of fluid in its body.
- Decrease in the thickness of egg shells.
- Stress.
- Damage to fish eggs, larvae, and young fish.
- Contamination of beaches where turtles breed, resulting in contamination of eggs, adult turtles, or newly hatched turtles.
- Damage to estuaries, coral reefs, seagrass, and mangrove habitats that are the breeding areas of many fish and crustaceans, interfering with their breeding.
- Tainting of fish, crustaceans, mollusks, and algae.
- Interference with a baleen whale's feeding system by tar-like oil, as this type of whale feeds by skimming the surface and filtering out water.
- Poisoning of young through the mother, as a dolphin calf can absorb oil through the mother's milk.

Since the quality of petroleum changes with time, animals covered in oil at the onset of a spill may be affected differently from animals encountering it at a later time. For example, oil may be more poisonous during the period immediately following the spill (due to volatiles), resulting in the uptake by animals of more noxious chemicals.

Weather conditions can influence the potential for oil to cause damage to the environment and wildlife (Votier et al., 2005). For example, warm seas and high winds result in lighter oils forming gases, reducing the amount of oil that stays in the water.

The impact of an oil spill on wildlife is also affected by where spilled oil reaches. For example, fur seal pups are affected more than adults by oil spills because they swim in tidal pools and along rocky coasts, whereas adults swim in open water where it is less likely for oil to linger over time. Dugongs (relatives of manatees) also feed on seagrass along the coast within their range and therefore are more affected by oil spills.

Another report by the Center for Biological Diversity brings up the potential of a disaster of an oil spill in the Arctic (Robertson, 2013). The Gulf of Mexico has by far the largest, best-equipped, most experienced oil spill-containment system in the nation. It has hundreds of experienced volunteer fishing boats at its disposal. The water is warm year round and relatively calm except in hurricane season. Wildlife rehabilitation and cleanup crews have access to a road system in close proximity to much of the shoreline. Yet, with all these advantages, the government and the oil industry were unable to contain the spill. There is also some question about the value of bioremediation. It appears that sometimes the cure can be just as stressful for the environment and its inhabitants as the cause, possibly more.

Imagine what would happen if a similar spill occurred in the Arctic—140 miles from land, in subzero temperatures, with miles of sea ice to hack through, ship-killing icebergs in all directions and darkness for 20h a day in the winter. According to authorities, it would be a disaster many magnitudes worse than what we are suffering in the Gulf of Mexico.

"There's no way to clean up a massive oil spill in the broken-ice conditions that prevail in these Arctic areas for much of the year," they say. "In fact, the ice-free drilling season is so short in the Arctic July to early October—that leaking oil from a similar accident there could continue to gush for an entire winter while efforts to drill a relief well were necessarily postponed":

> "In a huge victory for the Arctic, on May 27, 2010, President Obama announced he wouldn't allow Shell Oil to conduct exploration drilling in Alaska's Beaufort and Chukchi seas this summer—but we don't know what will happen in the next few years." With the need for energy, the victory is most probably a temporary one. In fact, "Obama proposed opening up both the Chukchi and Beaufort seas to additional offshore oil drilling in coming years. And BP continues to pursue high-risk drilling plans in the Beaufort Sea: Its 'Liberty' project is exempted from the Alaska offshore moratorium, as reported in the June 24 edition of The New York Times, due to the fact that the rig, which lies three miles offshore, is built onto an artificial island and [is] thus classified as onshore."

The Center for Biological Diversity also states that it is "imperative that drilling *never goes forward* to prevent an Arctic oil-spill catastrophe that could be even worse than the Gulf disaster, threatening the polar bear, Pacific walrus, ringed, spotted, bearded and ribbon seals, cetaceans like the North Pacific right whale, the bowhead whale, the Cook Inlet beluga, migratory birds and many other species." In addition, as we have learned through these pages, there are many organisms that never get mentioned that are adversely affected by such environmental disasters.

As I was in the process of reviewing this material, a public radio broadcast entitled *Shell Faces Pushback as Alaska Drilling Nears* reported on preparations to explore for oil off the north coast of Alaska (Van Halderen et al., 2014). They said that many Alaskans welcome the venture, which could lead to the development of trillions of dollars worth of oil. So now the Inuit Alaskans are losing touch with nature as well. It, like many facets of human life, is all about the money. But not everyone was convinced that Shell had all the environmental safeguards in place to protect areas of the pristine Arctic. I include that article here, mostly to show the attitude of people who need work and lack either a knowledge of the potential disaster drilling in the Arctic can bring or concern for the biota that will be devastated in case of an oil spill there:

> The federal government could soon give the final go-ahead for Royal Dutch Shell to begin drilling for oil in the Arctic Ocean. Shell has spent $4 billion since 2007 to prepare for this work, and is hoping to tap into vast new deposits of oil. But the plan to drill exploratory wells is controversial—opposed by environmental groups and some indigenous people as well.
>
> You can get a feel for this controversy by stepping aboard the Nanuq. Shell plunked down $100 million for this 300-foot-long, Arctic-class spill-response vessel. And unlike similar vessels that serve the oil industry in the Gulf of Mexico, there's no foul-smelling black smoke coming out of the stacks.
>
> "They retrofitted this vessel with clean-air technology just to keep those emissions down," says Shell scientist Michael Macrander. "All of our vessels have been worked on to keep our emissions down."
>
> That's not simply because Shell is big-hearted. The company's plans to drill in the Arctic last year were thwarted when environmental groups successfully petitioned the EPA to reject the company's original air-quality permits. Shell switched to low-sulfur fuel and installed particle scrubbers to meet the higher standard. So the fleet heading north is a lot cleaner than it would have been in 2011.
>
> Air isn't the only issue. Inupiat who live along the north coast of Alaska have been worried that oil exploration could scare away the whales they hunt. The company has taken steps to address those concerns.
>
> "That goes all the way to agreements with the Alaska Eskimo Whaling Commission to shut down during critical periods while they are hunting," Macrander says (Suhre, 1999).
>
> Shell says it will pull its drilling rig out of the Beaufort Sea for weeks during the summer whale hunt. And Macrander also notes that the Nanuq is painted blue and white because the Inupiat told Shell that those colors are less likely to scare whales.
>
> Those accommodations have tempered Inupiat opposition to the drilling. But they do not satisfy environmental groups. Everyone from Greenpeace to the Sierra Club to the Alaska Wilderness League continue to fight a pitched battle to block Shell.
>
> Richard E. Reanier, an independent archaeologist working for Shell Oil, uses GPS to record the location of the remains of a sod house along the Chukchi Sea coast near Wainwright, Alaska, in July 2011.
>
> "We're not ready scientifically, we're not ready in terms of having effective cleanup, and we're not ready in terms of having the regulatory oversight mechanism that we should have had," says Lois Epstein of the Wilderness Society in Anchorage.[5]
>
> Epstein says government agencies don't have enough technical staff to fully evaluate oil and gas operations (Cross and Baird, 2000). She says it helps a bit that the government has decided to put inspectors on the Arctic drilling rigs for the entire time they are drilling offshore. But that's not enough.

Another concern for environmental groups is that the federal government has been leasing tracts to oil companies without first stopping to figure out what parts of the Chukchi and Beaufort seas should be set aside as wildlife preserves. "Scientists are still trying to identify the most critical habitat for marine mammals," Epstein says.

"When we do identify areas that are highly productive in terms of calving and important in terms of raising young, we should set aside those areas to assure that whales and other marine mammals have that habitat," he says.

Scientific research has accelerated in the region over the past few years—funded primarily by Shell and other oil companies, and carried out largely by academic researchers. Macrander argues that they're making good progress in closing the data gaps that previous reviews have identified. But Epstein says development is simply happening too fast, in an area where so much is at stake, for the environment and for the native peoples.

"Yes, there's enormous pressure to go after the Arctic, but we're not so desperate right now that we have to do that," she says.

Environmental groups reach sympathetic ears in the Lower 48 states, but that's not the case in Alaska. Scott Goldsmith, an economics professor at the University of Alaska, Anchorage, says most Alaskans are excited that Shell is about to start drilling offshore.

"It could be one of the next big events that will keep the petroleum economy going for another generation or more," he says, adding that most Alaskans care more about the petroleum economy than they do about the pristine Arctic.

The economic concerns have been growing because oil production from Alaska's major sources has been in real decline. Drilling offshore could turn that around dramatically.

There are potentially 25 billion barrels of oil up there, arguably worth several trillion dollars. So it's no mystery why Shell has already invested $4 billion on this prospecting. The company will doubtless spend a lot more before oil could start to flow.

Pete Slaiby, Shell's vice president for Alaska, says the costs of doing business in Alaska are big, "but the volumes are substantial (Longwell, 2002). The only area in the U.S. with potentially more resource would be those remaining in the deep-water Gulf of Mexico."

Slaiby says if they do find rich pockets of oil, it will take at least a decade to drill production wells and to string pipeline to carry the oil to shore and on to market. And Shell isn't the only company interested. Conoco, Phillips, and Statoil hope to start drilling exploratory wells next year.

So will the Arctic Ocean eventually bristle with thousands of oil rigs, like the Gulf of Mexico? Slaiby doubts it. Modern technology requires far fewer oil-production platforms. And the harsh conditions will keep out everyone but the big guys, "so it will never, in my view, have that kind of industrial look that the Gulf of Mexico might have."

Some Inupiats argue that oil and gas exploration puts their traditional means of livelihood at stake. U.S. readiness for increased activity in the Arctic is limited, and it's not part of a key treaty. The land was set aside for oil production in the 1970s; now, some groups want part of it protected.

Since the BP blowout in the Gulf of Mexico, there has been enormous attention paid to the risks of offshore oil. The situation in Alaska is different from the Macondo—the water is only a few hundred feet deep, unlike the 5,000-foot depth in the Gulf, so it's less technically challenging.

On the other hand, the Arctic Ocean is so far from major ports and airports that it would be a major challenge to mount a large-scale spill response. The U.S. Coast Guard doesn't have vessels stationed along the North Slope of Alaska, though this summer, at least, the agency is planning to have a full-time presence in the region.

And with all the buildup to this drilling season, it may come as a surprise to learn that this is not, in fact, the first time Shell has drilled in the Beaufort and Chukchi seas north of Alaska.

"We drilled wells two decades ago, drilled them all successfully without any kind of major incident," Slaiby says. "And the greatest compliment I hear about that is one I hope to get at the end of 2012, and that is [people saying], 'We didn't know you guys were here.'"

Actually, there was one way that Shell wanted to make its presence felt—in creating jobs, and the political support that comes with them. Slaiby says these operations could ultimately generate 55,000 jobs—a huge number for sparsely populated Alaska.

Onboard the Nanuq spill cleanup vessel near Valdez, Robert Long is one of the nearly 200 Alaskans hired for spill cleanup duty. "The pay's good, the work's good here," says the 50-year-old Alaska native.

Are he and the members of his community concerned about an oil spill? "I'm sure anybody is," he says. "You know if there is [one], we'll be right there cleaning it up."

As we see in this chapter, oil spills are dangerous for all forms of life (Bostrom et al., 2014). When one realizes that we have mostly covered marine mammals and birds and that marine plants, fish, invertebrates, and colonial and unicellular organisms (all of which are part of food chains and complex food webs) also suffer from oil spills, it widens the effects to all marine creatures. Since at least half of the oxygen in the world is generated in the oceans, this is of particular importance to marine photoautotrophs (creatures that make their own food from sunlight, CO_2, and water) and organisms that depend on them for an aerobic environment.

Further, upon reaching the edges of a marine environment, precious mangrove environments and intertidal zones, areas that are extremely biodiverse, are affected, and this, in turn, affects terrestrial organisms (Jensen et al., 2008). In addition to the ecological significance of mangrove and intertidal destruction, the die-off of sea creatures halts marine-related businesses and affects the livelihood of numerous people who depend on coastlines.[1]

ESTABLISHING MARINE DEAD ZONES

As disastrous as these oil spills were, we may wonder whether they were the beginning of coastline destruction? Not by a long shot. Actually, the Gulf of Mexico, like many other areas in the world that lie close to human populations, was a deteriorating ecological disaster area long before the release of oil, and the roots of these disasters, also from the hands of humans, are on land.

The Mississippi River is a mighty river that receives drainage water from terrestrial and aquatic ecosystems throughout its course. Small and large streams and rivers throughout the Midwest drain extensive areas and flow into the Mississippi. Pesticides and fertilizers used by humans enter runoff which flows into the river, and they, along with vast amounts of other pollutants, are dumped into the Gulf of Mexico (Mateos, 2011). These chemicals, which have killed unknown numbers of terrestrial and aquatic organisms en route to the Gulf, have subsequently created a dead zone in Gulf waters off the coast of Louisiana (Fig. 16.2).

Some of the chemicals found in runoff promote the growth and reproduction of many marine microorganisms (Smith, 1987). Nitrogen-rich compounds found in fertilizer and waste material runoff provide nutrients that result in algal blooms.[2] Upon the termination of a bloom, there is a great dying of organisms that sink to the bottom of the water and decompose, depleting the water of oxygen. It occurs in similar waters around the world as well (Fig. 16.2).

The Northern Gulf of Mexico's hypoxic (oxygen-deficient) waters represent one of the Western Hemisphere's largest "dead zones"—areas where a lack of oxygen kills fish, crustaceans, and other marine life (Rabalais, 2011). The size of the zone varies but at its peak, it stretches along the inner continental shelf from the mouth of the Mississippi River westward to the upper Texas coast, covering about 7000 mi^2, an area as large as New Jersey. Long-term consequences to biodiversity, species abundance, and biomass in the Gulf are not yet known, but experience with other coastal dead zones has shown significant ecological deterioration and depleted fisheries.

The UN Environment Program's first Global Environment Outlook Year Book (GEO Year Book, 2003) reported 146 dead zones in the world's oceans where marine life could not be supported due to depleted oxygen levels. Some of these were as small as a square kilometer, but the largest dead zone covered 70,000 km^2 (27,000 mi). A 2008 study (just 5 years after the 2003 report) counted 405 dead zones worldwide.

In January, 2011, the World Resources Institute (WRI) and Virginia Institute of Marine Science (VIMS) identified 534 low-oxygen "dead zones" and an additional 228 sites worldwide exhibiting signs of marine eutrophication (areas with an overabundance of certain chemicals, e.g., fertilizers and pesticides). WRI has since discovered 13 additional sites that are already eutrophic and in danger of becoming dead zones, bringing the total number of coastal areas around the world known to be suffering from nutrient pollution at the time of this writing to 775.

The GEO Year Book reported that "Low oxygen levels recorded along the Gulf Coast of North America have led to reproductive problems in fish involving decreased size of their reproductive organs, low egg counts and lack of spawning. It might be expected that fish would flee this potential suffocation, but they are often quickly rendered unconscious and doomed. Slow moving bottom-dwelling creatures like clams, lobsters and oysters are unable to escape. All colonial animals are extinguished. The normal re-mineralization and recycling that occur among benthic (bottom) life-forms is stifled."

FIGURE 16.2 Red tide in the Gulf of Mexico often results from chemicals that originate from pesticides and fertilizers used throughout the farm belt on each side of the Mississippi River. Runoff from farmland is carried to the Mississippi River which eventually flows into the gulf, causing eutrophication of marine water and rapidly swelling algal populations. Upon dying, oxygen normally in the water is depleted (hypoxia), resulting in a die-off of many marine organisms. Die-offs such as these result in dead zones where very little life is left. In addition to a large dead zone between Mississippi and Texas in the Gulf of Mexico, there have been almost 800 dead zones reported around the world.

Thus, nitrogen-rich chemicals represent the primary cause of a serious planetary depletion of oxygen (hypoxia) in many parts of the ocean, especially in coastal zones around the world. The resulting lack of dissolved oxygen greatly reduces the ability of these areas to sustain oceanic organisms.

Fresh water habitats are suffering as well. About half of all the lakes in the United States are now said to be eutrophic (waters rich in phosphates, nitrates, and organic nutrients that promote a proliferation of plant life, especially algae), and oceanic dead zones near inhabited coastlines are increasing around the world. The production of eutrophic bodies of water and resulting dead zones is not acceptable if we want to continue living on planet Earth. In addition, even if eutrophication can be reversed, it would take decades before the accumulated nitrates in groundwater can be broken down by natural processes.

The high application rates of inorganic nitrogen fertilizers required to maximize crop yields that feed the high numbers of people now in the world, combined with the high solubilities of these fertilizers, lead to increased runoff into surface water as well as leaching into groundwater. The use of ammonium nitrate in inorganic fertilizers is especially damaging, as plants absorb ammonium ions preferentially over nitrate ions, while excess nitrate ions that are not absorbed dissolve (by rain or irrigation) into runoff or groundwater.

Half of the farms in the United States are located in the Mississippi River Basin (Changnon, 2009). The entire drainage basin empties into the gulf. Much of the nitrogen reaching the gulf is from agricultural fertilizers, with lesser amounts from residential fertilizers and other sources. Water in the 20,000-km (7728-mi^2) dead zone, extending from the mouth of the Mississippi River Basin to beyond the Texas border, has so little oxygen that very little marine life exists.

If human-accelerated eutrophication is not reversed, the entire coastal ecosystem ultimately may be changed (Hoagland and Franti, 2014). Sensitive species may be replaced by more tolerant and resilient species (if even they can withstand the toxic conditions), and biologically diverse communities may eventually be replaced by less diverse ones. Further, nutrient enrichment and the associated eutrophication in coastal waters is implicated in the presence of some harmful algal blooms, in which certain species of algae produce biotoxins (natural poisons, as in red tides) that can be transferred through the food web, potentially harming higher-order consumers, such as marine and terrestrial mammals, even humans.[3]

HUMAN AND OTHER ANIMAL BODY WASTES

Biological and chemical pollution does not stop with oil spills and the use of pesticides, herbicides, and commercial fertilizers. Where there are animals, there are animal waste products, and as the growth of industrial farming concentrates thousands of animals on increasingly fewer farms, it produces massive amounts of animal waste on relatively small plots of land (Burns and Raman, 2010). Dairy cows in confined feeding operations throughout the United States, for instance, produce more than 2 billion pounds of nitrogen from manure each year.

A 2002 agriculture census pointed out that there were 95.5 million cows and calves, as well as 60.4 million hogs and pigs in the United States, each producing wastes every day.[4] According to G. Koneswaran and D. Nierenberg (*Global Farm Animal Production and Global Warming: Impacting and Mitigating Climate Change*) (2008), about 56 billion land animals are reared and slaughtered for human consumption annually worldwide, and livestock inventories are expected to double by 2050, with most increases occurring in the developing world.

When an overabundance of wastes is produced at one place, there is no safe, cost-effective way to either use it productively or dispose of it. While government regulation and better waste management practices can make a difference and should be encouraged for existing farms, the problem of livestock waste will never end so long as we rely on concentrated industrial farms to produce our food.

According to Koneswaran and Neirenberg (2008), increasing numbers of animals in recent decades are "raised in intensive production systems in which chickens, pigs, turkeys and other animals are confined in cages, crates, pens, stalls and warehouse-like grow-out facilities. These production systems are devoid of environmental stimuli, adequate space, or means by which to experience most natural behaviors. Furthermore, because these industrialized, 'landless' facilities tend to produce more manure than can be used as fertilizer on nearby cropland, manure is instead distributed to a small, local landmass resulting in soil accumulation and runoff of phosphorus, nitrogen and other pollutants. Although extensive or pasture-based farming methods remain the norm in Africa and some parts of Asia, the trend in Latin America and Asia [where the human population is growing most rapidly] is to increasingly favor intensive production systems over more sustainable and more animal welfare–friendly practices."

Thus, disposing of organic wastes is difficult with our current policy on nonuse. Governmental policies on the nonuse of organic fertilizers on plant products for human consumption because of recent high counts of *Escherichia coli* has also put a damper on their depletion (Browman and Maskarinec, 2008).

Organic fertilizers (e.g., manures) usually improve biodiversity (especially soil life) and long-term productivity of soil, and they can represent a large depository for excess carbon dioxide. Nutrients in organic fertilizers increase the abundance of important soil organisms by providing organic matter and micronutrients for organisms such as fungal mycorrhiza, which function in a mutualistic fashion to aid plants in absorbing nutrients. Organic fertilizers also can drastically reduce external inputs of pesticides, energy, and other forms of fertilizer.

However, there are several disadvantages: (1) organic fertilizers often contain pathogens and other forms of parasites unless the fertilizers are treated to eliminate these organisms; (2) the nutrient content of organic fertilizer is variable, and their release in variable forms that are useable by plants may not occur at the right plant growth stage; (3) organic fertilizers are comparatively voluminous and can be too bulky to deploy the right amount of nutrients that are required for good plant growth; and (4) organic fertilizers at this time are more expensive to prepare for use (Möller and Schultheiß, 2014).

Comparing the impact of different types of animals, livestock statistics are often cited in terms of animal units. One animal unit, for instance, equals 1000 lb of the live weight of an animal. For example, 125 8-lb chickens make up 1 animal unit of chicken, and four 1250-lb cows equal 5 animal units of cattle. While these figures may appear a bit confusing, they provide us with some interesting statistics.

With reference to this system, one animal unit of broiler (meat) chickens produces an average of 14.97 tons of manure each year, fattened cattle 10.59 tons per year, and dairy cows 15.24 tons per year. In comparison, one animal unit of humans produces a mere 5.48 tons of waste per year. Thus, the majority of polluting waste materials originate from domestic animals. However, even this most definitely points directly to human overabundance because we are dependent upon this number of domestic animals to keep us alive.

The USDA estimates that more than 335 million tons of dry body wastes (the portion of waste remaining after water is removed) is produced annually on farms in the United States, representing almost a third of the total municipal and industrial waste produced every year. In addition, animal feeding operations annually produce about 100 times more manure than the amount of human sewage sludge processed in United States municipal wastewater plants (Chapinal et al., 2008). One dairy farm with 2500 cows produces as much waste as a city with around 411,000 residents. Unlike human waste, however, the law (in most cases) does not require that livestock waste be treated.

At farms where animals are allowed to graze on pasture, most of their manure is excreted directly onto the land, serving as fertilizer and recycling nutrients back into the soil (Mugwira et al., 2007). On industrial livestock farms, however, animals drop their manure in buildings where they live. From there, the manure must be cleaned out, transported, and stored, each step of which can negatively affect the environment. Simply cleaning out livestock houses can waste vast amounts of water. For instance, a dairy operation that utilizes an automatic flushing system can use up to 150 gallons of water per cow each day.

Manure is usually stored for many months, often in giant outdoor lagoons. As it decomposes, the manure emits harmful gases, such as hydrogen sulfide, methane, and ammonia. Meanwhile, these lagoons can leak or rupture, polluting the surrounding soil and water systems.

One study conducted by North Carolina State University in 1995 estimated that as many as 55% of the manure lagoons on hog farms in that state were leaking (Furuseth, 2010). Even without leaks, manure lagoons often overflow. Perhaps the best-known example occurred in 1999, in which the majority of North Carolina's manure lagoons spilled over into waterways during Hurricane Floyd, leading to widespread water contamination. North Carolina (at the time of this writing), like most states, required no treatment of animal waste.

Since manure is produced on factory farms in excess of what can safely be absorbed by the farm's soil, it is often shipped to neighboring farms for use as fertilizer. Unfortunately, manure is quite heavy, so transporting it both consumes large amounts of fuel (needed to power the trucks that haul it) while at the same time contributing to air pollution (due to emissions from the trucks that haul it).

Once the manure arrives at its destination, it is sprayed onto farm fields as fertilizer. Under the current system of animal production, however, there is always more manure available than can possibly be absorbed by the soil as fertilizer. In fact, studies show that between 1982 and 1997, as industrial agriculture grew, the United States experienced a 64% increase in the amount of manure that could not be absorbed by its soils (Lewthwaite, 2010). This practice is not only harmful to the soil, but can also result in contamination of human drinking water and vegetables grown for human consumption and thus lead to serious public health problems.

While many people believe animal manure is harmless, it, in truth, can be quite hazardous (Keener et al., 2013). Factory livestock facilities pollute the air by releasing over 400 separate gases, mostly due to the large amounts of manure they produce. The principal gases released are hydrogen sulfide, methane, ammonia, and carbon dioxide. Gases can be dangerous air pollutants that threaten both the environment and human health.

Methane and carbon dioxide, for instance, are important greenhouse gases. Nitrous oxides are also released in large quantities from farms through manure application and are among the leading causes of acid rain (Van der Weerden et al., 2014).

Risks of lagoon leakage, overflows, and illegal discharge of waste also pose a direct threat to the quality of soil and water systems. A report for the US Geological Survey documented over 1000 spills and dumps of animal waste in the 10 mid-western states it surveyed over the course of 3 years.

Manure from leaky lagoons or saturated farm fields has also been known to enter public water sources and infect humans with potentially harmful microbes. For example, a study of waterborne disease outbreaks from 1986 to 1998 conducted by the Centers for Disease Control demonstrated that in every case where pathogens could be identified, it most likely originated from livestock (Bicudo and Goyal, 2008).

Among the many nutrients usually present in high concentrations in animal waste are phosphorous and nitrogen, which are beneficial to the soils when the manure is added in small concentrations. However, the volume of manure usually found in lagoons and storage systems, and their very high concentrations of nutrients can cause a range of ecological problems, such as fish kills or a loss in biodiversity, when released into the environment and can affect human health when leached into drinking water.

Nitrogen in manure is tied up in its organic state until, through decomposition by nitrogen-fixing bacteria, it is converted to a soluble form (ammonium nitrate) (Raviv et al., 2013). When ammonium nitrate is mixed with water, nitrates can leach into groundwater systems and threaten the water quality.

According to the Environmental Protection Agency (EPA), drinking water with nitrate concentrations above 10 parts per million can cause developmental deficiencies in infants and death in severe cases due to oxygen deprivation (Zhang et al., 2009). Nitrates introduced into the body through affected water significantly reduce the blood's oxygen-carrying capacity and deprive the body of oxygen. High nitrate concentrations are also believed to have caused spontaneous abortions and possibly cancer.

Little known to the public, 3 million gallons of cow manure spilled from a ruptured tank on a 3000-head dairy farm in upstate New York during August of 2005, spilling into the Black River and polluting an area one-fourth the size of the Exxon Valdez oil spill. While the New York State Department of Environmental Conservation cited the farm for numerous environmental and permit violations, it is estimated that the spill still caused the deaths of 200,000 to 250,000 fish.

The storage of animal waste under industrial livestock facilities and in manure tanks also poses a direct health risk to both animals and humans. Since animal waste is often stored directly beneath the barns in which livestock live, livestock commonly die from poor ventilation that allows for the buildup of toxic gases inside confinement facilities.

In addition, manure pits have been known to claim the lives of farm workers. Between 1992 and 1997, at least 12 workers died due to asphyxiation by manure gases and drowning while trapped in manure lagoons (Beaver et al., 2008). The gases in livestock facilities can also pose other risks to workers; for example, methane is highly flammable, and if not vented properly from manure tanks, it can cause explosions. As pointed out earlier, it is also an important greenhouse gas.

Until recently, there has been very little regulation of animal wastes. Federal law changed in 2002 to require virtually all confined animal feeding operations to apply for National Pollutant Discharge Elimination System permits for their waste discharge (Collentine, 2007). In 2005, the rule was revised, requiring only confined operations that discharge waste (i.e., into streams, rivers, and lakes) to apply for a permit. However, these new regulations make it clear that Federal law prohibits any confined operations from discharging waste, even accidentally, without a permit, and they will face fines if this is violated. Additionally, confined operations applying for this permit are now required to submit a nutrient management plan with the permit application that is open for public review.

This new regulation makes nutrient management a Federal regulation (no longer leaving it up to the states to enforce) and requires a plan of action for the management of waste to ensure that no waste is discharged from the confined operation site. The EPA has the right to prosecute those who discharge animal wastes illegally under the Clean Water Act, although these cases are brought up infrequently.

Apart from regulation, there are some other innovations that may help control the potential problems associated with animal waste. Researchers have discovered that adding sodium carbonate—a mineral commonly found in laundry detergents—to manure can dramatically decrease the amount of harmful *E. coli* bacteria present. There are also feed additives for cattle—including one derived from a type of seaweed that is already widely used in human foods and cosmetics—that can significantly reduce the amount of this dangerous strain of *E. coli* in cattle manure.

Another proven and simple way to reduce the presence of *E. coli* in cattle manure is the method of sending them out to graze on pasture, and taking them off of industrial feed made of corn and other grains (Rasmussen and Casey, 2008). Of course, this would lead to further destruction of habitat to provide pastureland for such animals.

While feed additives are a creative way to address some problems, ultimately they do nothing to address the fact that too much waste is being produced in areas that are too concentrated. Eliminating *E. coli* bacteria does nothing to address the problems of harmful gases or the detrimentally high concentrations of manure (and therefore nitrogen and phosphorous) have on the environment and human health. While methane digesters can partially reduce the discharge of harmful gases, they cannot eliminate the solid waste that still must be stored and discharged, nor do they protect against leaks or overflows that can contaminate water supplies.

OTHER ANIMAL-RELATED PROBLEMS

There are many other problems associated with the large number of animals that are being used by humans, and although they do not relate directly to dominance and aggression, they point out the results of overpopulation by the most dominant organism on Earth. A very important summary of these problems may be found in the online article by G. Koneswaran and D. Neirenberg called *Global Farm Animal Production and Global Warming: Impacting and Mitigating Climate Change* (2008), where they point out that, "The farm animal sector is the single largest anthropogenic user of land, contributing to many environmental problems, including global warming and climate change. Immediate and far-reaching changes in current animal agriculture practices and consumption patterns are both critical and timely if greenhouse gases (GHGs) from the farm animal sector are to be mitigated."

Because of the importance of what Koneswaran and Neirenberg have to say, I have included the material here much in its entirety (although their references to other sources of information have been removed). In light of what is said, we should remember the numbers of animals necessary for human consumption, the amount of greenhouse gas production, and the deforestation required to provide pasture for their growth.

In reading the following material, we may also want to reevaluate our desires to cut down more trees to grow material with which we can produce biofuels. This will also contribute to climate change in a variety of ways.

"Although much evidence has been amassed on the negative impacts of animal agricultural production on environmental integrity, community sustainability," they say, "public health and animal welfare, the global impacts of this sector have remained largely underestimated and underappreciated. In a recent review of the relevant data, H. Steinfeld et al. (2006) calculated the sector's contributions to global greenhouse gas emissions and determined them to be so significant that—measured in carbon dioxide equivalent—the emissions from the animal agricultural sector surpass those of the transportation sector."

GLOBAL WARMING AND CLIMATE CHANGE

The three main (greenhouse gases, GHGs) are CO_2, methane (CH_4), and nitrous oxide (N_2O). Although most attention has been focused on CO_2, (it has been found that) methane and N_2O—both extremely potent GHGs—have greater global warming potentials (GWPs) than does CO_2. By assigning CO_2 a value of 1 GWP, the warming potentials of these other gases can be expressed on a CO_2-equivalent basis: CH_4 has a GWP of 23, and N_2O has a GWP of 296.

"Many impacts of global warming are already detectable. As glaciers retreat, the sea level rises, the tundra thaws, hurricanes and other extreme weather events occur more frequently and penguins, polar bears and other species struggle to survive, experts anticipate even greater increases in the intensity and prevalence of these changes as the 21st century brings rises in GHG emissions." According to a NASA report, "The five warmest years since the 1890s were 1998, 2002, 2003, 2004, and 2005. Indeed, average global temperatures have risen considerably, and the Intergovernmental Panel on Climate Change predicts increases of 1.8–3.9 °C (3.2–7.1 °F) by 2100. These temperature rises are much greater than those seen during the last century, when average temperatures rose only 0.06°C (0.12°F) per decade. Since the mid-1970s, however, the rate of increase in temperature rises has tripled." The latest report by the Intergovernmental Panel on Climate Change (IPCC) warns that climate change "could lead to some impacts that are abrupt or irreversible."

ANTHROPOGENIC INFLUENCES

"Although some natural occurrences contribute to GHG emissions, the overwhelming consensus among the world's most reputable climate scientists is that human activities are responsible for most of this increase in temperature." The IPCC concluded with high confidence that "anthropogenic warming over the last three decades has had a discernible influence on many physical and biological systems"[5]:

> Although transportation and the burning of fossil fuels have typically been regarded as the chief contributors to GHG emissions and climate change, a 2006 Food and Agriculture Organization report, *Livestock's Long Shadow: Environmental Issues and Options*, highlighted the substantial role of the farm animal production sector. Identifying it as "a major threat to the environment," the FAO found that the animal agriculture sector emits 18%, or nearly one-fifth, of human-induced GHG emissions, [which is] more than the transportation sector.

Carbon Dioxide Emissions From Pastured Animals

Koneswaran and Neirenberg state that regarded as the most important GHG, CO_2 has the most significant direct-warming impact on global temperature because of the sheer volume of its emissions. Of all the natural and human-induced influences on climate over the past 250 years, the largest is due to increased CO_2 concentrations attributed to burning fossil fuels and deforestation.

The animal agriculture sector accounts for about 9% of total CO_2 emissions, which are primarily the result of fertilizer production for feed crops, on-farm energy expenditures, feed transport, animal product processing and transport, and land use changes. Burning fossil fuels to produce fertilizers for feed crops may emit 41 million metric tons of CO_2 per year.

Vast amounts of artificial nitrogenous fertilizer are used to grow farm animal feed, primarily composed of corn and soybeans. Most of this fertilizer is produced in factories dependent on fossil-fuel energy. The Haber–Bosch process, which produces ammonia in order to create nitrogen-based artificial fertilizer, is used to produce 100 million metric tons of fertilizer for feed crops annually.

An additional 90 million metric tons of CO_2 per year may be emitted by fossil fuels expended for intensive confinement operations. Energy uses in these industrial facilities differ substantially from those in smaller-scale, extensive, or pasture-based farms. Although a large portion of the energy used for intensive confinement operations goes toward heating, cooling, and ventilation systems, more than half is expended by feed crop production, specifically to produce seeds, herbicides, and pesticides, as well as the fossil fuels used to operate farm machinery in the production of feed crops.

According to the FAO's estimates, CO_2 emissions from farm animal processing total several tens of millions of metric tons per year. The amount of fossil fuels burned varies depending on the species and type of animal product. For example, processing 1 kg of beef requires 4.37 mega-joules (MJ), or 1.21 kW-hours, and processing 1 dozen eggs requires >6 MJ, or 1.66 kW-hours.

That same 1 kg of beef may result in GHGs equivalent to 36.4 kg of CO_2, with almost all the energy consumed attributed to the production and transport of feed. About 0.8 million metric tons of CO_2 are emitted annually from the transportation of feed and animal products to the places where they will be consumed.

Farm animals and animal production facilities cover one-third of the planet's land surface, using more than two-thirds of all available agricultural land including the land used to grow feed crops. Deforestation, land degradation, soil cultivation, and desertification are responsible for CO_2 emissions from the livestock sector's use of land.

Animal agriculture is a significant catalyst for the conversion of wooded areas to grazing land or cropland for feed production, which may emit 2.4 billion metric tons of CO_2 annually as a result of deforestation. This sector has particularly devastated Latin America, the region experiencing the largest net loss of forests and greatest releases of stored carbon into the atmosphere, resulting from disappearing vegetation. One of the chief causes of Latin America's deforestation is cattle ranching.

Other important ecosystems are also threatened by increasing farm animal populations. Brazil's Cerrado region, the world's most biologically diverse savanna, produces half of the country's soy crops. As noted by the WWF, the region's animal species are competing with the rapid expansion of Brazil's agricultural frontier, which focuses primarily on soy and corn. Ranching is another major threat to the region, as it produces almost 40 million cattle a year.

Farm animal production also results in releases of up to 28 million metric tons of CO_2/year from cultivated soils. Soils, like forests, act as carbon sinks and store more than twice the carbon found in vegetation or in the atmosphere.

Human activities, however, have significantly depleted the amount of carbon sequestered in the soil, contributing to GHG emissions:

> Desertification, or the degradation of land in arid, semiarid, and dry subhumid areas, is also exacerbated and facilitated by the animal agriculture sector. By reducing the productivity and amount of vegetative cover, desertification allows CO_2 to escape into the atmosphere. Desertification of pastures due to animal agriculture is responsible for up to 100 million metric tons of CO_2 emissions annually.[6]

Nitrogen from Fertilizer and Food Production

Feeding the global population of livestock requires at least 80% of the world's soybean crop and more than one-half of all corn, a plant whose growth is especially dependent on nitrogen-based artificial fertilizers. Natural sources of fixed nitrogen, the form easily available as fertilizer for plants, are limited, necessitating artificial fertilizer production. Before the development of (a process called) the Haber–Bosch process, the amount of sustainable life on Earth was restricted by the amount of nitrogen made available to plants by bacteria and lightning. Modern fertilizer manufacturing, heavily reliant on fossil fuels, has taken a once-limited nutrient and made it available in massive quantities for crop farmers in the industrialized world and, increasingly, the developing world.

According to Elizabeth Holland, a senior scientist with the National Center for Atmospheric Research, "changes to the nitrogen cycle are larger in magnitude and more profound than changes to the carbon cycle…, but the nitrogen cycle is being neglected."

In addition, the co-chairs of the Third International Nitrogen Conference highlighted the role of farm animal production in the Nanjing Declaration on Nitrogen Management, a statement presented to the United Nations Environment Programme, recognizing that a growing proportion of the world's population consumes excess protein and calories, which may lead to human health problems. The associated production of these dietary proteins (especially animal products) leads to further disturbance of the nitrogen cycle:

> According to Vaclav Smil, a nitrogen cycle expert at the University of Manitoba, "we have perturbed the global nitrogen cycle more than any other, even carbon. Indeed, the overwhelming majority of all crops grown in the industrialized world are nitrogen-saturated, and overuse of nitrogen in crop production, nitrogen runoff into waterways, and the millions of tons of nitrogen found in farm animal manure threaten environmental integrity and public health."[7]

Methane

The animal agriculture sector is also responsible for 35–40% of annual anthropogenic methane emissions that result from enteric fermentation in ruminants and from farm animal manure. Methane emissions are affected by a number of factors, including the animal's age, body weight, feed quality, digestive efficiency, and exercise.

Ruminants emit methane as part of their digestive process, which involves microbial (enteric) fermentation. Although individual animals produce relatively small amounts of methane, the >1 billion ruminants reared annually amount to a significant methane source. Indeed, enteric fermentation generates about 86 million metric tons of methane emissions worldwide.[8]

Typically, cattle confined in feedlots or in intensive confinement dairy operations are fed an unnatural diet of concentrated high-protein feed consisting of corn and soybeans. Although cattle may gain weight rapidly when fed this diet, it can cause a range of illnesses. This diet may also lead to increased methane emissions. The standard diet fed to beef cattle confined in feedlots contributes to manure with a "high methane-producing capacity." In contrast, cattle raised on pasture, eating a more natural, low-energy diet composed of grasses and other forages, produce manure with about half of the potential to generate methane.

Farm animals produce billions of tons of manure, with confined farm animals in the United States alone generating about 500 million tons of solid and liquid waste annually. Storing and disposing of these immense quantities of manure can lead to significant anthropogenic emissions of methane and N_2O. For example, according to the Pew Center on Global Climate Change, farm animal manure management accounts for 25% of agricultural methane emissions in the United States and 6% of agricultural N_2O emissions. Globally, emissions from pig manure alone account for almost half of all GHG emissions from farm animal manure:

> Farm animal manure is the source of almost 18 million metric tons of annual methane emissions. Between 1990 and 2005 in the United States, methane emissions from dairy cow and pig manure rose by 50% and 37%, respectively. The US EPA traces this increase to the trend

toward housing dairy cows and pigs in larger facilities that typically use liquid manure management systems, which were first in use in the 1960s but are now found in large dairy operations across the United States and in some developing countries, as well as in most industrial pig operations worldwide.

Nitrous Oxide

Although 70% of anthropogenic emissions of N_2O result from crop and animal agriculture combined, farm animal production, including growing feed crops, accounts for 65% of global N_2O emissions. Manure and urine from farm animals, once deposited on the soil, emit N_2O; in the United States, a 10% rise in N_2O emissions between 1990 and 2005 can be traced, in part, to changes in the poultry industry, including an overall increase in the domestic stock of birds used for meat and egg production.

Conflict, Hunger, and Disease

As is the case with animal agriculture's impacts on soil, water, and air quality, the sector's contributions to climate change cannot be viewed in a vacuum. Climate change is having far-reaching consequences, perhaps most startlingly seen in growing conflicts among pastoral communities. Environmental degradation has been cited as one of the catalysts for ongoing conflicts in Darfur and other areas of Sudan, where the effects of climate change have led to untenable conditions. As temperatures rise and water supplies dry up, farmers and herders are fighting to gain and control diminishing arable land and water.

The UNEP (United Nations Environment Program) tied two of its critical concerns in Sudan—land degradation and desertification—to an explosive growth in livestock numbers. In addition to citing climate change as one factor that led to the Darfur conflict, United Nations Secretary-General Ban Ki-moon has noted that natural disasters, droughts, and other changes brought about by global warming are likely to become a major driver of war and conflict.

According to the IPCC, many areas already suffering from drought will become drier, exacerbating the risks of both hunger and disease. By 2020, up to 250 million people may experience water shortages, and, in some countries, food production may be cut in half. By 2050—the same year by which the FAO projects that meat and dairy production will double from present levels, primarily in the developing world—130 million people in Asia may suffer from climate change–related food shortages:

> Global temperature shifts may also hasten the speed at which infectious diseases emerge and reemerge. According to Francois Meslin of the World Health Organization, 'the chief risk factor for emerging zoonotic diseases is environmental degradation by humans, particularly deforestation, logging, and urbanization.' The clear-cutting of forests for soybean cultivation, logging, and other industries enables viruses to exploit such newly exposed niches.

Strategies and Next Steps

It has been stressed that "mitigating the animal agriculture sector's contributions to climate change necessitates comprehensive and immediate action by policy makers, producers and consumers. Enhanced regulation is required in order to hold facilities accountable for their GHG emissions. One critical step is accurately pricing environmental services—natural resources that are typically free or underpriced—leading to 'overexploitation and pollution.'"

Thus far, most mitigation and prevention strategies undertaken by the animal agriculture sector have focused on technical solutions. For example, researchers are investigating the reformulation of ruminant diets to reduce enteric fermentation and some methane emissions. One such remedy is a plant-based bolus, formulated to reduce excessive fermentation and regulate the metabolic activity of rumen bacteria to reduce methane emissions from both the animals and their manure (Flachowsky and Meyer, 2015).[9]

The USDA and US EPA assist in funding anaerobic digester projects domestically and abroad. These digesters, now in use at some large-scale intensive confinement facilities, capture methane from manure to use as a source of energy, but are typically not economically viable for small-scale farms.

In addition, producers are burning animal waste for fuel. The world's foremost pig producer, Smithfield Foods (Smithfield, VA), and one of the top poultry producers, Tyson Foods (Springdale, AR), are both using animal by-product fats to create biofuels.

McDonald's (Oak Brook, IL) and agribusiness giant Cargill (Wayzata, MN), which was supplying McDonald's with soy for use as chicken feed, recently entered into an agreement with Brazil's other chief soy traders. Engineered by international environmental organization Greenpeace, a 2-year moratorium was enacted in 2007 to prevent purchases of soy from Brazil's newly deforested areas.

As consumers increasingly favor more environmentally friendly products and techniques, reducing consumption of meat, eggs, and milk, as well as choosing more sustainably produced animal products, such as those from organic systems, may prove equally critical strategies. Indeed, organic farming has the potential to reduce GHG emissions and sequester carbon. Also, raising cattle for beef organically on grass, in contrast to fattening confined cattle on concentrated feed, may emit 40% less GHGs and consume 85% less energy than conventionally produced beef.

However, there remains an immediate need for more research regarding both technical and less technology-dependent strategies to record existing GHG emissions from individual production facilities and to provide lessons to producers and policy makers for reducing the climate-damaging impacts of animal agriculture.

Given the urgency for global action—calls echoed by scientists and world leaders alike—individual consumers must also participate. McMichael et al. put forth several recommendations, including the reduction of meat and milk intake by high-income countries as "the urgent task of curtailing global greenhouse-gas emissions necessitates action on all major fronts"; they concluded that, for high-income countries, "greenhouse-gas emissions from meat-eating warrant the same scrutiny as do those from driving and flying":

> As the numbers of farm animals reared for meat, egg, and dairy production increase, so do emissions from their production. By 2050, global farm animal production is expected to double from present levels. The environmental impacts of animal agriculture require that governments, international organizations, producers, and consumers focus more attention on the role played by meat, egg, and dairy production. Mitigating and preventing the environmental harms caused by this sector require immediate and substantial changes in regulation, production practices, and consumption patterns.

Based on this frightening summary of world environmental conditions that arise from the overabundance of agricultural animals and projections for the future, I would like to reemphasize that it is the high numbers of humans on Earth who require the animals that are causing these environmental problems (Unno, 2014). What appears to be a simple problem of raising cattle and other animals for meat and milk production has far-reaching results, involving deforestation, the production of massive amounts of greenhouse gases, and global climate change. However, the problem does not end there. In addition to making Earth changes to produce and maintain this large animal population, humans more directly affect the homeostasis of Earth in numerous other ways, including the production of their own wastes which result in additional environmental pollution and rising health concerns.

Disposal of Human Wastes

Land-based water pollutants that flow into streams, rivers, and eventually into marine waters range from petroleum wastes to pesticides, fertilizers, and excessive amounts of animal and industrial waste products. Marine waters also receive wastes directly from offshore activities, such as ocean-based dumping (e.g., from ships and offshore oil and gas operations) (Anderson and Gurnham, 2009; Ogg and Menczel, 1981).

One marine water pollutant, human sewage, largely consists of excrement from toilets, wastewater from bathing, laundry, and dishwashing, as well as animal and vegetable matter from food preparation that is disposed of through in-sink garbage disposal systems (Bolzonella et al., 2008). Because coasts around the world represent the most densely populated areas, the amount of sewage reaching marine waters is of particular concern because some substances contained in it can seriously harm ecosystems and pose a significant public health threat. In addition to nutrients that cause overenrichment in bodies of water that receive them, sewage carries an array of pathogens (disease-causing microbes).[10]

According to M.D. Sobsey et al. in an article entitled Pathogens in Animal Wastes and the Impacts of Waste Management Practices in their Survival, Transport and Fate (2012), "animal pathogens posing potential risks to human health include a variety of viruses (e.g., swine hepatitis E virus), bacteria (e.g., *Salmonella* species), and eukaryote parasites (e.g., *Cryptosporidium parvum*)," some of which are endemic in commercial livestock and difficult to eradicate from both the animals and their production facilities. Hence, pathogens in animal manure and other wastes pose potential risks to human and animal health both on and off animal agriculture production facilities if the wastes are not adequately treated and contained. There are also growing public health concerns about the high concentrations of antibiotic-resistant bacteria in agricultural animals resulting from the therapeutic and growth-promotion use of antibiotics in animal production.

The report reviews: (1) the types of pathogens potentially present in the manure of swine and other agricultural animals, (2) the levels of some important microbial pathogens and indicators for them that have been detected in animal wastes, (3) the potential for off-farm release or movement of pathogens present in manure and other wastes under current or proposed management practices, and (4) the extent to which these pathogens are reduced by currently used and candidate manure treatment and management technologies.

In parts of the United States, concern is increasing due to outbreaks of coliform bacteria, giardiasis, cryptosporidiosis, and hepatitis A. Some of these are bacteria, while others are viruses or organisms known as protistans. For example, *Giardia* is a genus of flagellated protistans that are parasitic in the intestines of humans and other animals. Another protistan, *Cryptosporidium parvum*, is a parasite that causes gastrointestinal illness in humans and is commonly isolated from HIV patients. Hepatitis A is an enteric virus that causes infectious hepatitis, and it can be transferred through contaminated water. Coliform bacteria live in soil or vegetation and in the gastrointestinal tract of many animals:

> These organisms enter water supplies from the disposal of waste into streams or lakes, or from runoff from wooded areas, pastures, feedlots, septic tanks, and sewage plants into streams or groundwater. In addition, coliform bacteria can enter an individual's house via backflow of water from a contaminated source, carbon filters, or leaking well caps that allow dirt and dead organisms to fall into the water.

Let us take a quick look at land-based pollutants. As pointed out earlier, "animal wastes from feedlots and other agricultural operations (e.g., manure-spreading on cropland) pose health concerns similar to those of human wastes mainly because of their microbial composition." Manure is also loaded with the eggs of parasitic flatworms and roundworms. Just as inland rivers, lakes and groundwater can be contaminated by pathogens, so can coastal waters. But it is not just pathogens that are a threat in animal wastes. Chemical and radioactive forms of pollution are dangerous as well.

The major types of marine pollutants from industrial sources can be generally categorized as petroleum, as well as hazardous, thermal, and radioactive wastes. Petroleum products are oil and oil-derived chemicals used for fuel, manufacturing, making plastics, and many other purposes. Hazardous wastes are chemicals that are toxic (poisonous at certain levels), reactive materials which are capable of producing explosive gases, corrosive materials which are able to even corrode steel, or ignitable materials which are flammable. Thermal wastes are heated wastewaters, typically from power plants and factories, where water is used for cooling purposes. Radioactive wastes contain chemical elements that have an unstable nucleus that will spontaneously decay and emit ionizing radiation.

Sewage is waste material that originates primarily from domestic, commercial, and industrial sources. In developed countries, such wastes typically are delivered either to onsite septic systems or to centralized sewage treatment facilities. In both methods, sewage is treated before it is discharged, either underground (in the case of septic tanks) or to receiving bodies of water (in the case of sewage treatment plants). Receiving bodies of water are typically streams, rivers, or coastal outlets.

While sewage treatment facilities are designed to accommodate and treat sewage from their service area, partly treated or even untreated sewage sometimes is accidently discharged into the environment. Factors responsible for such discharges include decayed infrastructure of the facility's holding tanks, facility malfunctions, or heavy rainfall events which overwhelm systems using combined sewers and storm-water drains:

> In areas lacking a sewage system, improperly designed or malfunctioning septic tanks can contaminate groundwater and surface water, including coastal waters (e.g., in the dead zone of the Gulf of Mexico off the coast of Texas, Louisiana, and Mississippi). In some developed areas, raw sewage continues to pour into harbors, bays, and coastal waters. In developing countries with no onsite or centralized sanitation facilities, no opportunity exists for any type of treatment, and human wastes go directly into surface waters, including marine coastal ecosystems.

Sewage Sludge

Disposal of biosolids, semisolid by-products of the sewage treatment process often referred to as sludge, is another source of ocean pollution by sewage-related waste. Historically, sludge in developed nations was disposed of in coastal waters: New York's 20 sewage treatment plants, for example, once disposed of their sludge offshore in a region known as the New York Bight. As recently as the 1980s, for example, the New York Bight was essentially lifeless due to oxygen depletion (hypoxia), caused largely by decades of sewage and sludge disposal. As of 2002, Halifax Harbor was still receiving a daily influx of raw sewage, creating serious ecological and public health concerns:

> Although today's environmental regulations in the United States prohibit this practice, the system is often abused, and sewage sludge is still typically disposed of in marine waters by some countries. Disease-causing microbes represent the primary human health risk in sewage-contaminated waters, and it is also the main cause of recreational beach closures.

Human Health

Sewage, particularly if partially treated or untreated, brings high microbe concentrations into the ocean. Human diseases can be caused by waterborne pathogens that contact the skin or eyes, or when they are accidentally ingested when water is swallowed. They can also be caused by pathogens found in the tissues of fish and shellfish that are consumed as seafood.

Beach pollution consequently is a persistent public health problem. Annually, thousands of swimming advisories and beach closings are experienced because high levels of disease-causing microbes are found in the water. Sewage often is responsible for these harmful microbial levels:

> Seafood contaminated by sewage-related pathogens sickens untold numbers of people worldwide. Regulatory agencies will close a fishery when contamination is detected, but many countries lack regulatory oversight or the resources to adequately monitor their fisheries.

Industrial Wastes

Industrial wastes primarily enter coastal waters from terrestrial (land-based) activities. Industries, like municipalities and other entities that generate wastes, dispose of many liquid wastes through wastewater systems (and ultimately to water bodies), whereas they dispose of their solid wastes in landfills.

The amount and type of industrial wastewater depends on the type of industry that releases it, its water and wastewater management, and the type of waste pretreatment (if any) before delivery to a wastewater (sewage) treatment plant. Because industrial waste frequently goes down the same sewers as domestic and commercial nonindustrial waste, sewage often contains high levels of industrial chemicals and heavy metals (e.g., lead, mercury, cadmium, and arsenic).

Chemicals not removed by wastewater treatment processes are discharged via the treated effluent to a receiving stream, river, or coastal outlet. Inland waters ultimately reach the ocean, carrying with them some residual chemicals that are not attenuated, stored, or degraded during their journey through the watershed. Other land-based sources of industrial pollutants in the ocean are pipeline discharges, transportation accidents, leaking underground storage tanks, and activities at ports and harbors. Intentional, illegal dumping in inland watersheds and in inland bodies of water also can deliver industrial wastes to drainage-ways, and ultimately to the ocean.

Some industries in coastal watersheds discharge their wastes directly into the ocean. As with industries located inland, these industries must first obtain a permit as part of the Clean Water Act. Industrial pollutants also can directly enter the ocean through accidental spills or intentional dumping at sea.

Wet and dry deposition of airborne pollutants is a sometimes overlooked, yet significant, source of oceanic pollution. For example, sulfur dioxide that originates from a factory smokestack begins as air pollution. Polluted air mixes with atmospheric moisture to produce airborne sulfuric acid which, like nitric oxide, falls on water and land as acid rain. This deposition can change the chemistry and ecology of an aquatic ecosystem. For example, polychlorinated biphenyls (PCBs), some of the major transport chemicals that end up in the ocean, enter it through airborne deposition:

> Industrial chemicals can adversely affect the growth, reproduction, and development of marine animals. Pollutants of various types are appearing not only in the Pacific, Atlantic, and Indian Oceans, and their marginal seas, but also in the more remote and once-pristine polar oceans. An array of contaminants has been found in the flesh of fish and marine mammals in polar regions. In addition to the environmental and ecological issues, there is growing concern over potential human health impacts in aboriginal communities whose residents depend on fish and marine mammals for daily sustenance.

A major public health concern is the safety of seafood as it relates to chemical pollution in waters used for commercial and recreational fishing and mariculture (a specialized branch of aquaculture involving the cultivation of marine organisms for food and other products in the open ocean, an enclosed section of the ocean or in tanks, ponds, or raceways that are filled with seawater). "Heavy metals (e.g., copper, lead, mercury, and arsenic) can reach high levels inside marine animals and subsequently be passed along in seafood to humans. A well-known case of human poisoning occurred in Japan, where one industry dumped mercury compounds into Minimata Bay from 1932 to 1968 (Lech and Goszcz, 2008). Methyl mercury that accumulated in fish and other animals was subsequently passed along to humans who consumed them. Over 3000 human victims and an unknown number of animals succumbed to what became known as "Minimata Disease," a devastating illness that affects the central nervous system":

> Monitoring by fisheries, environmental and public health agencies can prevent or minimize cases of human illness caused by chemical contaminants in seafood. Some shellfish-producing areas off US coasts have been either permanently closed or declared indefinitely off-limits by health officials as a result of this type of pollution. A large percentage of US fish and shellfish-consumption advisories are due to abnormally high concentrations of chemical contaminants in seafood.

Increasing Population

When we look into our forests and fields and recognize that we are setting aside wildlife preserves to protect part of our wildlife, we get the feeling that although there are problems in the world, there is plenty of land upon which to continue building and expanding the human population (Crompton, 2011). However, it is clear from the information provided in this chapter (based on reports by our leading scientists) that most of our problems stem directly or indirectly from human overpopulation, and we do not have to reach a point where all land is used by humans before we feel the effects of overpopulation.[11]

Population Growth and Environmental Carrying Capacity

Let us look further at the present picture of the world, its dominant inhabitants, and how the human animal utilizes its resources. During March, 2010, the United States Census Bureau reported Earth's human population to be 6.8 billion (Falek and Konner, 2014). It is currently over 7 billion. As our numbers increase, our resources and other biological (nonhuman) forms decrease their numbers.

According to an article at worldpopulationbalance.com, "current global population of over 7 billion is already two to three times higher than the sustainable level. Several recent studies show that Earth's resources are enough to sustain only about 2 billion people at a European standard of living. If all of the world's 7 billion people consumed as much as an average American, it would take the resources of over five Earths to sustainably support all of them. On average, each American uses nearly 20 acres of biologically productive land and water (biocapacity) per year." At an American standard of living and consumption, "Earth's 29.6 billion acres of biologically productive land and water could sustainably support only about 1.5 billion people."

The world's human population has been growing at a more-or-less continuous pace since the end of plague epidemics around the year 1400. Subsequently, humans commenced invading foreign lands and building populations in their new homes. We have gone through periodic spurts of population growth that resulted in what we should rightly consider an already-overpopulated planet. Although it is true that world population growth in recent years has shown some signs of slowing in certain areas, it is obvious that there are already way too many people and domestic animals in the world. Population growth is not slowing enough. Based on the damage we have already inflicted on our planet, it is quite apparent that we need to not only slow population growth down but show a negative growth rate.

The fastest rates of world population growth (increases above 1.8% per year) were during the 1950s and subsequently during the 1960s and 1970s. The 2008 rate of growth has almost been cut in half since its peak of 2.2% per year, which was reached in 1963.

On a worldwide scale, annual births (as reported in 2008) have reached a brief plateau of about 134 million per year since their peak of 163 million in the late 1990s. Some population experts expect population growth to remain relatively constant. However, with deaths around 57 million per year, and an expectation of an increase to 90 million by the year 2050, the world's population should reach 9 billion between 2040 and 2050.

The rapid increase in human population over the course of the 20th century has raised concerns about how many people the world can support (our environmental carrying capacity) (Shaofeng, 2013). And let us remember that with human population increases comes the need for an increase in forest destruction, use of fertilizers and agricultural animals, and all the associated environmental problems that accompany them. Some say we are approaching overpopulation status, while others have argued (and the data provided here strongly supports their arguments) that we have already long surpassed the overpopulation mark (as indicated earlier by the statement made by worldpopulationbalance.com). It is clear to the scientific community that the current population size, its degree of expansion, the pollution and damage that results from it, and an accompanying increase in usage of resources are linked to significant destruction of the world ecosystems, all resulting from overpopulation (Falek and Konner, 2014).

Increases in population, improved medical attention and drugs, comfort, and technological skills are all beneficial features to humans. Many of us are fortunate to be in a comfort zone of life, but living in a comfort zone and disregarding the consequences of overpopulation result in certain overbearing demands on the world by humans.

How comfortable do we need or want to be? We, as the most dominant animals that have ever lived on Earth, are forced to stretch the limits of what we can take and still attempt to maintain harmony. Let us examine some other indications that we are stretching and have already stretched these limits beyond our environmental (ecological) carrying capacity?

Genetically Modified Organisms and the World Population

Overpopulation has been on the minds of the world leaders for many years. In *War Against the Weak*, Edwin Black states that "in 1798, English economist Thomas Malthus published a watershed theory on the nature of poverty and the controlling socioeconomic systems at play. Malthus reasoned that a finite food supply would naturally inhibit a geometrically expanding human race. He called for population control by moral restraint. He even argued that in many instances charitable assistance promoted generation-to-generation poverty and simply made no sense in the natural scheme of human progress."

Thus, began the concept of eugenics (Susanne, 2014; Queiroz, 2014). "In the 1850s, agnostic English philosopher Herbert Spencer popularized a powerful new term: survival of the fittest. Through evolution," he said, "the fittest would naturally continue to perfect society. The unfit would naturally become more impoverished, less educated and ultimately die off." His conclusion was that "all imperfection must disappear."

Charles Darwin published his *Origin of Species* in 1859. He, of course, "was writing about a natural world, distinct from man. But it wasn't long before leading thinkers were distilling the ideas of Malthus Spencer and Charles Darwin into a new concept, bearing a name never used by Darwin himself: social Darwinism."

Unlike Darwinism, social Darwinism and eugenics led to thoughts of artificial selection in the human species, on the order of breeding domestic animals, incorporating the perfection of a superior race and the elimination of social misfits. While eugenicists were using techniques to kill and perform surgical sterilization, the war against the weak subsequently entered a new phase which, according to F. William Engdahl in *Seeds of Destruction*, accomplishes the same goals through the distribution of genetically modified plants (Rogers, 2011).

In spite of the enthusiasm of eugenicists and their attempts at the elimination of misfits, other genocides, and frequent wars, the world population continues to grow beyond its carrying capacity, possibly at three or four times the sustainable level outlined by worldpopulationbalance.com. Population control appears to be the only logical answer to the overpopulation dilemma, but it is only being carried out by countries that have become so overpopulated that they are at their absolute limits of sustainability and beyond (Lamé et al., 2013).

According to a recent article in Wikipedia (https://en.wikipedia.org/wiki/Human_overpopulation), some of the problems we are now facing with overpopulation and can expect to accompany further overpopulation will be:

- Inadequate fresh water for drinking as well as discharge treatment and effluent discharge.
- Depletion of natural resources, especially fossil fuels.
- Increased levels of air pollution, water pollution, soil contamination, and noise pollution.
- Loss of arable land and increase in desertification.
- Deforestation and loss of ecosystems that valuably contribute to the global atmospheric oxygen and carbon dioxide balance; about 8 million hectares of forest are lost each year.
- Changes in atmospheric composition and consequent global warming.
- Mass species extinctions.
- High infant and child mortality.
- Intensive factory farming to support large populations.
- Increased chance of the emergence of new epidemics and pandemics.
- Starvation, malnutrition, or poor diet with ill health and diet-deficiency diseases (e.g., rickets).
- Poverty coupled with inflation in some regions and a resulting low level of capital formation.
- Low life expectancy in countries with fastest growing populations.
- Unhygienic living conditions for many based upon water resource depletion, discharge of raw sewage, and solid waste disposal.
- Elevated crime rate due to drug cartels and increased theft by people stealing resources to survive.
- Conflict over scarce resources and crowding, leading to increased levels of warfare.
- Less personal freedom and more restrictive laws.

Is this what the world wants? Do we all desire to continue increasing our population beyond its sustainable limits, to face the problems outlined earlier, and have to depend on genetically modified organisms (GMOs) in an attempt to keep up with food demands?

Reviewing Our Situation

On a global scale, can we accommodate a continuously rising world population? According to the Center for Research on Globalization, "the number of starving people in the world has exceeded 1 billion. Every sixth earthling," they say, "is underfed."

This statement was issued by the UN Food and Agriculture Organization's Director-General, Jacques Diouf. There may be many reasons for this, but overproduction by humans, an inability to grow an adequate amount of food crops in many areas of the world, a dependence on GMOs, environmental degradation, and disregard for biodiversity are extremely important points to consider.

The director warned that "unless urgent measures are taken, the number of those starving may be staggering the world over by the year 2050. The recent St. Petersburg international grain summit took up food security and urgent moves to be made amid the current world crisis."

Because of food demands around the globe, humans have changed their growing habits over the years from utilizing small farms with organic fertilizers to huge corporate farms and artificial fertilizers. In order to keep up with demands, many standard crops have been genetically modified and represent a potential threat to our well-being.

Since recent *E. coli* scares, along with numerous encounters with a long list of culinary pathogens, farms have been forced to use artificial fertilizers over manures on crops that represent human foods. In addition to a lower food value in chemically synthesized inorganic fertilizers, such fertilizers (which have mostly developed during the industrial revolution) have been contaminating fresh and marine waters ever since (as indicated in foregoing statements involving eutrophication of fresh and marine waters).

But eutrophication is not our only problem. The presence of nitrate in groundwater causes other problems as well. Nitrate levels above 10 parts per million in groundwater, for instance, can cause a condition called "blue baby syndrome," which leads to hypoxia within the body (a deprivation of oxygen in tissues, which can lead to coma and death if not treated) (Thorpe and Shirmohammad, 2007).

Nitrogen-containing inorganic and organic fertilizers can cause soil acidification. This may lead to decreases in nutrient availability which must be offset by adding lime.

Uranium (a radioactive heavy metal) is a contaminant often found in phosphate fertilizers (Schnug et al., 2008). Eventually, it can build up to unacceptable levels in the soil and even in vegetable produce. The intake of uranium by adults is both from the ingestion of food and water and simply from breathing contaminated air.

Another highly radioactive isotope (polonium-210) is contained in phosphate fertilizers, absorbed through plant roots and stored in its tissues (Skwarzec et al., 2012). Tobacco, a plant product that is derived from plants fertilized by rock phosphates, contains this isotope, and alpha radiation emitted from it is estimated to cause about 11,700 lung cancer deaths each year worldwide.

Steel industry wastes, recycled into fertilizers for their high levels of zinc (an element that is essential for plant growth) sometimes include toxic metals, for example, arsenic, cadmium, chromium, lead, and nickel (Huang et al., 2013). The most common toxic elements in this type of fertilizer are arsenic, mercury, and lead. Fish-meal mercury has been a definite concern in certain areas.

The use of petroleum-derived fertilizers on a global scale emits significant quantities of greenhouse gas into the atmosphere (Viets and Lunin, 2009). Emissions of greenhouse gases come about through: (1) using animal manures and urea, which release ammonia, carbon dioxide, methane, and nitrous oxide in varying quantities; and (2) fertilizers that use ammonium bicarbonate or nitric acid result in emissions of nitrous oxide, ammonia, and carbon dioxide into the atmosphere. It has been found that excessive nitrogen fertilizer applications can also lead to pest problems by increasing the birth rate, longevity, and overall fitness of certain agricultural pests.

Global warming results from emissions of certain gases from fertilizer. Methane emissions from fields used in growing crops (especially rice paddy), for instance, have increased by the application of ammonium-based fertilizers. Since methane is a potent greenhouse gas, such emissions contribute greatly to global climate change.

By using nitrogen fertilizer, added at a rate of 1 billion tons per year to the already-existing amount of reactive nitrogen, nitrous oxide has become the third-most important greenhouse gas after carbon dioxide and methane. As pointed out earlier, it has a global warming potential 296 times larger than an equal mass of carbon dioxide, and it also contributes to stratospheric ozone depletion, which is a serious threat to all life forms on Earth. Even stored nitrous oxide adds to atmospheric emissions. Ammonia gas (NH_3) may also be emitted following an application of inorganic fertilizers.

Chemicals referred to as toxic persistent organic pollutants (POPs), for example, dioxins, polychlorinated dibenzo-p-dioxins (PCDDs), and polychlorinated dibenzofurans (PCDFs), have been detected in agricultural fertilizers and soil additives. Concentrations of up to 100 mg/kg of cadmium also contaminate the soil.[12]

There are various reasons for starvation in the world, and it would be incorrect to talk about it without discussing how foods are distributed around the world. In certain cases, grain supplies have been overproduced (McLaughlin, 1999). "Wheat stocks worldwide," for instance, "have doubled since 1996, soybean stocks have tripled and corn stocks have quadrupled. Midwest grain elevators are filled to bursting with the unsold portions of the wheat harvests of 1996, 1997, and 1998. When the 1999 wheat harvest comes in, an estimated 1 billion bushels—half the entire crop—will go unsold and have to be stored in temporary outdoor facilities."

Grains are at the base of our food chain. Much of it goes to growing cattle and other herbivorous animals that we eat. While we must admit that our culinary needs are dramatically affecting the numbers of fish and other nature-derived food stores important to feeding a growing population, an overabundance of grain implies that we possibly do have sufficient food to solve the needs of the starving segment of the world population.

What it really means is that the acquisition of food is based on money, and people who get the food belong to subpopulations in countries that have enough money to purchase it. Those in populations in undeveloped countries may either have no money and/or trading power and thus are not in a position to obtain the foods they need.

While it is important to understand this, it must not take our attention off the fact that while grain is abundant, we are nevertheless overharvesting fish and certain other aquatic species that we depend on, as well as obtaining other of Earth's resources and destroying ecosystems in the process. Assuming that we have adequate food stores and will always have them is a dangerous consideration.

The statement about the number of starving people in the world was issued by the UN Food and Agriculture Organization's Director-General, Jacques Diouf. In spite of the amount of stored grain in the United States, the director warned that "unless urgent measures are taken, the number of those starving may be staggering the world over by the year 2050. The recent St. Petersburg international grain summit took up food security and urgent moves to be made amid the current world crisis."

Because of food demands around the globe, humans have changed their growing habits over the years from utilizing small farms with organic fertilizers to huge corporate farms and artificial fertilizers. In order to keep up with demands, many standard crops have been genetically modified and represent a potential threat to our well-being.

There may be other considerations for the production of GMOs. Many of the foods we eat are GMOs. The companies that sell the seeds to grow crops patent their seeds. When farmers grow these crops, they are dependent upon the companies to supply the seeds. Many of the crops may be seedless or the seeds in the food crops are restricted from use because of the patents. Thus, farmers are required to purchase new seeds from these companies whenever they set up their crops.

We also have put ourselves in a position that does not allow us to readily grow organic foods. Organic foods are dependent upon the use of natural fertilizers. Laws have been passed to keep us from growing organically produced foods to be sold because of the possibility of transferring soil-dwelling pathogens to the general public. The only way we can be sure of getting foods that are grown organically is to grow them ourselves or purchase them from the few sources that are available.

Earth's Dwindling Resources

While people are starving and there is a likelihood that more will be starving in the coming years, the world is dramatically overusing its resources and overproducing certain unneeded materials for use. According to Tan Kin Lian, Chief Executive of NTUC Income, "The world is producing more [unnecessary] goods than is really needed."

Overproduction of unnecessary materials has led to the wasteful use of energy and resources and damage to the environment. Through competition, more people are working harder to produce goods that are not really needed. We have too much clothes, electronic equipment, and gadgets and too little time (due to long working hours) to enjoy them.

Value of Trees

Yet, there are many things that we either have an underabundance of or are running out of them. Take trees, for instance. In addition to clearing land for human farming and living space, one internet source lists 442 different products that are made from trees. The publishing field alone claims to use as much as 3 billion trees annually. This, of course, is changing due to the rapidly increasing field of epublishing.

Trees are important photosynthesizers and environmental filters (Saguaro, 2013). Having chlorophyll, absorbed light energy is converted to food which, in turn, is at the bottom of many food chains. Trees determine the type of environment that is found in a particular area (recall the mention of climax vegetation in Chapter 2). Climax vegetation, for instance, defines what other plants and animals are found in an area. They alter the environment by moderating climate, improving air quality, conserving water, and harboring wildlife. Climate control is obtained by moderating the effects of the sun, wind, and rain.

Radiant energy from the sun is absorbed or deflected by the leaves of deciduous trees in the summer and filtered by branches of deciduous trees in winter. Animals are cooler when they stand in the shade of trees and, therefore, are not exposed to direct sunlight.

Wind speed and direction can be affected by trees. The more compact the foliage on the tree or group of trees, the greater the influence of the windbreak. The downward fall of rain, sleet, and hail is initially absorbed or deflected by trees, which provide some protection for people, other animals, and buildings. Trees intercept water, store some of it, and reduce storm runoff and the possibility of flooding.

Dew and frost are less common under trees because less radiant energy is released from the soil in those areas at night. The roots of trees form a stabilizing feature to topsoil, keeping it from blowing away. Temperature in the vicinity of trees is cooler than it is away from trees. The larger the tree, the greater the cooling.

Air quality can be improved because of trees, shrubs, and turf. Leaves filter the air we breathe by removing dust and other particulates. Rain then washes the pollutants to the ground.

Through photosynthesis, carbohydrates are made and later used in the plant's structure and function. In this process, leaves also absorb other air pollutants, for example, ozone, carbon monoxide, and sulfur dioxide, and give off oxygen.

Trees are utilized for an astonishing array of products (Alexander and Grimshaw, 2012). Thus, trees are being lost on a global scale through deforestation (mostly for agriculture and housing, as well as the production of paper) and acid rain (mostly because of industrial pollution). It is not that we cannot change the loss of many trees and improve the environment. I have said in other publications that a change to epublishing alone can revolutionize the publishing industry and at the same time save billions of trees and help the economy in the process.[13]

As dominant, highly cognitive animals, we have the power to help the environment tremendously if we choose to think of the environment first rather than hang on to traditional methods of publishing. Then, consider the equipment, transportation, and power consumption required to get and process those trees, the delivery system necessary after the merchandise is produced, and the pollution generated in their production. We must make an effort to join the publishing revolution and cut down on the number of trees required by the publishing industry?

Overuse of Resources

Overuse of Earth's resources is revealed in a Wikipedia report, expressing energy consumption in terms that are beyond the levels most of us can comprehend: the "total worldwide energy consumption in 2008, for instance, was 474 EJ (474×10^{18} J), with 80–90% derived from the combustion of fossil fuels. This is equivalent to an average power consumption rate of 15 TW (1.504×10^{13} W)."

While much of the Earth's warmth is generated internally, most of the world's energy resources have initially been from the sun's rays hitting Earth. Some of that energy was received millions of years ago and preserved as fossil energy (e.g., coal and petroleum), and some is directly or indirectly usable (e.g., solar, wind, hydro-, and wave power).

Earth receives significant amounts of solar energy. The term solar constant is the amount of incoming solar electromagnetic radiation per unit area, measured on the outer surface of Earth's atmosphere, in a plane perpendicular to the rays.

This constant includes all types of solar radiation in addition to visible light (Yilmaz et al., 2007). Most deleterious electromagnetic waves are absorbed by the ozone layer. It is measured by satellite to be roughly 1366 W per m^2, but it fluctuates by about 6.9% during a year—from 1412 W/m^{-2} in early January to 1321 W/m^{-2} in early July, due to the Earth's varying distance from the sun, and by a few parts per 1000 from day-to-day. For the entire Earth, with a cross-section of 127,400,000 km^2, the total energy rate is 174 PW (1.740×10^{17} W), plus or minus 3.5%. This value represents the total amount of solar energy impinging upon Earth; about half of which (89 PW) reaches the Earth's surface (thanks to the ozone layer):

Estimates of remaining nonrenewable worldwide energy resources vary, with the remaining fossil fuels totaling an estimated 0.4 YJ (1 YJ = 10^{24} J) and the available nuclear fuel such as uranium exceeding 2.5 YJ. Fossil fuels range from 0.6 to 3 YJ if estimates of reserves of methane clathrates are accurate and become technically extractable. Mostly thanks to the Sun, Earth also has a renewable usable energy flux that exceeds 120 PW (8000 times the 2004 total usage), or 3.8 YJ/yr, dwarfing all nonrenewable resources.

Other Pollution

Use of nonrenewable fuels produces toxic pollutants, and pollutants result in an unhealthy atmosphere. According to a Verda Vivo report (2012), pollution is responsible for 40% of human deaths worldwide. And if pollution is killing humans, we must logically assume it is killing other organisms as well. Two international environmental groups, US-based Blacksmith Institute and Green Cross Switzerland, issued important information about the world's worst pollution problems (Mays, 2014).

Accordingly, the Top Ten list includes commonly discussed pollution problems, such as urban air pollution, as well as more overlooked threats, such as car battery recycling. The problems included in the report have a significant impact on human health worldwide, resulting in death, persistent illness, and neurological impairment for millions of people, particularly children. Many of these deaths and related illnesses could be avoided with affordable and effective interventions.

Blacksmith Institute's list of the World's Worst Pollution Problems (written verbatim from the Verda Vivo report) is unranked and includes:

- Indoor air pollution (IAP): adverse air conditions in indoor spaces. An estimated 80% of households in China, India, and sub-Saharan Africa burn biomass fuels in improperly ventilated spaces for their cooking energy. IAP contributes to 3 million deaths annually and constitutes 4% of the global burden of disease.
- Urban air quality: adverse outdoor air conditions in urban areas. The World Health Organization (WHO) estimates that 865,000 deaths per year worldwide can be directly attributed to outdoor air pollution. Leaded gasoline (in countries where it is still used) and the combustion of fossil fuels, especially coal and diesel fuel, play a major role in air pollution.
- Untreated sewage: untreated waste water. WHO estimates that 1.5 million preventable deaths per year result from unsafe water, inadequate sanitation, or hygiene.
- Groundwater contamination: pollution of underground water sources as a result of human activity. Fresh water for drinking makes up only 6% of the total water on Earth and only 0.3% is usable for drinking.
- Contaminated surface water: pollution of rivers or shallow-dug wells mainly used for drinking and cooking. Almost 5 million deaths in the developing world annually are due to water-related diseases, much of this being preventable with adequate supplies of safe water.
- Artisanal gold mining: small-scale mining activities that use the most basic methods to extract and process minerals and metals. Mercury amalgamation, a by-product of artisanal and small-scale mining affects up to 15 million miners and others, including 4.5 million women and 600,000 children.
- Industrial mining activities: larger-scale mining activities with excessive mineral wastes. Unless a major accident occurs, the effects are often chronic in nature and include irritation of eyes, throat, nose, and skin; diseases of the digestive tract, respiratory system, blood circulation system, kidney, and liver; a variety of cancers; nervous system damage; developmental problems; and birth defects.
- Metal smelting and other processing: extractive, industrial, and pollutant-emitting processes. Steel production alone accounts for 5–6% of worldwide, man-made CO_2 emissions.
- Radioactive waste and uranium mining: pollution resulting from the improper management of uranium mine tailings and nuclear waste. Of the ten largest producers of uranium, seven are in areas where industrial safety standards do not always correspond to the best industrial practices: Kazakhstan, Russia, Niger, Namibia, Uzbekistan, Ukraine, and China. There is no "safe" level of radiation exposure. High exposures can result in death within hours to days to weeks. Individuals exposed to nonlethal doses may experience changes in blood chemistry, nausea, fatigue, vomiting, or genetic modifications.
- Used lead acid battery recycling: smelting of batteries used in cars, trucks, and backup power supplies. Blacksmith Institute estimates that over 12 million people are affected by lead contamination from the processing of used lead acid batteries throughout the developing world.

Improvements to ratify these problems leaves us with our most important, overwhelming dilemma, that of additional humans in the world. What happens, for instance, if human technology solves these and other problems and the population increases because of it? Is it possible to ignore reproductive limitations and simultaneously provide a better life for everyone?

Solid Wastes

It has been estimated that each of us throws away more than 1200 pounds of trash per year, far more than people in most other countries (Ackerman, 2010). The typical American family discards 2460 pounds of paper, 540 pounds of metals, 480 pounds of glass, and 480 pounds of food scraps every year. Eighty percent of what is discarded ends up in landfills, resulting in a significant loss of energy and a loss of materials that if recycled could help intercept the destruction of environments. Only 10% of what is left gets recycled and the other 10% gets incinerated. The amount of wood and paper we throw away each year alone is enough to heat 50 million homes for 20 years.

Each gallon of gas used by a car contributes about 19 pounds of carbon dioxide into the atmosphere. For a single car driving 1000 miles a month, that adds up to 120 tons of carbon dioxide a year. This, of course, is affecting global climate change and contributing to planetary pollution.

About 110 million Americans live in areas with levels of air pollutants the federal government considers harmful (Dimitriou and Christidou, 2010). Americans dump 16 tons of sewage into their waters, every minute of every day. Americans throw away 25 billion Styrofoam coffee cups every year and 2.5 million plastic beverage bottles every hour. Americans throw away about 40 billion soft drink cans and bottles every year. Placed end to end, they would reach to the moon and back nearly 20 times (Byström and Lönnstedt, 2009).

Eighty-four percent of a typical household's waste, including food scraps, yard waste, paper, cardboard, cans, and bottles, can be recycled. Using recycled paper for one print run of the Sunday edition of The New York Times would save 75,000 trees. America's refrigerators use about 7% of the nation's total electricity consumption, the output of about 25 large power plants.

Unfortunately, some of our most common products cannot be recycled. An article in Green Living Tips points to our dependence on Styrofoam.

Styrofoam is a trademark of the Dow company, but the material itself is called polystyrene. Like so many other plastics, it is all around us—very commonly used in packing material as peanuts or expanded foam, in food trays, and a wide variety of other products—even [in] explosives, such as napalm and hydrogen bombs!

The bad news is…polystyrene is manufactured from petroleum. It is highly flammable, and a chemical called benzene, which is a known human carcinogen, is used in its production (Fishbein, 2008).

"Polystyrene foam, used commonly as padding in appliance packaging, takes an incredibly long time to break down in the environment and additionally, animals may ingest it which blocks their digestive tracts and ultimately causes starvation. This foam is also abundant in the Great Pacific Garbage Patch," an accumulation of plastic and other debris that has built up over time in certain areas of the Pacific Ocean because of the careless disposal of debris and the action of Pacific currents.

Given the nature of polystyrene, it is surprising that such an energy-intensive, oil-sucking, and toxic substance is allowed to be used as packaging for food; particularly for items such as meat where the food has direct contact with it. Nearly two dozen cities in the United States have banned the use of polystyrene for this purpose. Packaging and products containing polystyrene can usually be identified by a recycling triangle logo with the number 6 inside it stamped on the item (Maharana et al., 2007).

The article further points out that it is likely to be a very long time before the use of polystyrene is totally discontinued. "Unfortunately, many curbside recycling programs do not accept polystyrene and given its bulk, it can be difficult to store. Also, polystyrene is often recycled to be used in single-use products, such as more packing material, so it is important to get the word out about recycling this form of packaging."

"Some people choose to burn polystyrene in order to be rid of it, believing that as chloro-fluoro hydrocarbons were eliminated from expanded polystyrene over a decade ago, it was safe to do so." Recall that chlorofluoro hydrocarbons affect the ozone layer between the troposphere and stratosphere above the Earth which is important in screening out threatening electromagnetic waves (Fig. 6.3). "Without an ozone layer, there would be very little terrestrial life on the planet":

> The burning of polystyrene releases styrene gas that can affect the nervous system. Also, as it usually burns with a sooty flame, this indicates combustion is not complete and a complex mixture of toxic chemicals can be produced by the relatively low temperature of a backyard burn.

Much of the wastes described earlier come under the heading of municipal solid waste (MSW) or urban solid waste (Chettiparamb, 2013). MSW represents yet another form of material that has resulted in polluted environments. It is a type of human waste that includes predominantly household waste (domestic waste), along with the addition of commercial wastes collected by a municipality. They are in either a solid or semisolid form and generally exclude such products as industrial hazardous wastes. Residual wastes are waste materials that are left from household sources which contain materials that have not been separated for reprocessing (Flores et al., 2013).

Such waste materials include: (1) biodegradable waste, for example, food and kitchen waste, green waste, and paper; (2) recyclable material, for example, paper, glass, bottles, cans, metals, certain plastics, and other assorted materials; (3) inert waste, for example, construction and demolition waste, dirt, rocks, and debris; (4) composite wastes, for example, waste clothing, Tetra Paks, and waste plastics (such as toys); and (5) domestic hazardous waste (also called "household hazardous waste") and toxic waste, for example, medication, e-waste, paints, chemicals, light bulbs, fluorescent tubes, spray cans, fertilizer and pesticide containers, batteries, and shoe polish.

The disposal of wastes by land filling or land spreading is the ultimate fate of all solid wastes in contemporary times, no matter what their origin. A modern sanitary landfill is considered an engineered facility used for disposing solid wastes on land without creating unpleasant conditions or hazards to public health or safety, such as the breeding of insects and contamination of ground water (Sangodoyin, 2008).

If handled properly, municipal solid waste can be used to generate energy. Several technologies have been developed that make the processing of MSW for energy generation cleaner and more economical than ever before, including landfill gas capture, combustion, pyrolysis, gasification, and plasma arc gasification. While older waste incineration plants emitted high levels of pollutants, recent regulatory changes and new technologies have significantly reduced this concern (Waid, 2007).

EPA regulations in 1995 and 2000 under the Clean Air Act have succeeded in reducing emissions of dioxins from waste-to-energy facilities by more than 99% below 1990 levels, while mercury emissions have been reduced by over 90%. The EPA noted these improvements in 2003, citing waste-to-energy as a power source "with less environmental impact than almost any other source of electricity."

Recycling

Although recycling is a means of energy conservation, many people, institutions, and communities do not practice it (Overton, 1994). Recycling involves the processing of used materials into new products to prevent the waste of potentially useful materials. Recycling also reduces the consumption of fresh raw materials and reduces energy usage, air pollution (from incineration), and water pollution (from landfilling) by avoiding the need for "conventional" waste disposal. It also lowers greenhouse gas emissions when compared to virgin production.

Recyclable materials include many kinds of glass, paper, metal, plastic, textiles, and electronics. Although the composting or other reuse of biodegradable waste (such as food or garden waste) is not typically considered recycling, it does provide nutrients to soil and helps to reduce the amount of solid wastes requiring disposal and avoids the use of industrially produced fertilizers. Materials to be recycled are generally brought to a collection center or picked up from the curbside, sorted, cleaned, and reprocessed into new materials which are made available for manufacturing.

In a strict sense, recycling of a material would produce a fresh supply of the same material, for example, used office paper to more office paper or used foamed polystyrene to more polystyrene. However, this is often difficult or too expensive (compared with producing the same product from raw materials or other sources), so "recycling" of many products or materials involves their reuse in producing different types of materials (e.g., paperboard) instead.

Another form of recycling is based on salvaging certain materials from complex products, either due to their intrinsic value (e.g., lead from car batteries, or gold from computer components) or due to their hazardous nature (e.g., removal and reuse of mercury from various items) (Aitken and Murray, 2010).

Critics dispute the net economic and environmental benefits of some recycling over its costs and suggest that proponents of recycling often make matters worse and suffer from confirmation bias. Specifically, critics argue that the costs and energy used in collection and transportation detract from (and often outweigh) the costs and energy saved in the production process; also that the jobs produced by the recycling industry can be a poor trade for the jobs lost in logging, mining, and other industries associated with virgin production; and that materials such as paper pulp can only be recycled a few times before material degradation prevents further recycling. Proponents of recycling dispute each of these claims, and the validity of arguments from both sides has led to an enduring controversy.

However, with these examples, we must realize that we and our planet are in trouble, and that many of the problems we have are directly or indirectly connected with the amount of people and domestic animals that share Earth and its resources with other populations of organisms.

Our Dilemma

Being animals with incredible technological capabilities, we have discovered methods of using Earth's resources (including petroleum products that commenced their development as far back as 350 million years ago in Paleozoic oceans, fresh-water environments, and swamps). Our dilemma appears to relate to an inability to understand our position of dominance while maintaining some degree of empathy for the demise of Earth's other biotic forms and abiotic (nonliving) components. In addition, we appear to be living in a Catch-22 situation in which we require the things that are necessary for a respectable quality of life, but the processes of obtaining them are destroying the world we live in (Perelman, 2014).

Our complex contemporary lives and biodiversity have led us not to a perfect life but to a stressful, Earth-destructive one, and paradoxically, we continue to overexpress certain inherited animalistic traits (such as selfishness, increasing our numbers exponentially, and expressing our dominant and aggressive nature), as well as overusing resources available to us. Behaviorally, we have become a very demanding species, and to survive our devastating nature, we must become more serious about striving for the following goals:

- Decrease the numbers of individuals in our populations; exponential growth in any animal population leads to deleterious effects.
- Pay strict attention to our devastating relationship with the environment and its biotic components (including ourselves); except for certain environmentally conscious groups and new laws enacted by governments to reduce pollutants, we have ignored the respect we should give to our environment, fellow humans, and other organisms.
- Make amends for our lack of concern for the misuse and overuse of the planet's resources.

This is a very tall order. In spite of overwhelming evidence to the contrary, some may not agree that overpopulation is the root of our problems, but based on the figures presented here, how can it be denied? Dominant individuals who say we can sustain our populations without worrying about future problems are either in denial about the facts pointed out here or have ulterior motives for not concerning themselves with overpopulation and world destruction.

Even those who agree that it is the root of our problems think that accomplishing these recommendations borders on the impossible, but what choice do we have? Others (generally those who are nonscience oriented) are convinced that we can handle our ecosystem problems as they arise. However, what basis do they have to say this when we have not been able to handle them to this point, and we are plunging deeper into a pit of despair by increasing our populations further.

It is a common thought by many that because we are humans, the dominant form of life on the planet, we can alter our living conditions as we see fit (Totton, 2015). This may be true, to a degree, but at what expense? And while we are technologically superior to any other animals that ever existed on Earth, it may take more than technological expertise to save our species, other species, and planet Earth. It may, stretching our thought processes somewhat, also take a modicum of logic.

It is not that we, as a species, have not recognized our dilemma. Many scientists have pointed out the rocky road we may be headed for. Rachael Carson's book, *Silent Spring*, published in 1962 (as well as other books she has written and those of other authors), documented detrimental effects of pesticides on the environment, particularly on birds, and pointed out serious damage that we are causing to our environment.

She brought to the world's attention the deleterious effects that DDT was having on a wide variety of organisms, including thinner egg shell development in birds, its result in reproductive problems, and death. In *Silent Spring*, she also accused the chemical industry of spreading disinformation, as well as corruption between them and governmental officials.

If it was not for groups of people who have chosen to ignore these warnings, we may actually be able to decrease our populations and turn this thing around to live in harmony with the biotic and abiotic world that is left. If we keep up the pace toward destruction that we and people of other countries have followed for some time, at what stage of our development will we reach the point of no return and lose our chance to maintain our planet as a livable system that includes humans and many other life forms?

Other Indications of Our Destructive Nature

E.O. Wilson summarizes additional important statistics that world leaders should seriously consider (Williams, 2000). While many of the following examples from Wilson concern nonhuman species, they nevertheless represent an indication of a deteriorating world, the same world in which we live, and they show how we are using our dominance and aggressiveness to promote this deterioration.

- One-fifth of the world's bird species have been eliminated in the past two millennia, principally following human occupation of islands." According to a report by Bird Life International, "One in eight of the world's bird species is deemed globally threatened and the fortunes of 189 critically endangered species are now so perilous that they are at risk of imminent extinction. Some of these species," they say, "have not been sighted for many years and may already have succumbed."

Let us also remember that it is not only the loss of important species on Earth that we must ponder. Dwindling species indicate a deteriorating environment. Many of the problems we have discussed in this chapter stem from the destructive nature of humans and the vast number of individuals in our global population.

Wilson points out that "the colonization of tropical Pacific Islands by humans is believed to have led to the extinction of more than 2000 species of native birds. A 2004 report documents the extinction of 784 species of different types of organisms (including 338 vertebrates, 359 invertebrates and 87 plants) in the last 500 years."

Humans have overtaken from nature for a long time. As an example, certain Polynesian (Tahiti and New Zealand) and Hawaiian peoples have long created featherwork of amazing skill and artistry. These items were worn in battle and different state affairs to signify dominant status. Thus, cloaks and other feathered articles were treasured items of the ranking elite, red and yellow being the traditional colors for royalty. A great majority of feathers used in making these articles came from endemic birds, the Hawaiian honeycreepers. The 'i'iwi (*Vestiaria coccinea*) and the 'apapane (*Himatione sanguinea*) supplied most red feathers while the 'ō'ō (*Moho nobilis*) and mamo (*Drepanis pacifica*) provided feathers of yellow. Since the cloak of King Kamehameha consisted of about 450,000 yellow feathers of the mamo bird, found only on Hawai'i Island, it is not surprising that the collection of birds for such cloaks caused a reduction in population.

Wilson points out that on occasion, we receive a glimmer of hope from the scientific community that we can change to a more biologically friendly society. "Representatives of Bird Life International state that, 'there is cause for optimism." They point out that, "Conservation works. Around the world, dedicated conservationists… have orchestrated spectacular recoveries, bringing numerous species back from the brink. The message is clear, they say, that given sufficient resources and political will, species can be saved and the loss of biodiversity reversed."

That is the rub: political will. As the human population grows, will there be sufficient political will?

As indicated, it is not just bird species that are dwindling. Many other forms of life are being lost as well. In a World Conservation Union Red List, it has been said that, "Current extinction rates are at least 100 to 1000 times higher than natural rates found in the fossil record."

The World Conservation Union report concluded that, "humans are the main reason for most species' declines. Habitat destruction and degradation are the leading threats, but other significant pressures include overexploitation (for food, pets and medicine), introduced species (often through the pet trade and their subsequent release by pet owners), pollution and disease. Climate change is also increasingly recognized as a serious threat."

A recent *Science News* report stated the following about a decreasing species diversity in the United Kingdom: "Two new studies of UK flora and fauna offer some of the first comprehensive evidence that species diversity is decreasing in the United Kingdom. The findings support the hypothesis that the world is experiencing a mass extinction on par with the other five mass extinctions that have punctuated the history of life." As an example, "over 20 years, the ranges of about 70% of all the butterfly species in the United Kingdom declined to some degree—from a relatively small number of regional disappearances for some species to nationwide extinctions for a few others." If butterflies prove to be representative of insects as a whole, then "the world is indeed experiencing the extinction crisis many people have been suggesting and talking about for years."

Pesticides are often at the root of low butterfly numbers. Within recent years, I conducted butterfly counts on Sanibel Island, Florida, and could readily see the difference in species and individual numbers before and after the spraying of pesticides for mosquito and ceratopogonid fly control.

As another example, feral European honeybee populations in the United States have dropped about 90% in the past 50 years. It has been stated that they have been replaced in the Southwest by Africanized bees. In addition, managed honey bee colonies have dropped by about two-thirds.

This is especially bad news. While many other insects, as well as other animals, are important pollinators, honeybees rank #1 in this category. Most of the plants humans feed upon are honeybee-pollinated. Dwindling honeybee numbers is not a sustainable process.

- Population densities of migratory birds in the mid-Atlantic United States from the 1940s to the 1980s have dropped 50%, and many species have become locally extinct." According to an Environmental Protection Agency report on Basics of Bird Conservation in the United States by Lynne Trulio, "The shear decline in bird numbers tells us that we are damaging the environment through habitat fragmentation and destruction, pollution and pesticides, introduced species and many other impacts."
- About 20% of the world's freshwater fish species have either become extinct or they exist in a state of dangerous decline." Researchers have found that nearly 40% of freshwater fish species in North America are in trouble,

the number of species doubling since 1989. As many as 700 individual fish populations are threatened or endangered, up to 364 subspecies from 1989 reports. The study, published by the American Fisheries Society, found that 457 species might already be extinct.

- Relate these statements to the process of overfishing, as pointed out in the beginning of this chapter. The demands for fish to feed the growing masses of humans is a very destructive hit on biodiversity, and unless population control is instituted, it will only get worse.
- Through the introduction of exotic species (such as the Nile perch, Tilapia, and many other species) large numbers of indigenous fish species have been drastically reduced." According to a recent report by P. S. White, "exotic species invasions, referred to by one conservation biologist as the 'least reversible' of all human impacts, cause harm to economies (e.g., fisheries, wildlife populations, and tourism), the environment (e.g., in the form of broadcast of pesticides and herbicides), human health and well-being (e.g., allergic responses and the increase in fire severity in some landscapes), and aesthetics (e.g., the amount of mortality in vegetation)."
- These invasions," White says, "threaten biological diversity by causing population declines of native species and by altering key ecosystem processes like hydrology, nitrogen fixation and fire regime. The Earth is essentially a loaded gun of exotic species problems because: (1) evolution in isolation has produced continents with a similar range of environmental conditions but a very different array of species; and (2) species generally have an ability for exponential increase, particularly when removed from natural controls on their population growth. As a result, the problem is a global one. The exotic species problem," he says, "is neither trivial nor transitory. The human-caused mixing of formerly isolated biota stems from a failure to base decisions on the ecological and coevolutionary setting or organisms. We must employ many methods from our management tool box: eradication, containment, biocontrol, monitoring and, most importantly, prevention."
- As an example of exotics within the United States, according to the Florida Fish and Wildlife Conservation Commission, more than 500 nonnative fish and other wildlife species have been observed in Florida. Some have become invasive by causing harm to native species, posing a threat to human health and safety or causing economic damage.
- In the United States," Wilson says, "where the largest freshwater mollusk fauna in the world exists, species have long been in a steep decline from the damming of rivers, pollution and the introduction of alien mollusks and other aquatic animals."
- Whit Gibbons at the Savannah River Ecology Laboratory, states that, "Mollusks are arguably the most endangered major group of animals in the world. But few people realize the extent of their imperilment." As many as "291 of the 693 recorded extinctions of animal species since the year 1500 are mollusks. Virtually all have been freshwater or terrestrial species. In the last five centuries, the number of mollusks disappearing from the world has been more than all the mammals, birds, reptiles and amphibians combined that have gone extinct."
- In Tahiti, a number of local land snails were exterminated by the introduction of a single species of exotic carnivorous snail." *Euglandina rosea*, for instance, is a predatory snail that is known to have serious impacts when introduced to islands with native snail populations. While terrestrial, it has even been known to prey on aquatic species.
- The Center for Plant Conservation has revealed that out of about 20,000 plant species in the United States, between 213 and 228 are known to have become extinct." It has been estimated that 8323 plant and lichen species are now considered at risk.
- In the former Federal Republic in western Germany, 34% of 10,290 invertebrate species were classified as threatened or endangered in 1987.
- Collecting in Germany, Austria and the Netherlands has revealed a 40–50% loss in fungi species during the past 60 years.

While scientists (the most knowledgeable members of our society about the planet and its interactive components) point out with overwhelming evidence that humans have been a very destructive force in maintaining biodiversity and healthy ecosystems, developers (who are not as intimate with the environment or organisms which live in it) argue that biodiversity does not necessarily result in ecological stability (Sauri-Pujol, 2007). Their misunderstanding and approach is to ignore biodiversity and proceed with removing diverse forest ecosystems, developing all remaining wild areas for human use, disregarding overpopulation and premature extinction of other biota, and putting our destiny in the hands of human ingenuity. As E.O. Wilson put it in *The Diversity of Life* (1999), "What counts [in populations of contemporary humans] is food today, a healthy family, tribute for the chief, victory celebrations, rites of passage [and] feasts."

By all means, we must consider our basic needs (such as food, clean water, clean air, and shelter), but we must simultaneously coexist with other life forms by considering their needs as well. Dominance of all living creatures by humans and a lack of attention paid to a scientific approach toward survival and reproduction have by no means been overstated and overplayed. Wilson points out that it is clear that "we are living unsustainably by depleting and degrading the Earth's natural capital at an accelerating rate as our population and demands on the Earth's resources and life-sustaining processes increase exponentially."

The world's population in 2003 showed an increase of 7000 people an hour, an exponential growth of 1.2% per year (Höhler, 2007). Less than 2 billion people represent the total population of developed countries (Australia, countries of Europe, the United States, Canada, Japan, and New Zealand) while over 5 billion people inhabit the developing countries.

Much of the world population (especially in the developing countries) is struggling to maintain its existence, and such people are forced to degrade local ecosystems as a means of survival. In addition, while at least 9 million poor people die annually from malnutrition and infectious disease, it is in such areas and populations that most children are born.

Reproductivity is one of our most endearing animalistic qualities. However, because of our dominant nature in the world, it is imperative that we consider controlling it.

We are using up our nonrenewable resources and exceeding our renewable ones, leading to environmental degradation. We are polluting our terrestrial, fresh water, and marine environments, leading to the disruption of our life-support systems; wildlife, property, and human health damage; and an increase of dissonant sounds, unpleasant tastes, smells, and sights.

Forest Devastation

Wilson states that "the cutting of primeval forest and other disasters, fueled by the demands of growing human populations are the overriding threat to biological diversity everywhere." Some rather interesting data related to declining plant and animal diversity in what Wilson refers to as 18 of the world's hot spots are summarized as follows:

- The environment of a Mediterranean domain called the California Floristic Province in Oregon and California is being rapidly constricted by urban and agricultural development.
- South American vegetation is under pressure, especially by rural families, who rely on natural vegetation for fuel and livestock fodder.
- Forests in Colombia's coastal plain and low mountains are down to about three quarters of their original cover and are being destroyed at an accelerating rate.
- The rainforests of Ecuador, notable for their diversity of epiphytes, have been almost completely wiped out.
- At least 65% of the upland forests in western Amazonia have been cleared and converted into palm-oil plantations.
- Unique rain forests along the Atlantic coast of Brazil have been reduced to less than 5% of the original cover.
- 160,000 km^2 of unrestricted logging and slash-burn farming along the southwestern Ivory Coast of Africa have reduced native vegetation to about 16,000 km^2. Even land protected as National Parks is threatened by illegal logging and gold prospecting.
- Forests in the eastern arc of Tanzania are down to half their original cover and shrinking fast from the incursions of Tanzania's exploding population.
- By 1985, Madagascar forests were down to a third of the cover encountered by the first colonists. Due to encroachment because of rapidly accelerating population growth, most of the loss has occurred since 1950.
- Forests on the lower slopes of the Himalayas are down by two-thirds and are disappearing quickly through unregulated logging and conversion to farmland.
- About a third of the cover in the western Ghats of India are gone, the remainder disappearing at a rate of 2–3% a year.
- The wet forests of Sri Lanka have been reduced to slightly less than 10% of its original area.
- The richest repository of biological diversity in Peninsula Malaysia has been degraded. About half of the endemic tree species are now extinct or endangered.
- The forest cover in northwestern Borneo has been reduced by nearly a half, and most of the remaining trees have been consigned to timber companies.
- In the past 50 years, two-thirds of the forest in the Philippines have been cleared.

- The forests of New Caledonia, with 1575 species of plants, 8% of which are endemic, are being exploited through logging, mining, and burning.
- Since 1955, heathland vegetation west of the Nullarbor Plain in southwestern Australia has been reduced by half, mostly through agricultural conversion. Mining operations, introduction of exotic weeds, and wildfires are exerting continuous pressure on the area's species, and half of its species are now classified as rare or threatened.

Evidence provided in this chapter offers an eye-opening account of the influence of the world's most dominant animal, *Homo sapiens*. Is there any question that our population around the world is already much too large to maintain a homeostatic biotic system, that we are contributing to global mass extinction, overusing the resources that the world provides, and polluting the environment much faster than we can repair it?

Note

Superscript numerals appearing in this chapter refer to additional text/explanation given in the appendix.

CHAPTER

17

Attempts to Save the Natural World

We cannot return to what used to be, but we can restore, protect, and more effectively manage what we have. Sound science needs to be the foundation, and communication, education and public involvement, the cornerstones. **Elizabeth D. Purdum**, Florida Waters

All news about our relationship with planet Earth is not bad. The foregoing sections have not only depicted human problems, but they also have introduced a number of examples of certain attempts by humans to solve many of the problems for our growing demands. Many attempts are being made by individuals and special groups to minimize the use of resources, levels of pollution, and other environmental problems. Also, certain members of our population are seriously considering the importance of other living organisms and their needs, pointing out that there are definite positive attributes in some humans that are associated with our dominant nature. In addition, population increase has slowed somewhat. But are we doing enough and is it too late?[1]

What is unfortunate is that as the world becomes more complex, many humans grow further apart from the system that produced all life forms, including other members of the species to which they belong. Instead of living a mutualistic or even a commensalistic existence with our fellow organisms and planet Earth, we as a species appear to be more parasitic in our ways. Instead of contributing to world homeostasis or going to the roots of our problems, we find ourselves constantly attempting to compensate for our overwhelming misconduct. If it were not for the heroic efforts of certain people to save the natural world, we may have succumbed to a dismal existence and deteriorating Earth long ago (Wals, 1990).

Wikipedia provides a list of the more notable environmental organizations by organization type (intergovernmental, governmental or nongovernmental) and further subdivides the list by country. That there is such a list means that certain people around the world recognize the problems and care enough to do something about environmental and species abuse.

Since it would be impossible to adequately cover all of the organizations properly here, I have simply listed many of the organizations outside of the United States. For those organizations within the boundary of the United States, I have offered a more complete explanation. Interested readers may want to pursue the topic by going to the source. The following information is from that article. I apologize if I have left any organizations out.

WORLDWIDE INTERGOVERNMENTAL ORGANIZATIONS

Intergovernmental organizations that are interested in global problems, including the Intergovernmental Panel on Climate Change (IPCC), the United Nations Environment Programme (UNEP), the Earth System Governance Project, and the Global Environment Facility (GEF).

REGIONAL ORGANIZATIONS

Regional organizations in global problems, including the European Environmental Agency (EEA) and the Partnerships in Environmental Management for the Seas of East Asia (PEMSEA).

Dominance and Aggression in Humans and Other Animals
http://dx.doi.org/10.1016/B978-0-12-805372-0.00017-1

WORLDWIDE ENVIRONMENTAL PROTECTION GROUPS

The governments of many countries have ministries or agencies devoted to monitoring and protecting the environment: for example, Australia—Department of Sustainability, Environment, Water, Population, and Communities; Brazil—IBAMA; Canada—Environment Canada; Denmark—Danish Ministry of Climate and Energy; Germany–Federal Ministry for Environment, Nature Conservation, and Nuclear Safety; Hong Kong—Environmental Protection Department; India—Central Pollution Control Board (CPCB), Gujarat Pollution Control Board, and Ministry of Environment and Forests; Indonesia—Directorate General of Forest Protection and Nature Conservation; Ireland—Environmental Protection Agency; Ireland (Northern)—Northern Ireland Environment Agency; Isle of Man—Manx National Trust; Israel—Ministry of the Environment; Mexico—Secretariat of the Environment and Natural Resources; The Netherlands—Ministry of Housing, Spatial Planning and the Environment; New Zealand—Department of Conservation, Ministry for the Environment and Parliamentary Commissioner for the Environment; Nigeria—Kano State Environmental Planning and Protection Agency; Norway—Norwegian Ministry of the Environment, Norwegian Directorate for Nature Management, and Norwegian Pollution Control Authority; Philippines—Department of Environment and Natural Resources; Portugal—Ministry for Environment, Spatial Planning and Regional Development; Republic of China (Taiwan)—Environmental Protection Administration; Saudi Arabia—Saudi Environmental Society; Scotland—Historic Scotland, Scottish Natural Heritage, and Scottish Environment Protection Agency; Sweden—Ministry of the Environment; United Kingdom—Department for Environment, Food and Rural Affairs, English Heritage, Environment Agency, and Natural England; United States—United States Environmental Protection Agency, United States Fish and Wildlife Service, United States National Park Service, and Inter-Tribal Environmental Council (Native America); Wales—Cadw, Countryside Council for Wales, and Environment Agency Wales.

Wikipedia provides a more extensive international list of the most notable environmental organizations. Since the following organizations are within the United States, I have attempted to explain their activities somewhat.[2]

UNITED STATES ENVIRONMENTAL GROUPS

As pointed out earlier, many of us live a life that is quite pleasant (at least, for the moment). Thus the reader may wish to give credence to the aforementioned groups and many of the following US-based organizations that have been formed to terminate planetary abuse and help cure the planet's current ills.

An organization called *American Rivers*, for instance, founded in 1973, is dedicated to protecting and restoring rivers for people, fish, and wildlife. Another group, *Defenders of Wildlife*, functions in saving endangered wildlife, habitat, and biodiversity. The *Defenders Electronic Network* was established to save wildlife for future generations.

As a nonprofit law firm for the environment, *Earthjustice* protects people, wildlife, and natural resources by providing free legal representation to citizen groups to enforce environmental laws. *Environmental Defense* links science, economics, and law to create innovative, equitable, and cost-effective solutions to the most urgent environmental problems. As the United States voice of an international network spanning 70 countries, *Friends of the Earth* champion the thought of having a healthier and just world.

Greenpeace is the leading independent campaigning organization that uses nonviolent direct action and creative communication to expose global environmental problems and promote solutions that are essential to a green and peaceful future. As the bipartisan political voice for the national environmental community, the League of Conservation Voters Education Fund publishes the National Environmental Scorecard, which rates members of Congress on key environmental votes.

The *National Audubon Society* conserves and restores natural ecosystems, focusing on birds, other wildlife and their habitats for the benefit of humanity and Earth's biological diversity. With experience in media relations, issue campaign management, government affairs, federal environmental law and investigative research, the National Parks Conservation Association functions in protecting and enhancing America's National Park System for present and future generations.

The National Wildlife Federation provides education, inspiration, and action, helping people from all walks of life to restore and safeguard the world's wildlife and wild resources. The Natural Resources Defense Council uses law, science, and the support of its members to protect the planet's wildlife and wild places to ensure a safe and healthy environment for all people.

A group called Physicians for Social Responsibility works at the intersection of environment and health, striving to protect Americans through education and action about global warming, toxic materials, drinking water quality, and antibiotic resistance. The *National Environmental Trust* provides public education campaign and communications expertise on national environmental issues.

The *Sierra Club* works to inspire all Americans and every public official to protect the natural world that surrounds and sustains us. The *Ocean Conservancy* informs, inspires, and empowers people to act for the survival of marine organisms. The *State PIRGs* protect public health and the environment by organizing citizens against threats to clean air and water, to prevent global warming and preserve natural resources. The *Wilderness Society* works to protect America's natural environments.

The *Union of Concerned Scientists* is a nonprofit partnership of scientists and citizens combining rigorous scientific analysis, innovative policy development, and effective citizen advocacy to achieve practical environmental solutions. *World Wildlife Fund*, the global conservation network, is dedicated to preserving wildlife and wild lands through science, policy advocacy, and the *Conservation Action Network*.

Other federal and state conservation groups (some of which have already been mentioned) are working on attempts at restoring the natural environment. While it would be difficult to cover all of the activities involved with restoration, I have chosen one particular state and a resource that is of extreme value to all species. I quote from a 2004 report by Elizabeth D. Purdum called *Florida Waters, A Water Resource Manual from Florida's Water Management District*.

FLORIDA'S ATTEMPT TO RESTORE ITS NATURAL ENVIRONMENT

"In Florida for at least 14,000 years, human settlement has been shaped by water." In the initial stages of existence in Florida, "much of the world's water was frozen in glaciers." Florida was a dry, large, grassy prairie. Many present-day rivers, springs, and lakes had yet to be formed; even groundwater levels were far lower than they are today. Sources of freshwater were limited, and finding them was critical to the survival of the Paleoindians and the animals they hunted for food.

"About 9000 BC, glaciers melted, sea level rose and Florida's climate became wetter. By 3000 BC, when Florida's climate became similar to today's climate, people occupied almost every part of the present state. On average, more rain falls in Florida (135 centimeters or 53 inches) per year than in any other state in the nation besides Louisiana, which receives an average of 140 centimeters (55 inches). "

"Florida has two types of climate: humid subtropical in the northern two-thirds of the state and tropical savanna in the southern third and the Keys." Most of the state's water cannot be seen because it seeps beneath the ground through sand and gravel and flows through cracks and channels in underlying limestone. Florida has more available ground water in aquifers than any other state. Of the 84 first-magnitude springs (those that discharge water at a rate of 100 cubic feet per second or more) in the United States, 33 are in Florida—more than any other state.

"Water sources in Florida are expressed in springs, rivers, lakes, wetlands, and estuaries, in addition to the Atlantic Ocean and Gulf of Mexico." Ecosystems are in the form of pine flatlands, scrub, dry prairies, hardwood hammocks, swamps, marshes, lakes, rivers, dunes and maritime forests, mangroves, seagrass beds, and coral reefs.

Native Americans in South Florida investigated water flow and changed it to suit their needs. However, when human populations other than indigenous Americans first entered Florida, it was the beginning of extreme water use, control, and modification. Wetlands were drained for farms, groves, and houses. Canals were cut to facilitate drainage and to improve navigation. Floodwaters were held back with engineering works. Wastes were discharged without treatment into rivers, lakes, and coastal waters.

Beginning in the 1800s, many of Florida's natural systems were radically changed. Thousands of acres were drained for agriculture. Thousands more were drained for houses for the steady stream of new residents. Rivers were straightened and canals were dug for drainage and flood control to make travel easier for ships and barges. Rivers were dammed for hydroelectric power and to create lakes for recreation. Forests were cut and trees were tapped for turpentine and rosin. In northern Florida, centuries-old longleaf pine trees were replaced with acre upon acre of fast-growing slash pine. Farther south, ancient cypress were logged and the land left bare.

A serious consequence of the conversion of the natural Florida landscape to human uses has been the fragmentation of remaining natural habitats. Habitat fragmentation increases the amount of "edge" habitat. Although edges are desirable for some game species, such as deer and rabbits, and for some birds, such as song sparrows and cardinals, excessive amounts of edge are undesirable for interior forest dwellers. Edges of forests are also hotter and drier than the forests themselves and may become dominated by common weeds, whereas forest interiors are more diverse and support more rare species.

"Complexity and diversity tend to be hallmarks of unaltered systems, and this makes restoration very difficult. Like a broken eggshell, a fragmented and altered ecosystem that is put back together may never be as strong and resilient as the original." In spite of these challenges, attempts are currently being made throughout Florida to return the state's ecosystems to what they were before humans began to destroy them or at least improve them.

"The US Army Corps of Engineers and the South Florida Water Management District are embarking on the most ambitious ecosystem restoration ever undertaken in the United States. At an estimated cost of $7.8 billion, a 50-year plan provides the road map for reviving what was once an uninterrupted ecosystem from the Kissimmee River valley, through Lake Okeechobii, through the water conservation area and Everglades National Park to Florida Bay and the coral reefs." At the same time, highways are being widened to accommodate additional travelers and additional lands are being cleared for building and agricultural purposes.

"The Everglades landscape began to change in 1882 when Hamilton Disston attempted to channelize the Caloosahatchee and Kissimmee rivers. In 1904, modification of the south Florida environment accelerated when Napoleon Bonaparte Broward was elected governor of Florida on a promise to 'drain the Everglades.' Between 1905 and 1927, six major canals and channelized rivers were connected to Lake Okeechobee for drainage and navigation. People began to settle and farm newly drained land south and east of Lake Okeechobee." From those times, people, not nature, determined where and, to some degree, how much water would flow.

The Central and Southern Florida Flood Control Project opened vast areas for agriculture and urban development, making it possible for more and more people to live in south Florida. It did so at tremendous ecological cost to the Everglades. While the population of people in south Florida has risen from 500,000 in the 1950s to more than 6 million today, the number of wading birds in Everglades National Park has declined by 95%. Sixty-eight plant and animal species are threatened or endangered, and over 1.5 million acres are infested with invasive exotic plants. The current Everglades is only about half the size of the Everglades that existed 100 years ago.

"To improve natural area connectors and to enhance overland flow, more than 240 miles of levees and canals" are scheduled to "be removed from the Everglades. Portions of the Tamiami Trail" are scheduled to be rebuilt with bridges and culverts, allowing a more natural flow of water across the land into Everglades National Park. In the Big Cypress National Preserve, the levee that separates the preserve from the Everglades will be removed, restoring more natural overland water flow.

"Tampa Bay is Florida's largest open water estuary, with a surface area of nearly 100 square miles and a watershed of 2200 square miles." It is "the year-round home to more than 100 dolphins and a winter refuge for the endangered Florida manatees." It is also home to one of the highest populations of lancelets in the world, organisms that are similar to the invertebrates that started the line of evolution that led to all vertebrates, including humans.

"Beginning in 1950, the [human] population in the bay area began to soar." Subsequent "algal blooms and fish kills [became] common in the bay. Water [became] so murky that divers couldn't see their own hands. The biggest decline of the bay was [due to an overabundance of] nutrients, primarily nitrogen, from wastewater discharges and stormwater runoff. Goals of [an improvement] plan include restoring at least 2000 acres of coastal habitat and increasing seagrass beds to 40,000 acres."

Through the 1800s, there were over 400,000 acres of floodplain marsh in the Upper St. Johns River Basin. By the 1970s, nearly two-thirds of the floodplain marsh was lost, resulting in flooding, declines in water quality and decreases in fish and wildlife populations.

In 1954, following devastating flooding from hurricanes in the 1940s, Congress authorized construction of engineering works in the Upper St. Johns River Basin as part of the Central and Southern Florida Flood Control Project. Nearly, a century after they were first altered, 125,000 acres of marsh (many of which had been drained and converted to pastureland) have been restored.

Longleaf pine forests originally stretched from Virginia to eastern Texas, covering 6.9 million acres in Florida's upper peninsula and Panhandle regions. These forests are home to hundreds of species, including the federally endangered red-cockaded woodpecker and the declining gopher tortoise. Longleaf pine forests have one of the most diverse plant populations on Earth because of frequent lightening fires, which keep one species from outcompeting the other, and many longleaf pine forests are important groundwater recharge areas.

Destruction of longleaf pine forests began in earnest after the Civil War and has accelerated in the last 50 years. Since World War II, Florida's longleaf pine forests have been cut at an annual rate of 130,000 acres and largely replaced by single-species plantations of slash pine. These plantations do not support the diversity of the original sandhill and flatwoods communities[3]:

The Northwest Florida Water Management District is restoring thousands of acre as within its 16 counties, including many where longleaf pine once thrived. About 4.4 million longleaf pines have been planted on District lands, as well as 563,000 wire grass [plants].

In summing up her report on Florida's water, Purdum made a statement that could apply to any geographical area that has undergone restoration: "Florida once had extensive and highly productive ecosystems, many of which were altered and degraded by urban and agricultural development. Much of the activity resulted from a lack of knowledge concerning how ecosystems function, how they are interrelated and the ways in which they help sustain people. We cannot return to what used to be," she said, "but we can restore, protect and more effectively manage what we have. Sound science needs to be the foundation, and communication, education and public involvement, the cornerstones."

Such restoration projects are remarkable attempts at returning the ecosystems to what they were before humans began to destroy them. Add to this the numerous science teachers, nature centers and similar groups around the United States and other parts of the world that strive to make a connection between the wilderness and the contemporary human population and we have a formidable force of individuals who can influence the preservation of wilderness areas. Without these groups, nature would not have lasted as long as it has.

A NEED FOR MORE CONCERN

Impressive as these organizations and the people who belong to them are, they are not nearly enough for the preservation of natural areas and survival of Earth's species as long as our populations remain large. Because we are such powerful, aggressive, and dominant animals, people from all walks of life and from every country must learn to care about what happens to our biotic and abiotic world. Biologists have been saying this for years, and a number of important books have issued warnings about what can happen if we ignore environmental abuse and the delicate nature of Earth (Brenes and Winter, 2001).

Yet, large numbers of people (many of whom are not knowledgeable about environmental issues or the organisms that live in nature) are not interested in wildlife or the environment. Even with the educational opportunities available to learn about wilderness areas and the wildlife that inhabits them (including incredible wildlife movies on the Discovery channel), there often is a lack of concern by humans for the preservation of nature and natural resources.

From talking to students in colleges and universities, as well as people from all walks of life, I see a large segment of the population that has lost any connection with the Earth, other than making their way in life on a forgotten planet. In another book (*Making the Wind Sing, Native American Music and the Connected Breath*; Hermann, 2011), I quote a psychologist, Michael DeMaria who stated, "most people today suffer from depersonalization and derealization," causing them to feel disconnected or out of touch. One of the reasons, he says, is that they have lost their connection with nature. Many, possibly most, people in contemporary societies are quick to disregard the natural world and even resort to killing everything they do not understand. They seem to have somehow lost any connection with and empathy for life forms that we should coexist with rather than simply dominate.

It is true that people may have empathy for a young seal that is being beaten to death for its pelt or for domestic animals that are being abused, but they are often unaware of, unknowledgeable or unconcerned about the environmental atrocities that are being committed on Earth.

Many people are driven to different endeavors by instincts that influence their success in obtaining wealth and power (see Chapter 8). Many humans are dominant, competitive animals, and they sometimes will go to great lengths to obtain the things they desire in life, often at the expense of other humans, other organisms and the world. It has been obvious over the last few hundred years that we, as a species, are more interested in aggressively dominating each other and everything around us, including the Earth and its biota, and taking whatever we need to boost our personal ambitions (see Chapter 11). Let us look at some examples.

LOVE CANAL ENVIRONMENTAL DISASTER

In 1920, for instance, a company called Hooker Chemical had turned an area in Niagara Falls into a municipal and chemical disposal site. In 1953, the site was filled, and a thick layer of impermeable red clay was used to seal the dump, theoretically preventing chemicals from leaking out of the landfill.[4]

Despite warnings from Hooker Chemical, a city near the dumpsite eventually bought the site and proceeded to dig into the ground to develop a sewer system. In the process, they damaged the red clay cap that covered the dumpsite below, and blocks of homes and a school were built atop the site. The neighborhood was named Love Canal. An anonymous article by Wikipedia, called *Love Canal* documents the details of this environmental disaster (Robinson, 2002).

On the surface, Love Canal seemed like a regular neighborhood. The only thing that distinguished it from others was the strange odors that often hung in the air and an unusual seepage of unidentifiable substances noticed by inhabitants in their basements and yards. Children in the neighborhood often fell ill. Love Canal families regularly experienced miscarriages and birth defects.

An activist named Lois Gibbs noticed the high occurrence of illness and birth defects in the area and started documenting it. In 1978, newspapers revealed the existence of the chemical waste dump in the Love Canal area, and anticipating problems, Lois started petitioning for closing the school. In August, 1978, the claim succeeded when the NYS Health Department ordered closing of the school after a child suffered from some form of chemical poisoning.

Following an investigation, over 130 pounds of the highly toxic carcinogenic TCDD, a form of dioxin, were discovered. The 20,000 tons of waste present in the landfill appeared to contain more than 248 different types of chemicals. The waste mainly consisted of pesticide residues and chemical weapons research refuse.

Chemicals originating beneath Love Canal had escaped its confines and entered homes, sewers, yards, and creeks. Gibbs recommended that the 900 families that belonged to the community be moved away from the location, and President Carter eventually provided funds to move all the families to a safer area. Hooker's parent company was sued and settled for 20 million dollars.

Despite protests, some houses in Love Canal went up for sale some 20 years later, even though none of the chemicals had been removed from the dumpsite. It has been resealed and the surrounding area was cleaned and declared safe. Today, the Love Canal dumpsite is known as one of the major environmental disasters of the century.

The Love Canal tragedy largely resulted from human greed and a lack of environmental concern, instituted by developers who saw a lucrative opportunity. In the long run, results negatively affected the health of humans and other organisms. There are other cases as well. The following account of releasing environmental pollutants created similar problems.

THE BAIA MARE CYANIDE AND HEAVY METAL SPILL

Cyanide (CN) is used by workers in gold mines to purify gold that has been removed from rocks. At 10 PM on January 30, 2000, CN used in a gold mine in Baia Mare, Romania, overflowed into the Somes River and subsequently into the Tisza River. The spill resulted from a break in a dam that surrounded a settling basin. At least 100,000 cubic meters of water with extraordinarily high concentrations of CN and heavy metals (e.g., copper, zinc and lead) were released (Frentiu et al., 2008).

When Rumanian authorities were notified of the spill, they responded immediately, preventing any human deaths. The spill nevertheless did kill all aquatic plant and animal life for dozens of miles downstream. On February 12, it even impacted the Danube River, which receives water from the Tisza. The impact was eventually noticed in Hungary and Serbia, and inhabitants of Belgrade witnessed high numbers of dead fish flowing by. Up to 100 people, most of them children, were treated in hospitals after eating contaminated fish. The Romanian media entitled this environmental disaster "the largest since Chernobyl."

Environmental organizations suggest that large companies routinely take advantage of flexible environmental regulations in poor countries such as Romania, resulting in environmental disasters such as the one in Baia Mare. It is not surprising that owners of companies that may be responsible for such disasters often deny that there is a connection between their company and environmental disasters.

Following the spill, fishing was banned from the Tisza and the population was recommended not to use the water. Local residents have suffered from drinking-water shortages, and the fishing industry has suffered as well.

THE COMPLEX HUMAN

The human animal is very unusual and complex indeed. In spite of the good, sometimes altruistic acts that many humans and their organizations perform, other humans use their selfish, sometimes antisocial attributes to gain personal benefits, often at the expense of shattering homeostatic environmental systems and human lives. That is part of who we are as a species. Earlier chapters have brought out detailed accounts of other human attributes.

Note

Superscript numerals appearing in this chapter refer to additional text/explanation given in the appendix.

18

The Nature of Things

The cost of our neglect of Gaia could cause the greatest human tragedy in living memory because the Earth, in its own but not our interests, is moving into a new hot epoch. Within the next century, we may have to retreat from our homes and move to what were formerly plateaus of ice in the most extreme regions of the world. We will certainly be forced to give up many of the comforts of western living as food, water, and electricity supplies are threatened. Only the fittest—and the smartest—will survive. *James Lovelock, The Vanishing Face of Gaia, a Final Warning*

The preceding statement about Gaia (Earth) by James Lovelock is grim. Nevertheless, the message of a deteriorating Earth has been expressed by numerous writers who know a lot about the Earth and how it works. In spite of the credentials of these writers, their messages have not adequately impacted our political leaders. Thus, our concern comes to this: How serious should we be about what environmental scientists are saying about Earth and its most dominant species?

Based on the forgoing chapters, it is clear that humans have inherited the genetics of a long line of predecessor animals and express themselves accordingly. Comparisons with numerous other animals reveal our animalistic nature. And yet, humans have evolved to a point of possessing a brain that has taken most of them far from their initial animalistic world to an ultramodern state of existence. Many of the circumstances that plagued them as animals have all but been eliminated in many places in the world, at least for the time being. But while they continue to function animalistically toward many environmental pressures, there are new problems, problems that bring new types of stress (distress) to them and the world around them.

A look at their dominant and aggressive nature in terms of their genetics, chemistry, widely divergent alternate approaches to a moral, amoral, or immoral life, brain chemistry, reaction to variations in nutrition, everyday interactions with other conspecific individuals, religion, politics, and warfare reveals that humans are indeed animals of extreme complexity and capable of expressing themselves in widely divergent ways.

A large and complex brain, accompanying technological skills and complex interactions with one another have allowed humans to build a civilization so complex, demanding, and sometimes threatening that it may be difficult to maintain intraspecific and interspecific homeostasis in both the biotic and abiotic world. James Lovelock and others strongly feel that our chances for rectifying the harm that has been done to the planet are very slim.[1] Even if we could learn to cure our environmental ills, with the population at an all-time high, we, as Lovelock has done, may wonder where and how the human population will live if we are ever faced with catastrophic events on Earth.

None of us like to ponder catastrophic events, especially if they result in the death of numerous members of our population or the rise of new, substandard living conditions. Yet, it is something we must anticipate as a precautionary measure to survive as a species. We may stretch our imaginations, attempt to conceive adverse conditions, and wonder if, under pressure, we would make an attempt to flaunt our animalistic rights to survive and reproduce. Who would survive? Would it be an egalitarian process? Would it subject us to superdominant struggles to determine who is to survive? Where would we go?

Many, Lovelock says, may enter underground bunkers which are usually restricted to an elite few (selected dominants and their loyal followers). As delusional as this may sound, where to retreat in case of cataclysmic threats has been on the minds of humans for many years, including those in ancient populations. Rather than control our population and halt planetary abuse, we are willing to consider alternative, albeit substandard, conditions to avoid extinction.

Dominance and Aggression in Humans and Other Animals
http://dx.doi.org/10.1016/B978-0-12-805372-0.00018-3

Do such retreats actually exist in our world, are they extensive enough to accommodate large numbers of people, and are they open for occupancy by the general population? Some such survival areas may be governmental installations that are apparently extensive and generally available for a select few, subterranean railway systems, war-time bunkers (constructed to survive bombing attacks), personal installations that people with adequate funds have had constructed, subterranean mining installations (some of which have been constructed as underground cities), or vast tunnels that have existed for many years, some of which are beneath previous civilizations, for example, the Mayan and Egyptian wonders of the world.

At this point, you may be wondering if this kind of thinking is real or fictional, but read ahead and make your own decision (Allen, 2012). An anonymous report for *Survival Blog* lists some of the older governmental bunkers which are available for potential survivors:

1. Mount Weather, also known as the "Mount Weather Emergency Operations Center." The facility, located in the heart of Virginia, is considered a major relocation site for the highest level of civilian and military officials in case of national disaster.
2. The famed Cheyenne Mountain Complex, subject of much curiosity, is located at Cheyenne Mountain Air Force Station (CMAFS), a short distance from NORAD and USNORTHCOM headquarters at Peterson Air Force Base in Colorado Springs, Colorado. It is situated deep in 2000 feet of granite, and was designed to withstand a 5 megaton nuclear explosion up to 1.7 miles away.
3. Raven Rock, a facility that was constructed in 1950, when it was approved by President Harry S. Truman. It was meant for use as a relocation site for up to 3000 of the pentagon staff. Five, three-story buildings were originally built in the complex, with roughly 700,000 square feet of usable space.
4. Moscow metro underground cities and rail lines, a large subterranean system of secret train tunnels and bunkers beneath Moscow, most of which were built during the reign of Stalin (a despotic dominant) during the Cold war.
5. An Iron Mountain facility, which is not officially recognized as a survival bunker. Taking up 10,000 square feet of space in a 1000-acre limestone mine, this storage facility is massive and rumored to be one of the most secure locations in the world.
6. The Greenbrier bunker, a 120,000-square feet bunker which was constructed by the government beneath a luxury hotel.
7. The Burlington Bunker, constructed 100 feet beneath the surface or Corsham, was built by the British government in 1950 in case of a nuclear strike, and it was intended to be the Emergency Government War Headquarters. The shelter was designed to accommodate 6000 people for up to 3 months.
8. The Shanghai Complex, a very little known complex, is a million square foot bunker capable of housing up to 200,000 people.
9. Denver International Airport may overlay a bunker, for which there is tertiary evidence.
10. The Svalbard Global Seed Vault, located in the remote arctic of Norway, only 810 miles away from the North Pole, is designed to be a sort of Noah's Ark for humanity.

In *Ground Penetrating Radar Finds Hidden Cities*, Paul White (2012) points out that since the declassification of the new ground-penetrating radar, staggering data has emerged about complex and labyrinthine underground systems in various parts of the world. As an example, tunnels have been mapped under the Mayan pyramid complex at Tikal, which extend a full 800 km, reaching to the opposite side of the country.

Using SIRA radar in Egypt, an extraordinary subterranean complex was discovered beneath Egyptian pyramids. A vast megalithic metropolis, 15,000 years old, was found to reach several levels below the Giza plateau. "Complete with hydraulic underground waterways…massive chambers, the proportions of our largest cathedrals, with enormous statues, the size of the Valley of the Nile, [were discovered to be] carved in situ."

There are many other caves being used by humans around the world, some newly constructed and others of ancient age. According to Celeste Adams in *Cave Dwellers, The Magic of Living in the Earth* (2014), actively used "cave dwellings have been found all over the world." "A few notable caves," Adams points out, "include Goat's Hole at Paviland, South Wales, the Huapoc Canyon caves in Mexico, Kent's Cavern at Torquay in the south of Britain; Wookey Hole and many others in Britian; Lascaux and other caves painted by Cro-Magnon man in Les Eyzies, France; Altamira in northern Spain; Mount Carmel rock shelters in the Middle East; the Dravidian caves in southern India; Choukoutien in China; Devils Lair and Kennif Cave in Australia; Olduvai Gorge in Africa; and Cyrenaica, in Libya. The United States has Danger, Utah; Ventana, Arizona; and Bat and Sandia Caves in New Mexico."

In Northern China, according to a University of Washington report, there are more than 40 million people living in cave and pit dwellings. In the provinces of Shaanxi and Shanxi where the yellow Earth (called loess) is quite compacted, cave houses have been in use for centuries.

Many people have already taken advantage of underground facilities that were originally constructed for defensive purposes. "During World War I," for instance, "troops burrowed underground for protection, and later, whole cities descended into bomb shelters to escape aerial attacks. In the 1950s and 1960s, people began to build fallout shelters. Today, as the balance of world peace becomes more unstable, political leaders all over the world are building underground chambers for protection against terrorism, nuclear bombs and bioterrorism. Both the American government and the Taliban fighters are resorting to underground structures for protection and defense."

Within the last decade, the caves of Afghanistan have been on the minds of people. "The United States military effort in Afghanistan scoured a network of 50 caves in the Zawar Kili region of eastern Afghanistan. Many of them were high in the walls of cliffs and accessible only by experienced climbers. The [United States] military also tried to clean out caves in the Tora Bora area to the north of Zawar Kili, and it is thought that Al-Qaida and Taliban fighters may yet be hiding in other Afghan caves":

"Jack Shroder, a geologist and professor at the University of Nebraska who has a special interest in eastern Afghanistan, said in an interview that warriors of the ethnic Pashtun group in the region have been digging caves for hundreds—maybe thousands—of years. Quoting Shroder, Adams points out that "most of the caves in that area are manmade.""

Another example of cave dwelling was revealed by Adams in Cappadocia, "a barren and mountainous region in central Turkey." As Adams points out, "The winters are terribly cold and the summers are very hot. There is very little vegetation, and building materials are limited. Perhaps it was because of the harshness of the terrain that the people of Cappadocia started to carve out underground cities." On the other hand, she says, It also has been speculated that the underground cities were built for defense by Christian refugees escaping persecution. For whatever reason, it is estimated that there are 300 underground sites, and portions of these are still in use.

In the caves of Cappadocia, networks of passageways link family rooms with communal spaces (Weiskopf, 1990). Most of these underground cities are no longer in use and can be visited by tourists. These include Derinkuyu, which is the largest and has eight levels that people can view (though it is thought that there are 12 more levels that have not yet been excavated). It is estimated that there are some 600 doors to this underground city, hidden in surface dwellings, and courtyards. It is also thought that Derinkuyu is linked to Kaymak, another underground city just 9 km away:

"The subterranean lifestyle still exists in Cappadocia. Subterranean canals irrigate terraced farmland. And in villages such as Zelve and Soganl, semi-subterranean rooms are still in use. Locally grown potatoes and fruit are stored underground because the temperature remains constant in these underground rooms. There also are more than twenty thousand cones throughout the region made of volcanic rock that has been hollowed out and made into living quarters."

When one contemplates using such facilities, many of us may wish to ignore the thought as overkill. Yet, it has become clear that some of us are actually preparing for a doomsday event, Armageddon if you will, brought on by overpopulation and intolerance between dominant individuals who have all the earmarks of sociopaths, smart, highly destructive nuclear weapons, all with world-wide destruction capabilities (Steinbruner and Dyson, 2007).

But the question arises: Why are we preparing for doomsday when we could very well avoid it? Some of us are considering living in bunkers to survive and colonizing nearby planets. Our approach to solving our problems is somewhat like going to a doctor and getting pills to alleviate our symptoms when we should be investigating their roots.

A major part of our growing problem in the world is human overproduction, an animalistic trait which typically is found in all organismic populations. Unlike most animals, however, we humans have no methods for stabilizing our population other than killing one another or weak attempts at slowing our population growth.

Based on what has been said in this book and according to numerous well-informed writers, our planet is very ill. Should we feel this is an alarmist attitude? I do not think so. They are based on reports that point out that the Earth and its inhabitants do not form a homeostatic whole. While many of us currently lead a comfortable life, there are problems which have arisen because of human dominance and aggression that may, in the future, threaten our very existence.

A TRANSFORMING EARTH

An article in *The Independent* and a report by the American Association for the Advancement of Science (AAAS) summarize many of the problems humans have caused, leaving "an indelible mark on the planet." Earth, they point out, is currently "undergoing an unprecedented transformation."

Maps compiled from satellite images show that "almost a quarter of the Earth's surface has been entirely transformed, either by being covered over by roads and buildings or ploughed up for crops and pasture. Another quarter has been exploited to a lesser degree, but in a way that has completely altered its natural state."

A rapidly growing human population, rising economic expectations, a continual decline in natural resources, increasing pollution by industrialized countries, dispersal of exotic species around the world, a disregard for the lives of other organisms, an increasing fear of terroristic factions, and the threat of nuclear war are leading (to the possibility of) a crisis of epic proportions.

"We have become a force of nature comparable to volcanoes or cyclical variations in the Earth's orbit," the AAAS report warns. "As we enter the third millennium, the destiny of the planet is in our hands as never before; yet, they are inexperienced hands. We are modifying ecosystems and global systems faster than we can understand the changes and prepare responses to them."

Humans are perhaps the most successful species in the history of life on Earth in terms of domination and modifying their environment. From a few thousand individuals, some 200,000 years ago, our cosmopolitan population size passed the 1 billion mark around 1800, 6 billion in 1999, and 7 billion in 2011. Our levels of consumption and the scope of our technologies have grown in parallel with, and in many ways outpaced, our numbers. But our success is showing signs of overreaching itself, of threatening the key resources on which we depend. Today, our impact on the planet has reached a truly massive scale. In many fields, our ecological footprint far outweighs the impact of all other living species combined.

"We have transformed approximately half the land on Earth for our own uses—around 11% each for farming and forestry, and 26% for pasture, with at least another 2 to 3% for housing, industry, services, and transport. The area used for growing crops has increased by almost six times since 1700, mainly at the expense of forest and woodland." Add to this a gross disregard by most of our dominant leaders for the planet and its inhabitants, as was pointed out in Chapter 16, and we have a situation that is most difficult to correct:

> "Past attempts to estimate global land use have been hindered by the lack of full geographical coverage, which is not a problem with a polar-orbiting satellite," Bromley said. "By getting this eye-in-the-sky view, you can prove that cropland is far more extensive than anybody recognized."[2]

According to *The Independent* report, the atlas shows the extent to which soil erosion has affected a substantial part of the Earth's surface that has gone under cultivation and has subsequently been abandoned. "Worldwide, an estimated 12 million hectares of cropland falls out of use for this reason each year. Economists have estimated the value of this lost soil, in terms of nutrients and water-holding capacity, at about $400 billion." Not utilizing our organic materials to rebuild the soils makes the problem worse.[3]

Freshwater has also been degraded. "Chronic or acute water shortage is increasingly common in many countries with fast-growing populations, becoming a potential source of conflict. The distribution of water resources around the globe is highly unequal. Canada, [for instance], has more than 30 times as much water available to each citizen of China."

"Today, it is estimated that 31 countries with 8% of the world's population—mostly in Africa and the Middle East—have water shortages. By 2025, the figure is likely to have risen to 48 countries and 35% of the population. The crisis is likely to be worsened by the deteriorating quality of water, polluted by industrial wastes, [pesticides] and sewage discharges, and spreading diseases such as cholera and schistosomiasis."[4] The AAAS report concludes that humans have:

- regulated the flow of about two-thirds of all rivers on Earth, creating artificial lakes and altering the ecology of existing lakes and estuaries;
- fished two-thirds of marine fisheries to the limit or beyond and altered ecologies of many marine species. In 100 years, we have destroyed half of the coastal forests and irrevocably degraded a tenth of the coral reefs;
- contributed 50% more to the nitrogen cycle than all natural sources combined, leading to the impoverishment of forest soils and forest death, and at sea to the development of toxic algal blooms and expanding dead zones devoid of oxygen;
- released toxic metals into the biosphere through mining and processing that would otherwise have remained safely locked in stone;
- had an incalculable effect on biodiversity. The 484 animal and 654 plant species recorded as extinct since 1600 are only "the tip of a massive iceberg";

- become a major force of phyletic and environmental confusion, not just for the "new" species we artificially breed and genetically engineer, but for the thousands of species we introduce as exotics to indigenous populations and thousands of others whose habitats we modify and destroy, consigning many to extinction:

"In this unprecedented situation, the need to be fully aware of what we are doing has never been greater," the report says. "We need to understand the way in which population, consumption and technology create their impact, to review that impact across the most critical fields and to find ways of using our understanding of the links to inform policy."[5]

Assuming these writers are correct in their predictions, we may wonder what we must do to avoid further complications. With the many problems that exist on Earth, there is no denying that the human population is already beyond its carrying capacity and we rightly should be concerned about it. But humans have had their sights on overpopulation and its inherent problems for quite some time, and the population has increased in spite of early knowledge of what population growth could do to sustainability (Mandani, 1972; Tomlinson, 1975).

As early as 300 BC, Indian political philosopher, Kautilya (350–283 BC), had written about it. Plato (427–347 BC) and Aristotle (384–322 BC) theorized that Greek cities should be small enough for efficient administration and citizen participation but large enough for defensive purposes. They recommended procreation and immigration if the population was too small and emigration to colonies if the population grew too large. Aristotle reasoned that excess population growth would result in poverty, and that poverty was the cause of sedition and evil. Confucius (551–478 BC) and other Chinese philosophers suggested that "excessive growth would reduce output per worker, repress levels of living and engender strife."

Tertullian (AD 160–220) described famine and war as factors that can prevent overpopulation: "The scourges of pestilence, famine, wars and earthquakes have come to be regarded as a blessing to overcrowded nations" (Neurath, 1994).

Attitudes toward human reproductivity before and during the Middle Ages mostly supported the biblical command: "Be ye fruitful and multiply." During the 16th and 17th centuries, both pros and cons of population growth were debated from different viewpoints. Religious dominants like Martin Luther stated that "God makes children. He is also going to feed them." This may have been a righteous approach at one time, but it is a dangerous assumption in contemporary times. The various reasons why human population growth must be curtailed are evident throughout this book.

In spite of potential hardships arising because of increasing population size, populations sometimes increase for purely selfish reasons. Some philosophers, for instance, recommended having large populations because they favored more production and increased exports, leading to greater wealth. This is largely the approach by contemporary world leaders. While religious figures pointed out that population should not increase beyond its food supply, they expressed that "the greatness of a city rests on the multitude of its inhabitants and their power" (Neurath, 1994).

In 1798, Thomas Malthus published *An Essay on the Principle of Population* in which he stated that "positive checks on exponential population growth, e.g., disease, war, disaster, famine and genocide, would ultimately save humanity from itself" and that human misery was an "absolute necessary consequence" of overpopulation. He advocated the use of "moral restraint" or voluntary abstinence to slow the population growth rate. He further stated that we need compulsory birth regulation by adding "temporary sterilants to water supplies or staple food" (Knudsen, 2006).

Once the world's philosophers recognized serious problems with exponential population growth, their thoughts automatically became directed at various means of keeping the population in check, for example, contraception and abortion, and such subjects are well documented in early history. Both the Ebers Papyrus (1550 BCE) (Carpenter et al., 2006) and the Kahun Papyrus (1850 BCE) (Mesopotamia and Ancient Egypt) document the use of honey, acacia leaves, and lint to be placed in the vagina to block sperm. The Kahun Gynecological Papyrus (1850 BCE) describes various contraceptive pessaries, including acacia gum, which research has confirmed to have spermacidal qualities and is still used in contraceptive jellies (Stevens, 1974). Other contraceptive methods mentioned in the papyrus include the application of gummy substances to cover the uterine cervix, a mixture of honey and sodium carbonate applied to the inside of the vagina, and a pessary made from crocodile dung. Prolonged lactation of up to 3 years was also used for birth control purposes in ancient Egypt (Bonte and Van Balen, 1969).

Withdrawal, or coitus interruptus, is mentioned as a method of contraception in the *Book of Genesis* (Onan "spills his seed" on the ground so as to not father a child with his deceased brother's wife Tamar).

In ancient Greece, plants with contraceptive properties were used from the 7th century BCE onwards and documented by numerous ancient writers, for example, Hippocrates (Farnsworth et al., 1975). The botanist Theophrastus, documented the use of silphium (*Ferula* or *Thapsia* species), a plant well known for its contraceptive and abortifacient properties. Asafoetida, a plant extract with similar properties to silphium, was also used for its contraceptive properties. Other plants commonly used for birth control in ancient Greece included Queen Anne's lace, willow, date palm, pomegranate, pennyroyal, artemisia, myrrh, and rue. Since some of these plants are toxic, ancient Greek documents recommend using safe dosages. Research has confirmed the birth control properties of many of these plants. Queen Anne's lace, for instance, has postcoital antifertility properties and is still employed as a birth control measure in India. Like the Greeks, ancient Romans used this plant for both contraception and abortion.

In the 7th century BC Far East, the Chinese physician Master Tung-hsuan documented both coitus reservatus and coitus obstructus to prevent the release of semen during intercourse. Sun Ssu-mo documented an oil and quicksilver preparation that was heated together for one day and taken orally (Middleberg, 2003).

In India, a variety of birth control prescriptions, mainly made up of herbs and other plants, are listed in the 12th century *Ratirahasya* ("Secrets of Love") and the *Ananga Ranga* ("The Stage of the God of Love") (Middleberg, 2003). Indians also used a potion made from powdered palm leaf and red chalk, as well as pessaries made of honey, ghee, rock salt, or the seeds of the palasa tree.

In late 9th to early 10th century Islam, the Persian physician Muhammad ibn Zakariya al-Razi documented coitus interruptus, preventing ejaculation and the use of pessaries to block the cervix as birth control methods. His descriptions included elephant dung, cabbages, and pitch, used alone or in combination (McTavish, 2007). During the same period, Ali ibn Abbas al-Majusi documented the use of pessaries made of rock salt for women for whom pregnancy may be dangerous. In the early 10th century, Abu Ali al-Hussain ibn Abdallah ibn Sina, included a chapter on birth control in his medical encyclopedia *The Canon of Medicine*, listing 20 different methods of preventing conception (Middleberg, 2003).

In spite of recommendations by philosophers in world populations, religious leaders influenced humans to procreate. In medieval Western Europe, any efforts to halt or prevent pregnancy were deemed immoral by the Catholic Church (Cuomo, 2010). However, women of the time continued to use a number of birth control measures such as coitus interruptus, inserting lily root and rue into the vagina, and infanticide after birth (McTavish, 2007).

Knowledge of herbal abortifacients and contraceptives to regulate fertility decreased in the Early Modern period— John M. Riddle attributed this to attempts of European states to "repopulate" Europe after dramatic losses following the plague epidemics that started in 1348 (Heinsohn and Sterga, 1999, 2004).

In 1484, witches were accused of having "slain infants yet in the mother's womb" (abortion) and of "hindering men from performing the sexual act and women from conceiving" (contraception). Witches were also accused of infanticide and having the power to steal men's penises (Broedel, 2004).

Barrier methods such as use of condoms have been around at least for a thousand years, but until recently they were primarily aimed at preventing sexually transmitted diseases. While between 6 and 9 billion condoms are currently sold every year, the first use of condoms was recorded about a thousand years ago (Donadio and Goodstein, 2010). Cave paintings dating back to the year AD 200 depict condom use, the earliest known visual evidence of their use. In the 1500s, an Italian doctor by the name of Gabrielle Fallopius suggested that linen sheath condoms be used to protect against syphilis, a venereal disease which resulted in a deadly epidemic at that time in history.

Very little is known about the early condoms in history, and some controversy exists about the origin of condoms even at later times. It is believed by some that farmers in Condom, France, began using sheep guts as condoms in the 1640s, which is possibly the origin of the lambskin condom. Others believe that in the 1660s, the term "condom" was coined when Charles II was given oiled sheep intestines to use as condoms by a physician named, Dr. Condom. Still others insist that the "condom" came from the Latin word *condus* which simply means "vessel."

Casanova in the 18th century was one of the first reported for using what, at the time, was called "assurance caps" to prevent impregnating his mistresses (Fryer, 1965; Dingwall, 1953).

In 1855, rubber was introduced as a component of condoms. At that time, men were advised that the rubber version could be washed and reused until they crumbled. Subsequently, the single-use, latex condom was born. By World War II, latex condoms were mass produced and given to troops globally. In the 1950s, the latex condom was improved by making them thinner, tighter, and lubricated. Also, the reservoir tip was introduced that collects semen in the end, decreasing the risk of leakage and unintentional pregnancy.

The emergence of HIV as a sexually transmitted disease has placed condoms in the reproductive mainstream. Experts agree that condoms are the best way outside of abstinence to avoid HIV.

Thoughts about avoiding pregnancy were also taking inventurous physicians into the realm of the female reproductive system. The history of intrauterine devices (IUDs) dates back to the early 1900s. In 1909, Richard Richter developed the first IUD made from silkworm gut, which was further developed and marketed in Germany, but it was not widely used (Wyndham, 2012). Unlike IUDs of today, early IUDs crossed both the vagina and the uterus, causing a high incidence of pelvic inflammatory disease in a time period when gonorrhea was endemic.

Ernst Gräfenberg, another German physician, created the first Ring IUD, made of silver filaments. His work was suppressed during the Nazi regime, when contraception was considered a threat to Aryan women (Wyndham, 2012). He moved to the United States, where his colleagues H. Hall and M. Stone took up his work after his death and created the stainless steel Hall-Stone Ring. A Japanese doctor named Tenrei Ota also developed a silver or gold IUD called the Precea or Pressure Ring.

Jack Lippes promoted the use of IUDs in the United States in the late 1950s. At this time, thermoplastics, which can bend for insertion and retain its original shape, subsequently became the material used for IUDs. Lippes also devised the addition of the monofilament nylon string, which facilitates IUD removal. The trapezoid shaped Lippes Loop IUD became one of the most popular first generation IUDs. In the following years, many different shaped plastic IUDs were invented and marketed, including some poorly designed devices that caused bacterial infection.

The invention of the copper IUD in the 1960s brought with it the T-shaped design used by most modern IUDs. US physician, Howard Tatum, determined that the T-shape would work better with the shape of the uterus, which forms a T when contracted, and he predicted this would reduce rates of IUD expulsion. Together, Tatum and others discovered that copper could be an effective spermicide, and they developed the first copper IUD. Improvements led to the creation of the currently preferred copper IUD (Wyndham, 2012).

The hormonal IUD was also developed in the 1960s and 1970s. Initially, the goal was to mitigate the increased menstrual bleeding associated with copper and inert IUDs. The first model was conceived of by Antonio Scommengna and created by Tapani J.V. Luukkainen, but the device only lasted through 1 year of use (Hammerstein, 1987). A commercial hormonal IUD, which is used even in 2016, was also developed by Luukkainen and released in 1976 (Wyndham, 2012). The manufacturer of this model became the target of multiple lawsuits over allegations that Bayer failed to adequately warn users that the IUD could pierce the uterus and move to other parts of the body (Blue et al., 2002).

Birth control became an important political issue in Britain during the 19th century, based on statements by Thomas Malthus, whose ideas came to carry great weight in British political debate in the 19th century. Malthusians were in favor of limiting population growth and began actively promoting birth control through a variety of groups. The power of free choice by women was simultaneously coming into vogue. The term "voluntary motherhood" was coined by feminists in the 1870s as a political critique of "involuntary motherhood" and as expressing a desire for women's emancipation (Gordon and Speroff, 2002). Yet, advocates for voluntary motherhood disapproved of contraception, arguing that women should only engage in sex for the purpose of procreation (Gordon and Speroff, 2002).

In contrast, the overall birth control movement advocated contraception to allow sexual intercourse as desired without the risk of pregnancy (Gordon and Speroff, 2002). By emphasizing "control," leaders of the birth control movement argued that women should have control over their own reproduction, a decision that was closely tied to the emerging feminist movement.

The Malthusian League, which was established in 1877, promoted the education of the public about the importance of family planning and advocated the elimination of penalties against the promoters of birth control (Simms, 1977). It was initially founded during the "Knowlton trial" of Annie Besant and Charles Bradlaugh in July 1877 (D'arcy, 1977), who were prosecuted for publishing Charles Knowlton's *Fruits of Philosophy* which explained various methods of birth control. Besant and Bradlaugh wrote that it was "...more moral to prevent the conception of children, than, after they are born, to murder them by want of food, air and clothing." The trial of Bradlaugh and Besant triggered a wave of public interest in contraception, and sales of Knowlton's book surged (McLaren, 1978).

Starting in the 1880s, birth rates began to drop steadily in industrialized countries, as women married later and families in urban living conditions increasingly favored having fewer children. This trend was particularly acute in the United Kingdom, where birth rates declined from almost 35.5 births per 1000 in the 1870s to about 29 per 1000 by 1900. While the cause is uncertain, the 29% decline within a generation shows that birth control methods used by Victorian women were effective. Many women were educated about contraception and how to avoid pregnancy. While the rhythm method was not yet understood, condoms and diaphragms made of vulcanized rubber were reliable and inexpensive (Draznin, 2001).

In the United States, contraception had been legal throughout most of the 19th century, but in the 1870s, the Comstock Act and various state laws outlawed the distribution of information about safe sex and the use of contraceptives. Margaret Sanger and Otto Bobsein popularized the phrase "birth control" in 1914 (Meyer, 2004). Sanger was mainly active in the United States, but had gained an international reputation by the 1930s.

Sanger established a short lived birth control clinic in 1916, which was shut down just nine days later. She was arrested for distributing contraceptives and went on trial (Lepore, 2011). Here as well, the publicity surrounding the arrest, trial, and appeal sparked birth control activism across the United States and earned the support of numerous donors, who would provide her with funding and support for future endeavors. She went on to found the first birth control league in America in 1921.

The first permanent birth control clinic was established in Britain in 1921 by the birth control campaigner Marie Stopes, in collaboration with the Malthusian League. Stopes, who exchanged ideas with Sanger, wrote a book *Married Love* on birth control in 1918; it was eventually published privately due to its controversial nature (Greer, 1984; Rose, 1992). The book was an instant success, requiring five editions in the first year and elevating Stopes to a national figure. Its success was followed up with *Wise Parenthood: A Book for Married People*, a manual on birth control, published later that year (Hall, 1977). She originally tried to publicize her message through the dissemination of pamphlets in the slums of East London, but this approach failed to work, as the working class was too mistrustful of well-intentioned meddlers at the time.

In 1921, after years of planning, Stopes and her husband Humphrey Verdon Roe opened the Mothers' Clinic in Holloway, North London (Hall, 1977). The clinic, run by midwives and supported by visiting doctors, offered mothers birth control advice and taught them the use of a cervical cap. Later in the same year, Stopes founded the *Society for Constructive Birth Control and Racial Progress*, a support organization for the clinic. Her clinic made contraception acceptable during the 1920s by framing it in scientific terms, and they subsequently gained an international reputation. The Malthusian League opened up a second clinic shortly afterward (Wyndham, 2012).

These two clinics "opened up a new period in the history of the movement aimed at the emancipation of women from their slavery to the reproductive function." Although the clinic helped few patients in 1921, "the year was one of the most important in the whole history of birth control simply because of their very existence" (Clive and Suithers, 1970).

Throughout the 1920s, Stopes and other feminist pioneers, including Dora Russell and Stella Browne, played a major role in breaking down taboos about sex and increasing knowledge, pleasure, and improved reproductive health. Stopes was particularly influential in helping emerging birth control movements in a number of British colonies (Blue et al., 2002). In 1930, the National Birth Control Council was formed. Stella Browne's initial activism was limited to giving speaking tours across the country, providing information on birth control, women's health problems, tribulations related to puberty and sex education (Hall, 1977).

In 1929, Browne began to openly appeal for the legalization of abortion (Hall, 1977). In 1930, the Birth Control Conference assembled 700 delegates and was successful in bringing birth control and abortion into the political sphere—three months later, the Ministry of Health allowed local authorities to give birth control advice in welfare centers.

The societal acceptance of birth control required the separation of sexual activity from procreation, making birth control a highly controversial subject in some countries at some points during the 20th century. Birth control also became a major theme in feminist politics. As an example, reproductive issues were cited as examples of women's powerlessness to exercise their rights. Starting in the 1930s and intensifying in the 1960s and 1970s, the birth control movement advocated the legalization of abortion and large-scale education campaigns about contraception by governments (Gordon, 2002). In a broader context, birth control became an arena for conflict between liberal and conservative values, resulting in questions about family, personal freedom, state intervention, religion in politics, sexual morality, and social welfare (Gordon, 2002).

Gregory Pincus and John Rock, with help from the Planned Parenthood Federation of America, developed the first birth control pills in the 1950s, which became publicly available in the 1960s (Poston, 2010). Medical abortion became an alternative to surgical abortion, with the availability of prostaglandin analogs in the 1970s and the availability of mifepristone in the 1980s (Kulier et al., 2011).

In 1965, a Connecticut law prohibiting the use of contraceptives was deemed a violation of the constitutional "right to marital privacy." In 1972, rights to possess and use contraceptives became available to unmarried couples.

In France, the 1920 Birth Law contained a clause that criminalized dissemination of birth-control literature (Soubiran, 1969). That law, however, was annulled in 1967, thus authorizing contraception, which was followed in 1975 with the Veil Law. Women fought for reproductive rights, and they helped end the nation's ban on birth control in 1965 (Hunt et al., 2009). In 1994, 5% of French women aged 20 to 49, who were at risk of unintended pregnancy, did not use contraception (Laurent and Leridon, 1998).

The availability of contraception in the Republic of Ireland was illegal in the Irish Free State (later the Republic of Ireland) from 1935 until 1980, when it was legalized with strong restrictions. This reflected Catholic teachings on sexual morality. In Italy, women gained the right to access birth control information in 1970 (Hunt et al., 2009).

In the Soviet Union, birth control was made readily available to facilitate social equality between men and women. Alexandra Kollontai, USSR commissar for public welfare, promoted birth control education for adults. A recent, well-studied example of governmental restriction of birth control in order to promote higher birth rates is found in Ceauşescu-era Romania (Kligman, 1998; Lataianu, 2001). The surge in births resulting from Decree 770 led to great hardships for children and parents, matched by an increase in illegal abortions. In Eastern Europe and Russia, natality fell abruptly after the dissolution of the Soviet Union.

With the new rights people obtained toward controlling their reproductive output, improved methods of sanitation and the impact of modern medicine, population growth is leveling off (demographic transition) in certain areas of the world, but it is increasing in other parts. Increased attention to family planning, sexual education, along with more women entering the corporate world, is resulting in slower population growth, reportedly because women are choosing to put off having children until later in life, having less children or no children at all. On the other hand, an increase in reproductivity in undeveloped countries is contributing to a significant rise in population.

We know that the human population has already reached a point in which it is causing environmental disorder. Throughout this book are numerous examples of ways humans have influenced environmental change because of overpopulation. While some investigators see little chance of saving the world, there are others who have not lost faith in humans. As pointed out by R. Engleman (2012), there are steps we can take toward halting population growth and even decreasing it, but it must be implemented with some degree of responsibility on the behalf of world citizens. He outlines nine possible strategies that could promote an environmentally sustainable existence:

- Provide universal access to safe contraceptive options for both sexes. Sexual partners must have easy access to a range of safe, effective and affordable contraceptive methods, along with information on how to use them. There are many birth control options available, including pills, patches, injections, diaphragms, condoms and copper, and hormone-infused IUDs.
- Guarantee education through secondary school for all people, especially females.
- Eradicate gender bias from law, economic opportunity, health, and culture. Women must be allowed to make their own decisions about their reproductive lives and be free from fear of coercion or pressure from partners, family, and society.
- Offer age-appropriate sexuality education to all students.
- Terminate all policies that reward parents financially based on the number of children they have.
- Integrate lessons on population, environment, and development into school curricula at multiple levels.
- Put prices on environmental costs and impacts.
- Adjust to an aging population instead of boosting child-bearing through government incentives and programs.
- Convince dominants within our societies to commit to stabilizing population growth through the exercise of human rights and development.

Engleman suggests that if all of the strategies in the preceding list were to be put into effect, global populations would most likely peak and subsequently undergo a gradual decline, resulting in a more sustainable Earth. But that's the rub: People must take overpopulation and its negative impact on Earth seriously before we can see promising results. And while citizens may be convinced, politicians must be made aware as well. Of course, implementation may be negatively influenced by religious and cultural beliefs. The Roman Catholic Church, for instance, has continued to oppose abortion, contraception and sterilization as a general practice, specifically with regards to policies developed for population control.

With modern medicine curing diseases like never before, people are living longer, in spite of the rapidly evolving capabilities of pathogenic microbes. Our reproductive rate is high. With agriculture producing food like never before and modern technology providing protection and comfort like never before, many humans feel that they are experiencing the comforts of modern living like never before. But our comforts are real and will they last? Are they laced with certain threats involving an increase in the consumption of genetically modified foods and the loss of worldly resources that will cause friction among our planetary brethren? And are we willing to give up these comforts because religious and political dominants tell us that we cannot or should not control our population.

According to A. Srikanthan and Reid (*Religious and Cultural Influences on Contraception*) (2008) and F. Campbell (*Birth Control and the Christian Churches*) (1960), religious adherents vary widely in their views on birth control. Some religious leaders disagree with their respective religions and find that their own opinions of the use of birth control differ from the beliefs of their religious dominants. In addition, many grapple with the ethical dilemma of what is conceived as the moral thing to do, according to their faith, versus personal circumstance, reason, and choice.

Among Christian denominations today, there are a large variety of positions toward contraception. The Roman Catholic Church has disallowed artificial contraception for their followers as far back as one can historically trace. In early history, a population increase was an acceptable thing to do, but under modern conditions of overpopulation and all the ailments that accompany it, population increases are destroying the planet. Contraception was also officially disallowed by non-Catholic Christians until 1930 when the Anglican Communion changed its policy. Most Protestant groups subsequently came to accept the use of modern contraceptives as a matter of biblically allowable freedom of conscience.

What will be easier for our future generations, continued or even better comforts in life or an overcrowded, sick Earth with fewer resources and polluted environment? I think the answer is obvious, but what do we intend to do about it?

WHAT THIS BOOK HAS ATTEMPTED TO DO

The Great Game of Life, Aggression and Dominance in Humans and Other Animals is an attempt to describe what humans are, where they obtained some of the traits that characterize them as both animal and human and what they have done and are doing to planet Earth and its co-inhabiting species. Thus, emphasis is placed on understanding and protecting the environment and its biota and the growing human population.

To readers who say that Lovelock and others who bring up negative facts about how humans have interacted with Earth and its ecosystems are alarmists, my response to them is this: Somehow, among all the scientific investigations that have been carried out around the planet and the enormous amount of data that have been produced from them, the entire point has been missed.[6]

The skeptical reader's first question may be: Is information provided by the scientific community reliable? To answer this, let us analyze what scientists do. The overwhelming goal of a scientific investigation, any scientific investigation, is to seek the truth about whatever subject is being investigated. In seeking the truth, scientists with vast knowledge about their subjects' test and retest hypotheses repeatedly and analyze abundant data, attempting to find out what is behind the subjects they are investigating. Other scientists often repeat investigations that have already been carried out by earlier investigators, in order to test their hypotheses, and final statements are based on comparative analyses.

The information provided in the forgoing chapters has been gleaned from the results of some of those investigations. What has been said in each chapter is based on what science and the extraordinary people who have carried out the investigations feel is the truth.

CRITICISM OF SCIENCE BY ANTISCIENCE PERSONALITIES

As we might expect, there are people who claim that scientific investigations are not necessary and sometimes untruthful. Often, statements made by such people have little basis, but they are made and often influence part of the population to be skeptical toward scientific findings (Swartz and Lederman, 2008).

Consider what the world would be like if we did not have scientific investigations. We would live under substandard conditions, walk naked, and barefooted through grasslands and forests (assuming that some such habitats would continue to exist), become heavily parasitized like most animals that overpopulate their ecosystems, use outdoor waste-disposal systems or spread our wastes randomly through the environment, freeze in the winter and become overheated in the summer (unless we live in caves).

The electronic world would not exist. We would be unable to watch TV or use electronic devices of any kind, and many other conveniences that have been bestowed upon us. Cyberspace and do-anything telephones would not exist. Is this how people who talk negatively about science live? I do not think so.

There are antiscience movements in human populations around the world which criticize the workings of science. Many of these movements are groups of people who abhor the thought that humans and other animals evolve rather than are created. Religious groups often are intolerant of the findings of science. More oddly, like prejudiced individuals who are intolerant of others who have a different ethnicity, skin color or came from "the other side of the track," they are often intolerant of other religious groups as well. Recall that the concept of evolution was ridiculed from its inception and ridicule continues today by people who do not understand or want to understand the basic nature of natural selection. Other criticism is directed at scientific findings that may affect lucrative endeavors.

According to an anonymous article in a Flathead Area Secular Humanist Association publication called the *Top US Antiscience Groups*, antiscience sentiments are especially seen in groups of people who are associated with the following topics. I quote them verbatim so as not to destroy their precise meaning:

- Climate Change
  Heartland Institute—Largely, a mouthpiece for big-oil interests that spews misinformation about climate science in order to keep big-oil's profit margin high.
- Evolution
  Answers in Genesis—a nonprofit Christian apologetics ministry with a particular focus on supporting Young Earth creationism and a literal interpretation of the Book of Genesis.
  The Discovery Institute—a nonprofit public policy think tank based in Seattle, Washington, best known for its advocacy of intelligent design. Founded in 1990, the institute describes its purpose as promoting "ideas in the common sense tradition of representative government, the free market, and individual liberty." Its Teach-the-Controversy campaign aims to teach creationist anti-evolution beliefs in United States public high school science courses alongside accepted scientific theories, positing that a scientific controversy exists over these subjects.
- Health
  Age of Autism—Promoting an imaginary link between vaccinations and autism.
  Huffington Post (Health Sections)—HuffPo aggressively promotes worthless alternative medicine such as homeopathy, detoxification, and the thoroughly debunked vaccine-autism link. In 2009, Salon.com published a lengthy critique of HuffPo's unscientific (and often exactly wrong) health advice, subtitled "Why bogus treatments and crackpot medical theories dominate 'The Internet Newspaper.'" HuffPo's tradition is neither new nor just a once-in-a-while thing.
  Mercola—promoting a wide range of bogus health products and information.
  Natural News—a Website founded and owned by self-proclaimed "health ranger," Mike Adams. According to Jeff Black, in *Rousseau's Critique of Science: A Commentary on the Discourse on the Sciences and the Arts* (2009), the site promotes almost every sort of medical woo known, though it specializes in promoting vaccine hysteria and quack cancer medicine. The site also promotes conspiracy theories concerning modern medicine, geared to gain sympathy for alternative medicine. Mission is based on paranoid theories about the links between government, food industry and big-pharma, spews inaccurate information about a wide range of health issues.
- General Science
  Answers in Genesis (mentioned earlier)—a nonprofit Christian apologetics ministry with a particular focus on supporting Young Earth creationism and a literal interpretation of the Book of Genesis.
  Character Health—their statement is, "We believe in the direct creation by God of the physical universe, all spirit beings, man and lower forms of life, without the process of evolution, and also that the early chapters of Genesis are literal and accurate history. God sustains all creation but exists in no necessary relationship to it."
  Conservapedia—promoted to homeschoolers and their parents, it is a great place to get the wrong answer in any branch of science.
  Prison Planet/Info. Wars—the kind of site that tries to come up smelling legit by mixing a teaspoon of scientific accuracy with a dump truck load of 100% crap.
- Other
  9/11 Truth.org—despite overwhelming evidence to the contrary, 9/11 Truth continues to promote a conspiracy theory in which the American government was behind the attacks of 9/11.
  Prison Planet or Info Wars (mentioned earlier)—Vague worldwide plot (New World Order) to control or kill law abiding citizens.

As indicated in the statements, antiscience sentiments issue from a variety of people and organized groups. General philosophical expressions against scientific studies have stemmed from both ancient and contemporary individuals. On the less-critical end, Jean-Jacques Rousseau (*Discourse on the Arts and Sciences*) (1973), for instance, claimed that the sciences are not necessarily bad. Following his not-so-bad approach to science, he also states that personages like René Descartes (dubbed the Father of Modern Philosophy), Francis Bacon (advocate and practitioner of the scientific method during the scientific revolution) (Irving, 2006) and Isaac Newton (regarded by philosophers as one of the most influential scientists of all time and a key figure in the scientific revolution) should be held in high regard (Ward, 1978).

At the other extreme, philosophers like Paul Feyerabend, regard science as an epistemological anarchism. According to him, there are no useful and exception-free methodological rules governing the progress of science or the growth of knowledge. Feyerabend believes that to think science can or should operate according to universal and fixed rules is unrealistic, pernicious and detrimental to science itself.

In making these statements, Feyerabend shows a gross lack of knowledge and understanding of science, especially in the biological sciences, the methods employed and the people who carry it out. Firstly, his comments are not new to antiscientific sentiments. Some Romantic philosophers felt that the intellect can never be separated from the passions of life which embody it. For example, they state that truth and justice cannot be accounted for at the expense of emotion and desire. This, I would think, depends on the type of emotion and desires we are talking about.

As we know, emotion can be expressed in what we may refer to as a "normal" way or by our darkest personalities who have immoral desires which threaten other humans. In the Romantic era, rationality was thus seen by some as a threat to human individuality and creativity, and it is treated in the same fashion by certain individuals today. Friedrich Nietzsche (1844–90), for instance, believed that truth (i.e., scientific thought) is merely a cultural necessity, that there is no moral truth and no scientific natural truth that governs our existence.

Certain sciences are intrinsically more apt than others to make concrete statements. Chemistry and physics, for instance, are sciences for which certain formulas and laws can be recognized because of the intrinsic nature of the science. In most cases, chemical and physical events are predictable. While these sciences are often important to biological concepts, it is especially difficult to apply many rules and laws established in these fields of science to biology because biological systems naturally change through time (commencing with mutations), and thus they have tremendous variation, each population and each individual sometimes reacting to its surroundings in a different way. The process of natural selection weeds out the unfit and results in population changes over time.

Let us look at the field of science in a more positive light. Scientific investigations attempt to analyze all variables that influence its outcome. To do otherwise is not good science. As pointed out earlier, experiments are carried out by individuals who wish to determine the true nature of the subject on which they are working, and experiments are generally replicated exhaustively until they are written up and published.

Many investigations are rather minor, but they are used to further subsequent investigations on the same or similar subjects. When papers are written for publication, they most often present a comparative view, showing a depth to formulating hypotheses. When they are published, they are subjected to peer evaluation and represent answers to questions that the investigator feels are correct. If they show signs of irrational thought, they are rejected. If they show questionable results, another scientist may choose to repeat the experiment to prove or disprove them. At times, certain scientific studies are major breakthroughs, but they are often based on numerous other experiments that lead the way to the breakthroughs.

Scientific breakthroughs during the 19th and 20th centuries (and continued into this century) have demonstrated the extraordinarily innovative nature of humans and were so numerous that it would be impossible to list them here. Scientific contributions to the world may be found in books such as Bryan Bunch and Alexander Hellemans' *The History of Science and Technology: A Browser's Guide to the Great Discoveries, Inventions and the People* (2005); James McClellan and Harold Dorn's *Science and Technology in World History: An Introduction* (2006); and National Geographic's *Concise History of Science and Inventions: An Illustrated Time Line* (2009).

In a 2013 Wikipedia article called *Science*, "modern science" was defined as a way of pursuing knowledge. Over the course of the 19th century, the word "science" became increasingly associated with the scientific method itself, as a disciplined way to study the natural world, including such fields as physics, chemistry, geology, and biology.

Scientific output attracts a multitude of other people from all walks of life, some of which may have other-than-honest uses for information that has been obtained in scientific experiments. At times, when entrepreneurs recognize a lucrative use for the results of scientific experiments, scientific concepts are sometimes used by them for reasons other than for what they were originally designed, in both negative and positive ways.

Yet, there are scientific results that may have been developed under the direction of governments that were used for immoral reasons, such as the development of bombs and other weapons to be used for destructive purposes, but many of those projects were later used for the benefit of humans. The extensive use of scientific innovation during the wars of the last century, for instance, led to the space race, increased life expectancy and the nuclear arms race, giving a widespread public view of the importance of modern science in various walks of life.

Anyone criticizing scientific research for projects like the one just mentioned should realize that many scientific research projects do not have anything to do with governments, lucrative endeavors or sociopaths. They are simply projects carried out by scientists to obtain answers of a scientific nature.

The field of biology was initially introduced to the human mind at the time of the great philosophers, Socrates, Plato, and Aristotle. Modern biological science got its most important boost with the work of Charles Darwin (Fig. 18.1). When he published On the Origin of Species by Means of Natural Selection, or the Preservation of Favored

FIGURE 18.1 Charles Darwin. Modern biological science got its most important boost with the work of Charles Darwin. When he published *On the Origin of Species by Means of Natural Selection, or the Preservation of Favored Races in the Struggle for Life* (1859), establishing descent with modification as the prevailing evolutionary explanation of biological complexity, his theory of natural selection provided a rational explanation of how species originated and changed over time. In subsequent years, scientists have tested and retested his concepts (as investigators in any science should), adding new information and challenging old thoughts, and the concept remains strong, although modified, modern interpretations coming under the title of neodarwinism. *Source: chrisdorney/Shutterstock.com*

Races in the Struggle for Life (1859), establishing descent with modification as the prevailing evolutionary explanation of biological complexity, his theory of natural selection provided a rational explanation of how species originated and changed over time. In subsequent years, scientists have tested and retested his concepts (as investigators in any science should), adding new information and challenging old thoughts, and the concept remains strong, although modified, modern interpretations coming under the title of neodarwinism.

But that is the rub for both creationists and evolutionists. Evolution and creationism are conflicting thoughts, and intolerants on either side of the fence feel the need to criticize one another. Of prime concern in the ongoing argument is the thought that creationists are determined to believe that humans and other animals were created by God and put on Earth in the form that they are presently found in, in 4004 BC, basing their conclusions on scriptures that were written by humans during periods of inspiration. Scientists, on the other hand, base their evidence of evolution on aging rocks, determining the age of fossils found in aged rocks, understanding how the Earth changes, studies of extinct and extant organisms, DNA similarities and differences and rational thought processes involving mutations and natural selection.

With the impact of modern humans on the world, it has been argued most recently that the ultimate purpose of science is to make sense of the nature of human beings. In *Consilience*, for instance, Wilson stated that "The human condition is the most important frontier of the natural sciences." In its own small way, this is also what the present volume attempts to touch upon.

In spite of what science critics say, there is no doubt that the scientific world has contributed greatly to the human world, including an astonishing amount of information about Earth and its inhabitants. Within, on and around Earth, there is an almost inexhaustible supply of biotic material for scientific research, including such fields as human anatomy, physiology, embryology, genetics, and behavior.

Over the years, life for humans has steadily improved. Consider the fields of medicine, psychology and the multitude of modern devices that most of us have to make life more pleasurable. There is hardly a device or concept that makes modern life more pleasurable that does not have a scientific connection.

Due to the collaborative work of thousands upon thousands of scientists throughout the world, we have answers to many of the questions that have been asked through the ages, especially since the middle and late 1800s. An enormous range of scientific literature has been and continues to be published. Scientific journals communicate and document the results of research carried out in colleges, universities, and various other research institutions, serving as an archival record of science. Many additional magazines are written in a more popular style which bring scientific information to the general public.

Since the publication by J.D. Watson and F.H. Crick on the structure of DNA (Bates et al., 1977), an unfathomable number of hypotheses about how evolution works have issued forth from the scientific community, many of which have helped us to understand the process better. Many of the questions that Darwin was not able to answer because of his limited knowledge at the time are now answered or further investigated because of advancements in numerous biological studies, especially in genetics and molecular biology. Repeated efforts eventually either refute the findings of earlier science or support it.

Following the work of Darwin and his contribution to understanding variation between animals on the different islands of the Galopogos, the study of biogeography gained tremendous popularity with the work of Alfred Russel Wallace in the mid-to-late nineteenth century. He first studied the Amazon River extensively and then investigated the Malay Archipelago (the islands located between the mainland of Southeast Asia and Australia). In his research, he examined the flora and fauna and determined what came to be known as Wallace's Line—a line that divides Indonesia from the Australian region—and the distribution of organisms found there. Those closer to Asia were said to be more related to Asian animals while those close to Australia were more related to Australian animals. Such studies are influenced by a knowledge of isolation and population concepts, mutations and natural selection. Because of his extensive early research, Wallace is often called the "Father of Biogeography" (Wilson, 2013).

Following Wallace were a number of other biogeographers who also studied the distribution and isolation of species. Most of those researchers looked at history for explanations, thus making it a descriptive field. In 1967 though, Robert MacArthur and E.O. Wilson published *The Theory of Island Biogeography* (1967). Their book changed the way biogeographers looked at species and made the study of environmental features of that time important to an understanding of their spatial patterns. As a result, island biogeography and the fragmentation of habitats caused by island development and continental drift became popular because it was possible to explain plant and animal patterns on islands and the various land masses that moved over great periods of time. The study of habitat fragmentation in biogeography then led to the development of conservation biology and landscape ecology.

Discoveries in fundamental science can be world changing. For example, a well-documented presentation in Wikipedia summarized the following research topics with the impact they had on the world:

- Research: The strange orbit of Mercury (1859) and other research leading to special (1905) and general relativity (1916).
  Impact: Satellite-based technology such as GPS (1973), satnav and satellite communications.
- Research: Radioactivity (1896) and antimatter (1932).
  Impact: Cancer treatment (1896), nuclear reactors (1942) and weapons (1945), PET scans (1961) and medical research (via isotopic labeling).
- Research: Germ theory (1700).
  Impact: Hygiene, leading to decreased transmission of infectious diseases; antibodies, leading to techniques for disease diagnosis and targeted anticancer therapies.
- Research: Static electricity and magnetism (1600).
  Impact: All modern electronics, including electric lighting, television, electric heating, magnetic tape, loudspeaker, plus the compass and lightning rod.
- Research: Crystallography and quantum mechanics (1900).
  Impact: Semiconductor devices (1906), hence modern computing and telecommunications, including the integration with wireless devices: the mobile phone.
- Research: Diffraction (1655).
  Impact: Optics, hence fiber optic cable (1840s), modern intercontinental communications and cable TV and the Internet.
- Research: Photovoltaic effect (1839).
  Impact: Solar cells (1833), hence solar power, solar powered watches, calculators, and other devices.

- Research: Radio waves (1887).
  Impact: Radio had become used in innumerable ways beyond its better-known areas of telephony, and broadcast television (1927) and radio (1906) entertainment. Other uses included: emergency services, radar (navigation, and weather prediction), sonar, medicine, astronomy, wireless communications, and networking. Radio waves also led researchers to adjacent frequencies, such as microwaves, used worldwide for heating and cooking food.
- Research: Vaccination (1798).
  Impact: Leading to the elimination of most infectious diseases from developed countries and the worldwide eradication of smallpox.

These are but a few of the leading breakthrough in science over the years. We must admit that all scientific investigations do not terminate in breakthroughs that lead to improvements in human health and the comforts that we experience in our contemporary world, and we must also understand that there have been mistakes made by scientists. All humans make mistakes. However, scientists attempt to avoid mistakes. To criticize science for making mistakes is an approach to investigation that reeks of ignorance, often expressed by people who are dominated by mythological beliefs or armchair philosophers who do not understand the dedication and honest nature of individuals in the various scientific fields or the practices used by scientists. Scientists are people, often dedicating their lives to science, and they may occasionally make mistakes. So do philosophers and anyone else who is investigating the unknown. But scientists strive to apply rational thinking and repeat their investigations until they find what they feel is the truth.

Rousseau goes so far as to say that political influence can cultivate sciences to great benefit and that morality's corruption is mostly because of society's bad influence on scientists. Firstly, science is not generally cultivated by politicians unless it benefits them (the politicians) personally. In fact, politicians often ignore scientific data (including data that exposes the declining conditions in an overpopulated world) unless they are put under pressure to recognize it, in spite of its possible potential importance to an improvement of the environment or collectively to the human species. They are usually untrained and uninterested in the sciences, and their avoidance of facts about increasingly unclean environments, the air, the water and many other problems associated with an overpopulated world is beyond their comprehension.

Understanding these subjects and acting on ways to cope with them are important in propelling humans into a homeostatic future. It is my opinion that in this day and age, when environment and planetary homeostasis is of prime importance, people who are destined to be governmental leaders should be forced to take courses on environmental, planetary stability and sustainability. But, of course, the impact of using such courses to help the environment would depend on the personality of the individuals taking it. Sociopathic dominants may somehow use the new information for their benefit.

Many issues and the way that they are handled damage the relationship between science, the media and the political community. As a very broad generalization, for instance, many politicians who do make an effort to use science in societal improvements have a tendency to seek certainties and facts while scientists typically offer probabilities and caveats, not wishing to comment on subjects that have poor foundation. However, that is the nature of science, and scientists are reluctant to make a statement without adequate study.

In addition, politicians' abilities to express themselves in the mass media frequently distort the scientific understanding for the public. Examples in Britain include the controversy over the MMR inoculation, and the 1988 forced resignation of a Government Minister, Edwina Currie, for revealing the high probability that battery farmed eggs were contaminated with *Salmonella*.

Science is subjected to many additional political atrocities. John Horgan, Chris Mooney and researchers from the United States and Canada, for instance, have described Scientific Certainty Argumentation Methods (SCAMs), where an organization or think tank makes it their only goal to cast doubt on supported science because it conflicts with political agendas. Hank Campbell and microbiologist Alex Berezow have described "feel-good fallacies" used in politics, where politicians frame their positions in a way that makes people feel good about supporting certain policies even when scientific evidence shows there is no need to worry or there is no need for dramatic change on current programs. Such is the mark of alternate human behavior.

Science can be split into two functional groups: (1) a group that is interested in pure (basic) science and in which basic research is carried out (pure science); and (2) a group that is interested in problems of economic importance (applied science). While the former approach to science is funded mostly by governmental agencies that recognize that all forms of research are important, the latter group is sometimes supported by businesses that most likely will benefit from the research by making or improving products that can be used for a variety of purposes. The data that is obtained in a scientific manner may be used in many ways by people who are not scientists, even in unethical ways. Thus this is not a criticism of the scientific work that was carried out but with the way the information was subsequently utilized by nonscientists.

Many antiscience activist and pseudoscientific groups are dominated by misguided philosophers and intolerant religious dominants who find fault with human behavior and seek perfection in the human species. Based on what has been said in earlier chapters and knowing something about the diversity of human (and all animal) behavior, this is an impossible goal.

They often ridicule the medical profession as a scientific group for using what they feel is the right medication for an ailment they are treating. This is not a fair judgment because medication is dispersed based on what is known about the illness and medication at the time. Once additional knowledge is obtained about the illness and the medication for which it was used, recommendations about what to do change. In addition, in order to make judgments about humans, their thoughts and behaviors, we must understand the differences between humans in their specific fields.

Medical researchers themselves, for instance, may have a scientific background and may use scientific methods, but many people in the medical field, including the doctors who dispense medications and are so important to the well-being of humans in modern society, are not scientists and do not carry out scientific research. Their training and their beliefs may depend on scientific discoveries, but each person may vary as to how they relate to scientific reasoning. Their views on certain biological subjects, in fact, may be in direct contrast to those of the scientific community.

Anyone who finds fault with the results of scientific investigations may want to look into themselves for a reason for their criticism. Scientists in most cases (especially in university environments) do not knowingly falsify their results. They stand behind them and attempt to understand them on the basis of what is known and what is logical. Maybe disbelief stems from an ignorance of what the scientific approach is. Maybe it is a fear of losing their connection with their more mythological beliefs. Maybe it is self-preservation and greed that makes people ignore them or treat them as insignificant.

For instance, is it monetary rewards, lack of concern for Earth and its inhabitants or lack of understanding that make a person or a nation want to continue destroying natural habitats to build something upon? Is it ignorance, selfish desire for large families, a nonbelief of the facts that science has obtained or a lack of concern for problems humans and other biotic populations have that make individuals want to continue propagating?

Are the same reasons behind a lack of understanding that planet Earth is the only planet we can live on comfortably, that we must preserve it at any cost or suffer the consequences? Is it selfish political power, dominant status or both that make one want an ever-growing population and a continuous supply of followers for sociopathic dominants to influence? Is it a lack of concern, procrastination or lack of understanding that keep us from recycling everything that is possible to recycle?

Is it a throw-down from earlier evolution or religious intolerance that some humans are overwhelmingly intolerant, prejudiced, domeering and war-inclined animals, willing to commit genocide and war upon others of our species who are different from us? Or are we simply animals who have grown technologically beyond our abilities and are caught up in a catch-22 situation that we can't seem to escape from? What alternatives do we have? Can we fix the ills of the planet? It has been suggested that we can if we focus on our problems and ignore our irrational desires. But will we do that?

While some people and organizations are trying to counter the destructive nature of humans, marine and fresh water habitats are still polluted. The air we breathe is often polluted. Because of our numbers, we require an enormous number of domestic animals that need an extraordinary amount of pastureland and food to survive. Their fecal material is so abundant, we cannot use it all, and it, along with commercial fertilizers and pesticides are polluting terrestrial environments, our drinking water, rivers, and oceans. And yet, we do not use natural organic materials to replenish nutrients to the soil and possibly produce forms of energy in the process. We throw trash away that can be recycled, and it ends up in landfills that are toxic to the environment or in the ocean as a gigantic floating garbage dump. Many species, including humans, are suffering and dying because of this.

Acid rain has increased. Toxic and greenhouse gases are increasing. Resources of all types are dwindling. Exotic species are being introduced to every land mass in the world. We continue to have oil spills and rely on fossil fuels for our energy when there are alternate forms of energy that are nonpolluting and available. Have we been responsible to this point for curing the planet's ills?

We are the cause of mass extinctions. At times, it is due to fragmentation or removal of populations in the process of taking and modifying land. In many cases, modification of natural habitats is associated with our demands for food, which, in turn, is related to overpopulation.

OUR ULTIMATE FATE

After looking at the social human animal as we have done throughout the chapters of this book, what can we say about our species, its worthiness as an animal and its destiny as a human that has not already been said in the foregoing chapters or in other contemporary publications? It is certain that other writers, who have been cited in the foregoing chapters, writers who are well attuned to the affairs of the human species and condition of Earth, have emphasized the complexity of human nature and what this rather strange animal has contributed to or taken from the Earth. Thus, those and similar publications offer a multitude of important facts, hypotheses and thoughts that we should pay attention to, not ignore. To understand our species, truly understand it, its dominance and aggressive components, it is worth perusing all forms of literature about them with an open mind.

Humans have reached a point in evolution that has been unmatched by any other species, extinct or extant, ever. We define ourselves in elite humanistic terms, often ignoring the fact that we, like all other species, are the recipients of genes that have been handed down through many forms of organisms by a process we refer to as natural selection. As we would expect from any species with extreme diversity in cognitive abilities but has evolved from a long line of simpler, more primitive ancestors, the human animal is an amalgamation of features from a primitive animalistic past and a more modern technological and almost god-like one.

In recognizing who we are, we should acknowledge that we have arrived at this time in cosmological and biological evolution as an elite brainy species that has been subjected to numerous natural selective events over millions of years, some of which have threatened our survival and reproduction. Just as any surviving animal that has been subjected to repeated trials and tribulations over time, we now stand as descendant mutants who have repeatedly experienced Nature's natural selective processes, facing physical and population threats and subsequently solving them in due course.

While many of us are no different from pre-human species in our approach to the world around us, certain of us are sensitive to the care of both human and nonhuman populations, as well as to the Earth itself. Care is not an animalistic trait. It is a humanistic one, a very important one for a species that controls its environment and world affairs. And yet, our leaders, those who express dominance and aggression most aptly, often practice a more selfish existence and choose to ignore the trials and tribulations of world populations until it is too late to deal with them in a humanistic way. Genocide and mass killing has been our answer to overpopulation and aggressive expressions of territoriality rather than controlling our reproductive rate. Within recent years, we have possibly added complications stemming from genetically modified foods and possible mass sterilization.

Based on what we know, it appears obvious that problems that threaten our existence pretty much relate to our dominant and aggressive nature and inability for many of us to recognize the importance of overpopulation. Biologists and eco-friendly people from many other fields best understand the demise of nonhuman organisms in an overpopulated environment: Food and other resources dwindle, living conditions deteriorate, fighting with one another escalates and populations experience heavy parasitism, greater stress and a sudden population crash. These are concepts that have originated in the minds of our earliest philosophers but have not halted exponential human population growth.

As *The Great Game of Life, Dominance and Aggression in Humans and Other Animals* has shown us, the human species has multiple problems in a contemporary world, but who really knows what our most threatening problems will be in the future? In *The Coming Plague, Newly Emerging Diseases in a World Out of Balance*, author Laurie Garrett (1994) states that "Unpurified drinking water, improper use of antibiotics, local warfare, massive refugee migration and changing social and environmental conditions around the world have fostered the spread of new and potentially devastating viruses and diseases—HIV, Lassa, Ebola and others."

Johathan M. Mann, who contributed the Preface to Garrett's book, states that "The world has rapidly become much more vulnerable to the eruption and, most critically, to the widespread and even global spread of both new and old infectious diseases." Infectious diseases are illnesses that are caused by a biological agent such as bacteria, viruses, and fungi.

Infectious diseases have always been a danger to the human race, but as with other animal populations, infectious diseases are increasingly easy to transmit in overpopulated areas. These diseases can spread from people to people throughout the world. In addition, there are often reservoir animals which can harbor the diseases, keeping them within a threatening distance, and it can be subsequently transferred back to humans. Of course, this is yet another way to control overpopulation.

"A new and heightened vulnerability for the human global population is not mysterious," Garrett says. "Increases in the worldwide movement of people, goods, and ideas is the driving force behind the globalization of disease," especially when people visit certain hot spots where such diseases are most prevalent. Take the Ebola virus from West Africa, for instance, which has the highest case-fatality rate of the group of viruses to which it belongs, averaging 83% since the first outbreaks in 1976. Fatality rates up to 90% have been recorded in one epidemic (2002–03).

People not only travel more in a contemporary world, they travel much more rapidly and go to many more places than ever before. The tourism business, in fact, often offers exotic, out-of-the-way locations as perks to their traveling clients.

A person who contracts an infectious disease and harbors life-threatening microbes may be symptomless in the early stages of parasitism. They can easily board a jet plane and be on another continent before or when the symptoms of illness reveal themselves. The jet plane upon which such a person travels suddenly becomes a zone of threat to other travelers, as well, and the disease is off and running. Garrett mentions that, "Few habitats on the globe remain truly isolated or untouched, as tourists and other travelers penetrate into the most remote and previously inaccessible areas in their search for new vistas, businesses, or recreational areas."

Intolerance between people and quarrels over resources also affect our population homeostasis with microbial threats. On the last page of Garrett's book, she quotes an American journalist, L.F. Stone, as saying that because of global unrest and lack of concern for the environment, "Either we learn to live together [with other people] or we die together." Garret's last statement is, "While the human race battles itself, fighting over [increasingly] crowded turf and scarcer resources, the advantage moves to the microbes' court. We humans represent a viable feeding source for them, and as opportunists, they (as most pathogenic microbes) will overwhelm us if we do not take overpopulation and parasitism by infectious microbes seriously. It's either that or we brace ourselves for the coming plague."

Why is it that we do not accept these signs of planetary distress for the human animal as well as for populations of other organisms? Is it because we feel God will take care of our problems or direct us to the ultimate life on Earth, along with eternal bliss no matter what we do, or is it simply that many of us do not care about what happens when we are gone? Lacking training in the sciences, we can expect that most people are happy in their ignorance of what is happening to Earth and its inhabitants. To put it in more recognized forms, "ignorance is bliss" and "not knowing is better than knowing and worrying."

What about our children, their children and the generations to come? Have we gone sour on our pledge to provide them with a homeostatic world? Haven't biologists and psychologists, the people who understand animalistic and humanistic concepts best, said enough about how animalistic humans are and how they react to environmental threats?

Dominance and aggression in humans run our world and threaten our relationship with both other organisms and fellow humans when it infringes on the topics of intolerance and immoral behavior. An overabundance of our species stresses natural populations of all organisms, the availability of resources and relationships with one another.

Natural selection, a process that helps populations of organisms reach their survival and reproductive goals, has been the mechanism that has shaped all populations through the eons of time since life began on Earth about 4 billion years ago. It has been a problem-solver for many extant populations, helping them to adjust to changing conditions, but it also has been an eliminator of populations that cannot cope. Populations that lack the capacity (the diversity) to change become extinct, and as biological, chemical and geological evidence has shown us, most of the life forms that have graced the Earth with their presence have become extinct.

As the ultimate thinking and spiritual animal, our cognitive and innovative brain has become the product of incredible changes over millions of years from a poorly developed, almost unrecognizable nervous mass at the anterior end of a dorsal hollow nerve cord to a device of extreme complexity which has helped us to become what and who we are. Are we going to treasure the habitable zone in space that we have inherited and the phyletic success that got us to where we are, or are we going to forfeit our achievements to extinction, like so many species have done in the past?

While we may recognize our importance as individuals to humankind, to ourselves and to others we relate to, love and care about, our presence and status on Earth is biologically nothing more than a stage of evolution, caught in a span of time (a Goldilocks moment) that seems (in our mere lifetimes) to be undergoing little change. Yet, over greater spans of time, as with all living organisms, we have undergone enormous change and will continue to change or become extinct, and we will never escape from our biological roots because we remain as biological entities. In the end, no matter how we take care of our species, our Earth and our Earth-bound co-inhabitants and how we adjust to changing conditions, it will be natural selection that determines our ultimate fate.

It is not the strongest species that survive, nor the most intelligent, but the ones most responsive to change. *Charles Darwin*

Note

Superscript numerals appearing in this chapter refer to additional text/explanation given in the appendix.

Appendix

CHAPTER 1

Defining Dominance and Aggression

1. One of the main points brought out in this book concerns the retention of animalistic traits by humans, although they have changed significantly over time in their cognizant abilities and technological skills. As pointed out in Chapter 7, most features demonstrated by humans that have a dominant and aggressive nature have been inherited from a long line of predecessors and their survival and reproductive strategies. Their combined cognizant nature and deceptive abilities often have created an environment of nontrust or unpredictability in some humans, showing that some of their behavior is suited to humans alone. Yet, deceptive behavior in most nonhuman animals is clearly deceptive, and there are cases of extreme deception, as pointed out in these chapters (Plumwood, 2000).

2. Genetically modified plants and their role in human society have been discussed by Black (2012), Engdahl (2007), and others. Since the chemistry of modified foods consumed by humans is poorly understood, we may later learn that genetically modified foods produce numerous deleterious effects.

3. A second significant topic to dwell upon concerning the human population is its increasing size and what this has to do with degrading both the biotic and abiotic world. Most of the tribulations facing the human species are either directly or indirectly related to overpopulation. In *Ecological Psychology: Creating a More Earth-Friendly Human Nature*, G.S. Howard states that "Our world now faces several ecological threats (e.g., global warming, ozone depletion, acid rain, deforestation, soil poisoning, water pollution, and desertification) that could make life in the 21st century nightmarish. All of these threats are human produced. Thus, alterations in humans' thoughts, actions, and lifestyles must be important parts of the problems' solutions. The twin engines of ecological destruction are human overpopulation and overconsumption (i.e., unsustainable lifestyles). Both of these aspects of human nature are malleable to social/political and psychological interventions. Psychologists should become involved in dealing with impending ecological threats, since ecological degradation will be one of the leading causes of human pain and suffering in the next century.

4. To understand human behavior, all fields of study must collaborate and recognize the terminology and investigations in other fields. To do this, we must recognize humans as animals who have not only progressed to an advanced stage of cognizance and technological expertise, but as organisms that display numerous animalistic traits (Haslam et al., 2008).

5. Being social has its benefits in survival and reproduction: dominant individuals have a chance to be the prime genetic contributors to the next and subsequent generations, and the species is able to use cooperative defense to protect members of the group to which they belong. On the other hand, species that are not social have developed other means of self-preservation (De Catanzaro, 1986).

CHAPTER 2

Traits of Dominant Animals

1. While haplometrotic species generally live haplometrotic lives, with some exceptions, species that we refer to as pleometrotic sometimes exist haplometrotically. Whether queens of these species function as haplometrotic or pleometrotic depends on the degree of dominance and tolerance females demonstrate. As pointed out in this chapter, the success of a nest not only depends on the establishment of a dominance hierarchy but also on how tolerant the alpha female is with respect to subordinate cofoundresses. In certain respects, the process of pleometrosis is quite similar to tropical wasp species that establish their nest as a group (group-founding species). With *Polistes annularis*, fertile females return to the nest site they emerged from during the previous

year after hibernating through the winter. They do not reuse the nest but utilize the site to interact in agonistic confrontations which result in establishing a dominance hierarchy. A nest, which is controlled by a dominant female, is subsequently constructed nearby. Group founding is similar in that cofoundresses move directly from their parent nest to a new nest site. There is no hibernation period.

2. Queens are usually capable of defending their nest from certain intruders that approach it from the substrate to which the nest is attached (such as ants, but they are poor defenders against larger predators (Hermann et al., 2017). We have been bitten by queens, even though they possess a sting. Moth and wasp parasitoids are persistent in approaching the nest, resulting in a nervous appearance in defending wasps, and they are usually successful in parasitizing nests, especially during the ergonomic period.

3. Tolerance is defined as the ability or willingness to tolerate something, in particular the existence of opinions or behavior that one does not necessarily agree with (Oxford Dictionary). In the context of this chapter, tolerance pertains to the queen's tolerance toward other members of her colony. In other social animal groups, it relates to the tolerance an alpha individual has toward subordinates in a group. Tolerance by an alpha individual is important in maintaining group homeostasis.

4. Marmots are large squirrels in the genus *Marmota*, of which there are 15 species (Thorington and Hoffman, 2005). The yellow-bellied marmot, *Marmota flaviventris* is found in southwestern Canada and western United States. Marmots have a harem-polygynous mating system in which the male reproduces with two or three mates at the same time. Female offspring tend to stay in the area around their home. Male offspring typically leave when they are yearlings and will defend one or more females (Linzey and Hammerson, 2008).

5. Marmosets are New World monkey species of the genera *Callithrix*, *Cebuella*, *Callibella*, and *Mico*. All four genera belong to the family Callitrichidae (Rylands and Mittermeier, 2009). Most marmosets are about 20 cm (8 in.) long. Relative to other monkeys, they show some apparently primitive features, for example, possessing claws rather than nails and tactile hairs on their wrists. They lack wisdom teeth, and their brain has some primitive features. They exhibit germline chimerism, which is not known to occur in nature in any primates other than callitrichids. Ninety-five percent of marmoset fraternal twins trade blood through chorionic fusions (Ross et al., 2007), making them hematopoietic chimeras. They live in family groups of 3–15, consisting of one to two breeding females, an unrelated male, their offspring, and occasionally extended family members and unrelated individuals. Their mating systems are highly variable and can include monogamy, polygyny, and polyandry. In most species, fraternal twins are usually born, but triplets are known. Like other callitrichines, marmosets are characterized by a high degree of cooperative care of the young and some food sharing and tolerated theft. Adult males, females other than the mother and older offspring, participate in carrying infants (i.e., they function as alloparents). Most groups scent mark and defend the edges of their ranges, but it is unclear if they are truly territorial, as group home ranges greatly overlap.

6. Diamond and Bond (2003) point out that "although social play is broadly distributed among mammals, it is infrequently encountered in other vertebrate taxa. It is, however, displayed in a fully realized and complex form in several groups of birds. Unambiguous accounts of social play have been recorded from 13 species of parrots, seven species of corvids and several hornbills and Eurasian babblers." These investigators "conducted an analysis of the avian play literature, testing for differences between avian taxa, as well as for correlations between play complexity, brain size and age of first reproduction. Corvids were far more likely to show social object play than parrots. Corvids, parrots and hornbills had larger relative brain sizes than would be predicted from a class-level allometric regression, but brain size was not associated with the complexity of social play among genera within taxa. Play complexity within parrots and corvids was, however, significantly associated with the age of first reproduction. The likelihood of complex social play appears to increase when delayed reproduction is accompanied by persisting relationships between adults and post-fledging juveniles. The adaptive significance of social play in birds thus offers intriguing parallels to similar analyses in mammals."

7. Reeve et al. state that "recent evolutionary models of reproductive partitioning within animal societies (known as 'optimal skew,' 'concessions' or 'transactional' models) predict that a dominant individual will often yield some fraction of the group's reproduction to a subordinate as an incentive to stay in the group and help rear the dominant's offspring. These models quantitatively predict how the magnitude of the subordinate's 'staying incentive' will vary with the genetic relatedness between dominant and subordinate, the overall expected group output and the subordinate's expected output if it breeds solitarily." They report that "these predictions accord remarkably well with the observed reproductive partitioning between conesting dominant and subordinate queens in the social paper wasp *Polistes fuscatus*. In particular, the theory correctly predicts that (1) the dominant's share of reproduction, i.e. the skew, increases as the colony cycle progresses and (2) the

skew is positively associated both with the colony's productivity and with the relatedness between dominant and subordinate. Moreover, aggression between foundresses positively correlated with the skew, as predicted by transactional but not alternative tug-of-war models of societal evolution." Thus, their "results provide the strongest quantitative support yet for a unifying model of social evolution."

8. Francis (1988) states that "it is frequently assumed that variation in dominance status is in large part a function of differences in levels of aggressiveness. It is argued that the large role ascribed to aggression with respect to dominance success results from a conceptual error, in particular, the temptation to conflate the two constructs. There may be an association between aggression and social dominance, not because levels of aggression determine dominance success, but because aggression is a manifestation of dominance. It is further argued that the analysis of dominance relationships exclusively in terms of individual attributes such as aggression is flawed. Social dominance is a relational quality, which could be profitably studied from a sociological perspective."

9. Several good books on human personality discuss variations in behaviors: Bandura, 1999; Eysenck, 2013; Zuckerman, 1991.

10. Hoegl and Gemuenden (2001) state that "an extensive body of literature indicates the importance of teamwork to the success of innovative projects. This growing awareness, that 'good teamwork' increases the success of innovative projects, raises new questions: What is teamwork, and how can it be measured? Why and how is teamwork related to the success of innovative projects? How strong is the relationship between teamwork and various measures of project success such as performance or team member satisfaction?" Their article "develops a comprehensive concept of the collaboration in teams, called Teamwork Quality (TWQ). The six facets of the TWQ construct, i.e., communication, coordination, balance of member contributions, mutual support, effort, and cohesion, are specified. Hypotheses regarding the relationship between TWQ and project success are tested using data from 575 team members, team leaders, and managers of 145 German software teams. The results of the structural equation models estimated show that TWQ (as rated by team members) is significantly associated with team performance as rated by team members, team leaders and team-external managers. However, the magnitude of the relationship between TWQ and team performance varies by the perspective of the performance rater, i.e., manager vs. team leader vs. team members. Furthermore, TWQ shows a strong association with team members' personal success (i.e., work satisfaction and learning)."

11. Spence (1985) states "that contemporary theories of achievement and achievement motivation are rooted in individualism and may have validity primarily for American and similar cultures. Concerns about the erosion of the work ethic and the destructive aspects of individualism are discussed."

12. According to Mace et al. (1993), "the field of applied behavior analysis has developed an impressive technology for reducing the myriad aberrant behaviors engaged in by individuals with developmental disabilities. Most interventions," they say, "apply basic behavioral principles (e.g., positive negative reinforcement, punishment and stimulus control) to discourage aberrant responses and promote adaptive behavior."

13. According to Mitchell and Thompson (1986), "deception is a formal property of certain interactions and as such it encompasses a variety of manifestations." In order to provide evidence for a systematic examination of varying types of deception, these authors "combined examples of deceptions based on mimicry with those based on imitation and imagination."

14. Natoli and De Vito (1991) found that a linear dominance hierarchy of male and female feral cats living in a large urban colony was based on the outcome of agonistic encounters among males. They stated that their results "did not correlate with copulatory success. Courting males did not fight around the female in oestrus. The optimal mating strategies of male and female cats conflict: females would do best to copulate with more than one male, whereas males should monopolize the female and guard her from other males. In this study, however, females mated polygamously, but males did not attempt to monopolize females." Possible explanations for this obscure male behavior are presented.

15. As an example of a genuine facad in human society, Mohr (2013) states by "examining the concepts and examples of leadership within the white supremacist movement, an image of harmful leadership begins to develop. These leaders were found to use leadership strategies such as creating purpose, community building, follower focus and meaning making, empowerment and motivation and trust and respect to gain followers. These strategies were combined with detrimental approaches to leadership that included pursuing and abusing power, developing a sense of self-importance, using fear and having a sense of isolation that contributed to a harmful leadership style." Harmful leadership and its implications for leaders and followers in more mainstream organizations are explored by Mohr.

16. The reader is directed to M. J. Connelly's book (2009), *Fatal Misconception: The Struggle to Control World Population*. Connelly states that "the idea of population control is at least as ancient as Plato's Republic, which described how a 'Guardian' class could be bred to rule, the unfit left to die, and everyone sold the same myth that political inequality reflected the natural order of things. It has been argued that some kind of population policy is common to every culture. Most have been pro-natalists, in that they taught people by means more or less subtle—from tax breaks to witch hunts—to be fruitful and multiply." Connelly mentions that "controlling the population of the world is a modern phenomenon." It was at the "end of the nineteenth century, as rival empires seized the few remaining regions that had withstood European settlers, that observers began to see trends in fertility, mortality and migration as interconnected."

CHAPTER 3

Significance of Comparative Studies

1. Most nonbiologists are unfamiliar with the mechanisms of natural selection and evolution, and many who follow a religion are reluctant to delve into its meaning. Most biologists, on the other hand prefer to disclose factual information about how organisms evolved and find beauty in the role mutations play in organismic diversification and how natural selection results in population change (Enquist and Arak, 1994).
2. Both DNA (deoxyribonucleic acid) and RNA (ribonucleic acid) are composed of a series of chemicals called nucleotides (Chargaff, 2012). While DNA is double stranded and represents a blueprint for the making of peptides, forms of single-stranded RNA (messenger and transfer) represent the force and codes important in arranging amino acids for the making of peptides. Ribosomal RNA is important in linking the amino acids together. Forms of RNA also play a role in catalyzing certain biological reactions, controlling certain genetic expressions, and playing a role in sensing and responding to cellular signals.
3. Sexual reproduction involves the production of male and female gametes (e.g., spermatozoa and ova) and their fusion to produce a zygote. It thus requires meiotic divisions to supply haploid gametes. Asexual reproduction is a form of reproduction which does not require gametes. Binary fission by prokaryotes, cellular divisions in eukaryotes, and growing viable clones are examples of asexual reproduction (Bengtsson, 2003).
4. Human embryos pass through these stages in the process of gestation that takes about 9 months, but the process of evolution to develop each stage took millions upon millions of years. The various organisms that currently represent these various developmental stages never progressed to higher levels of development (Cheverud et al., 1983).
5. The development of the amniotic egg in reptiles was a major step taken by vertebrates in their invasion of terrestrial environments. Rather than being deposited in an aquatic environment like their predecessors, amniotic eggs provided the aquatic environment necessary for the development of an embryo. Humans, like birds and other mammals, retain the amniotic egg in their development, although the yolk sac (for embryonic food) and allantois (for the storage of embryonic wastes) have been replaced by a placenta. In a journal article called *Has the Importance of the Amniote Egg Been Overstated*, J. Skulan (2000) states that "the evolution of the amniote egg is commonly regarded as an important milestone in the history of the vertebrates, an innovation that completed the transition from aquatic to fully terrestrial existence by permitting eggs to be laid away from standing water. This view derives ultimately from the recapitulationist theories of Haeckel, and rests on the assumption that extant frogs and salamanders are good models for the reproductive habits of early tetrapods and the ancestors of the amniotes. It also assumes that it is more difficult to lay eggs on land than in water, and that the amniote egg is an adaptation to the physical rigors that eggs encounter in terrestrial environments. Taken together, these assumptions comprise what may be termed the 'Haeckelian framework' for the origin of vertebrate terrestriality. Several independent lines of evidence suggest that the assumptions of the Haeckelian framework are false. There appear to be no theoretical reasons to assume that the evolution of terrestrial egg-laying was difficult, or required a structure as elaborate as the amniote egg. The physical conditions eggs encounter in the terrestrial environments where they are actually laid are quite mild. Land may in fact be an easier place to lay eggs than water. In addition, analysis of the distribution of key reproductive character states among vertebrates provides no evidence that the 'typical amphibian' reproductive mode is primitive for tetrapods. Amniotes are as likely as frogs or salamanders to retain primitive reproductive character states."

CHAPTER 4

Social Nonprimate Animals

1. According to J.U.M. Jarvis et al. (1994), studies of vertebrate and invertebrate sociality have proceeded more or less independently. "Interest in the evolution of eusociality in insects has focused mostly upon intrinsic or genetic selective factors, while analysis of vertebrate cooperative breeding has drawn attention to extrinsic or ecological selective factors. The African mole-rats (Bathyergidae) are of particular interest since, like many families of social insects, they exhibit a range of sociality from solitary through to truly social or eusocial species. They therefore provide a means of comparing cooperatively breeding vertebrates and invertebrates." Jarvis cites numerous authors who have contributed articles on the subject of vertebrate sociality.

2. While most of my work on wasp behavior was carried out on *Polistes annularis*, I and my graduate students researched all of the species that were local in the Athens, Georgia, area. The climate in NE Georgia is quite cold in the winter, but there is a relatively long summer season when compared to more northern states. Most wasp behavior in that part of Georgia was haplometrotic, with a single founding queen. As we go south, cold seasons are mild compared to those in more northern locations, and wasps often stay on the nest throughout the winter period, and predation and parasitism are often more harsh, forcing species to be more pleometrotic and group founding.

3. While this paragraph on birds has been placed immediately behind reptiles because of their genetic link, birds are endothermic like mammals, and some are quite advanced behaviorally. Both birds and mammals arose from a reptile lineage (Feduccia, 1999).

4. The point to be made here is that agonistic behavior in most cases is not life threatening unless both individuals are high on the dominance scale. In most cases, initial threats establish a position of dominance in the hierarchy for every competing individual, and subsequent interactions are subtler, involving body positions and gazes. This is more-or-less a universal phenomenon in dominance reactions between intraspecific agonists. A lack of life-threatening agonistic behaviors is of more value to each species by building tolerance between survivors and eliminating the nontolerant alphas from the population (Enquist and Leimar, 1987).

CHAPTER 5

From Whence We Came: Primates

1. It is difficult to present an absolute up-to-date table of events to naming and classifying hominines since new fossils are being uncovered at a relatively rapid rate, and new techniques are providing new means of investigating them (Washburn, 2013).

2. We could say this about any organisms (that they are not the same as the species they came from) because evolution is an ongoing process that affects every population of organisms (Andersson and Andersson, 1999).

3. Evolution of the mammalian brain from a reptile precursor is little understood. Aboitiz et al. (2002) state that "the isocortex is a distinctive feature of the mammalian brain, with no clear counterpart in other amniotes. There have been long controversies regarding possible homologues of this structure in reptiles and birds. The brains of the latter are characterized by the presence of a structure termed dorsal ventricular ridge (DVR), which receives ascending auditory and visual projections, and has been postulated to be homologous to parts of the mammalian isocortex (i.e., the auditory and the extrastriate visual cortices). Dissenting views, now supported by molecular evidence, claim that the DVR originates from a region termed ventral pallium, while the isocortex may arise mostly from the dorsal pallium (in mammals, the ventral pallium relates to the claustroamygdaloid complex). Although it is possible that in mammals the embryonic ventral pallium contributes cells to the developing isocortex, there is no evidence yet supporting this alternative. The possibility is raised that the expansion of the cerebral cortex in the origin of mammals was a product of a generalized dorsalizing influence in pallial development, at the expense of growth in ventral pallial regions. Importantly, the evidence suggests that organization of sensory projections is significantly different between mammals and sauropsids. In reptiles and birds, some sensory pathways project to the ventral pallium and others project to the dorsal pallium, while in mammals sensory projections end mainly in the dorsal pallium." The authors suggest "a scenario for the origin of the mammalian isocortex which relies on the development of associative circuits between the olfactory, the dorsal and the hippocampal cortices in the earliest mammals."

4. During the period in which the species of *Ardipithecus* reigned (6–2 million years ago), the human brain was similar in size to those of a chimpanzee. Even though they were primitively bipedal, they spent time climbing in trees. Toward the end of this time, humans commenced making primitive tools and hunting. Protein intake had increased with meat-eating habits. From 2 million to 800,000 years ago, dispersal to new areas commenced, and the human brain began to increase in size. Humans in the *Australopithecus* and *Paranthropus* groups increased their proficiency in tool making and began to use fire. From 800,000 to 200,000 years ago, large climate shifts and more demands resulted in a relatively rapid increase in brain size and complexity, allowing them to cope with environmental changes and a more emotional existence (Tattersall, 1999).

5. Ledley (1982) states that "the birth of a child with a caudal appendage resembling a tail generates an unusual amount of interest, excitement and anxiety. There is something seemingly unhuman about the presence on a human infant of a 'tail' like the tails found on other primates. It is incongruous; it violates our sense of anthropocentricity, and it raises issues that involve not only teratology and embryology but also our view of ourselves and our place in evolution. The human tail has long been an object of scientific curiosity." Studies of developing mice by K. D. Economides et al. (2003) have shown that "tail bud outgrowth is a process similar to limb outgrowth, where a ventral ectodermal ridge (VER) has a similar function as the apical ectodermal ridge (AER) on the fates of mesenchymal cells within the tail bud (Goldman et al., 2000). The caudal neural tube, the caudal somites and the neural crest derived caudal spinal ganglia are all derived from the tail bud. This process is well conserved in vertebrates regardless of whether or not the final adult form has a tail, but little is known about the genes involved in determining the final length and pattern of the adult tail." They concluded that "Hoxb13 mutations cause an overgrowth of the caudal spinal cord and tail vertebrae [while] 5′ Hoxgene may act as an inhibitor of neuronal cell proliferation, an activator of apoptotic pathways in the SNT and as a general repressor of growth in the caudal vertebrae."

6. It is not so much the complexity of the base structures in nucleic acids, but the arrangement of nucleotides to form genes that represent the codes for assembling mRNA, tRNA, and peptides.

7. Thorpe et al. state that "human bipedalism is commonly thought to have evolved from a quadrupedal terrestrial precursor, yet some recent paleontological evidence suggests that adaptations for bipedalism arose in an arboreal context. However, the adaptive benefit of arboreal bipedalism has been unknown." Here we show that it allows the most arboreal great ape, the orangutan, to access supports too flexible to be negotiated otherwise. Orangutans react to branch flexibility like humans running on springy tracks, by increasing knee and hip extension, whereas all other primates do the reverse. Human bipedalism is thus less an innovation than an exploitation of a locomotor behavior retained from the common great ape ancestor.

8. This is a case where there are important behavioral differences between two different populations of genomic brethren, relating to dominance and aggression in the use of tools.

9. Throughout time, different populations of humans were on their way to speciation, developing anatomical and behavioral differences due to isolation. Even today, we can see these features. However, any progress toward speciation was slowed or discontinued when humans began to significantly increase the size of their population and travel around the world, resulting in the breeding of different population members and the blending of anatomical and behavioral features.

CHAPTER 6

The Human Animal

1. There were numerous forms of humans and subhumans that led to the rise of *Homo sapiens*. See the geological time table (Table 5.1).

2. Both speciation and extinction are natural events in all groups of organisms. When major environmental changes occur, new conditions often put stress on populations to change or become extinct. Cataclysmic environmental changes occurred during the five major mass extinctions since the Cambrian explosion. Species diversification generally follows mass extinctions, along with a change in the dominant forms of organisms on Earth.

3. The Earth was a molten mass in the initial stages of its formation. Even as it cooled, high amounts of carbon dioxide were present, along with noxious gases, and oxygen was in poor supply. It was not until about 3.5 billion years after its formation when cyanobacteria began releasing oxygen into the atmosphere and the subsequent Great Oxygen Event up to about 2.5 billion years ago that a catastrophic change occurred on Earth, changing an anoxic environment into an aerobic one. Robert M. Hazen precisely documents these events in *The Story of Earth, The First 4.5 Billion Years from Stardust to Living Planet*.

4. We take for granted the type and amount of energy we receive from the sun in the form of electromagnetic waves. Electromagnetic waves generated in the sun's core include ionizing radiation in the form of alpha rays, beta rays, neutron rays, X-rays, and gamma rays, as well as nonionizing electromagnetic radiation, including visible light, infrared waves, microwaves, radio waves, very-low-frequency waves, extremely low-frequency waves, ultraviolet waves, and black body waves. Prior to the build-up of an ozone layer between the troposphere and stratosphere (due to a lack of oxygen on Earth), many of the sun's electromagnetic waves reached Earth, making Earth unfit for terrestrial organisms. R. M. Hazen points out several types of extant organisms that do not require the sun's energy, but most organisms on Earth today are dependent upon the sun for the commencement of food chains and production of oxygen. When photosynthesizing prokaryotes and eukaryotes began releasing oxygen as a waste gas some 3.5 billion years ago and into the Great Oxidation Event 2.5 billion years ago, ultraviolet B and C rays reacted with it to produce O_3 (ozone), and terrestrial organisms were relatively safe from the most destructive waves. The electromagnetic waves that reach Earth, and our planet's size and position from the sun all provide the proper conditions for life as we know it.

5. Early philosophers were under the influence of religious domination and thus considered the Earth as a changeless planet, and organisms were believed to be changeless as well. It took almost 2000 years for the human mind to grasp the significance of DNA and the relationships between organisms on Earth. Robert M. Hazen points out that Earth remained relatively static in some respects during a period referred to as *The Boring Billion, The Mineral Revolution*, about 3.7–2.7 billion years ago:

> If nothing else, the mysterious, not-so-boring billion teaches us that Earth has the potential to settle into an occasional temporary stasis, a benign balance of its many competing forces. Gravity and heat flow, sulfur and oxygen, water and life can find and maintain a stable equilibrium for hundreds of millions of years.

6. While there are several ways that life can be defined, for example, the cell theory which states that all organisms are composed of cells, biologists recognize that the true nature of living organisms is its ability to replicate itself. Replication of a cell involves DNA replication and cell division. Prokaryotes replicate a single, ringed chromosome and subsequently undergo a division referred to as binary fission. Eukaryote organisms replicate their DNA and often undergo genetic changes during mitosis and meiosis.

7. Endosymbiosis was a process in which cellular inclusions, for example, chloroplasts and mitochondria, originated as separate organisms and somehow ended up in certain cells of the organisms that have them. Support for this theory is found in the overall similarity of these organelles to prokaryotes, including their anatomy, number and placement of their single, ringed chromosome, and their duplication by binary fission.

8. Cyanobacteria has been recognized as the first large group of organisms on Earth to function as photoautotrophs. This feature developed in this group of organisms about 3.5 billion years ago, and it has been this and derived groups of organisms that began supplying Earth with oxygen. Following the Great Oxygen Event, Earth has been more or less an aerobic planet.

9. It is natural for all species to overproduce their offspring (generally referred to as an expression of biotic potential). However, environmental carrying capacity (the maximum number of individuals that an environment can support, based on such ecosystem features as food availability, predation, and parasitism) generally cuts the population back to a size that works best with the ecosystem's resources). While humans do not have natural predators, we kill one another, especially when we vie for natural resources. With increasing populations, food abundance will eventually become insufficient and parasitism will become more prevalent.

10. As we can see, speciation and extinction are natural processes that happen in all groups of organisms. However, mass extinctions are caused by catastrophic events. Prior to the rise of *Homo sapiens*, catastrophic events were mostly in the form of environmental changes or asteroid bombardment. Photosynthetic organisms were responsible for a world change in the Great Oxygen Event, but humans may be the first to experience extinction at their own hand while understanding the reason behind their demise.

CHAPTER 8

Human Nature

Its Beginning and Philosophical Base

1. While this chemical apparently existed in placoderms (bony fishlike organisms at the root of the chordate line) and brachiopods (primitive clam-like organisms from Paleozoic seas) over 500 million years ago, it apparently

did not significantly show up in other vertebrates until the ossification of certain skeletal components in cartilaginous fish. The human endoskeleton incorporates both cartilage and ossified tissue into it.

2. Hair, a filamentous biomaterial consisting of the protein keratin, has been described as one of the defining characteristics of mammals. In the human body, lanugo of prenatal individuals is later replaced by vellus hair (fine, often considered a juvenile characteristic) and terminal hairs (androgenic hairs) at various parts of the body. It has therefore been suggested that the presence of vellus hairs in adult humans may be a neotenic feature.

As in other mammals, hair color in humans ranges from brown and black, due to a predominance of eumelanin, to blond and red, due to a predominance of pheomelanin. Other colors result from various combinations of these two types of melanin, or a loss of melanin, for example, in gray or white hair.

Hair in humans may be found growing on most parts of the body, except for areas of glabrous skin, for example, on the palms and soles of the hands and feet and on the lips. While there are numerous explanations for the importance of hair for mammals in general (e.g., thermoregulation, camouflage, mating rituals, protection, mechanoreception, parent/child relationships, and ectoparasitism), its origin is not clear. It probably arose in the common ancestor of mammals, the synapsids, about 310 million years ago.

It has been said that the hairlessness of humans may be due to the loss of functionality in the pseudogene KRTHAP1, a genetic component that is important in the production of keratin, possibly in the early human lineage as far back as 240,000 years ago. Thus, genetic research may provide important information on the rise and loss of hair in the human body. It has been found that mutations in the gene HR can result in complete hair loss.

Of about 5000 extant species of mammals, only a few have lost most of their hair, including elephants, hippopotamuses, rhinoceroses, walruses, pigs, cetaceans, and naked mole rats. Humans are the only primate species that has undergone significant hair loss.

Based on what we know about other great apes, it is likely that early hominines possessed extensive hairiness in a similar fashion to their closest relatives. It has been estimated that humans underwent a change in which they developed numerous eccrine sweat glands over the surface of the body at about the time they were moving from forested areas to more open savanna, about 2.5 million years ago, possibly as a means to facilitate sweating in order to cool the body.

In *Evolution of Bipedality and Loss of Functional Body Hair in Hominids*, Peter Wheeler suggested that "a need for decreased body hair originated as a response to climate change that began approximately 3 million years ago. It was at that time that Earth entered a period of global cooling. Much of the lush, cooler wooded forests were replaced by dry grassland savanna, a habitat that promoted a change in the human cooling system. Humans kept the head hair which [blocks] harmful UV rays, and long terminal hairs in the axillary and pubic areas helped dissipate heat in those regions."

3. The production of milk is poorly understood. Milky secretions are not uncommon in the animal kingdom. The discus fish (*Symphysodon aequifasciata*) feeds its offspring with epidermal mucus secretions. Therapsids (like cynodonts) secreted a nutrient-rich milk-like substance. Pigeons and other birds produce substances referred to as crop milk. Many of these substances appear closely related to prolactin. Even tsetse flies (the flies that are vectors of trypanosomes in Africa) and at least one cockroach species are known to produce milky secretions. In mammals, mammary glands are thought to arise from certain sweat glands, for example, glands of adults called apocrine glands. Modern mammals, including monotremes (egg-laying mammals), marsupials (pouched mammals) and placentals (a group to which humans belong) all have mammary glands. One of the differences in these three groups of mammals is in the way that milk is distributed to the young. In monotremes, there are no teats, and the young lap milk up from the mother's body in the vicinity of the exit ducts. Marsupials and placentals have teats for the young to latch on, those of the former being inside a pouch and those of the latter generally being somewhere in the thoracic area. The number of teats varies among mammals, most mammals having a number of them. It is in the simian primates that the number of teats have been reduced to two.

4. The rise of humans from a chimpanzee-like hominid involved movement from forest to savanna, increase in omnivory, a dramatic change in the endoskeleton to an upright position and bipedal locomotion, increased tool-making skills, a brain size increase, increased social skills, and a dramatic change in body hair.

5. Deforestation to provide increased amounts of land for growing food and genetic engineering to produce increased yields has played a role in keeping up with a growing human population, often at the loss of natural habitat and the organisms that exist in it.

CHAPTER 9

Alternate Human Behavior

1. Dominant status and altruism are most often opposites. Dominance, especially despotic (super) dominance, and selfishness often accompany one another.
2. Alternate, chemical-dependent behaviors may sometimes be corrected by medication.
3. The reader should keep in mind that while psychologists label closely related conditions under separate titles, seeming to suggest separate, well-defined categories of personality, humans are nonetheless biologically driven, and thus there is variation that may cause overlap in one or more of the categories.
4. The difficulty in understanding the variation in behavioral expressions demonstrated in humans and other animals comes from not knowing genetic predisposition and the history of emotional and physical trauma during the formative years.
5. B. S. McEwen (1998) states that the physiological systems activated by stress, developed over 60 years ago, can not only protect and restore but also damage the body. He refers to the long-term effect of the physiological response to stress as the allostatic load, which is critical to survival. "Through allostasis, the autonomic nervous system, the hypothalamic–pituitary–adrenal (HPA) axis and the cardiovascular, metabolic, and immune systems protect the body by responding to internal and external stress. The price of this accommodation to stress can be allostatic load, which is the wear and tear that results from chronic overactivity or underactivity of allostatic systems.
6. Many people possess certain personality disorders listed here but lead a relatively normal existence. The degree or type of disorder may be expressed in different ways, and the past history of the individual experiencing them may only slightly or strongly affect them. A combination of all such personality traits can lead to a dangerous lifestyle.

CHAPTER 10

The Chemical, Physical, and Genetic Nature of Dominance, Aggression and Aberrant Behavior

1. The basic chemicals in Earth's early environment which are thought to have influenced the generation of a living form are: methane (CH_4), ammonia (NH_3), water (H_2O), hydrogen sulfide (H_2S), carbon dioxide (CO_2) or carbon monoxide (CO), and phosphate (PO_4^{3-}). Molecular oxygen (O_2) and ozone (O_3) were either rare or absent. The current hypothesis in the progression of chemical change that led to life, if it did indeed arise on Earth, is that there were three stages: (1) the origin of biological monomers; (2) the origin of biological polymers; and (3) the evolution from molecules to cells.

    According to J. D. Bernal (*The Origin of Life*), evolution may have commenced early, possibly between stages 1 and 2 in the preceding scheme, utilizing particular organic chemicals. The synthesis of organic chemicals on early Earth were probably derived in conjunction with one of the following: (1) energy sources, for example, ultraviolet waves or electrical discharges; (2) delivery by extraterrestrial objects, for example, carbonaceous meteorites (chondrites); or (3) organic synthesis that was driven by impact shocks.

    Various experiments in recent years have delved into how these chemicals could generate life forms, including: (1) a now-famous Miller–Urey experiment in which a reduced mixture of gases was used to form organic monomers; (2) the spontaneous formation of peptide structures under conditions that might plausibly have existed early in Earth's environment, proposed by Sidney W. Fox; (3) a clay model, suggested by A. Graham Cairns-Smith, postulating that complex organic molecules arose gradually on a preexisting, nonorganic platform, for example, silicate crystals in solution; (4) a subsurface model, proposed by Thomas Gold in 1970, that life may have first developed within the Earth, possibly involving nanobes (filamental structures that are smaller than bacteria but which may contain DNA). On this latter point, it has been reasonably well established that microbial life is plentiful on, above, and in Earth, even up to 5 km below the surface, in the form of archaean extremophiles. These primitive organisms do not need the sun for an energy source and synthesize their energy from the chemical environment in which they live.

    The actual chemicals of life are mostly large macromolecules that belong to four groups: (1) carbohydrates, the monomers of which are simple sugars; (2) lipids that are seen in various hydrocarbons (phospholipids) that are common in the plasma membranes of living cells; (3) proteins, the monomers of which are amino acids;

and (4) nucleic acids, the monomers of which are nucleotides. Nucleotides are composed of phosphate, pentose sugar, and a nitrogenous base.

Biologists agree that self-organization and self-replication are the hallmarks of living systems. As an early molecule that could function in this capacity, RNA (ribonucleic acid) works both as an enzyme and as a relay system in the process of protein synthesis. It may very well have been the first genetic material. DNA (deoxyribonucleic acid), a double-stranded molecule with chemical components that are mostly exactly like those of RNA, possibly arose from or assumed its genetic role from RNA. Supporting evidence comes from the following findings: (1) the structure of DNA and RNA are quite similar, the main differences being that DNA is a double-stranded molecule instead of single-stranded like RNA, and there is one nucleotide in RNA that is not in DNA (uracil) and one in DNA that is not in RNA (thymine); (2) some RNAs function as enzymes; (3) some viruses use RNA for heredity; and (4) many of the most fundamental parts of the cell require RNA.

Proteins are the workhorses of cellular functions and are made by using the genes and codes of DNA and RNA. As a container for these macromolecular chemicals, lipids formed into bilayers early in the rise of a living organism, just as they do naturally. Once organic chemicals are enclosed in such a membrane, more complex biochemistry is then possible.

Each cell, whether in a unicellular organism or multicellular one, is a unit of life that must eat, process foods, and eliminate wastes. During a cell's activities, it must break down chemicals (catabolic processes) and build them up (anabolic processes), depending on its needs. In essence, a single cell has to do everything we do but at the cellular level. On the other hand, multicellular life is complicated by the fact that they make up larger organisms and in attempting to undergo metabolic functions, cells must differentiate to assist the movements of substances into, through and out of the various organs and organ systems that accompany the multicellular body. Life on Earth has indeed become complex as the number of cells have increased in multicellular organisms.

What about other hypotheses? Could life have arisen somewhere else? Panspermia is a concept developed by Fred Hoyle, that life developed elsewhere in the universe and arrived on Earth in a dormant form (e.g., spores), possibly being spread by meteorites. It has been suggested that if panspermia had anything to do with the presence of the first life forms on Earth, the origin of simple life forms could have arisen on Mars, since its atmosphere was a more appropriate location to start life than early Earth and minerals containing elements like boron and molybdenum (which are more common on Mars than on Earth) could assist in certain stages in the formation of a simple life form.

2. The human brain is indeed a complex structure. According to the physicist Sir Roger Penrose, "compared to the complexity of a brain, a galaxy is just an inert lump." While its basic anatomy is complex in itself, researchers agree that many of our high-level abilities are carried out by extensive brain networks linking many different areas of the brain. They suggest it may be the structure of these extended networks more than the size of any isolated brain region that is critical for cognitive functioning.

3. In a very real sense, we are all mutants, products of millions of years of genetic change that has led to tremendous variation in all organismic populations. Because of this variation, populations that best coped with environmental change succeeded in surviving and passing their genes on to their offspring. Thus, mutations can give an advantage to individuals that are faced with environmental hardships, but many mutations can result in features that represent a threat to survival and reproduction.

4. Genetic disorders in humans represent an illness caused by one or more abnormalities in the genome, especially a condition that is congenital. Most genetic disorders are quite rare and affect one person in every several thousands or millions.

All genetic disorders are not heritable, that is, passed down from the parents' genes to their offspring. In nonheritable genetic disorders, defects may be caused by new mutations or changes to the DNA. In such cases, the defect will only be heritable if it occurs in the germ line (not in the somatic cells). The same disease, such as some forms of cancer, may be caused by an inherited genetic condition in some people, by new mutations in other people or mainly by environmental causes in still other people. Whether, when, and to what extent a person with the genetic defect or abnormality will actually suffer from the disease is almost always affected by environmental factors and events in the person's development.

Some categories that cover mutations in humans are autosomal dominants, autosomal recessives, X-linked dominants, X-linked recessives and mitochondrial (maternal inheritance).

5. Road rage is a form of human aggressive behavior. The term originated in the United States in 1987–1988 when a rash of freeway shootings occurred on the 405, 110, and 10 freeways in Los Angeles, California. According to Paul Eberle (*Terror on the Highway*) and others, road rage may be defined as aggressive or angry behavior

exhibited by a driver of an automobile or other road vehicle. Such behavior, stemming from interactions between people who may be influenced by other stresses in life, often includes rude gestures, verbal insults, deliberately driving in an unsafe or threatening manner or making threats. Road rage can lead to altercations, assaults, and collisions that result in injuries and even deaths. It can be thought of as an extreme case of aggressive driving and intolerance between one human and another.

More than 300 cases of road rage annually in the United States have ended with serious injuries or even fatalities—1200 incidents per year, according to the AAA Foundation study, and rising yearly throughout the 6 years of the study that examined police records nationally. Studies have found that individuals with road rage are predominantly young (33.0 years of age on average) and male (96.6%).

The National Highway Traffic Safety Administration points out that road rage comes about when "an individual commits a combination of moving traffic offenses so as to endanger other persons or property; an assault with a motor vehicle or other dangerous weapon by the operator or passenger(s) of one motor vehicle on the operator or passenger(s) of another motor vehicle caused by an incident that occurred on a roadway"(National Highway Traffic Safety Administration). This definition makes an important distinction between a traffic offense and a criminal offense.

As our population increases, there are also increasing accounts of rage between people in parking lots, especially on holidays (e.g., Christmas) when people undergo stressful buying sprees. Such exhibitions are the result not only of delinquent behavior but reactions by people who are intolerant and suffering from personal distress.

Many in the field of psychoanalysis say this definition does not cover the entirety of road rage. Psychologists are now considering a new label for road rage, a form of mental illness or a combination of emotional responses culminating in a pattern of behavior or syndrome, implying that the behavior may be outside the control of the perpetrator.

6. Darke (1998) critically reviews the literature on the "reliability and validity of self-reported drug use, criminality and HIV risk-taking among injecting drug users." The author states that the literature shows respectable reliability and validity of self-reported behaviors when compared to biomarkers, criminal records, and collateral interviews." It concludes that the self-reports of drug users are sufficiently reliable and valid to provide descriptions of drug use, drug-related problems, and the natural history of drug use. Also see the reports by Koob and Bloom (1988) and Markou et al. (1998).

7. In a national child care survey, S. L. Hofferth et al. (1991) address the large gap in the understanding "of the employment patterns of mothers and the care of their children while they work." They present new data "on forms of care used for infants, toddlers and school-age children, as well as previously unknown national data on how parents find programs, what alternatives are available, what child care arrangements cost and how parents juggle employment and the care of their children."

CHAPTER 11

Dominance and Aggressiveness in the Workplace

1. Mention of the various personalities in human society has not covered the topic of indolence, the act of being lazy, which is a disinclination to activity or exertion despite having the ability to do so. It is a condition in which certain individuals cannot perform up to their innate potential because of "output failures."

In *Evolution of Self*, L. F. Seltzer defines laziness not as a lack of mobility but of motivation to embark upon and follow through on life's various challenges and difficulties. A phenomenon found in all societies, indolence is poorly understood but may be caused in part by a variety of biological, neurological, and psychological deficits stemming from both innate and learned circumstances.

Indolent people, Seltzer says, lack a sense of self-efficacy, the conviction that if we put our mind to something, we will be effective with it. People who are indolent also lack self-discipline (stemming from unresolved self-doubts and a lack of self-esteem), a lack of interest in whatever an endeavor may be, ambivalence (lacking faith that the action will be worth the effort), a fear of failure, fear of refusal or rejection, sense of discouragement, hopelessness, futility and possibly other fears and an attitude of pessimism, cynicism, hostility, or bitterness.

Lethargic avoidance, Seltzer states, has "virtually everything to do with depression." Depression is characterized by a diminished interest or pleasure in all, or almost all, activities. Underlying this self-restraining orientation, according to Seltzer, is undischarged anger (or rage) from past disappointments, which propels such people into resistant negativity. Because of the depth of past psychological wounds, indolent people are left disheartened, disillusioned, and disenchanted, causing them to refuse to do what otherwise they might achieve without much difficulty.

CHAPTER 12

Dominance in Religion

1. Why do you believe in God? Michael Shermer (Who believes in God—and why?) had been asking people that question for most of his adult life. In 1998, he and Frank Sulloway presented the query in a more official format—along with the question "Why do you think other people believe in God?"—in a survey given to 10,000 Americans. Just a few of the answers they received are as follows:

    A 22-year-old male law student with moderate religious convictions (a self-rated five on a nine-point scale), who was raised by very religious parents and who today calls himself a deist, writes, "I believe in a creator because there seems to be no other possible explanation for the existence of the universe." Shermer and Sulloway found that "other people believe in God to give their lives [more of a] purpose and meaning."

    A 43-year-old male computer scientist and Catholic with very strong religious convictions (a nine on the nine-point scale) had a personal conversion experience, where he had what he believed was a direct contact with God. This conversion experience and ongoing contacts in prayer, he said, formed the only basis for his faith. According to Shermer and Sulloway, other people believe in God because of (1) their upbringing, (2) the church environment and (3) a hope for this contact.

    A 36-year-old male journalist and evangelical Christian with a self-rated eight in religious conviction wrote: "I believe in God because to me there is ample evidence for the existence of an intelligent designer of the universe." Yet, "others accept God out of a purely emotional need for comfort throughout their life and use little of their intellectual capacity to examine the faith to which they adhere."

    A 40-year-old female Catholic nurse with very strong religious convictions (a nine on the nine-point scale) stated that "I believe in God because of the example of my spiritual teacher who believes in God and has unconditional love for people and gives so completely of himself for the good of others. And since I have followed this path, I now treat others so much better." On the other hand, she writes that "I think people initially believe in God because of their parents and unless they start on their own path—where they put a lot of effort into their spiritual part of their life—they continue to believe out of fear."

    When Shermer and Sulloway completed their initial study, they undertook a more extensive analysis of all the written answers people provided in their survey. In addition, they inquired about family demographics, religious background, personality characteristics, and other factors that contribute to religious belief and skepticism. They discovered that the seven strongest predictors of belief in God are: (1) being raised in a religious manner; (2) parents' religiosity; (3) lower levels of education; (4) being female; (5) a large family; (6) lack of conflict with parents; (7) being younger.

2. Spider Woman is an important diety in the mythology of several Native American tribes, including the Navajo, Keresan, and Hopi tribes. In most cases, she is associated with the emergence of life on Earth (as indicated in this chapter). She helps humans by teaching them survival skills. Spider Woman also teaches the Navajos the art of weaving. Before weavers sit down at the loom, they often rub their hands in spider webs to absorb the wisdom and skill of Spider Woman.

    In the Navajo creation story, Spider Woman (Na'ashjéii asdzáá) helps the warrior twins Monster Slayer and Child of Water find their father, the Sun. The Keresan say that Spider Woman gave the corn goddess Iyatiku a basket of seeds to plant.

    According to the Hopi, at the beginning of time, Spider Woman controlled the underworld, the home of the gods, while the sun god, Tawa, ruled the sky. Using only their thoughts, they created the Earth between the two other worlds. Spider Woman molded animals from clay, but they remained lifeless, so she and Tawa spread a soft white blanket over them, said some magic words, and the creatures began to move. Spider Woman then molded people from clay. To bring them to life, she clutched them to her breast and, together with Tawa, sang a song that made them into living beings. She divided the animals and people into the groups that inhabit the Earth today. She also gave men and women specific roles. Women were to watch over the home and men to pray and make offerings to the gods.

Another Hopi myth says that Tawa created insect-like beings and placed them in the First World. Dissatisfied with these creatures, Tawa sent Spider Woman to lead them, first to the Second World and then to the Third World, where they turned into people. Spider Woman taught the people how to plant, weave, and make pottery. A hummingbird gave them fire to help them warm themselves and cook their food. However, when sorcerers brought evil to the Third World, Spider Woman told the people to leave for the Fourth World. They planted trees to climb up to the Fourth World, but none grew tall enough. Finally, Spider Woman told them to sing to a bamboo plant so that it would grow very tall. She led the people up the bamboo stalk to the Fourth World, the one in which the Hopi currently live.

3. According to religious beliefs, the origin of the Bible is God. This poses a dilemma because whether prophets actually were in touch with God cannot be checked, so a question arises about whether prophets or writers of holy scriptures actually were in touch with a spiritual being.
4. The controversy between science and religion still exists, although some have worked out a compromise between the two. It is difficult to understand how such a compromise is possible, for it blends what biologists feel is the truth with undeniable myth. Scientists, especially biological scientists believe in the evolution of species from a prelife form, for example, a protobiont-like organism and the progression of life forms through the process of natural selection, without the intervention of a diety. To them, this is the most logical explanation for the rise and progression of life. In addition, it is an ongoing process, a process which has never stopped, continues to go on, and will continue to go on until all life is eliminated.

To a biologist, there is beauty and logic in the connections between all organisms (e.g., similar DNA structure throughout all life forms), the process of mutation, the pathways taken by populations through the process of natural selection due to environmental stresses placed on them over time for 3.5–4 billion years. This process, of course, does not entertain the mythological concept of a supreme being that has been behind any of the events that have occurred over 4 billion years since Earth and the solar system arose. To them, the universe and all of its events are based on natural consequences, based on scientific explanations.

A person who chooses to follow a religious pathway in life, generally ignores the findings of science, although there is ample backing for the concept of evolution and its various ramifications, finding more solace in a deity who has been responsible for everything that happens and has happened in the universe, including the making of humans, creatures that they believe are not part of the animal kingdom. The deity they choose to follow and communicate with is often believed to be a just and rewarding god who they can count on to lead them through life.

People who follow a religious path believe in scriptures as historical records of events that took place and were recorded by authentic scribes. Thus, the written scriptures are used as proof of the existence of a deity who is in charge of human destiny.

It is difficult for a biological scientist to understand how people with religious interests can rely on scriptures for the belief in a deity, especially when the process of evolution is so logical and has such backing from the fossil record. However, most nonscientists do not understand the process of evolution or care to understand it, fearing that it would take the beauty away from their current beliefs.

The commandments, listed in the Old Testament and based on holy scripture, are basically rules composed by early prophets to follow in leading a moral life. It is claimed to be the word of God. Most humans would follow these rules, no matter what their religion or lack of a religion. Certain humans, however (the ones with personality disorders) choose to disregard them and live their life according to their own rules. In many different ways, they disrupt the lives of other people in the population, as covered in this book:

1. Thou shalt have no other gods before me (God is the only god).
2. Thou shalt show tolerance towards our fellow humans who do not believe in a god or who believes in gods other than the one we believe in.
3. Thou shalt not make unto thee any graven image, or any likeness of anything that is in heaven above, or that is in the Earth beneath, or that is in the water under the earth.
4. Thou shalt not bow down thyself to them, nor serve them: for I the Lord thy God am a jealous God, visiting the iniquity of the fathers upon the children unto the third and fourth generation of them that hate me; and shewing mercy unto thousands of them that love me, and keep my commandments. (Don't make idols.)
5. Thou shalt not take the name of the Lord thy God in vain; for the Lord will not hold him guiltless that taketh his name in vain. (Don't swear falsely in God's name.)
6. Remember the Sabbath day, to keep it holy. Six days shalt thou labour, and do all thy work: But the seventh day is the Sabbath of the Lord thy God: in it thou shalt not do any work, thou, nor thy son, nor thy daughter, thy

manservant, nor thy maidservant, nor thy cattle, nor thy stranger that is within thy gates. For in six days the Lord made heaven and earth, the sea, and all that in them is, and rested the seventh day: wherefore the Lord blessed the Sabbath day, and hallowed it. (Keep the Sabbath.)

7. Honour thy father and thy mother: that thy days may be long upon the land which the Lord thy God giveth thee. (Honor one's parents.)
8. Thou shalt not kill. (Don't murder anyone.)
9. Thou shalt not commit adultery. (Don't commit adultery.)
10. Thou shalt not steal. (Don't steal.)
11. Thou shalt not bear false witness against thy neighbour. (Don't bear false witness.)
12. Thou shalt not covet thy neighbour's house, thou shalt not covet thy neighbour's wife, nor his manservant, nor his maidservant, nor his ox, nor his ass, nor any thing that is thy neighbour's. (Don't covet what others have.)

The commandments are inadequate rules for populations in a modern, overpopulated world, a world in which humans are utilizing more of the planet's resources than can be replenished, killing the Earth's coinhabitants at record speed, not taking care of our families and often raising children in inappropriate environments. Other rules of a contemporary nature that may help the human and other animal populations, as well as the Earth, are:

- Thou shalt not overpopulate the Earth. Instead thou shalt investigate a proper carrying capacity and abide with the findings, using whatever moral way necessary for keeping the population in check. In making this decision, we shalt not depend on criteria that have a mythological basis. In the process, we shall maintain homeostatic habitats and hopefully deem it unnecessary to make more room for agricultural purposes in order to feed an excessive number of humans on Earth.
- Thou shalt not destroy natural habitat beyond the needs of a nonthreatening homeostatic human carrying capacity.
- Thou shalt not pollute the Earth or its water supply. Instead thou shalt determine the best way to recycle whatever can possibly be recycled. In addition, thou shalt encourage all homeowners to recycle their organic materials and other waste materials.
- Thou shalt not construct containment facilities to store human and animal wastes. Also, thou shalt not dispose of such wastes in the ocean or any other aquatic habitat. Instead, thou shalt utilize such wastes to produce energy and fertilizers, and recycle the remainder back to Earth.
- Thou shalt not use pesticides unless absolutely necessary within human homes to keep household pests at bay. Even in homes, thou shalt refrain from polluting our home environment with pesticides, using extremely low dosages or none at all.
- Thou shalt use natural organic fertilizers to fertilize our agricultural plants rather than commercial fertilizes and make efforts to plow them into the ground and contain such fertilizers within the limits of growing areas. Thou shalt also investigate and use other forms of food production, for example., hydroponics, aquaculture, or aquaponics whenever possible.
- Thou shalt spend time and energy into producing a form of energy that can be used to run vehicles without using petroleum products and thus avoiding polluting the environment.
- Thou shalt show respect toward our fellow humans, practicing tolerance whenever possible and loving them as we would like to be loved. Thou shalt not use deceptive means against our fellow humans or Earthly coinhabitants, using truth and logic whenever possible and functioning as one tolerant species with many variations.
- Thou shalt show respect toward all forms of life just as we respect our fellow humans, never killing them because of cravings or for misguided reasons. Instead, thou shalt learn as much as possible about our Earthly coinhabitants so that we may share Earth with them in a peaceful and homeostatic manner.
- Thou shalt live in harmony with all other creatures lower than humans, never overfishing or overhunting them. Instead, we shalt learn their ways so that we can help them when they need it.
- Thou shalt show tolerance toward fellow men, women, and children under all circumstances, counting on tolerance and understanding to maintain homeostasis in the home and in our communities.
- Thou shalt put emphasis on raising children in a favorable environment, without emotional or physical trauma, providing knowledge for them to understand Earth and their responsibilities to keep it homeostatic.
- Thou shalt not consume mind-altering drugs or substances in such a way that will harm our bodies or harm our fellow humans in any way.
- Thou shalt include learning about the environment and all of its creatures in the curricula of all political majors and all other people on Earth, exposing them to these commandments so that they will contemplate the importance of maintaining homeostatic ecosystems and tolerance among humans.

CHAPTER 13

Dominance in Politics

1. According to R. J. Brulle (*Politics and the Environment*), environmental concerns have been part of the U.S. political agenda for nearly 150 years. As early as 1864, the U.S. Congress debated the proper use of national lands, and, motivated by press accounts of the logging of Giant Sequoia trees, decided to protect Yosemite Valley for aesthetic reasons.

   Since then, as industrial and environmental impacts have risen in tandem, environmental politics has expanded its range over an increasingly wide spectrum of political action, ranging from local level land use decisions to global controls over CO_2 emissions. Thus, the study of environmental politics encompasses a range of issues across virtually all political arenas. Using a wide variety of intellectual tools, ranging from legal studies to geospatial analysis, the literature on environmental politics has expanded into an immense field.

   While the environment has been the subject of concern during this period, the effect of human population size has been less of a concern, although the higher the population size is, the greater is the impact on environmental degradation. More emphasis is put on ridiculing birth control practices than considering the effects of a continuously climbing human population in other parts of the world.

   Human population control is the practice of artificially altering the rate of growth of a human population. Historically, human population control has been implemented with the goal of increasing the rate of population growth. In the period from the 1950s to the 1980s, concerns about global population growth and its effects on poverty, environmental degradation, and political stability led to efforts to reduce population growth rates. In the 1980s, tension began to grow between population control advocates and women's health activists who advanced women's reproductive rights as part of a human rights–based approach, as pointed out in Lara Knudsen's Reproductive Rights in a Global Context. Growing opposition to the narrow population control focus led to a significant change in population control policies during the early 1990s.

   Paul R. Ehrlich, a U.S. biologist and environmentalist, published *The Population Bomb* in 1968, advocating stringent population control policies. http://en.wikipedia.org/wiki/Antinatalistic_politics–cite_note-15. His central argument on population is as follows: Like a cancer is an uncontrolled multiplication of cells; the population explosion is an uncontrolled multiplication of people. Treating only the symptoms of cancer may make the victim more comfortable at first, but eventually he dies—often horribly. Ehrlich states that a similar fate awaits a world with a population explosion if only the symptoms are treated.

   We must shift our efforts from treatment of the symptoms to the cutting out of the cancer. The operation will demand many apparent brutal and heartless decisions. The pain may be intense. But the disease is so far advanced that only with radical surgery does the patient have a chance to survive.

   In his concluding chapter, Ehrlich offered a partial solution to the "population problem," "[We need] compulsory birth regulation ... [through] the addition of temporary sterilants to water supplies or staple food. Doses of the antidote would be carefully rationed by the government to produce the desired family size."

   Ehrlich's views came to be accepted by population control advocates in the United States and Europe in the 1960s and 1970s. Since Ehrlich introduced his idea of the "population bomb," overpopulation has been blamed for a variety of issues, including increasing poverty, high unemployment rates, environmental degradation, famine, and genocide. http://en.wikipedia.org/wiki/Antinatalistic_politics—cite_note-Knudsen_2006_2. E2.80.933–12. In actuality, most if not all of the environmental problems that have been brought out in this book have a connection with population size, some directly and some indirectly.

   In a 2004 interview, Ehrlich reviewed the predictions in his book, and found that while the specific dates within his predictions may have been wrong, his predictions about climate change and disease were valid. Ehrlich continued to advocate for population control and coauthored the book *The Population Explosion*, released in 1990 with his wife Anne Ehrlich.

2. One of the problems with making a choice of a political figure is that most of us are unaware of what to look for, especially, since politicians train themselves to debate and create deception. That they lie about the various subjects with which they appear to be concerned sometimes upsets their audience (as it should), many people learn to expect lies and other forms of deception in the political arena.

   Kevin DeLapp (*Weapons of Mass Deception: The Uses and Abuses of Honesty in Politics*) (1991) states that "any of us react with trigger-finger sensitivity to any perceived instance of politicians lying—at least when the politician is a member of a rival political party! Leaks, cover-ups, marketing spin, flip-flopping on issues, all of these get reduced to the level of bald, straight-faced perjury. In our more everyday moral lives, however, we usually make finer and more charitable discriminations. Perhaps it is the pace or the medium of modern politics that makes for this double standard? Or, perhaps there is something genuinely unique about the way we should view deception in politics?

It seems we deceive ourselves when we righteously assert that we never want our politicians to deceive us. As philosopher Ruth Grant has put it, "We do not really believe that our leaders will never tell a lie, and I doubt we would be simply proud if they were scrupulously honest when it cost us something. We often hurl around accusations of dishonesty more for rhetorical and political purposes, than as honest expressions of moral disapproval."

"Much of our outrage at perceived political lies can be understood by looking at modern reactions to Plato's (in)famous defense of deception in his dialogue *Republic*. In constructing his utopia, Plato describes a creepy superking who is given the power (indeed, the obligation) to deceive the ignorant masses who could not otherwise handle the truth. Plato's 'Noble Lie,' as he called it, makes some cultural sense when we appreciate the general atmosphere of anxiety regarding the possibility of illusion and disguise—themes dominant in ancient Greek art and mythology—in which he was writing. But for the most part, we today do not share Plato's cosmic suspicion of appearances or his pessimism about the capacity to figure things out for ourselves. Hence our indignation about being condescendingly lied to by our politicians.

The ancient Chinese thinker Han Feizi (early 3rd-century BCE) might offer a more accessible and realistic model for our otherwise schizophrenic attitudes to political deception. Han Feizi enjoins politicians to "hide their hearts" and keep their intentions "closed and dark." Partly, this is a gamble to smoke-out corrupt lobbyists. And partly "it's a technique for the politician to safe-guard [his or] her own identity as the personification of the community. If a politician is too forthright about [his or] her personal likes and dislikes, that can tempt lobbyists to try to bribe or intimidate [him or] her by playing on those likes."

Han Feizi, after all, subscribed to a view of human nature according to which most people by default tend to be motivated by lazy and profiteering considerations. While Plato viewed most people as intellectually lazy but morally pretty decent, for Han Feizi, most people can be quite intellectually discerning, but are morally self-centered. The two thinkers' accounts of political deception are predicated on that difference in moral psychology: Plato thinks we wouldn't understand the truth, whereas Han Feizi thinks we'd understand just fine, but that we'd abuse it for short-sighted personal gain.

Notably, though, Han Feizi claims that strategic deception can also be a valuable strategy for political advocacy. Han Feizi relates the story of a politician named Sunjin who managed to convince an otherwise intractable ruler to abandon a wasteful building project by stacking eggs on top of one another in order to indirectly showcase the project's foolishness. Sunjin's action was in some sense "deceptive," for he had to lie to get an audience with the ruler in the first place and then evade the ruler's questions as he performs his egg-stacking. But such tactical deception was necessary to terminate a wasteful political policy. Thus, Han Feizi recognizes that certain deceptions are essential to the art of persuasion, which in turn is at the heart of constructive policy-making and governance.

Furthermore, to those who would call for total transparency in government, Han Feizi references a man who self-righteously calls himself "Honest Gong," but who is so brutally and indiscriminately truth-telling that he makes a federal case about even the smallest perceived untruth. Why should we allow the self-proclaimed Honest Gongs in our own society to disrupt the constructive—albeit more indirect and sometimes frustratingly slow—political efforts of the Sunjins out there? To diagnose every evasion and deception as if it were a Platonic "noble lie" and then to yell "You lie!" to the face of any would-be Sunjin is to conflate truthfulness with truth-bullying.

Han Feizi also helps us remember that politicians should probably not be viewed in the same way as we view our friends. Evasiveness would be an undesirable trait in a friend, after all, but politicians are not our friends (at least not qua politicians): they operate in an arena in which rapidly-changing facts make for shifting truths, in which every statement needs to be checked against the threat of de-contextualization, and in which different audiences have different motives and may hear quite different messages.

Do we ascribe to this way of thinking? Do we expect dishonesty, evasiveness, and outright lying from a person who we have chosen to accomplish feats that will help our populations deal with immoral subjects?

CHAPTER 14

Human Aggression: Killing and Abuse

1. It is stated by certain individuals that we all are capable of murder if subjected to certain stressful circumstances. D'Cruze et al. (2013) have "explored the social construction of the murderer in the context of interpersonal murder and explored the social construction of the murderer over time in an effort to understand how such constructions search for ways of rendering the murderer pathological: not like us."

CHAPTER 16

Are We Our Own Worst Enemy?

1. We should consider why humans allow environmental abuse to occur. Is the world so large that we do not notice such abuse, is it human ignorance that we do not understand the results of abuse or is it human complacency or greed that leads us onward without regard for what we are doing to the environment?

2. Nitrogen in the environment is useful at certain levels because once fixed into a form suitable for organisms to use, it promotes life. When living things die, nitrogen is recycled for use again. However, in overabundance, nitrogen (as well as other chemicals) can cause populations to grow so rapidly that they overpopulate their environment and subsequently suffer a rapid decline. In marine waters, it is typical for dinoflagellates to form blooms as a result of high nitrogen content. The process of dying requires oxygen which becomes unavailable to other organisms, and they die off as well.

3. Examples of eutrophication and hypoxia are clear cases of environmental abuse at the hand of humans, and it is associated with population size. The more humans and domestic animals that are on the planet, the more the population will face environmental disasters.

4. There are other animals as well. Cats, dogs, horses, sheep, goats, and others are producing more wastes than humans and thus represent a greater threat to environmental degradation. However, even though they represent a threat to the planet, through land lost to produce them, the unsettling amounts of their waste materials and the consequent forms of environmental degradation, humans depend on them for food and pleasure.

5. According to Ehhalt et al. (2001) in their and cited works, "substantial, preindustrial abundances for CH_4 and N_2O are found in the tiny bubbles of ancient air trapped in ice cores." Both gases have large, natural emission rates, which have varied over past climatic changes but have sustained a stable atmospheric abundance for the centuries prior to the Industrial Revolution. Emissions of CH_4 and N_2O due to human activities are also substantial and have caused large relative increases in their respective burdens over the past century. The atmospheric burdens of CH_4 and N_2O over the next century will likely be driven by changes in both anthropogenic and natural sources. A second class of greenhouse gases—the synthetic HFCs, PFCs, SF_6, CFCs, and halons—did not exist in the atmosphere before the 20th century. CF_4, a PFC, is detected in ice cores and appears to have an extremely small natural source. The current burdens of these latter gases are derived from atmospheric observations and represent accumulations of past anthropogenic releases; their future burdens depend almost solely on industrial production and release to the atmosphere. Stratospheric H_2O could increase, driven by in situ sources, such as the oxidation of CH_4 and exhaust from aviation, or by a changing climate. Tropospheric O_3 is both generated and destroyed by photochemistry within the atmosphere. Its in situ sources are expected to have grown with the increasing industrial emissions of its precursors: CH_4, NO_x, CO, and VOC. In addition, there is substantial transport of ozone from the stratosphere to the troposphere (see also Section 4.2.4). The effects of stratospheric O_3 depletion over the past three decades and the projections of its recovery, following cessation of emissions of the Montreal Protocol gases, has been recently assessed.

6. There is a direct relationship between CO_2 levels and global warming. Recall statements in other parts of this book, which mention conditions on Earth during its formative years. With abundant CO_2, the Earth was too hot to support life. The more wastes we produce, the more CO_2 we will have in the environment. With a growing human population, there will be more demands for agricultural animals and a consequent rise in CO_2. With the greater demand for agricultural animals, there will be an increased demand for agricultural land and the cutting of forests with trees that absorb CO_2 and help clean the air. Increased numbers of animals will also result in an increase of wastes that will flow through the land and into ponds, lakes, streams, and rivers,

polluting ground water and eventually reaching marine waters, causing eutrophication and hypoxia and further destruction of marine biota.

7. Considering the points made in this chapter about animal wastes, it is evident that the environment is becoming degraded because of it. And while there are many people who say there remains a lot of land available for human use, environmental degradation refutes this idea. Yes, there is more land to use, but what are we doing to the planet with the populations of humans and animals at their present level? Dwindling resources and loss of species diversification indicate a planet in trouble.

8. Combining the threat of eutrophication and chemical pollution of our waters, CO_2 and other GHG build-up, global warming, limited amounts of nutrients in plants required for human consumption, a dwindling number of trees and reduction in the consequent number of other biota, we can see the dilemma we are faced with, a situation that cannot be rectified without a reduction in the global population of humans.

9. Methane is an excellent gas to use as an energy source. Natural gas, which comes from within the Earth, is used for a variety of reasons, but methane generated in manure and the guts of animals is not utilized in most cases, and it passes directly into the environment. While methane is not the most significant greenhouse gas on a per-molecule level, it is significant because of its abundance. Methane production, of course, is another indication of an overpopulated Earth because the farm animals that produce most of the methane are being grown to feed humans. It is a vicious cycle. The more humans, the more animals required to feed them. The more animals there are, the more methane is produced the more natural habitat must be taken to provide pastureland.

10. An overabundance of animals in a wild population often results in a population collapse due mostly to parasitic infections. While the contemporary human world appears to be faring better than the ancient world in terms of serious parasite attack, parasites, like any other organisms, undergo natural selection, resulting in a population-wide immunity to chemicals used against them. Occasionally, the human population is threatened by parasite attack, and with increased population size, the human population will be increasingly faced with mass die-off caused by parasites.

11. There is a direct relationship between biotic potential (the maximum number of offspring a species can produce) and environmental carrying capacity (the maximum number of individuals a population can support). If we watch nature closely, we can already see the effects of spring elevations in food production in bodies of water. Plant material begins to grow, herbivores have more to eat and begin reproducing, predators do the same, each species producing more offspring than the environment can support. Due to certain factors (e.g., environmental chemistry, predation, and parasitism), the numbers of individuals in these populations is cut back as part of ecosystem homeostasis. Bringing in exotic organisms or removing the predators alter the balance and cause certain of its members to die or dramatically increase in numbers, and homeostasis is lost.

12. M. Hutton (1983) quantifies "the major sources of cadmium in the European Community and assessing the relative significance of such inputs to the environmental compartments, air, land and water. The methodology involved identification of potential sources of cadmium, including natural processes, as well as those associated with human activities." He follows his introductory comments with a review of emission studies of these processes and subsequent estimation of an emission factor for each source. He applies the emission factor "to the most recent production or consumption data for the process in question to obtain an estimate of the annual discharge. The steel industry and waste incineration, followed by volcanic action and zinc production, are estimated to account for the largest emissions of atmospheric cadmium in the region. Waste disposal results in the single largest input of cadmium to land; the quantity of cadmium associated with this source is greater than the total from the four other major sources—coal combustion, iron and steel production, phosphate fertilizer manufacture and use, and zinc production. The characterization of cadmium inputs to aquatic systems is incomplete but of the sources considered, the manufacture of cadmium-containing articles accounts for the largest discharge, followed by phosphate fertilizer manufacture and zinc production."

13. Such trees are often loblolly (*Pinus taeda*) and slash pines (*Pinus elliottii*) which grow fast and can be harvested periodically. Such trees represent maintained climax vegetation, and the areas that support them are not allowed to reach climax status. Thus, populations of other organisms that would be associated with climax trees never become established.

CHAPTER 17

Attempts to Save the Natural World

1. It is easy to become negative about the rising number of people in the world and increasingly bad conditions that result from a rising human population. There are numerous people in the world who are concerned and

attempting to alleviate the problems we have created. The solving of environmental problems stems from concerned citizens and filters through the population's dominance hierarchy to governmental leaders who may not understand or care about such problems. Eventually, though, laws may be changed to improve the environment or bring it back into homeostasis. With increases in the number of humans in the world and the animals necessary to keep them alive, there is a point at which homeostasis will not be possible.

2. With a dwindling interest in the biota and environment by many people in the world, individuals who make up the organizations in this chapter, the world and its biota would be worse off.

3. With humans degrading the environment, considerable care of wildlife areas will be necessary to maintain some degree of environmental homeostasis. Yet, with compounding problems, population fragmentation and the introduction of exotic species, any environmental crises may result in population crashes and extinction.

4. There is constant environmental pressure that results from human greed and human necessity. It is often the case that promoters of land use care about the money the sale of land will bring and not about what will happen to the land and the biota that is found upon it.

CHAPTER 18

The Nature of Things

1. Curing the planet of its ills is particularly challenging since it is not a focal point for contemporary society or the government to which it belongs. Without recognizing and accepting our problems, there will be threatening circumstances that the human species must face in the future.

While the rate of population growth has been reduced somewhat since the 1980s, it has been growing continuously since the end of the Black Death outbreaks in Europe. While population growth is slowing can be taken as a good sign, it means little if it has already surpassed the environmental carrying capacity. Much damage has already been done, and as long as the population shows any growth at all, Earth and its biota will suffer the consequences.

The human population rate varies throughout the world, but there are some important hot spots of excessive population growth. The United Nations, for instance, has expressed concern on continued excessive population growth in sub-Saharan Africa. Yet, China and India are countries that have the most outstanding problems with overpopulation, and they have suffered personally from it.

In an Urban Progress article by Alon Tal, called *Overpopulation Is Still the Problem*, it was pointed out that overpopulation "remains the leading driver of hunger, desertification, species depletion and a range of social maladies across the planet. Conservative estimates report that China's most recent food crisis, between 1958 and 1961, led to the starvation of over 20 million people, in part due to the erosion of China's natural capital. Uncontrolled human fertility led to a depletion of the land's fertility. Previous famines were worse. Over the years, hundreds of millions died a horrible death [from] hunger. Their misery should teach a sobering lesson [to all of us] about insouciant disregard for the balance between human numbers and natural resources."

As Tal further states, overpopulation is not just about food shortages and human suffering. It is clear that the collapse in global biodiversity is also linked to [human] overpopulation. "China, Mexico and Brazil have been singled out as extreme cases of species loss. Brazil's [human] population grew four-fold during the past 60 years; there is little wonder that the Amazon is feeling the pressure. Mexico and China's growth is comparable."

Most environmental scientists would agree that "it is time to realize that there is a trade-off between 'quality of life' and 'quantity of life.' In a planet with limited resources—sustainable growth is an oxymoron. Of course, humanity could all shift to vegan diets, forgo national parks and crowd in a few more billion people, hoping that new levels of efficiency will allow us to survive. But it is well to ask if this really is the kind of world that we want for us and future generations. There is much we can do to reduce the suffering caused by human population growth, but recognizing that overpopulation is a perilous problem constitutes a critical first step."

As of March 31, 2014, the United States Census Bureau estimated the world's human population to be 7.153 billion. Most contemporary estimates for the carrying capacity of the Earth under existing conditions are between 4 and 16 billion. A complete analysis of this range in estimated population size would be very interesting. I would suspect environmentalists to be at the low end and politically influenced groups at the other.

Whatever estimate is used, problems of an environmental nature strongly indicate that the human environmental carrying capacity has already been long surpassed, and the recent rapid increase in human population represents a serious threat to the planet. Based on estimates, the human population is expected to reach between 8 and 10.5 billion between the year 2040 and 2050. In May 2011, the United Nations increased the medium variant projections to 9.3 billion for 2050 and 10.1 billion for 2100. By then, we should all be at each other's throats.

Based on what has been said in this and other books, most or all of our environmental problems are due to human overpopulation. Citizens (the reproductive source of overpopulation) and governments should seriously consider that overpopulation is the number one global problem that will lead to the human species' demise, in spite of what religious and political leaders say to the contrary.

Readers interested in assisting in making and maintaining the world as a homeostatic system may look into the many organizations that deal with environment and biota that are listed in Chapter 17. In addition to the many organizations that are concerned about the environment, the reader should be aware of the Environmental Protection Agency (EPA) and its activities, a governmental organization that caters to certain environmental problems. The EPA was established on December 2, 1970. The following material was taken from an article by Shelly Barclay at the Life 123 website called *Environmental Protection Agency Facts.*

"The EPA was established to address issues with environmental law on a federal level. It operates under legislation from Congress and, because of this, is often accused of having political motives." If the EPA is so concerned with environmental problems, one should question why the government does not get involved with human overpopulation as the leading cause of environmental problems.

"The headquarters of the EPA is located in Washington, D.C. The agency also has offices and laboratories spread throughout the United States. Jurisdiction is split into 10 regions. Each of these regions covers multiple states:

> The Environmental Protection Agency enforces specific legislation involving clean air, clean water and clean soil. Other entities and departments, such as the Department of Agriculture, have jurisdiction as it pertains to their issues. However, the Environmental Protection Agency has more general jurisdiction and often finds itself handling industry issues.

Recommendations and Laws

On top of handling environmental law enforcement, the Environmental Protection Agency conducts research and launches initiatives that study current negative impacts as well as prevent further negative impact on the environment. For example, EnergyStar and WaterSense are programs that are not law but encourage use of energy and resource-efficient products. The agency tests a variety of media to gauge the impact of certain practices on the environment and suggest changes to curtail said impact where it is found:

> While the Environmental Protection Agency is not necessarily dealing with public health issues directly, they do deal with the materials that cause environmental health problems. For example, the Clean Air Act enables the EPA to handle cases where the mishandling of dangerous products leads to health concerns. Such was the case in 2012 when an Illinois man was sentenced to 10 years in prison for mishandling asbestos cleanup.

2. Cropland and space for growing animals to be butchered for human consumption are the major reasons for destroying habitats, and water use sources are utilized for these animals as well.
3. Not only can animal excrement be used to supply fertilizer to plants, it can be used as a relatively clean fuel. There are farms that are run entirely on methane gas which is produced from manure (often referred to as gobar gas). When research is carried out to utilize manure to obtain both natural fertilizer for plants and energy that can also be used, the process will be an important step in helping the environment and providing good nutrition for humans.
4. Natural animal populations often suffer from heavy parasitism when they are dense. Population density and parasitism go hand in hand toward the downfall of natural populations and will also affect human populations as the density increases.
5. Understanding that overpopulation is our main concern is part of the battle to regain homeostasis on Earth. The other part is taking action. The most likely dominant personalities that could lead such a program are political leaders, but they usually lack adequate concern to set goals in that direction.
6. It is easy to disregard environmental and behavioral problems in the world, hoping that they will somehow go away, but they do not, and population density has a lot to do with these conditions.

References

Aboitiz, F., Montiel, J., Morales, D., Concha, M., 2002. Evolutionary divergence of the reptilian and the mammalian brains: considerations on connectivity and development. Brain Research Reviews 39 (2), 141–153.

About Health, 2016. Fromm's Five Character Orientations. About Health. http://psychology.about.com/od/theoriesofpersonality/fl/Fromms-Five-Character-Orientations.htm.

Abrams, R., 2012. Why Do We Fish. Fishing Lines. Florida Fish and Wildlife Conservation. myfwc.com/media/316636/SW_FishingLines_WhyFish.pdf.

Ackerman, F., 2010. Waste management: taxing the trash away. Environment: Science and Policy for Sustainable Development 34 (5), 2–43.

Actman, J., 2015. From Trees to Tigers, Case Shows Cost of Illegal Logging. Wildlife Watch. http://news.nationalgeographic.com/2015/11/151110-timber-russian-far-east-illegal-logging-siberia/.

Adams, M.D., Celniker, S.E., Holt, R.A., et al., 2000. The genome sequence of Drosophila melanogaster. Science 287 (5461), 2185–2195.

Adams, J., 2005. North America during the Last 150,000 Years. http://www.esd.omi.gov/projects/qen/nercNorthAmerica.html.

Adams, C., 2014. The Magic of Living in the Earth. http://www.spiritofmaat.com/archive/apr2/trogs.htm.

Adkins-Regan, E., 2005. Hormones and Animal Social Behavior. Princeton University Press.

Agnarsson, I., Aviles, L., Coddington, J.A., Maddison, W.P., 2006. Sociality in theridiid spiders: repeated origins of an evolutionary dead end. Evolution 60 (11), 2342–2351.

Aiello, L., Dean, C., 1990. An Introduction to Human Evolutionary Anatomy. Elsevier Academic Press.

Aiello, L., Dean, C., 2006. An Introduction to Human Evolutionary Anatomy. Elsevier Academic Press.

Aigner, D., 1985. Hitler's ultimate aims – a programme of world dominion? In: Koch, H.W. (Ed.), Aspects of the Third Reich. MacMillan.

Aitken, J., Murray, A., 2010. Crash repair in the UK: reusing salvaged parts in car repair centres. International Journal of Logistics Research and Applications 13 (5), 359–372.

Albert, D.J., Walsh, M.L., Jonik, R.H., 1993. Aggression in humans: what is its biological foundation? Neuroscience & Biobehavioral Reviews 17 (4), 405–425.

Alberti, L.B., 1956. On Painting (J.R. Spencer, Trans.). Yale University Press.

Alcock, J., 2001. Animal Behavior, third ed. Sinauer Associates.

Alcock, J., 2001. Animal Behavior: An Evolutionary Approach, seventh ed. Sinauer Associates, Inc.

Alexander, W.P., Grimshaw, S.D., 2012. Treed regression. Journal of Computational and Graphical Statistics 5 (2), 156–175.

Alexander, R.D., 1990. Special Publication. How Did Humans Evolve? Reflections on the Uniquely Unique Species, vol. 1. Museum of Zoology, University of Michigan, pp. 1–38.

Allchin, R., 1995. The Archaeology of Early Historic South Asia: The Emergence of Cities and States. Cambridge University Press.

Allchin, B., 1997. Origins of a Civilization: The Prehistory and Early Archaeology of South Asia. Viking.

Allee, W.C., 1926. Studies in animal aggregations: causes and effects of bunching in land isopods. Journal of Experimental Zoology 45 (1), 255–277.

Allee, W.C., 1927. Animal aggregations. The Quarterly Review of Biology 2 (3), 367–398.

Allen, J.G., 2012. Underground structures of the cold war: the world below. Defense and Security Analysis 28 (4), 368–369.

Alroy, J., 2008. Dynamics of origination and extinction in the marine fossil record. Proceedings of the National Academy of Sciences of the United States of America 105 (Suppl. 1), 11536–11542.

Alter, S., 2004. Elephas Maximus: A Portrait of the Indian Elephant. Penguin Books.

Altman, I., 1975. The Environment and Social Behavior: Privacy, Personal Space, Territory and Crowding. Brooks/Cole Publishing Company.

American Psychiatric Association, 1994. Diagnostic and Statistical Manual of Mental Disorders, fourth ed. Washington.

Andelman, D.A., 2008. A Shattered Peace: Versailles 1919 and the Price We Pay Today. J. Wiley.

Anderson, C.A., Bushman, B.J., 2002. Human aggression. Annual Review of Psychology 53, 37–51.

Anderson, D., Gurnham, C.F., 2009. Disposal of sludge solids from food industry waste treatment. Critical Reviews in Food Technology 3 (1), 27–87.

Anderson, C., Kilduff, G.J., 2009. Why do dominant personalities attain influence in face-to-face groups? The competence-signaling effects of trait dominance. Journal of Personality and Social Psychology 96 (2), 491–503.

Anderson, J.D., 2004. Inventing Flight: The Wright Brothers and Their Predecessors. Johns Hopkins University Press.

Andersson, J.O., Andersson, S.G., 1999. Genome degradation is an ongoing process in Rickettsia. Molecular Biology and Evolution 16 (9), 1178–1191.

Andersson, M., 1984. The evolution of eusociality. Annual Review of Ecology and Systematics 165–189.

Andreopoulos, G.J., 1994. Genocide: Conceptual and Historical Dimensions. University of Pennsylvania Press.

Andrews, P.J., 1983. The Natural History of Sivapithecus. Advances in Primatology, pp. 441–463.

Andrews, D.G., 2001. Neuropsychology. Psychology Press.

Anonymous, 2012a. 10 Impressive Survival Bunkers. Survival Blog. http://www.survival-spot.com/survival-blog/10-impressive-doomsday-bunkers/.

Anonymous, 2012b. Love Canal. http://en.wikipedia.org/wiki/Love_Canal.

Anonymous, 2013. Top US Antiscience Group, Flathead Area Secular Humanist Association. http://www.flatheadsecular.com/anti-science-groups/.

Anonymous, 2014a. Petroleum – Oil and Natural Gas. Energy4Me. http://www.energy4me.org/energy-facts/energy-sources/petroleum/.

Anonymous, 2014b. What Respect Really Means in a Relationship. TwoOfUs.org.

Anstey, P.R., 2003. The Philosophy of John Locke: New Perspectives. Routledge.

Antai-otong, D., 2009. Minimizing violence in the chemically dependent client: treatment considerations. Journal of Addictions Nursing 10 (1), 28–33.

Anton, S.C., 2003. Natural history of *Homo erectus*. American Journal of Physical Anthropology 122, 126–170.

Antonovsky, A., 1979. Health, Stress, and Coping. Jossey-Bass Publishers.

Appleby, R.S., 2006. Between Americanism and Modernism; John Zahm and theistic evolution. In: Critical Issues in American Religious History: A Reader. Baylor University Press.

Archer, I., Ferris, J.R., Herwig, H.H., Travers, T.H.E., 2008. World History of Warfare, second ed. University of Nebraska Press.

Archer, J., 1991. The influence of testosterone on human aggression. British Journal of Psychology 1, 1–28.

Ardrey, R., 1966. The Territorial Imperative. Atheneum.

Argyris, C., 1957. Personality and Organization: The Conflict Between System and the Individual. Harper & Row, pp. 47–54.

Ariely, J., 2012. The (Honest) Truth about Dishonesty, How We Lie to Evryone – Especially Ourselves. Harper Collins Publishers.

Arrow, K., Bolin, B., Costanza, R., et al., 1995. Economic growth, carrying capacity and the environment. Ecological Economics 15 (2), 91–95.

Ascalone, E., 2007. Mesopotamia: Assyrians, Sumerians, Babylonians (Dictionaries of Civilizations; 1). University of California Press.

Ash, P.J., Robinson, D.J., 2010. The Emergence of Humans: An Exploration of the Evolutionary Timeline. John Wiley & Sons.

Ashton, T.S., 1948. Some statistics of the industrial revolution in Britain. The Manchester School 16 (2), 214–234.

Ashton, E.H., 1975. Prosimian biology. Journal of Anatomy 120, 393.

Authority, A.M.S., 2005. Major Oil Spills in Australia. Australian Maritime Safety Authority.

Avery, M., 2014. A Message from Martha: the Extinction of the Passenger Pigeon and Its Relevance Today. Natural History.

B&N, 2005. Great Battles of the Ancient World. Teaching Company Limited Partnership.

Baier, K., 1990. Egoism. In: Singer, P. (Ed.), A Companion to Ethics. Blackwell.

Bailey, T., Launay, F., Sullivan, T., 2000. Health Issues of the International Trade of Falcons and Bustards in the Middle East: The Need for Regional Monitoring and Regulation? National Avian Research Center, Environmental Research and Wildlife Development Agency. http://www.falcons.co.uk/mefrg/PDF/tom1.pdf.

Baker, S.R., 2015. Measuring economic policy uncertainty. Nat. Bureau of Econ. Res. http://www.nber.org/papers/w21633.

Baker, R.L., Mwamachi, D.M., Audho, J.O., Aduda, E.O., Thorpe, W., 1998. Resistance of galla and small east African goats in the sub-humid tropics to gastrointestinal nematode infections and the peri-parturient rise in faecal egg counts. Veterinary Parasitology 79 (1), 53–64.

Baker, J.O., Bader, C.D., Hirsch, K., 2014. Descration, moral boundaries and the movement of law: the case of Westboro Baptist Church. Deviant Behavior 36 (1), 42–67.

Ballard, W., 2011. Predator-prey relationships. In: Hewitt, D.G. (Ed.), Biology and Management in White-Tailed Deer. CRC Press.

Bandura, A., 1999. A social cognitive theory of personality. In: Cervone, D., Shodam, Y. (Eds.), Handbook of Personality. Guilford Press, pp. 154–196.

Bandura, A., Walters, R.H., 1963. Social Learning and Personality Development. Holt Rinehart and Winston.

Barbault, R., Sastrapadja, S.D., 1995. Generation, Maintenance and Loss of Biodiversity. Global Biodiversity Assessment. Cambridge University Press, pp. 193–274.

Barclay, S., 2014. The Environmental Protection Agency Facts. http://www.life123.com/home-garden/green-living/green-homes/environmental-protection-agency-facts.shtml?o=2800&qsrc=999&ad=doubleDown&an=apn&ap=ask.com.

Barker, K.N., Flynn, E.A., et al., 2002. Medication errors observed in 36 health care facilities. Archives of Internal Medicine 162 (16), 1897–1903.

Barnard, A., 1998. The foraging spectrum: diversity in hunter-gatherer lifeways. American Ethnologist 25 (1), 36–37.

Barnard, A.J., 2004. Hunter-Gatherers in History, Archaeology and Anthropology. Berg.

Barnes, A., Ephross, P.H., 1994. The impact of hate violence on victims: emotional and behavioral responses to attacks. Social Work 39 (3), 247–251.

Barnett, R., Yamaguchi, N., Barnes, I., Cooper, A., 2006. The origin, current diversity and future conservation of the modern lion (*Panthera leo*). Proceedings of the Royal Society B: Biological Sciences 273 (1598), 2119–2125.

Barrett, J., Abbott, D.H., George, L.M., 1993. Sensory cues and the suppression of reproduction in subordinate female marmoset monkeys, *Callithrix jacchus*. Journal of Reproduction and Fertility 97, 301–310.

Barroso, I., Gurnell, M., Crowley, V.E., et al., 1999. Dominant negative mutations in human PPARgamma associated with severe insulin resistance, diabetes mellitus and hypertension. Nature 402 (6764), 880–883.

Barton, R.A., Dunbar, R.I., 1997. Evolution of the social brain. Machiavellian Intelligence II: Extensions and Evaluations 2, 240.

Bates, R.H.T., Lewitt, R.M., Rowe, C.H., Day, J.P., Rodley, G.A., 1977. On the structure of DNA. Journal of the Royal Society of New Zealand 7 (3), 273–301.

Bauer, M., Reynolds, S., 2009. Under Siege: Life for Low-income Latinos in the South: A Report by the Southern Poverty Law Center. Southern Poverty Law Center.

Baughman, M., 2003. National Georgraphic Reference to the Birds of North America. National Geographic Society.

Bauman, M.D., Toscano, J.E., Mason, W.A., Lavenex, P., Amaral, D.G., 2006. The expression of social dominance following neonatal lesions of the amygdala or hippocampus in rhesus monkeys (*Macaca mulatta*). Behavioral Neuroscience 120 (4), 749–760.

Bauman, R., 1999. Human Rights in Ancient Rome. Routledge Classical Monographs.

Baumann, P., 2000. Equity and Efficiency in Contract Farming Schemes: The Experience of Agricultural Tree Crops. Overseas Development Institute.

Beaver, R., Beaver, M.S., Field, W.E., 2008. Summary of documented fatalities in livestock manure storage and handling facilities – 1975–2004. Journal of Agromedicine 12 (2), 3–23.

Beck, A.T., 1975. Cognitive Therapy and Emotional Disorders. International Universities Press.

Beckman, J.S., Koppenol, W.H., 1996. Nitric oxide, superoxide, and peroxynitrite: the good, the bad, and ugly. American Physiological Society 40 (5), 1424–1437.

Beebe, S.A., Beebe, S.J., Redmond, M.V., 2008. Interpersonal Communication, fifth ed. Pearson Education.

Begon, T.H., 2006. Ecology: From Individuals to Ecosystems, fourth ed. Blackwell Publishing.

Bell, J.M., 1977. Stressful life events and coping methods in mental-illness and-wellness behaviors. Nursing Research 26 (2), 136–141.

Benchley, P.J., Harper, D.A.T., 1998. Paleoecology: Ecosiplems, Environments and Evolution. Chapman & Hall.

Bengtsson, B.O., 2003. Genetic variation in organisms with sexual and asexual reproduction. Journal of Evolutionary Biology 16 (2), 189–199.

Benis, A.M., 1986. A model of human personality traits based on Mendelian genetics. In: American Assn. for the Advancement of Science, Pub. 86-5, Washington, D.C., p. 124 (abstract).

Benjamin, J., 2008. Human personality traits. Molecular Genetics and the Human Personality 333.

Bennett, N.C., Faulkes, C.G., 2000. African Mole-Rats: Ecology and Eusociality. Cambridge University Press.

Benton, D., 1988. Hypoglycemia and aggression: a review. International Journal of Neuroscience 1 (3–4), 163–168.

Benzinger, K., Console, R.P., 2014. Catching the Red Flags: How the Animal Cruelty Legacy Impacts Public Safety. CU Independent. http://cuindependent.com/2013/05/02/catching-the-red-flags-how-the-animal-cruelty-legacy-impacts-public-safety/.

Berger, E.H., Riojas-Cortez, M., 2004. Parents as Partners in Education: Families and Schools Working Together. Merrill.

Berger, M.L., 2001. The Automobile in American History and Culture: A Reference Guide. Greenwood Publishing Group.

Bergman, T.J., Beehner, J.C., Cheney, D.L., Seyfarth, R.M., 2003. Hierarchical classification by rank and kinship in baboons. Science 302 (14), 1234–1236.

Bergstrom, C.T., Godfrey-Smith, P., 1998. On the evolution of behavioral heterogeneity in individuals and populations. Biology and Philosophy 13 (2), 205–231.

Berlanstein, L.R. (Ed.), 1992. The Industrial Revolution and Work in Nineteenth-Century Europe. Routledge.

Bernal, J.D., 1967. The Origin of Life. World Publishing.

Berner, W., Berger, P., Hill, A., 2003. Sexual sadism. International Journal of Offender Therapy and Comparative Criminology 47 (1), 383–395.

Bettinger, R.L., 1991. Hunter-Gatherers: Archaeological and Evolutionary Theory. Plenum Press.

Bicudo, J.R., Goyal, S.M., 2008. Pathogens and manure management systems: a review. Environmental Technology 24 (1), 115–130.

Biello, D., August 2009. The Origin of Oxygen in Earth's Atmosphere. Scientific American.

Birdsell, J.B., 1986. Some predictions for the pleistocene based on equilibrium systems among recent hunter gatherers. In: Lee, R., DeVore, I. (Eds.), Man the Hunter. Aldine Publishing Co.

Bjorklund, D.F., Pellegrini, A.D., 2002. The Origins of Human Nature: Evolutionary Developmental Psychology. American Psychological Association, US.

Black, E., 2004. Banking on Baghdad. John Wiley.

Black, J.J.S., 2009. Rousseau's Critique of Science: A Commentary on the Discourse on the Sciences and the Arts. Lexington Books.

Black, E., 2012. War Against the Weak: Eugenics and America's Campaign to Create a Master Race. Dialog Press.

Blair, J., Mitchell, D., Blair, K., 2005. The Psychopath: Emotion and the Brain. Blackwell Publishing.

Blanchard, D.C., Blanchard, R.J., 2010. Offensive and defensive aggression. Encyclopedia of Behavioral Neuroscience 484–489.

Blatberg, C., 2000. From Pluralist to Patriotic Politics: Putting Practice First. Oxford University Press.

Bloch, J., Silcox, M.T., Boyer, D.M., Sargis, E.J., 2006. New Paleocene skeletons and the relationship of plesiadapiforms to crown-clade primates. PNAS 104 (4), 1159–1164.

Blockmans, W.P., Hoppenbrouwers, C.M., 2014. Introduction to Medieval Europe 300-1500. Routledge. http://books.google.com/books?id=ixR2QgAACAAJ&dq=introduction+to+medieval+europe&hl=en&sa=X&ei=Y0E6UpDCAs3LqAH2z4AI&ved=0CD4Q6AEwAw.

Blomberg, O., 2011. Concepts of cognition for cognitive engineering. International Journal of Aviation Psychology 21 (1), 85–104.

Blue, G., Bunton, M.P., Croizier, R.C., 2002. Colonialism and the Modern Worlds: Selected Studies. M. E. Sharpe.

Boal, J.G., 2006. Social recognition: a top down view of cephalopod behavior. Life & Environment 56 (2), 69–79.

Bocquet-Appel, J.P., 2011. When the world's population took off: the springboard of the neolithic demographic transition. Science 333 (6042), 560–561.

Bodmer, W., Bonilla, C., 2008. Common and rare variants in multifactorial susceptibility to common diseases. Nature Genetics 40, 695–701.

Boehm, C.H., 1999. Hierarchy in the Forest: The Evolution of Egalitarian Behavior. Harvard University Press.

Boehm, C., 2012. Moral Origins: The Evolution of Virtue, Altruism and Shame. Basic Books.

Boesch, C., Boesch, H., 1990. Tool use and tool making in wild chimpanzees. Folia Primatologica 54, 86–99.

Boesch, C., Boesch-Achermann, H., 2000. The chimpanzees of the Tai forest. Behavioral Ecology and Evolution 192.

Bolton, B., 1995. A New General Catalogue of the Ants of the World. Harvard University Press.

Boltwood, B., 1907. The ultimate disintegration products of the radio-active elements. Part II. The disintegration products of uranium. American Journal of Science 23 (4), 77–88.

Bolzonella, D., Pavan, P., Battistoni, P., Cecchi, F., 2008. The under sink garbage grinder: a friendly technology for the environment. Environmental Technology 24 (3), 349–359.

Bonte, M., Van Balen, H., 1969. Prolonged lactation and family spacing in Rwanda. Journal of Biosocial Science 1 (02), 97–100.

Booth, C.L., 1990. Evolutionary significance of ontogenetic colour change in animals. Biological Journal of the Linnean Society 40 (2), 125–163.

Borgerhoff Mulder, M., Schacht, R., 2012. Human Behavioural Ecology. Wiley Online Library.

BornFree, 2003. The dirty side of the exotic animal pet trade. Animal Issues 34 (2) Summer.

Bostrom, A., Walker, A.H., Scott, T., Pavia, R., Leschine, T.M., Starbird, K., 2014. Human and ecological risk assessment. An International Journal 21 (3), 581–604.

Botstein, D., Chervitz, S.A., Cherry, J.M., 1997. Yeast as a model organism. Science 277 (5330), 1259–1260.

Bourke, J., 1999. An Intimate History of Killing: Face to Face Killing in 20th Century Warfare. Basic Books.

Bowles, S., Gintis, H., 2011. A Cooperative Species: Human Reciprocity and Its Evolution. Princeton University Press.

Bowman, M.W., 2000. Castles in the Air: The Story of the B-17 Flying Fortress Crews of the U.S. 8th Air Force. Potomac Books.

Boyd, R., Silk, J.B., 2003. How Humans Evolved, third ed. Norton.

Boysen, S., 2009. The Smartest Animals on the Planet. Firefly Books.

Bradbury, S., Williams, J., 2006. New labour, racism and 'new' football in England. Patterns of Prejudice 40 (1), 61–82.

Brainerd, C.J., Reyna, V.F., Wright, R., Mojardin, A.H., 2003. Recollection rejection: false-memory editing in children and adults. Psychological Review 110 (4), 762–784.

Brandon-Jones, D., Eudey, A.A., Geissmann, T., Groves, C.P., Melnick, D.J., Morales, J.C., Shekelle, M., Stewart, C.B., 2004. Asian primate classification. International Journal of Primatology 25 (1), 100.

Brandt, R.B., 1979. A Theory of the Good and the Right. Clarendon Press.

Brank, E.M., Hoetger, L.A., Hazen, K.P., 2012. Bullying. Annual Review of Law and Social Science 8 (1), 213–230.

Bratsis, P., 2014. Political corruption in the age of transnational capitalism: from the relative autonomy of the state to the white man's burden. Historical Materialism 22 (1), 105–128. https://www.academia.edu/5949120/Political_Corruption_in_the_Age_of_Transnational_Capitalism_From_the_Relative_Autonomy_of_the_State_to_the_White_Mans_Burden.

Brattstrom, B.H., 1974. The evolution of reptilian social behavior. American Zoologist 14 (1), 35–49.

Brenes, A., Winter, D.D., 2001. Earthly dimensions of peace: the Earth charter peace and conflict. Journal of Peace Psychology 7 (2), 157–171.

Brewin, C.R., Andrews, B., Valentine, J.D., 2000. Meta-analysis of risk factors for posttraumatic stress disorder in trauma-exposed adults. Journal of Consulting and Clinical Psychology 68 (5), 748–766.

Briggs, D., Crother, P.R., 2008. Palaeobiology II. John Wiley & Sons, p. 600.

Britannica, 2007. Traditional Chinese Medicine and Endangered Animals. Britannica Advocacy for Animals. http://advocacy.britannica.com/blog/advocacy/2007/10/traditional-chinese-medicine-and-endangered-animals/.

Broad, S., Mulliken, T., Roe, D., 2003. The nature and extent of legal and illegal trade in wildlife. The Trade in Wildlife: Regulation for Conservation 3–22.

Broad, C.D., 1971. Egoism as a Theory of Human Motives, in His Broad's Critical Essays in Moral Philosophy. George Allen and Unwin.

Brody, H., 2001. The Other Side of Eden: Hunter-Gatherers, Farmers and the Shaping of the World. North Point Press.

Broedel, H.P., 2004. The Malleus Maleficarum and the Construction of Witchcraft: Theology and Popular Belief. Manchester University Press, p. 34.

Brosnan, S.F., 2006. At a crossroads of disciplines. Social Justice Research 19 (2), 218–227.

Brotheridge, C.M., Keup, L., 2005. Barnyard democracy in the workplace. Team Performance Management 11 (3/4), 125–132.

Brower, M.C., Price, B.H., 2001. Neuropsychiatry of frontal lobe dysfunction in violent and criminal behavior: a critical review. Journal of Neurology, Neurosurgery & Psychiatry 71, 720–726.

Browman, M.G., Maskarinec, M.P., 2008. Environmental aspects of organics in selected coal conversion solid wastes. Journal of Environmental Science and Health: Environmental Science and Engineering, Part A 17 (5), 737–766.

Brown, D., 1991. Human Universals. McGraw-Hill.

Brulle, R.J., 2010a. From environmental campaigns to advancing the public dialogue: environmental communication for civic engagement. Environmental Communication: A Journal of Nature and Culture 4 (1), 82–98.

Brulle, R.J., 2010b. Politics and the environment. In: Leicht, Kevin T., Craig Jenkins, J. (Eds.), The Handbook of Politics: State and Civil Society in Global Perspective. Springer Publishers.

Brulle, R.J., 2014. Politics and the Environment. http://www.pages.drexel.edu/~brullerj/Environmental%20Politics%20Chapter.pdf.

Brunner, B., 2015. The World's Most Notorious Despots. Fact Monster. http://www.factmonster.com/spot/topdespots1.html.

Brym, R.J., Araj, B., 2012. Are suicide bombers suicidal? Studies in Conflict & Terrorism 35 (6), 432–443.

Buckley, G., 1997. A new species of Purgatorius (Mammalia; Primatomorpha) from the lower Paleocene Bear Formation, Crazy Mountains Basin, south-central Montana. Journal of Paleontology 71, 149–155.

Budge, E.A.W., 1969. The Gods of the Egyptians, Studies in Egyptian Mythology, vol. 2. Dover Publications, Inc.

Bunch, B., Hellemans, A., 2005. The History of Science and Technology: A Browser's Guide to the Great Discoveries, Inventions and the People. Houghton Mifflin Co.

Burke, J., 2010. Triple Murder in India Highlights Increase in 'Honour Killings'. The Guardian. http://www.theguardian.com/world/2010/jun/25/triple-murder-india-honour-killings.

Burland, T.M., Bennett, N.C., Jarvis, J.U.M., Faulkes, C.G., 2002. Eusociality in African mole-rats: new insights from patterns of genetic relatedness in Damaraland mole-rat (Cryptomys damarensis). The Royal Society, Proceedings B 269, 1495.

Burns, R.T., Raman, D.R., 2010. Animal Waste: Treatment. Encyclopedia of Agricultural, Food and Biological Engineering, second ed.

Burston, D., 1991. The Legacy of Erich Fromm. Harvard University Press.

Buskirk, R.E., 1981. Sociality in the Arachnida. In: Hermann, H.R. (Ed.), Social Insects, vol. II. Academic Press, pp. 281–367.

Buss, A.H., 1961. The Psychology of Aggression. John Wiley & Sons, Inc.

Buttel, F.H., Humphrey, C.R., 2002. Sociological theory and the natural environment. In: Dunlap, R.E., Michelson, W. (Eds.), Handbook of Environmental Sociology. Greenwood Press, pp. 33–69.

Butts, D.P., Espelie, K.E., Hermann, H.R., 1991. Cuticular hydrocarbons of four species of social wasps in the subfamily Vespinae: Vespa crabro L., Dolichovespula maculata (L.), Vespula squamosa (Drury) and Vespula maculifrons (Buysson). Comparative Biochemistry and Physiology Part B: Comparative Biochemistry 99 (1), 87–91.

Byström, S., Lönnstedt, L., 2009. The economic and environmental impact of paper recycling. Critical Reviews in Environmental Science and Technology 27 (1), 193–211.

Cahoone, L., 2012. Wild Business: a Philosopher Goes Hunting. North Dakota Humanities Council. http://www.nd-humanities.org/mags/OSTmag_sum10.pdf.

Caldwell, H.K., Young, W.S., 2006. Oxytocin and vasopressin: genetics and behavioral implications. In: Handbook of Neurochemistry and Molecular Neurobiology: Neuroactive Proteins and Peptides, third ed. Springer, pp. 573–607.

Cameron, S.A., 1993. Multiple origins of advanced eusociality in bees inferred from mitochondrial DNA sequences. Proceedings of the National Academy of Sciences 90 (18), 8687–8691.

Campbell, N.A., Reece, J.B., 2002. Biology, sixth ed. Benjamin Cummings.

Campbell, F., 1960. Birth control and the Christian churches. Population Studies 14 (2), 131–147.

Campbell, D.B., 2003. Greek and Roman Siege Machinery 399 BC-AD 363. Osprey Publishing.

Campbell, D.B., 2005. Siege Warfare in the Roman World. Osprey Press. pp. 18, 33, 67.

Campbell, B.G., 2009. Human Evolution: An Introduction to Mans Adaptations. Aldine Transaction.

Canasova, J., 1980. Public Religions in the Modern World. The University of Chicago Press.

Canham, H., 2010. Group and gang states of mind. Journal of Child Psychotherapy 28 (2), 113–127.

Cardinal, S., Danforth, B.N., 2011. The antiquity and evolutionary history of social behavior in bees. PLOS One 6 (6), e21086Bibcode:2011PLoSO...621086C. http://dx.doi.org/10.1371/journal.pone.0021086. PMC 3113908. PMID 21695157.

Carey, B., 2006. Top Predators Key to Ecosystem Survival, Study Shows. Live Science. http://www.livescience.com/4171-top-predators-key-ecosystem-survival-study-shows.html.

Carpenter, F.M., Hermann, H.R., 1981. Antiquity of sociality in insects. In: Hermann, H.R. (Ed.), Social Insects, vol. 1. Academic Press, pp. 81–89 (Chapter 2).

Carpenter, S.M., et al., 2006. The Ebers Papyrus, vol. 3. Bard College.

Carpenter, J., 1996. Distributional checklist of species of the genus *Polistes* (Hymenoptera: Vespidae; Polistinae, Polistini). American Museum of Natural History Number 3188.

Carr, J., 1972. Robespierre: The Force of Circumstance. St. Martin's Press.

Carrasco, P., 2009. The civil-religious hierarchy in Meso-American communities: pre-Spanish background and colonial development. American Anthropologist 63 (3), 483–497.

Carrington, R., 1958. Elephants: A Short Account of Their Natural History, Evolution and Influence on Mankind. Chatto & Windus.

Carson, R., 1962. Silent Spring. Houghton Mifflin Co.

Carter, P., 1976. Mao. Oxford University Press.

Cartmill, M., Smith, F.H., 2011. The Human Lineage. John Wiley & Sons.

Cartwright, D.S., Howard, K.I., Reuterman, N.A., 1970. Multivariate analysis of gang delinquency: II. Structural and dynamic properties of gangs. Multivariate Behavioral Research 5 (3), 303–323.

Cartwright, M., 2013. Greek Religion. Ancient History Encyclopedia.

Case, J., 2007. Competition: The Birth of a New Science. Hill and Wang.

Cases, O., Seif, I., Grimsby, J., et al., 1995. Aggressive behavior and altered amounts of brain serotonin and norepinephrine in mice lacking MAOA. Science 268 (5218), 1763–1766.

Cashdan, E., 1994. A sensitive period for learning about food. Human Nature 5 (3), 279–291.

Cashmore, A.R., 2010. The Lucretian swerve: the biological basis of human behavior and the criminal justice system. Proceedings of the National Academy of Sciences 107 (10), 4499–4504.

Caspi, A., Roberts, B.W., 2001. Personality development across the life course: the argument for change and continuity. Psychological Inquiry 12 (2), 49–66.

Cavalli-Sforza, L.L., Menozzi, P., Piazza, A., 1994. The History and Geography of Human Genes. Princeton University Press.

Cavanaugh, W.T., 2009. The Myth of Religious Violence, Secular Ideology and the Roots of Modern Conflict. Oxford University Press.

Cela-Conde, C.J., Ayala, F.J., 2003. Genera of the human lineage. Proceedings of the National Academy of Sciences 100 (13), 7684–7689.

Ceumern-Lindenstjerna, I.A., Brunner, R., Parzer, P., Fiedler, P., Resch, F., 2002. Borderline personality disorder and attentional biases. Theoretical models and empirical findings. Fortschritte der Neurologie-Psychiatrie 70 (6), 321–330.

Chagnon, N.A., 1967. Yanomamö – the fierce people. Natural History 77, 22–31.

Chamberlain, M.I., Leopold, B.D., Conner, L.M., 2003. Space use, movements and habitat selection of adult Bobcats (*Lynx rufus*) in Central Mississippi. The American Midland Naturalist 149 (2), 395–405.

Chamovitz, D., 2012. What a Plant Knows: A Field Guide to the Senses. Scientific American/Farrar, Straus and Giroux.

Champman, R.N., 1928. The quantitative analysis of environmental factors. Ecology 9 (2), 111–122.

Chang, J., Halliday, J., 2005. Mao: The Unknown Story. Jonathan Cape.

Changnon, S.A., 2009. The historical struggle with floods on the Mississippi River basin. Water International 23 (4), 263–271.

Chapinal, N., Ruiz-de-la-Torre, J.L., Cerisuelo, A., et al., 2008. Feeder use patterns in group-housed pregnant sows fed with unprotected electronic sow feeder (Fitmix). Journal of Applied Animal Welfare Science 11 (4), 319–336.

Chapman, L.J., Chapman, J., 1982. Test results are what you think they are. In: Kahneman, D., et al. (Ed.), Judgment Under Uncertainty: Heuristics and Biases. Cambridge University Press, pp. 238–248.

Chargaff, E. (Ed.), 2012. The Nucleic Acids. Elsevier.

Charlesworth, B., Morgan, M.T., Charlesworth, D., 1993. The effect of deleterious mutations on neutral molecular variation. Genetics 134 (4), 1289–1303.

Chase, J.M., Leibold, M.A., 2003. Ecological Niches: Linking Classical and Contemporary Approaches. University of Chicago Press.

Chase, K., 2003. Firearms: A Global History to 1700. Cambridge University Press.

Chatterjee, H.J., Simon, Y.W., Ho, I., Barnes, I., Groves, C., 2000. Estimating the phylogeny and divergence times of primates using a supermatrix approach. Evolutionary Biology 9 (1), 259.

Chatterton, E.K., 1915. Sailing Ships and Their Story: The Story of Their Development from the Earliest Times to the Present Day. J.B. Lippincott Company.

Chen, F.C., Li, W.H., 2001. Genomic divergences between humans and other hominoids and the effective population size of the common ancestor of humans and chimpanzees. American Journal of Human Genetics 68 (2), 444–456.

Chesler, P., 2010. Worldwide trends in honor killings. Middle East Quarterly, Spring 17 (2), 3–11.

Chettiparamb, A., 2013. Solid waste: municipal. In: Encyclopedia of Environmental Management. http://dx.doi.org/10.1081/E-EEM-120046102.

Cheverud, J.M., Rutledge, J.J., Atchley, W.R., 1983. Quantitative genetics of development: genetic correlations among age-specific trait values and the evolution of ontogeny. Evolution 895–905.

Chiriboga, C.A., 2003. Fetal alcohol and drug effects. Neurologist 8 (6), 267–279.

Chivian, E., Bernstein, A., 2008. Sustaining Life: How Human Health Depends on Biodiversity. Oxford Univ. Press.

Christensen, I., Haug, T., Oien, N., 1992. Seasonal distribution, exploitation and present abundance of stocks of large baleen whales (Mystic3eti) and sperm whales (*Physeter microcephalus*) in Norwegian and adjacent waters. ICES Journal of Marine Science 49 (3), 341–355.

Christie, D., Viner, R., 2005. Adolescent development. BMJ 330 (7486), 301–304.

Churchland, P.S., 2011. Braintrust: What Neuroscience Tells Us about Morality. Princeton University Press.

Cialdini, R.B., Kallgren, C.A., Reno, R.R., 1991. A focus theory of normative conduct: a theoretical refinement and reevaluation of the role of norms in human behavior. Advances in Experimental Social Psychology 24, 201–234.

Cialdini, R.B., 2003. Crafting normative messages to protect the environment. Current Directions in Psychological Science 12, 105–109.

Cincotta, R.P., Engelman, R., 2000. Nature's Place Human Population Density and the Future of Biological Diversity. Population Action International, Hanover.

Ciochon, R., Fleagle, J., 1987. Primate Evolution and Human Origins. Benjamin/Cummings.

Clack, J.A., 2009. The fin to limb transition: new data, interpretations and hypotheses from paleontology and developmental biology. Annual Review of Earth and Planetary Sciences 37 (1), 163–179.

Clark, C.W., 1973. Profit maximization and the extinction of animal species. The Journal of Political Economy 950–961.

Clarke, T.A., 1970. Territorial behavior and population dynamics of a pomacentrid fish, the garibaldi. Hypsypops rubicunda. Ecological Monographs 40, 189–212.

Clarke, F.M., Faulkes, C.G., 1998. Hormonal and behavioural correlates of male dominance and reproductive status in captive colonies of the naked mole-rat, *Heterocephalus glaber*. Proceedings of the Royal Society B: Biological Sciences 265 (1404), 1391–1399.

Clarke, E., Reichard, U.H., Zuberbühler, K., 2006. The syntax and meaning of wild gibbon songs. PLoS One 1 (1), 73.

Clemens, W., 1974. *Purgatorius*, an early paromomyid primate. Science 184 (4139), 903–905.

Clements, F.E., 1916. Plant Succession: An Analysis of the Development of Vegetation. Carnegie Institution of Washington.

Cline, A., 2015. Ethics & Morality: Philosophy of Behavior, Choice and Character, about Religion. http://atheism.about.com/od/ philosophybranches/p/Ethics.htm.

Clutton-Brock, J., Wilson, D.E., 2002. Mammals, Smithsonian Handbooks.

Clutton-Brock, T.H., Brotherton, P.N.M., et al., 2001. Contributions to cooperative rearing in meerkats. Animal Behaviour 61 (4), 705–710.

Cochran, G., Harpending, H., 2009. The 10,000 Year Explosion: How Civilization Accelerated Human Evolution. Basic Books.

Cohen, J., 1995. How Many People Can the Earth Support? W. W. Norton & Co.

Cohen, J.E., 2003. Human population: the next half century. Science 302 (5648), 1172–1175.

Coie, J.D., Dodge, K.A., Damon, W.E., et al., 1998. Aggression and antisocial behavior. In: Handbook of Child Psychology, fifth ed. John Wiley & Sons, Inc., pp. 779–862.

Coleman, F.C., Figueira, W.F., Ueland, J.S., Crowder, L.B., 2004. The impact of United States recreational fisheries on marine fish populations. Science 305 (5692), 1958–1960.

Collentine, D., 2007. Composite market design for a transferable discharge permit (TDP) system. Journal of Environmental Planning and Management 49 (6), 929–946.

Collier, P., Hoeffler, A., 2002. On the incidence of civil war in Africa. Journal of Conflict Resolution 46 (1), 13–28.

Coltman, D.W., Bancroft, D.R., et al., 1999. Male reproductive success in a promiscuous mammal: behavioural estimates compared with genetic paternity. Molecular Ecology 8 (7), 1199–1209.

Condie, K.C., 1997. Plate Tectonics and Crustal Evolution, fourth ed. Butterworth-Heinemann.

Connelly, M.J., 2009. Fatal Misconception: The Struggle to Control World Population. Harvard University Press.

Corn, T.J., 2003. Invasive Species. The Burlington Free Press.

Corsini, R.J., 1999. The Dictionary of Psychology. Brunner/Mazel.

Cowan, R., 2007. Roman Battle Tactics 109BC – AD313. Osprey Press.

Cowley, W.H., 1931. The traits of face-to-face leaders. The Journal of Abnormal and Social Psychology 26 (3), 304.

Cowlishaw, G., Mendelson, S., Rowcliffe, J., 2005. Evidence for post-depletion sustainability in a mature bushmeat market. Journal of Applied Ecology 42 (3), 460–468.

Cranbrook, E., Payne, J., Leh, C.M.U., 2008. Origin of the elephants, *Elephas maximus* L., of Borneo. Sarawak Museum Journal 63 (84), 95–125.

Crane, M.T., 2014. Losing Touch with Nature, Literature and the New Science in Sixteenth-Century England. John Hopkins University Press.

Crespi, C.L., 1995. Xenobiotic-metabolizing human cells as tools for pharmacological and toxocological research. Advances in Drug Research 26, 180–237.

Cressman, R., 1995. Evolutionary game theory with two groups of individuals. Games and Economic Behavior 11 (2), 237–253.

Cribb, R., Gilbert, H., Tiffin, H., 2014. Wild Man from Borneo: A Cultural History of the Orangutan. University of Hawai'i Press.

Crompton, J.L., 2011. Empirical evidence of the contributions of park and conservation lands to environmental sustainability: the key to repositioning the parks field. World Leisure Journal 50 (3), 154–172.

Cronk, L., Chagnon, N., Irons, W., 1999. Adaptation and Human Behavior: An Anthropological Perspective. Aldine De Gruyter.

Cronk, L., 1991. Human behavioral ecology. Annual Review of Anthropology 20, 25–53.

Cross, R., Baird, L., 2000. Technology is not enough: improving performance by building organizational memory. MIT Sloan Management Review 41 (3), 69.

Crouwel, J., 2013. Studying the six chariots from the tomb of Tutankhamun – an Update. In: Veldmeijer, A.J., Ikram, S. (Eds.), Chasing Chariots: Proceedings of the First International Chariot Conference (Cairo 2012). Sidestone Press, p. 74.

Cruz, J.M., 2010. Central American maras: from youth street gangs to transnational protection rackets. Global Crime 11 (4), 379–398.

Cummins, D.D., 1996. Dominance hierarchies and the evolution of human reasoning. Minds and Machines 6 (4), 463–480.

Cuomo, A., 2010. Birth control. In: O'Reilly, A. (Ed.), Encyclopedia of Motherhood. Sage Publications, pp. 121–126.

D'Errico, F., Zilhão, J., Julien, M., Baffier, D., Pelegrin, J., 1998. Neanderthal acculturation in Western Europe? A critical review of the evidence and its interpretation. In: The Neandeerthal Problem and the Evolution of Human Behavior, pp. 1–44.

Dailykos, 2014. Unstoppable Human Population Growth to About 10 Billion Will Cause Ecosystem Decline or Collapse. http://www.dailykos. com/story/2014/10/28/1339730/-Unstoppable-Human-Population-Growth-to-about-10-Billion-Will-Cause-Ecosystem-Decline-or-Collapse.

D'arcy, F., November 1977. The Malthusian League and Resistance to Birth Control Propaganda in Late Victorian Britain. Population Studies 31, 429–448.

Darke, S., 1998. Self-report among injecting drug users: a review. Drug and Alcohol Dependence 51 (3), 253–263.

Darwin, C., 1859. On the Origin of Species by Means of Natural Selection, or the Preservation of Favoured Races in the Struggle for Life. John Murray.

Darwin, C., 1871. The Descent of Man, and Selection in Relation to Sex. John Murray.

Davin, D., 2013. Mao: A Very Short Introduction. Oxford UP.

Davis, J.J., 1985. Evangelical Ethics: Issues Facing the Church Today. Presbyterian and Reformed Pub. Co.

Dawkins, R., 2006a. The God Delusion. Houghton Mifflin Company.

Dawkins, R., 2006b. The Selfish Gene. Oxford University Press.

D'Cruze, S., Walklate, S.L., Pegg, S., 2013. Murder. Routledge.

De Catanzaro, D., 1986. A mathematical model of evolutionary pressures regulating self-preservation and self-destruction. Suicide and Life-Threatening Behavior 16 (2), 166–181.

De Grouchy, J., 1987. Chromosome phylogenies of man, great apes, and old world monkeys. Genetica 73 (1–2), 37–52.

De Heinzelin, J., Clark, J.D., White, T., Hart, W., Renne, P., WoldeGabriel, G., Beyne, Y., Brba, E., 1999. Environment and behavior of 2.5-million-year-old Bouri hominids. Science 284 (5414), 625.

De Lapp, K., 1991. Weapons of Mass Deception: The Uses and Abuses of Honesty in Politics. Psychology Today. http://www.psychologytoday.com/blog/the-love-wisdom/201206/weapons-mass-.

De Souza, D.V., 2013. Is aggressive behaviour biologically or environmentally based? Conflict and Health 7, 13.

De Vries, H., Jeroen, A.N., Stevens, M.G., Vervaecke, H., 2006. Measuring and testing the steepness of dominance hierarchies. Animal Behaviour 71 (3), 585–592.

De Waal, F., 2005. Our Inner Ape. Riverhead Books.

Deane, C.D., 1944. The Broken-Wing Behavior of the Killdeer. The Auk, pp. 243–247.

Debruyne, R., 2005. A case study of apparent conflict between molecular phylogenies: the interrelationships of African elephants. Cladistics 21 (1), 31–50.

Decker, S.H., Melde, C., Pyrooz, D.C., 2013. What do we know about gangs and gang members and where do we go from here? Justice Quarterly 30 (3), 369–402.

Deloria, V., 1974. Behind the Trail of Broken Treaties: An Indian Declaration of Independence. Delacorte Press.

Dennis, G.T., 1984. Maurice's Strategikon. Handbook of Byzantine Military Strategy. University of Pennsylvania Press.

Derue, D.S., Nahrgang, J.D., Wellman, N., Humphrey, S.E., 2011. Trait and behavioral theories – of leadership: an integration and meta-analytic test of their relative validity. Personnel Psychology 4 (1), 7–52.

Devlin, A.S., Arneill, A.B., 2003. Health care environments and patient outcomes a review of the literature. Environment and Behavior 35 (5), 665–694.

De Vries, M.K., 2003. Entering the Inner Theatre of a Despot: The Rise and Fall of Saddem Hussein. INSEAD.

Diamond, J., Bond, A.B., 2003. A comparative analysis of social play in birds. Behaviour 140 (8), 1091–1115.

Dictionary of Psychology (Online), 2014.

Dillehay, et al., 2010. Early Holocene coca chewing in northern Peru. Antiquity 84 (326), 939–953.

Dimitriou, A., Christidou, V., 2010. Pupils' understanding of air pollution. Journal of Biological Education 42 (1), 24–29.

Dingwall, E.J., 1953. Early contraceptive sheaths. British Medical Journal 1 (4800), 40–41.

Dixson, A.F., 1981. The Natural History of the Gorilla. Weidenfeld & Nicholson.

Dobbing, J., Sands, J., 1973. Quantitative growth and development of human brain. Archives of Disease in Childhood 48, 757–767.

Donadio, R., Goodstein, L., November 23, 2010. After Condom Remarks, Vatican Confirms Shift. New York Times.

Donaldson, S., Kymlicka, W., 2011. Zoopolis: A Political Theory of Animal Rights. Oxford University Press.

Dorn, L.D., Biro, F.M., 2011. Puberty and its measurement: a decade in review. Journal of Research on Adolescence 21 (1), 180–195.

Dornhaus, A., Chittka, L., 2005. Bumble bees (*Bombus terrestris*) store both food and information in honeypots. Behavioral Ecology 16 (3), 661–666.

Doyle, J.A., Donoghue, M.J., 1986. Seed plant phylogeny and the origin of angiosperms: an experimental cladistic approach. Botanical Review 52, 321–431.

Doyle, J., January 1995. Black Market Orchids: A Global Underground Smuggling Network May Drive Some Rare Species Into Extinction. San Francisco Chronicle, p. 6.

Draper, T.W., 1995. Canine analogs of human personality factors. Journal of General Psychology 122 (3), 241–252.

Draznin, Y.C., 2001. Victorian London's Middle-Class Housewife: What She Did All Day (#179). Contributions in Women's Studies. Greenwood Press, pp. 98–100.

Drews, C., 1993. The concept and definition of dominance in animal behavior. Behaviour 125 (3), 283–313.

Drews, C., 2002. Attitudes, knowledge and wild animals as pets in Costa Rica. Anthrozoös 15 (2), 119–138.

Dryden-Edwards, R., 2016. Rape and Sexual Assault. MedicineNet.com. http://www.medicinenet.com/rape_sexual_assault/article.htm.

Dryzek, J.S., 1983. Ecological rationality. International Journal of Environmental Studies 21 (1), 5–10.

Dugatkin, L.A., Earley, R.L., 2003. Group fusion: the impact of winner, loser and bystander effects on hierarchy formation in large groups. Behavioral Ecology 73 (3), 290–298.

Dunbar, R.I.M., Shultz, S., 2007a. Understanding primate brain evolution. Philosophical Transactions of the Royal Society of London. Series B, Biological Sciences 362, 649–658.

Dunbar, R.I.M., Shultz, S., 2007b. Evolution in the social brain. Science 317, 1344–1347.

Dunbar, R.I., 2002. The social brain hypothesis. Foundations in Social Neuroscience 5 (71), 69.

Dunham, A.E., 2008. Battle of the sexes: cost asymmetry explains female dominance in lemurs. Animal Behaviour 76, 1435–1439.

Dunlap, R.E., Michelson, W., 2002. Handbook of Environmental Sociology. Greenwood Press.

Dunn, K., Orellana, S., Singh, S., 2009. Legislative diversity and social tolerance: how multiparty systems lead to tolerant citizens. Journal of Elections, Public Opinion and Parties 19 (3), 283–312.

Dunne, M.S., Humphreys, Leach, F., 2006. Gender violence in school in the developing world. Gender and Education 18 (1), 75–98.

Dunn-Walters, D.K., Dogan, A., Boursier, L., MacDonald, C.M., Spencer, J., 1998. Base specific sequences that bias somatic hypermutation deduced by analysis of out of frame genes. The Journal of Immunology 160, 2360–2364.

Dupain, J., Van Krunkelsven, E., Van Elsacker, L., Verheyen, R.F., 2000. Current status of the bonobo (*Pan paniscus*) in the proposed Lomako Reserve (Democratic Republic of Congo). Biological Conservation 94 (3), 265–272.

De Ruiter, P.C., Wolters, V., Moore, J.C., 2005. Dymanic Food Web, Multispecies Assemblages, Ecosystem Development and Environmental Change. Academic Press.

East, M.L., Hofer, H., 2001. Male spotted hyenas (*Crocuta crocuta*) queue for status in social groups dominated by females. Behavioral Ecology 12 (5), 558–568.

East, M.L., Hofer, H., 2002. Conflict and cooperation in a female-dominated society: a reassessment of the "hyperaggressive" image of spotted hyenas. Advances in the Study of Behavior 31, 1–30.

Easton, C., 1979. The Ecology of Burning Beetles: Necrophorus: *Coleoptera*. University of Glasgow.

Eberle, P., 2006. Terror on the Highway. Prometheus Books.

Economides, K.D., Zeltser, L., Capecchi, M.R., 2003. Hoxb13 mutations cause overgrowth of caudal spinal cordand tail vertebrae. Developmental Biology 256 (2), 317–330.

Edgar, B., 2012. Why Do Politicians Lie? TED Weekends. The Huffington Post. http://www.huffingtonpost.com/rev-bob-edgar/why-politicians-lie_b_2104104.html.

Ehhalt, D., et al., 2001. Atmospheric Chemistry and Greenhouse Gases. Pacific Northwest National Laboratory(No. PNNL-SA-39647).

Ehret, B.D., Gray, W.D., Kirschenbaum, S.S., 2000. Special section: contending with complexity: developing and using a scaled world in applied cognitive research. Human Factors: The Journal of the Human Factors and Ergonomics Society 42 (1), 8–23.

Ehrlich, P.R., Ehrlich, A.H., 1991. The Population Explosion. Simon & Schuster.

Ehrlich, P.R., Ehrlich, A.H., 2008. The Dominant Animal: Human Evolution and the Environment. Island Press.

Ehrlich, P.R., Ehrlich, A.H., 2009. The population bomb revisited. Electronic Journal of Sustainable Development 1 (3), 63–71.

Ehrlich, P.R., Dobkin, D.S., Wheye, D., 1988. A Field Guide to the Natural History of North American Birds. Simon and Schuster.

Ehrlich, P.R., Dobkin, D.S., Wheye, D., 1988. Birder's Handbook. Touchstone.

Ehrlich, P.R., Dobkin, D.S., Wheye, D., Naeem, S., Pimm, S.L., 1994. The Birdwatcher's Handbook: A Guide to the Natural History of the Birds of Britain and Europe: Including 516 Species that Regularly Breed in Europe. Parts of the Middle East and North Africa. Oxford University Press.

Ehrlich, P.R., 1968. The Population Bomb. Ballantine Books.

Eickwort, G.C., 1981. Presocial insects. In: Hermann, H.R. (Ed.), Social Insects, vol. 2. Academic Press, pp. 199–280.

Eisner, T., 2005. For Love of Insects. Belknap Press.

El-Hani, C.N., Queiroz, J., Stjernfelt, F., 2010. Firefly femmes fatales: a case study in the semiotics of deception. Biosemiotics 3 (1), 33–55.

Elliott, F.A., 1988. Violence: a product of biosocial interactions. The Bulletin of the American Academy of Psychiatry and the Law 16 (2), 131–143.

Ellis, L., 1995. Dominance and reproductive success among nonhuman animals: a cross-species comparison. Ethology and Sociobiology 16 (4), 257–333.

Emery, N.J., Seed, A.M., von Bayern, A.M.P., Clayton, N.S., 2007. Cognitive adaptations of social bonding in birds. Philosophical Transactions of the Royal Society B: Biological Sciences 362 (1480), 489–505.

Emlen, J.M., 1968. Batesian mimicry: a preliminary theoretical investigation of quantitative aspects. American Naturalist 235–241.

Engdahl, F.W., 2007. Seeds of Destruction: The Hidden Agenda of Genetic Manipulation. Global Research, Canada.

Engleman, R., 2012. State of the World 2012: Moving Toward Sustainable Prosperity. Island Press/Center for Resource Economics.

Enquist, M., Arak, A., 1994. Symmetry, beauty and evolution. Nature 372 (6502), 169–172.

Enquist, M., Leimar, O., 1987. Evolution of fighting behaviour: the effect of variation in resource value. Journal of Theoretical Biology 127 (2), 187–205.

Erdal, D., Whiten, A., 1996. Egalitarianism and machiavellian intelligence in human evolution. In: Mellars, P., Gibson, K. (Eds.), Modeling the Early Human Mind. Cambridge MacDonald Monograph Series.

Erwin, T., et al., 2005. Mapping beta-diversity for beetles across the Western Amazon basin: A preliminary case for improving inventory methods and conservation strategies. Proc. Cal. Acad. Sci. 56 (7), 72–85.

Esmay, M.L., 1969. Principles of Animal Environment. CAB Direct.

Espelie, K.E., Hermann, H.R., 1990. Surface lipids of the social wasp *Polistes annularis* (L.) and its nest and nest pedicel. Journal of Chemical Ecology 16 (6), 1841–1852.

Estlund, D., 2011. Human nature and the limits (if any) of political philosophy. Philosophy & Public Affairs 39 (3), 207–237.

Evernden, N., 2006. The Natural Alien, Humankind and Environment. University of Toronto Press.

Extinction, 2008. Humans 'Almost Became Extinct in 70,000 BC'. The Telegraph. http://www.telegraph.co.uk/news/science/science-news/3340777/Humans-almost-became-extinct-in-70000-BC.html.

Eysenck, H.J., 2013. The Structure of Human Personality (Psychology Revivals). Routledge.

Falek, A., Konner, M.J., 2014. World population prospects: the impact of ecological and genetic factors on human population growth in the 21st century. Global Bioethics 12 (1–4), 31–41.

Falk, A., Fischbacher, U., 2006. A theory of reciprocity. Games and Economic Behavior 54 (2), 293–315.

FAO (Food and Agriculture Organization of the United Nations), 2006. Livestock a Major Threat to the Environment: Remedies Urgently Needed. http://www.fao.org/newsroom/en/news/2006/1000448/index.html.

Farnsworth, N.R., et al., 1975. Potential value of plants as sources of new antifertility agents. Journal of Pharmaceutical Sciences 64 (4), 535–598.

Faulkes, C.G., Abbott, D.H., 1997. The physiology of a reproductive dictatorship: regulation of male and female reproduction by a single breeding female in colonies of naked mole-rats. Cooperative Breeding in Mammals 302–334.

Fawcett, W.B., 1987. Communal Hunts, Human Aggregations, Social Variation and Climatic Change: Bison Utilization by Prehistoric Inhabitants of the Great Plains (Paleoecology) (Doctoral Dissertation). Univ. of Massachusetts, Amherst.

Feduccia, A., 1999. The Origin and Evolution of Birds. Yale University Press.

Felbab-Brown, V., 2013. The Illegal Trade in Wildlife in Southeast Asia and Its Links to East Asian Markets. Brookings. http://www.brookings.edu/research/articles/2013/04/05-illegal-trade-wildlife-southeast-asia-east-asian-markets-felbabbrown.

Feldhamer, G.A., Thompson, B.C., Chapman, J.A., 2004. Wild Mammals of North America. Johns Hopkins University Press, pp. 769–770.

Fernando, P., Vidya, T.N.C., Payne, J., Stuewe, M., Davison, G., Alfred, R.J., Andau, P., Bosi, E., Kilbourn, A., Melnick, D.J., 2003. DNA analysis indicates that Asian Elephants are native to Borneo and are therefore a high priority for conservation. PLoS Biology 1 (1), 1371.

Fernea, E., 1985. Women and the Family in the Middle East: New Voices of Change. University of Texas Press, pp. 258–269.

Ferriere, R., Michod, R.E., 2011. Inclusive fitness in evolution. Nature 471 (7339), E6–E8.

Ferson, S., Ginzburg, L.R., 1996. Different methods are needed to propagate ignorance and variability. Reliability Engineering & System Safety 54 (2), 133–144.

Fichtel, C., Kappeler, P.M., 2009. Human universals and primate symplesiomorphies: establishing the lemur baseline. In: Kappeler, P.M., Silk, J.B. (Eds.), Mind the Gap: Tracing the Origins of Human Universals. Springer, pp. 395–426 (Chapter 19).

Fimrite, P., 2015. Sardine Population Collapses, Prompting Ban on Commercial Fishing. SFGATE. http://www.sfgate.com/bayarea/article/Sardine-population-collapses-prompts-ban-on-6197380.php.

Fishbein, L., 2008. Analysis of carcinogenic and mutagenic aromatic amines: an overview. Toxicological & Environmental Chemistry 3 (2), 145–168.

Fisher, S.E., Reason, J.E., 1988. Handbook of Life Stress, Cognition and Health. John Wiley & Sons.

Fiske, A.P., 1992. The four elementary forms of sociality: framework for a unified theory of social relations. Psychological Review 99 (4), 689.

Flachowsky, G., Meyer, U., 2015. Challenges for plant breeders from the view of animal nutrition. Agriculture 5 (4), 1252–1276.

Flood, G.D., 1996. An Introduction to Hinduism. Cambridge University Press, p. 16.

Flores, R.A., Shanklin, C.W., Loza-Garay, M., Wie, S.H., 2013. Quantification and characterization of food processing wastes/residues. Compost Science & Utilization 7 (1), 63–71.

Florian, S., 2015. The Rhetoric of Surf: A Lexical and Archetypal Migration of Los Angeles Counterculture into Popular Culture. Project Essays, California State University, p. 154.

Flynn, G., 1999. Stop toxic managers before they stop you!. Workforce-Costa Mesa 78, 40–46.

Fonberg, E., 1988. Dominance and aggression. International Journal of Neuroscience 41 (3–4), 201–213.

Foremski, T., 2010. Computer crime, United States laws and law enforcement. International Review of Law, Computers & Technology 6 (1), 121–127.

Fortey, R., 2000. Trilobite! Eyewitness to Evolution. Alfred A. Knopf.

Fortey, R., 2012. Horseshoe Crabs and Velvet Worms: The Story of the Animals and Plants that Time Has Left Behind. Vintage Books.

Foster, J.B., 2000. Marx's Ecology: Materialism and Nature. NYU Press.

Fox, L.R., 1975. Cannibalism in natural populations. Annual Review of Ecology and Systematics 6, 87–106.

Fox, J., 2008. A World Survey of Religion and the State. Cambridge University Press.

Francis, R.C., 1988. On the relationship between aggression and social dominance. Ethology 78 (3), 223–237.

Fraser, I., 1998. Hegel and Martz: The Concept of Need. Edenbergh University Press.

Freeman, R.B., 1977. The Works of Charles Darwin: An Annotated Bibliographical Handlist, second ed. Wm Dawson & Sons Ltd.

Frentiu, T., Ponta, M., Levei, E., Gheorghiu, E., Benea, M., Cordos, E., 2008. Preliminary study on heavy metals contamination of soil using solid phase speciation and the influence on groundwater in Bozanta-Baia Mare Area, Romania. Chemical Speciation & Bioavailability 20 (2), 99–109.

Fridlund, A.J., 1994. Human Facial Expression. Academic Press.

Friedman, M., et al., 1986. Alteration of type A behavior and its effect on cardiac recurrences in post myocardial infarction patients: summary results of the recurrent coronary prevention project. American Heart Journal 112 (4), 653–665.

Friedman, M., 1996. Type A Behavior: Its Diagnosis and Treatment. Plenum Press (Kluwer Academic Press), p. 31.

Frodi, A., 1975. The effect of exposure to weapons on aggressive behavior from a cross-cultural perspective. International Journal of Psychology 10 (4), 283–292.

Frost, M., Andersen, T., Gossiel, F., Hansen, S., et al., 2011. Levels of serotonin, sclerostin, bone turnover markers as well as bone density and microarchitecture in patients with high bone mass phenotype due to a mutation in Lrp5. Journal of Bone and Mineral Research 26 (8), 1721–1728.

Fry, D.P., 2005. The Human Potential for Peace: An Anthropological Challenge to Assumptions about War and Violence. Oxford University Press, NY.

Fryer, P., 1965. The Birth Controllers. Secker & Warburg.

Furuichi, T., 1997. Agonistic interactions and matrifocal dominance rank of wild bonobos (Pan paniscus) at Wamba. International Journal of Primatology 18 (6), 855–875.

Furuseth, O.J., 2010. Restructuring of hog farming in North Carolina: explosion and implosion. The Professional Georgrapher 49 (4), 391–403.

Futuyma, D.J., 2005. Evolution. Sinauer Associates.

Gacono, C.B., et al. (Ed.), 2007. The Handbook of Forensic Rorschach Psychology. Lawrence Erlbaum, p. 80.

Gahlinger, P.M., 2004. Illegal Drugs: A Complete Guide to Their History, Chemistry, Use, and Abuse. Penguin Books.

Gallant, R., 1990. The Peopling of the Planet Earth. Macmillian Publishing Company.

Ganor, B., 2002. Defining terrorism: is one man's terrorist another man's freedom fighter? Police Practice and Research 3 (4), 287–304.

Garrett, L., 1994. The Coming Plague, Newly Emerging Diseases in a World Out of Balance. Panguin Books.

Gatlin, J.U., Wysocki, A., Kepner, K., 2007. Understanding Conflict in the Workplace. EDIS document HR 024, Dept. Food and Resource Economics. Florida Coop. Ext. Ser., Inst. Food and Agric. Sci., Univ. of Florida.

Gaukroger, S., 2001. Francis Bacon and the Transformation of Early-Modern Philosophy. Cambridge University Press.

Geen, R.G., 1990. Human aggression. In: Donnerstein, E.D. (Ed.), Milton Keynes. Open University Press.

Geisler, N.L., Turek, F., 2004. Kant's agnosticism: should we be agnostic about it? I don't have enough faith to be an atheist. Crossway 59–60.

Geist, H.J., Lambin, E.E., 2002. Proximate causes and underlying driving forces of tropical deforestation. BioScience 52 (2), 143–150.

Genn, R., Lerman, A., 2010. Fascism and racism in Europe: the report of the European parliament's committee of inquiry. Patterns of Prejudice 20 (2), 13–26.

Gesquiere, L.L., et al., 2011. Life at the top: rank and stress in wild male baboons. Science 333 (6040), 357–360.

Ghosh, K., Shen, E.S., Arey, B.J., López, F.J., 1998. Global model to define the behavior of partial agonists (bell-shaped dose-response inducer) in pharmacological evaluation of activity in the presence of the full agonist. Journal of Biopharmaceutical Statistics 8 (4), 645–665.

Gibbons, A., 1998. Ancient island tools suggest Homo erectus was a seafarer. Science 279 (5357), 1635–1637.

Gilbert, S.L., Dobyns, W.B., Lahn, B.T., 2005. Genetic links between brain development and brain evolution. Nature Reviews Genetics 6, 581–590.

Gillian-Flynn, G., 1999. Stop toxic managers before they stop you. Workforce 44–46.

Gillooly, J.F., Hou, C., Kaspari, M., 2010. Eusocial insects as superorganisms. Communicative & Integrative Biology 3 (4), 360–362.

Giustiniani, V., 1985. Homo, Humanus, and the meanings of humanism. Journal of the History of Ideas 46 (2), 167–195.

Godinot, M., 2006. Lemuriform origins as viewed from the fossil record. Folia Primatologica 77 (6), 446–464.

Goldman, N., Anderson, J.P., Rodrigo, A.G., 2000. Likelihood-based tests of topologies in phylogenetics. Syst. Biol. 49 (4), 652–670.

Goldstein, D.B., 2010. Common genetic variation and human traits. UC Shared Journal Collection 1696–1698.

Goodman, M., Tagle, D.A., Fitch, D.H., Bailey, W., Czelusniak, J., Koop, B.F., Benson, P., Slightom, J.L., 1990. Primate evolution at the DNA level and a classification of hominoids. Journal of Molecular Evolution 30 (3), 260–266.

Gordon, L., 2002. The Moral Property of Women, A History of Birth Control Politics in America. University of Illinois Press.

Gordon, J.D., Speroff, L., 2002. Handbook for Clinical Gynecologic Endocrinology and Infertility. Lippincott Williams & Wilkins.

Gore, A., 1992. Earth in the Balance. Houghton Mifflin Company.

Gray, D., 2015. Gang crime and the media in late nineteenth-century London. Cultural and Social History 10 (4), 559–575.

Greer, G., 1984. Sex and Destiny. Secker and Warburg, p. 306.

Griffiths, A., Wessler, J.F., Susan, R., Carroll, S.B., Sean, B., Doebley, J., 2012. 2: Single-Gene Inheritance, Introduction to Genetic Analysis, tenth ed. W. H. Freeman and Company, p. 57.

Grigorenko, E.L., Sternberg, R.J., 2003. The nature nurture issue. In: Slater, A., Bremner, G. (Eds.), An Introduction to Developmental Psychology. Blackwell.

Grosberg, R.K., Strathmann, R.R., 2007. The evolution of multicellularity: a minor major transition? Annual Review of Ecology, Evolution, and Systematics 38, 621–654.

Grosby, S., 1994. The verdict of history: the inexpungeable tie of primordiality – a response to Eller and Coughlan. Ethnic and Racial Studies 17 (1), 164–171.

Groves, C., 2002. A history of gorilla taxonomy. In: Gorilla Biology: A Multidisciplinary Perspective, vol. 3, pp. 15–34.

Groves, C.P., 2005. Family galagidae. In: Wilson, D.E., Reeder, D.M. (Eds.), Mammal Species of the World: A Taxonomic and Geographic Reference, third ed. Johns Hopkins University Press, pp. 123–127.

Grunwald, M., 2006. The Swamp: The Everglades, Florida, and the Politics of Paradise. Simon & Schuster.

Grusky, D.B., Takata, A.A., 1992. Social Stratification. The Encyclopedia of Sociology. Macmillan Publishing Company, pp. 1955–1970.

Gudykunst, W.B., Matsumoto, Y., 1996. Cross-cultural variability of communication in personal relationships. Communication in Personal Relationships Across Cultures 19–56.

Guetzloe, E., 1997. Origins of Violence and Aggression. www.ttac.odu.edu/articles/origins.html.

Guntrip, H., 1995. Personality Structure and Human Interaction: The Developing Synthesis of Psychodynamic Theory. Karnac Books.

Guthrie, W.K.C., 1968. The Greek Philosophers from Thales to Aristotle. Routledge.

Gutteridge, A., Thornton, J.M., 2005. Understanding nature's catalytic toolkit. Trends in Biochemical Sciences 30 (11), 622–629.

Guy, F., Lieberman, D.E., Pilbeam, D., et al., 2005. Morphological affinities of the *Sahelanthropus tchadensis* (late Miocene hominid from Chad) cranium. PNAS 102 (52), 18836–18841.

Gwynne, S.C., 2010. Empire of the Summer Moon. Simon and Schuster, Inc.

Hales, R.E., Zatazick, D.F., 1997. What is PTSD? American Journal of Psychiatry 154, 143–145.

Hall, R., 1977. Passionate Crusader. Harcourt, Brace, Jovanovich.

Hall, B.K., 2007. Fins into Limbs: Evolution, Development and Transformation. University of Chicago Press.

Hall, M.P., 2011. The Secret Teachings of All Ages: An Encyclopedic Outline of Masonic, Hermetic, Qabbalistic and Rosicrucian Symbolical Philosophy. Pacific Publishing Studio.

Hames, R., 2001. Human Behavioral Ecology. International Encyclopedia of the Social & Behavioral Sciences. Elsevier Science Ltd.

Hamilton, C., Layton, R.H., Rowley-Conwy, P., 2001. Hunter-gatherers: An Interdisciplinary Perspective, vol. 13. Cambridge University Press.

Hamilton, M.J., Milne, B.T., et al., 2007. The complex structure of hunter–gatherer social networks. Proceedings of the Royal Society of London B: Biological Sciences 274 (1622), 2195–2203.

Hamilton, W.D., 1963. The evolution of altruistic behavior. American Naturalist 97 (896), 354–356.

Hamilton, V., 1988. In the Beginning: Creation Stories from Around the World. Harcourt, Inc.

Hammersley-Fletcher, L., Brundrett, M., 2005. Leaders on leadership: impressions of primary school head teachers and subject leaders. School Leadership & Management 25 (1), 59–75.

Hammerstein, P., 1981. The role of asymmetries in animal contests. Animal Behaviour 29 (1), 193–205.

Hammerstein, J., 1987. Contraception: An overview. Amer. J. Obstetrics & Gynecology 157 (4/2), 1020–1023.

Hanlon, R.T., Messenger, J.B., 1996. Cephalopod Behaviour. Cambridge University Press.

Hardt, J., 2012. The Righteous Mind, Why Good People Are Divided by Politics and Religion. Pantheon Books.

Hare, R.D., 1993. Without Conscience: The Disturbing World of the Psychopaths Among Us. The Guilford Press.

Harff, B., 2003. No lessons learned from the holocaust? Assessing risks of genocide and political mass murder since 1955. American Political Science Review 97 (1), 57–73.

Harlow, W.F., Brantley, B.C., Harlow, R.M., 2011. BP initial image repair strategies after the *Deepwater Horizon* spill. Public Relations Review 37 (1), 80–83.

Harmon, R.B., Rosner, R., Owens, H., 1998. Sex and violence in a forensic population of obsessional harassers. Psychology, Public Policy, and Law 4, 236.

Hart, B., 2000. Conflict in the Workplace. Behavioral Consultants. http://www.excelatlife.com/articles/conflict_at_work.htm.

Hartman, W.L., 1972. Lake Erie: effects of exploitation, environmental changes and new species on the fishery resources. Journal of the Fisheries Research Board of Canada 29 (6), 899–912.

Hartwig, W., 2011. Primate evolution. In: Campbell, C.J., Fuentes, A., MacKinnon, K.C., Bearder, S.K., Stumpf, R.M. (Eds.), Primates in Perspective, second ed. Oxford University Press, pp. 19–31 (Chapter 3).

Harvey, P., 2000. An Introduction to Buddhist Ethics. Cambridge University Press.

Haslam, N., Kashima, Y., Loughnan, S., Shi, J., Suitner, C., 2008. Subhuman, inhuman, and superhuman: contrasting humans with nonhumans in three cultures. Social Cognition 26 (2), 248–258.

Hass, H., 1971. Human Animal: The Mystery of Man's Behavior. Hodder & Stoughton Ltd.

Hastings, P.J., Lupski, J.R., Rosenberg, S.M., et al., 2009. Mechanisms of change in gene copy number. Nature Reviews Genetics 10 (8), 551–564.

Hatcher, J.M., Dunn, M.A., 2011. Parasites in Ecological Communities. Cambridge University Press.

Hausfater, G., Hrdy, S.B. (Eds.), 1984. Infanticide: Comparative and Evolutionary Perspectives. Hawthorne.

Hawking, P., 1996. A Brief History of Time. Bantam Books.

Hawks, J., 2013. How has the human brain evolved? Scientific American Mind 24 (3).

Hayaki, H., 1983. The social interactions of juvenile Japanese monkeys on Koshima Islet. Primates 24 (2), 139–153.

Haynes, K.F., Gemeno, C., Yeargan, K.V., Millarand, J.G., Johnson, K.M., 2002. Aggressive chemical mimicry of moth pheromones by a bolas spider: how does this specialist predator attract more than one species of prey? Chemoecology 12 (2), 99–105.

Hazen, R.M., 2012. The Story of Earth: The First 4.5 Billion Years, from Stardust to Living Planet. Viking, New York.

Hecht, E., 1967. Chemical nature of human brain thromboplastin. Nature 212, 197–198.

Heinrich, J., Boyd, R., Bowles, S., Camerer, C., Fehr, E., Gintis, H., 2004. Foundations of Human Sociality: Economic Experiments and Ethnographic Evidence from Fifteen Small-Scale Societies. Oxford University Press.

Heinrich, B., 1989. Ravens in Winter. Summit Books.

Heinrich, B., 2006. Mind of the Raven. HarperCollins Publishers.

Heinsohn, G., Steiger, O., 1999. Birth control: the political-economic rationale behind Jean Bodin's Demonomanie. History of Political Economy 31 (3), 423–448.

Heinsohn, G., Steiger, O., 2004. Witchcraft, Population Catastrophe and Economic Crisis in Renaissance Europe: An Alternative Macroeconomic Explanation. Discussion Paper. University of Bremen.

Heng, H.H.Q., 2009. The genome-centric concept: resynthesis of evolutionary theory. BioEssays 31 (5), 512–525.

Henke, W., Hardt, T., Tatttersall, I., 2007. Handbook of Paleoanthropology: Phylogeny of Hominids. Springer, pp. 1527–1529.

Henshilwood, C.S., Marean, C.W., 2003. The origin of modern human behavior: critique of the models and their test implications. Current Anthropology 44, 627–651.

Hermann, H.R., Dirks, T.F., 1975. Biology of *Polistes annularis* (Hymenoptera: Vespidae): Spring behavior. Psyche 82 (1), 97–108.

Hermann, H.R., Kelting, T., Capobianco, P., 2017. Warning behavior expressed by three species of polistine wasps (Hymenoptera: Vespidae: Polistinae). Florida Entomologist (in press).

Hermann, H.R., 1971. Sting autonomy, a defensive mechanism in certain social hymenoptera. Insectes Sociaux 18 (2), 111–120.

Hermann, H.R., 1979. Insect sociality – an introduction. In: Hermann, H.R. (Ed.), Social Insects, vol. 1. Academic Press, New York, pp. 1–33 (Chapter 1).

Hermann, H.R., 1979. Sociality defined. In: Hermann, H.R. (Ed.), Social Insects. Academic Press, pp. 1–33.

Hermann, H.R., 1984a. Defensive Mechanisms in Social Insects. Praeger Scientific.

Hermann, H.R., 1984b. Defensive mechanisms: general considerations. In: Hermann, H.R. (Ed.), Defensive Mechanisms in Social Insects. Praeger Scientific, pp. 1–31 (Chapter 1).

Hermann, H.R., 1986. Selforganization in insects. In: Fox, S.W. (Ed.), Selforganization. Adenine Press, pp. 123–142.

Hermann, H.R., 2011. Making the Wind Sing, Native American Music and the Connected Breath. Masterwork Books.

Herrenkohl, R.C., 2005. The definition of child maltreatment: from case study to construct. Child Abuse and Neglect 29 (5), 413–424.

Heter, T.S., 2006. Sartre's Ethics of Engagement: Authenticity and Civic Virtue. Continuum.

Hicks, J.W., 2005. Fifty Signs of Mental Illness: A Guide to Understanding Mental Health. Yale University Press.

Hider, R.C., Kong, X., 2013. Iron: effect of overload and deficiency. In: Interrelations between Essential Metal Ions and Human Diseases. Metal Ions in Life Sciences. Springer, pp. 229–294 (Chapter 8).

Higman, B.W., 1996. Slave Populations of the British Caribbean. The Press, University of the West Indies, pp. 1807–1834.

Hill, K., Kaplan, H., 1999. Life history traits in humans: theory and empirical studies. Annual Review of Anthropology 28, 397–430.

Hingley, R., 1974. Joseph Stalin: Man and Legend. McGraw-Hill.

Hinton, H.E., 1973. Natural Deception. Duckworth.

Hipp, J.R., Tita, G.E., Boggess, L.N., 2009. Intergroup and intragroup violence: is violent crime an expression of group conflict or social disorganization? Criminology 47 (2), 521–564.

Hoagland, K.D., Franti, T.G., 2014. Eutrophication. In: Encyclopedia of Natural Resources: Water.

Hochner, B., Shomrat, T., Fiorito, G., 2006. The octopus: a model for a comparative analysis of the evolution of learning and memory mechanisms. The Biological Bulletin 210 (3), 308–817.

Hoegl, M., Gemuenden, H.G., 2001. Teamwork quality and the success of innovative projects: a theoretical concept and empirical evidence. Organization Science 12 (4), 435–449.

Hofferth, S.L., et al., 1991. National Child Care Survey, 1990. Urban Institute. http://webarchive.urban.org/publications/204604.html.

Hoffman, P.F., 1999. The break-up of Rodinia, Birth of Gondwana, True Polar Wander and the Snowball Earth. Journal of African Earth Sciences 17, 17–33.

Hoffstetter, R., 1974. Phylogeny and geographical deployment of the primates. Journal of Human Evolution 3 (4), 327–350.

Hofmann, H.A., Fernald, R.D., 2001. What cichlids tell us about the social regulation of brain and behavior. Journal of Aquaculture and Aquatic Sciences 9, 17–31.

Hofstede, G., 1997. Cultures and Organizations: Software of the Mind. McGraw Hill.

Hofstede, G.H., 2001. Culture's Consequences: Comparing Values, Behaviors, Institutions and Organizations across Nations. Sage Publications.

Höhler, S., 2007. The law of growth. Distinktion: Scandinavian Journal of Social Theory 8 (1), 45–64.

Holbrook, M.B., Hirschman, E.C., 1982. The experiential aspects of consumption: consumer fantasies, feelings, and fun. Journal of Consumer Research 132–140.

Holdren, J.P., Ehrlich, P.R., 1974. Human Population and the global environment: population growth, rising per capita material consumption, and disruptive technologies have made civilization a global ecological force. American Scientist 282–292.

Holland, H.D., 2006. The oxygenation of the atmosphere and oceans. Philosophical Transactions of the Royal Society: Biological Sciences 360, 903–915.

Hölldobler, B., Wilson, E.O., 1998. Journey to the Ants: A Story of Scientific Exploration. Belknap Press.

Hollingshead, A.B., 2011. Four factor index of social status. Yale Journal of Sociology 8, 21–52.

Holmberg, A.R., 1969. Nomads of the Long Bow: The Sirionó of Eastern Bolivia. Natural History Press.

Holmes, M.M., et al., 1996. Rape-related pregnancy: estimates and descriptive characteristics from a national sample of women. American Journal of Obstetrics and Gynecology 175, 320–324.

Holmes, F.L., 1987. Lavoisier and the Chemistry of Life: An Exploration of Scientific Creativity. University of Wisconsin Press.

Honeycutt, R., 1992. Naked mole-rats. American Scientist 80, 43–53.

Hooper, R., 2007. Marmosets May Carry Their Sibling's Sex Cells. New Scientist.

Horowitz, D.L., 1985. Ethnic Groups in Conflict. University of California Press.

Houpt, K.A., Law, K., Martinisi, V., 1978. Dominance hierarchies in domestic horses. Applied Animal Ethology 4 (3), 273–283.

Howard, G.S., 1997. Ecological Psychology: Creating a More Earth-friendly Human Nature. University of Notre Dame Press.

Huang, C.-J., Wang, S.-J., Wu, F., Zhu, P., Zhau, Z.-H., Yi, J.-M., 2013. The effect of waste slag of the steel industry on pulverized coal combustion. Energy Sources, Part A: Recovery, Utilization and Environmental Effects 35 (20), 1891–1897.

Huesmann, L.R., Moise-Titus, J., Podolski, C., Eron, L.D., 2003. Longitudinal relations between children's exposure to TV violence and their aggressive and violent behavior in young adulthood: 1977–1992. Developmental Psychology 39, 201–221.

Huggins, E.J., Jordan, W.C., Bourke, A.F., 2012. Reproductive conflict in bumblebees and the evolution of worker policing. Evolution 66 (12), 3765–3777.

Hughes, T., 2003. Neurology of swallowing and oral feeding disorders: assessment and management. Journal of Neurology, Neurosurgery & Psychiatry 74 (90003), 48.

Henshilwood, C.S., d'Errico, F., Yates, R., et al., 2002. Emergence of modern human behavior: middle stone age engravings from South Africa. Science 295 (5558), 1278–1280.

Humboldt, W., 1988. On Language: The Diversity of Human Language-Structure and Its Influence on the Mental Development of Mankind. Cambridge University Press.

Hunt, L., Martin, T.R., et al., 2009. third ed. The Making of the West: Peoples and Cultures, vol. C. Bedford/St. Martin's.

Hutchings, J., 1996. Spatial and temporal variation in the density of northern cod a review of hypotheses for the stock's collapse. Canadian Journal of Aquatic Science 53, 943–962.

Hutton, M., 1983. Sources of cadmium in the environment. Ecotoxicology and Environmental Safety 7 (1), 9–24.

Illegal Trade, 2013. High Demand for Animal Parts Driving Illegal Trade. Eco-Business. http://www.eco-business.com/news/high-demand-animal-parts-driving-illegal-trade/.

Investigative Staff, 2015. Betrayal: The Crisis in the Catholiic Church. Boston Globe.

Irving, S., 2006. In a pure soil': colonial anxieties in the work of Francis Bacon. History of European Ideas 32 (3), 249–262.

Isichei, E.A., 1997. A History of African Societies to 1870. Cambridge University Press.

Izard, C.E., 1978. Human Emotions. Springer.

Jack, K.M., 2003. Explaining variation in affiliative relationships among male white-faced capuchins (Cebus capucinus). Folia Primatol 74, 1–16.

Jackson, J.B.C., Coates, A.G., 1986. Life cycles and evolution of clonal (modular) animals. Philosophical Transactions of the Royal Society B: Biological Sciences 313 (1159), 7–22.

Jackson III, J.J., 2012. Why Do We Hunt? International Council for Game and Wildlife Conservation. http://www.cic-wildlife.org/index.php?id=18.

Jacobson, N.S., 1992. Behavioral couple therapy: a new beginning. Behavior Therapy 23, 493–506.

James, D.V., Farnham, F.R., 2003. Stalking and serious violence. Journal of the American Academy of Psychiatry and the Law 31, 432–439.

James, S.R., 1989. Hominid Use of Fire in the Lower and Middle Pleistocene: A Review of the Evidence. Current Anthropology, vol. 30 (1). University of Chicago Press, pp. 1–26.

Jamieson, G.S., 1993. Marine invertebrate conservation: evaluation of fisheries over-exploitation concerns. American Zoologist 33 (6), 551–567.

Janečka, J.E., Miller, W., Pringle, T.H., Wiens, F., Zitzmann, A., Helgen, K.M., Springer, M.S., Murphy, W.J., 2007. Molecular and genomic data identify the closest living relative of primates. Science 318 (5851), 792–794.

Jarvis, J.U.M., Bennett, N.C., 1993. Eusociality has evolved independently in two genera of bathyergid mole-rats—but occurs in no other subterranean mammal. Behavioral Ecology and Sociobiology 33 (4), 253–260.

Jarvis, J.U., O'Riain, M.J., Bennett, N.C., Sherman, P.W., 1994. Mammalian eusociality: a family affair. Trends in Ecology & Evolution 9 (2), 47–51.

Jarvis, J.U.M., 1981. Eusociality in a mammal: cooperative breeding in naked mole-rat colonies. Science 212 (4494), 571–573.

Jeanne, R.L., 1980. Evolution of social behavior in the Vespidae. Annual Review of Entomology 25 (1), 371–396.

Jenkins, R.K., Keane, A., Rakotoarivelo, A.R., Rakotomboavonjy, V., Randrianandrianina, F.H., Razafimanahaka, H.J., Ralaiarimalala, S.R., Jones, J.P., 2011. Analysis of patterns of bushmeat consumption reveals extensive exploitation of protected species in eastern Madagascar. PLoS One 6 (12), e27570.

Jensen, J.R., Narumalani, S., Weatherbee, O., Murday, M., Sexton, W.J., Green, C.J., 2008. Coastal environmental sensitivity mapping for oil spills in the United Arab Emirates using remote sensing and GIS technology. Geocarto International 8 (2), 5–13.

Joe, K.A., 1997. Getting into the gang: methodological issues in studying ethnic gangs. Substance Use & Misuse 32 (12–13), 1961–1966.

Johnson, A.W., Earle, T.K., 2000. The Evolution of Human Societies: From Foraging Group to Agrarian State. Stanford University Press.

Johnson, J.G., Smailes, E.M., Cohen, P., Brown, J., Bernstein, D.P., 2000. Associations between four types of childhood neglect and personality disorder symptoms during adolescence and early adulthood: findings of a community-based longitudinal study. Journal of Personality Disorders 14 (2), 171–187.

Johnson, R.A., 1976. Management, Systems, and Society: An introduction. Goodyear Pub. Co., pp. 148–242.

Johnson, S.C., 2005. North American Youth Gangs: Patterns and Remedies. The Heritage Foundation. http://www.heritage.org/Research/Testimony/North-American-Youth-Gangs-Patterns-and-Remedies.

Johnson, C.N., 2009. Ecological consequences of late quaternary extinctions of megafauna. In: Proceedings of the Royal Society of London B: Biological Sciences, rspb-2008.

Jones, C.M., 2005. Genetic Environmental Influences on Criminal Behavior. http://personalityresearch.org/papers/jones.html.

Jones, R., 2012. Neurogenetics: what makes a human brain? Nature Reviews Neuroscience 13 (10), 655.

Jordan, P., 2001. Neanderthal, Neanderthal Man and the Story of Human Origins. Sutton Publishing.

Joyce, G.H., 1913. Revelation. Catholic Encyclopedia. Robert Appleton Company.

Juang, R.M., Morrissette, N.A., 2008. François Duvalier. Africa and the Americas: Culture, Politics and History. ABC-CLIO, pp. 391–393.

Judge, T.A., Higgins, C.A., Thoresen, C.J., Barrick, M.R., 2006. The big five personality traits, general mental ability and career success across the life span. Personelle Psychology 52 (3), 621–652.

Juzhong, Z., Kuen, L.Y., September 2005. The Magic Flutes. Natural History, pp. 43–47.

Kahn, C., 1998. Plato and the Socratic Dialogue: The Philosophical Use of a Literary Form. Cambridge University Press.

Kak, S.C., 1991. The Honey Bee Dance Language Controversy. The Mankind Quarterly, pp. 357–365.

Kandel, E.R., Schwartz, J.H., Jessel, T.M., 2000. Principles of Neural Science. McGraw-Hill Professional.

Kaplan, J., 2003. Dreams and realities in cyperspace: White Aryan resistance and the world church of the creator. Patterns of Prejudice 37 (2), 139–155.

Kaplan, J., 2010. Dreams and realities in cyberspace: White Aryan resistance and the world church of the creator. Patterns of Prejudice 37 (2), 139–155.

Karnik, N.S., Popma, A., Blain, R.J.R., et al., 2008. Personality correlates of physiological response to stress among incarcerated juveniles. Z Kinder Jugendpsychiatr Psychother 36 (3), 185–190.

Kasting, J.F., 1993. Earth's early atmosphere. Science 259 (5097), 920–926.

Keener, H.M., Elwell, D.L., Ekinci, K., Hoitink, H.A.J., 2013. Composting and value-added utilization of manure from a sweine finishing facility. Compost Science & Utilization 9 (4), 312–321.

Keesing, F.M., 1966. Cultural Anthropology: The Science of Custom. Holt, Rinehart and Winston.

Keil, A., Sachser, N., 1998. Reproductive benefits from female promiscuous mating in a small mammal. Ethology 104 (11), 897–903.

Keller, E.F., 1985. Reflections on Gender and Science. Yale University Press.

Keller, E.A., 2001. Active Tectonics: Earthquakes, Uplift and Landscape, second ed. Prentice Hall.

Kellerman, B., 2004. Bad Leadership: What It Is, How It Happens, Why It Matters. Harvard Business Press.

Kendall, J., 2012. Wild at Home: Exotic Animals as Pets. National Geographic Animal Intervention. http://channel.nationalgeographic.com/wild/animal-intervention/articles/wild-at-home-exotic-animals-as-pets/.

Kent, C.F., Zayed, A., 2013. Evolution of recombination and genome structure in eusocial insects. Communicative & Integrative Biology 6 (2), 845–858.

Kent, S.B., 2009. Total chemical synthesis of proteins. Chemical Society Reviews 38 (2), 338–351.

Keown, D., 1992. The Nature of Buddhist Ethics. Macmillan.

Kerridge, E., 2006. The Agricultural Revolution. Routledge.

Keverne, E.B., Martel, F.L., Nevison, C.M., 1996. Primate brain evolution: genetic and functional considerations. Cross Mark Proceedings B. http://rspb.royalsocietypublishing.org/content/263/1371/689.short.

Kiernan, B., 1985. How Pol Pot Came to Power: A History of Communism in Kampuchea, 1930–1975. Verso.

Kimbel, W.H., Johanson, D.C., Rak, Y., 1994. The first skull and other new discoveries of Australopithecus afarensis at Hadar, Ethiopia. Nature 368, 449–451.

King, B., 2007. Evolving God: A Provocative View on the Origins of Religion. Doubleday Publishing.

Kingsbury, C.P., 1849. An Elementary Treatise on Artillery and Infantry. GP Putnam.

Kious, W.J., Tilling, R.I., 2001. This Dynamic Earth: The Story of Plate Tectonics, online ed. Geological Survey, U.S.

Kippenberg, H.G., 2011. Violence as Worship: Religious Wars in the Age of Globalization. Stanford University Press.

Kleinman, R.E., Murphy, J.M., Michelle, L., Pagano, M., Wehler, C.A., Regal, K., Jellinek, M.S., 1998. Hunger in children in the United States: potential behavioral and emotional correlates. Pediatrics 101 (1), 3.

Kligman, G., 1998. The Politics of Duplicity: Controlling Reproduction in Ceausescu's Romania. University of California Press.

Kling, A., Steklis, H.D., 1976. A neural substrate for affiliative behavior in nonhuman primates. Brain Behav Evol 13, 216–238.

Klopfer, P.H., 1969. Habitats and Territories: A Study of the Use of Space by Animals. Basic Books.

Kluger, J., 2007. What makes us moral. Time Magazine 170 (23), 54–60.

Knudsen, L., 2006. Reproductive Rights in a Global Context. Vanderbilt University Press.

Koizumi, M., Mamung, D., Levang, P., 2012. Hunter-gatherers' culture, a major hindrance to a settled agricultural life: the case of the Penan Benalui of East Kalimantan. Forests, Trees and Livelihoods 21 (1), 1–15.

Kolar, C.S., Lodge, D.M., 2001. Progress in invasion biology: predicting invaders. Trends in Ecology and Evolution 16 (4), 199–204.

Komro, K.A., Flay, B.R., Bingehang, F., Zelli, A., Rashid, J., Amuwo, S., 1999. Urban pre-adolescents report perceptions of easy access to drugs and weapons. Journal of Child & Adolescent Substance Abuse 8 (1), 77–90.

Kondrashov, A.S., 1988. Deleterious mutations and the evolution of sexual reproduction. Nature 336, 435–440.

Koneswaran, G., Nierenberg, D., 2008. Global farm animal production and global warming: impacting and mitigating climate change. Environ Health Perspect 116 (5), 578–582.

Koob, G.F., Bloom, F.E., 1988. Cellular and molecular mechanisms of drug dependence. Science 242 (4879), 715–723.

Koppel, H.L., 2015. Endangered Species Report #30: Eastern Box Turtle. http://www.wildlifewatchers.org/esReports/report30.html.

Koshkarova, V.L., Koshkarov, A.D., 2004. Regional signatures of changing landscape and climate of northern central Siberia in the Holocene. Russian Geology and Geophysics 45 (6), 672–685.

Kosinski, R.A., Zaremba, M., 2007. Dynamics of the Model of the Caenorhabditis elegans Neural Network. Acta Physica Polonica B 38 (6), 2201.

Kossowska, M., de Zavala, A.G., Kubik, T., 2010. Stereotype images of terrorists as predictors of fear of future terrorist attacks. Behavioral Sciences of Terrorism and Political Aggression 2 (3), 179–197.

Kozorovitskiy, Y., Gould, E., 2004. Dominance hierarchy influences adult neurogenesis in the dentate gyrus. The Journal of Neuroscience 24 (30), 6755–6759.

Kramer, A.E., 2008. Trade in Mammoth Ivory, Helped by Global Thaw, Flourishes in Russia. New York Times.

Krause, V., Singer, J.D., 2001. Minor powers, alliances and armed conflict: some preliminary patterns. In: Small States and Alliances, pp. 15–23.

Krebs, J.R., Sjolander, S., Sjolander, S., 1992. Konrad Zacharias Lorenz. Biographical Memoirs of Fellows of the Royal Society 38, 210.

Kronauer, D.J., Pierce, N.E., 2011. Myrmecophiles. Current Biology 21 (6), R208–R209.

Kulier, R., Kapp, N., et al., 2011. Medical methods for first trimester abortion. Cochrane Database of Systematic Reviews 11, 2207.

Lamb, H.H., Woodroffe, A., 1970. Atmospheric circulation during the last ice age. Quaternary Research 1 (1), 29–58.

Lamé, G., Lequien, M., Pionnier, P.A., 2013. Interpretation and limits of sustainability tests in public finance. Applied Economics 46 (6), 616–628.

Landes, D.S., 1969. The Unbound Prometheus: Technological Change and Industrial Development in Western Europe from 1750 to the Present. Press Syndicate of the University of Cambridge.

Lane, R., 1997. Murder in America: A History. Ohio State University Press, Columbus.

Langston, L., Berzofsky, M., Krebs, C., McDonald, H., 2010. Victimizations Not Reported to the Police, 2006–2010. U.S. Department of Justice Office of Justice Programs Bureau of Justice Statistics. http://www.bjs.gov/content/pub/pdf/vnrp0610.pdf.

Larsen, R.J., Buss, D.M., 2008. Personality Psychology: Domains of Knowledge About Human Nature, third ed. McGraw-Hill.

Lataianu, M., 2001. Proceedings of the Euroconference on Family and Fertility Change in Modern European Societies: Explorations and Explanations of Recent Developments. Max Planck Institute for Demographic Research. http://www.demogr.mpg.de/Papers/workshops/010623_paper25.pdf.

Laurent, T., Leridon, H., 1998. Contraceptive practices and trends in France. Family Planning Perspectives 30 (3), 114.

Laursen, L., Bekoff, M., 1978. Loxodonta africana. Mammalian Species 92 (92), 1–8.

Lavers, C., 2009. The Natural History of Unicorns. William Morris.

Lawlor, L., 2007. This Is Not Sufficient: An Essay on Animality and Human Nature in Derrida. Columbia University Press.

Lawrence, A., Lawrence, L.R., 2015. When Your Body Talks, Listen!. Allco Publishing.

Lawrence, J.G., Ochman, H., 1998. Molecular archaeology of the *Escherichia coli* genome. Proceedings of the National Academy of Sciences of the United States of America 95 (16), 9413–9417.

Lawrence, J.L., 1991. The United States and the Genocide Convention. Duke University Press.

League, A.D., 2001. Anti-Defamation League. New York.

Leakey, M.G., Feibel, C.S., McDougall, I., et al., 1998. New specimens and confirmation of an early age for *Australopithecus anamensis*. Nature 393, 62–66.

Lech, T., Goszcz, H., 2008. Poisoning from aspiration of elemental mercury. Clinical Toxicology 44 (3), 333–336.

Ledley, F.D., 1982. Evolution and the human tail: a case report. New England Journal of Medicine 306 (20), 1212–1215.

Lee, R.B., DeVore, I. (Eds.), 1968. Man the Hunter. Aldine de Gruyter.

Lee, Y.K., 2002. Building the chronology of early Chinese history, Asian perspectives. Journal of Archaeology for Asia and the Pacific 41.

Lehrman, S., 2002. The Virtues of Promiscuity. Alternet News & Politics. http://www.alternet.org/story/13648/the_virtues_of_promiscuity.

Lehrmann, D.J., Ramezan, J., Bowring, S.A., et al., 2006. Timing of recovery from the end-Permian extinction: geochronologic and biostratigraphic constraints from South China. Geology 34 (12), 1053–1056.

Lemberg, R., 2009. Yosemite Park and Region: Decrease in Snowpack Affecting the State's Water, Trees and Bears. National Geographic.

Lennings, C.J., Amon, K.L., Brummert, H., Lennings, N.J., 2010. Grooming for terror: the internet and young people. Psychiatry, Psychology and Law 17 (3), 424–437.

Leonard, K.I., 2006. Immigrant Faiths: Transforming Religious Life in America. Rowman Altamira.

Leopold, A. 328.

Lepore, J., 2011. Birthright: What's Next for Planned Parenthood? New Yorker.

Leslie, J., 1996. The End of the World: The Science and Ethics of Human Extinction. Routledge.

Levi, L., 1987. Stress in the Modern World. UNESCO Courier.

Lewis, J., Christopher, J., 1989. Childhood trauma in borderline personality disorder. American Journal of Psychiatry 146 (4), 490–495.

Lewthwaite, G.R., 2010. Wisconsin and the Waikato: a comparison of dairy farming in the United States and New Zealand. Annals of the Association of American Geographers 54 (1), 59–87.

Lieberman, D.E., Shea, J.J., 1994. Behavioral differences between archaic and modern humans in the Levantine Mousterian. American Anthropologist 96 (2), 300–332.

Lieberman, D.E., 2001. Another face in our family tree. Nature 410, 419–420.

Lin, N., Michener, C.D., 1972. Evolution of sociality in insects. Quarterly Review of Biology 131–159.

Lincove, D., 2005. Encyclopedia of Wars. Reference & User Services Quarterly 45 (1), 84–86.

Linnaeus, C., 1758. Systemae naturae. A photographic facsimile of the first volume of the tenth edition (1758). In: Regnum Animale. Trustees of the British Museum, London.

Linzey, A.V., Hammerson, G., 2008. Marmota Flaviventris. IUCN Red List of Threatened Species. Version 2014.3. International Union for Conservation of Nature.

Lipman-Blumen, J., 2005. The Allure of Toxic Leaders: Why We Follow Destructive Bosses and Corrupt Politicians – And How We Can Survive Them. Oxford University Press.

Littlefield, J., Ozanne, J.L., 2011. Socialization into consumer culture: hunters learning to be men. Consumption Markets & Culture 14 (4), 333–360.

Livshits, G., Kobyliansky, E., 1991. Fluctuating asymmetry as a possible measure of developmental homeostasis in humans: a review. Human Biology 63 (4), 441–466.

Lloyd, J.E., 1986. Firefly communication and deception: "oh, what a tangled web". In: Deception: Perspectives on Human and Nonhuman Deceit, pp. 113–128.

Long, J.A., C Young, G., Holland, T., Senden, T.J., Fitzgerale, E.M., 2006. An exceptional Devonian fish from Australia sheds light on tetrapod origins. Nature 444 (7116), 199–202.

Longwell, H.J., 2002. The future of the oil and gas industry: past approaches, new challenges. World Energy 5 (3), 100–104.

Lykken, D.T., 1995. The Antisocial Personalities. Lawrence Erlbaum Associates.

MacArthur, R.H., Wilson, E.O., 1967. The Theory of Island Biogeography. Princeton University Press.

Mace, F.C., Lalli, J.S., Lalli, E.P., Shea, M.C., 1993. Functional Analysis and Treatment of Aberrant Behavior. Springer, US, pp. 75–99.

MacKinnon, S., Silverman, S., 2005. Complexities Beyond Nature and Nurture. The University of Chicago Press.

MacLachlan, M., Carr, S.C., McAuliffe, E., 2010. The Aid Triangle: Recognizing the Human Dynamics of Dominance, Justice and Identity. Fernwood Pub.

MacLean, P.D., 1985. Evolutionary psychiatry and the triune brain. Psychological Medicine 15 (02), 219–221.

MacPhee, R.D.E., Sues, H.D., 1999. Extinctions in Near Time: Causes, Contexts and Consequences. Springer Science & Business Media.

Maestripieri, D., 2000. Mother nature: a history of mothers, infants, and natural selection. Animal Behaviour 59 (4), 895–896.

Maestripieri, D., 2007. Macachiavellian Intelligence: How Rhesus Macaques and Humans Have Conquered the World. University of Chicago Press.

Maestripieri, D., 2012. Games Primates Play: An Undercover Investigation of the Evolution and Economics of Human Relationships. Basic Books.

Maharana, T., Negi, Y.S., Mohanty, B., 2007. Recycling of polystyrene. Polymer-plastics. Technology and Engineering 46 (7), 729–736.

Malmberg, T., 1980. Human Territoriality: Survey of Behavioural Territories in Man with Preliminary Analysis and Discussion of Meaning, Mouton.

Malthus, T.R., 1798. An Essay on the Principle of Population. Oxford World's Classics reprint.

Mandal, F.B., 2010. Textbook of Animal Behaviour. PHI Learning, p. 47.

Mandani, M., 1972. The Myth of Population Control: Family Caste and Class in an Indian Village. Monthly Reader. 173 pp.

Mann, J., Connor, R.C., Tyack, P.L., et al. (Eds.), 1999. Cetacean Societies: Field Study of Dolphins and Whales. University of Chicago.

Mann, C.C., 2011. 1491, New Revelations of the Americas Before Columbus. Vintage Books.

Manson, J.H., Wrangham, R.W., et al., 1991. Intergroup aggression in chimpanzees and humans. Current Anthropology 369–390.

Markou, A., Kosten, T.R., Koob, G.F., 1998. Neurobiological similarities in depression and drug dependence: a self-medication hypothesis. Neuropsychopharmacology 18 (3), 135–174.

Marks, M., 2009. Infanticide. Psychiatry 8 (1), 10–12.

Marozzi, J., 2004. Tamerlane: Sword of Islam, Conqueror of the World. HarperCollins.

Marston, W.M., 1929. Emotions of Normal People. Taylor & Francis Ltd.

Maslow, A.H., 1954. Instinct Theory Reexamined, Motivation and Personality. Harper & Row.

Masters, R.D., Hone, B.T., Doshi, A., 1998. Environmental pollution, neurotoxicity, and criminal violence. Environmental Toxicology 13–48.

Mateos, L., 2011. Fertilizer and pesticide leaching: irrigation management. In: Encyclopedia of Water Science, second ed.

Matsumura, S., 1999. The evolution of "egalitarian" and "despotic" social systems among macaques. Primates 40 (1), 23–31.

Matsuoka, Y., Furuyasshiki, T., Yamada, K., et al., 2005. Prostaglandin E receptor EP1 controls impulsive behavior under stress. Proceedings of the National Academy of Sciences 102 (44), 16066–16071.

Matthiopoulos, J., Harwood, J., Thomas, L., 2005. Metapopulation consequences of site fidelity for colonially-breeding mammals and birds. Journal of Animal Ecology 74, 716–727.

May, H., 2000. On Socrates. Wadsworth/Thomson Learning, p. 20.

Mayhew, P.J., Jenkins, G.B., Benton, T.G., 2007. A long-term association between global temperature and biodiversity, origination and extinction in the fossil record. Proceedings of the Royal Society B 275 (1630), 47–53.

Mays, W., 2014. The best and the worst of possible worlds. Journal of the British Society for Phenomenology 10 (1), 49–59.

Mazur, A., Booth, A., 1998. Testosterone and dominance in men. Behvioral and Brain Sciences 21, 70–79.

Mazur, A., 2005. Biosociology of Dominance and Deference. Rowman & Littlefield Publishes Inc.

Mazur, A., 2015. Power affects performance when the pressure is on: evidence for low-power and high-power life. Personality and Social Psychology Bulletin 41 (5), 726–735.

McBrearty, S., Brooks, A., 2000. The revolution that wasn't: a new interpretation of the origin of modern humans. Journal of Human Evolution 39, 453–563.

McBurnett, K., Lahey, B.B., Rathouz, P.J., Loeber, R., 2000. Low salivary cortisol and persistent aggression in boys referred for disruptive behavior. Archives of General Psychiatry 57 (1), 38–43.

McClellan, J.E., Dorn, H., 2006. Science and Technology in World History: An Introduction. The Johns Hopkins Univ. Press.

McCrae, R.R., Costa Jr., P.T., 1997. Personality trait structure as a human universal. American Psychologist 52 (5), 509.

McElroy, S.L., 1999. Recognition and treatment of DSM-IV intermittent explosive disorder. The Journal of Clinical Psychiatry 60 (15), 12–16.

McEwen, B.S., Stellar, E., 1993. Stress and the individual: mechanisms leading to disease. Archives of Internal Medicine 153 (18), 2093–2101.

McEwen, B.S., 1998. Protective and damaging effects of stress mediators. New England Journal of Medicine 338 (3), 171–179.

McGrath, A.E., 2003. Christian Theology: An Introduction. Blackwell.

McGue, M., Bouchard, T.J., 1998. Genetic and environmental influences on human behavioral differences. Annual Review of Neuroscience 21, 1–24.

McLaren, A., 1978. Birth Control in Nineteenth-Century England. Taylor & Francis.

McLaughlin, M., 1999. Crisis of overproduction devastating American agriculture. World Socialist Web Site, wsws.org/en/articles/1999/08/farm-a)&.html.

McMahan, C.D., Chakrabarty, P., Sparks, J.S., Smith, W.L., Davis, M.P., 2013. Temporal patterns of diversification across global cichlid biodiversity (Acanthomorpha: Cichlidae). Plos One 8 (8), e71162.

McMichael, P., 2011. Development and Social Change: A Global Perspective. Sage Publications.

McNeely, J.A., 2003. Biodiversity, war and tropical forests. Journal of Sustainable Forestry 16 (3–4), 1–20.

McPeek, M.A., Gavrilets, S., 2006. The evolution of female mating preferences: differentiation from species with promiscuous males can promote speciation. Evolution 60 (10), 1967–1980.

McTavish, L., 2007. Contraception and birth control. In: Robin, D. (Ed.), Encyclopedia of Women in the Renaissance: Italy, France and England, pp. 91–92.

Mead, J.G., Brownell, R.L., 2005. Order Cetacea. In: Mammal Species of the World: A Taxonomic and Geographic Reference, third ed. Johns Hopkins University Press, pp. 723–743.

Mead, M., 2014. Coming of Age in Samoa: A Psychological Study of Primitive Youth for Western Civilisation. William Morrow Paperbacks.

Meade, C., 1999. The Effects of Substance Abuse on the Development of Children: Educational Implications. Teachnology.

Meiertöns, H., 2010. The Doctrines of US Security Policy – An Evaluation under International Law. Cambridge University Press.

Meltzer, D.J., 2009. First Peoples in a New World: Colonizing Ice Age America. University of California, Berkeley.

Menski, W., 2007. Hinduism. In: Morgan, P., Lawton, C. (Eds.), Ethical Issues in Six Religious Traditions, second ed. Columbia University Press.

Meyer, J.E.W., 2004. Any Friend of the Movement: Networking for Birth Control, 1920–1940. Ohio State University Press.

Michael, G., 2007. Rahowa! A history of the World Church of the Creator. Terrorism and Political Violence 18 (4), 561–583.

Middleberg, M.I., 2003. Promoting Reproductive Security in Developing Countries. Springer, p. 4.

Miles, V.J., 2012. Boys of the Cloth. Hamilton Books.

Miller, N.E., 1965. Chemical coding of behavior in the brain stimulating the same place in the brain with different chemicals can elicit different types of behavior. Science 148 (3668), 328–338.

Miller, D., 1997. Sir Karl Raimund Popper, C. H., F. B. A. 28 July 1902–17 September 1994: Elected F.R.S. 1976. Biographical Memoirs of Fellows of the Royal Society 43, 369–410.

Miller, G.T., 1993. Environmental science: sustaining the earth. In: Wadsworth Biology Series. Wadsworth Publishing Company.

Miller, T.C., 1993. The duality of human nature. Politics and the Life Sciences 221–241.

Miller, J.G., 1994. Cultural diversity in the morality of caring: individually oriented versus duty-based interpersonal moral codes. Cross-Cultural Research 28, 13–39.

Miller, G.T., 2004. Environmental Science, Working with the Earth. Thomson Learning, Inc.

Minkel, J.R., 2006. Humans and Chimps: Close but Not That Close. Scientific American, pp. 12–19.

MIT, 2015. 638 Primary Personality Traits – Ideonomy.

Mitchell, R.W., Thompson, N.S., 1986. Deception, Perspectives on Human and Nonhuman Deceit. State of New York University Press.

Mohr, J.M., 2013. Wolf in sheep's clothing: harmful leadership with a moral Façade. Journal of Leadership Studies 7 (1), 18–32.

Møller, A.P., Mousseau, T.A., 2013. Assessing effects of radiation on abundance of mammals and predator–prey interactions in Chernobyl using tracks in the snow. Ecological Indicators 26, 112–116.

Möller, K., Schultheiß, U., 2014. Chemical characterization of commercial organic fertilizers. Archives of Agronomy and Soil Science 61 (7), 989–1012.

Monaghan, J., Just, P., 2000. Social & Cultural Anthropology. Oxford University Press.

Montefiore, S.S., 2005. Stalin: The Court of the Red Tsar. Vintage.

Moore, S.E., DeMaster, D.P., 1997. Cetacean habitats in the Alaskan Arctic. Journal of Northwest Atlantic Fishery Science 22, 55–69.

Moore, B., 1993. Social Origins of Dictatorship and Democracy, Lord and Peasant in the Making of the Modern World. Beacon Press.

Moore, C., 1995. Betrayal of Trust: The Father Brendan Smyth Affair and the Catholic Church. Marino.

Moore, D.S., 2003. The Dependent Gene: The Fallacy of Nature vs. Nurture. Henry Holt.

Moore, J.R., 2009. Why religious education matters: the role of Islam in multicultural education. Multicultural Perspectives 11 (3), 139–145.

Morell, V., 2011. Killer whales earn their name. Science 331 (6015), 274–276.

Morgan, W., 2002. Origin and history of the earliest thematic apperception test. Journal of Personality Assessment 79 (3), 422–445.

Morris, D., 1967. The Naked Ape. McGraw-Hill Book Company.

Morris, D., 2009. The Human Zoo. Random House.

Morrison, K.D., Junker, L.L. (Eds.), 2002. Forager-Traders in South and Southeast Asia: Long Term Histories. Cambridge University Press.

Moynihan, M., Rodaniche, A.F., 1982. The behavior and natural history of the Caribbean reef squid Sepioteuthis sepioidea. With a consideration of social, signal and defensive patterns for difficult and dangerous environments. Advances in Ethology 25, 1–151.

Mugwira, L.M., Nyamangara, J., Hikwa, D., 2007. Effects of manure and fertilizer on maize at a research station and in a smallholder (peasant) area of Zimbabwe. Communications in Soil Science and Plant Analysis 33 (3–4), 379–402.

Mulder, M.B., Schacht, R., 2012. Human Behavioural Ecology. Nature Encyclopedia of Life Sciences.

Mullen, P.E., Pathé, M., Purcell, M., 2000. Stalkers and Their Victims. Cambridge University Press.

Mullen, M.G., 2011. Joint Operation Planning. US Government Publication. http://www.dtic.mil/doctrine/new_pubs/jp5_0.pdf.

Muller, M.N., Kahlenberg, S.M., Thompson, M.E., Wrangham, R.W., 2007. Male coercion and the costs of promiscuous mating for female chimpanzees. Proceedings of the Royal Society of London B: Biological Sciences 274 (1612), 1009–1014.

Munn, C.A., 1986. The deceptive use of alarm calls by sentinel species in mixedspecies flocks of neotropical birds. Deception: Perspectives on Human and Nonhuman Deceit 169–176.

Murray, C., 2004. Human Accomplishment: The Pursuit of Excellence in the Arts and Sciences, 800 B.C. to 1950. HarperCollins Publishers.

Nagel, T., 1978. The Possibility of Altruism. Princeton University Press.

Namka, L., 1995a. Conflict Over Values. http://www.angriesout.com/agreetob.htm.

Namka, L., 1995b. Incorporating group social skills training, shame release and play therapy with a child who was sexually abused. International Journal of Play Therapy 4 (1), 81–98.

NASA (National Aeronautics and Space Administration), 2005. Warmest Year in Over a Century. http://www.nasa.gov/vision/earth/environment/2005_warmest.html.

National Geographic, 2009. National Geographic Concise History of Science and Invention: An Illustrated Time Line. National Georgraphic.

National Geographic, 2015. Are We Along, and Other Mysteries of Space. Time Inc. Books.

Natoli, E., De Vito, E., 1991. Agonistic behaviour, dominance rank and copulatory success in a large multi-male feral cat, Felis catus L., colony in central Rome. Animal Behaviour 42 (2), 227–241.

Neurath, P., 1994. From Malthus to the Club of Rome and Back. M. E. Sharp.

Nickerson, C., Borchers, A., 2012. How Is Land in the United States Used? A Focus on Agricultural Land. http://www.usda.gov/.

Nielsen, C., 2012. The authorship of higher chordate taxa. Zoologica Scripta 41 (4), 435–436.

Nijhuis, M., October 2014. When the Snows Fail. National Geographic.

Nixon, M., Young, J.Z., 2003. The Brains and Lives of Cephalopods. Oxford University Press.

Nockleby, A., 2008. Animal Abuse and Its Link to Violent Criminals. ALDF's Online Content Manager. http://aldf.org/blog/animal-abuse-and-its-link-to-violent-criminals/.

Normile, D., February 2013. Japan's Scientific Whaling: An Expensive Proposition. Scienceinsider.

Norton, A., Zehner, O., 2008. Which half is mommy? Tetragametic chimerism and trans-subjectivity. Women's Studies Quarterly, Fall/Winter 106–127.

Nowak, et al., 2010a. The evolution of eusociality. Nature 466, 26.

Nowak, M.A., Tarnita, C.E., Wilson, E.O., 2010b. The evolution of eusociality. Nature 466 (7310), 1057–1062.

Nowak, R.M., Russell, A., Rylands, A.B., Konstant, W.R., 1999. Walker's Primates of the World. Johns Hopkins University Press, Baltimore.

Nuwer, R., October 2014. The Black Market Trade for Endangered Animals Flourishes on the Web. Newsweek.

Odegard, V.H., Schatz, D.G., 2006. Targeting of somatic hypermutation. Nature Reviews Immunology 6 (8), 573–583.

Oehlberg, B., 2014. Activities for Children Living in a Stressful World. Redleaf Press.

Ogg, R.N., Menczel, J.H., 1981. Disposal of hazardous wastes: problems and pitfalls. Journal of the Air Pollution Control Association 31 (2), 127–132.

Oil Spills, 2015. Oil Spills and Disasters. Infoplease. http://www.infoplease.com/ipa/A0001451.html.

O'Leary, K.J., Sehgal, N.L., Terrell, G., Williams, M.V., 2012. Interdisciplinary teamwork in hospitals: a review and practical recommendations for improvement. Journal of Hospital Medicine 7 (1), 48–54.

Olden, J.D., Hogan, Z.S., Vander Zanden, M.J., 2007. Small fish, big fish, red fish, blue fish: Size-biased extinction risk of the world's freshwater and marine fishes. Global Ecology and Biogeography 16 (6), 694–701.

Olson, P.J., 2006. The public perception of 'cults' and 'new religious movements'. Journal for the Scientific Study of Religion 45 (1), 97–106.

Olson, E.T., 2008. An argument for animalism. In: Martin, R., Barresi, J.J. (Eds.), Personal Identity. Blackwell.

Oprea, M., 1999. Antibody Repertoires and Pathogen Recognition: The Role of Germline Diversity and Somatic Hypermutation (Thesis). University of Leeds.

Osborne, M.J., Rubinsteun, A., 1994. A Course in Game Theory. MIT Press.

Otten, W., Puppe, B., Kanitz, E., Schön, P.C., Stabenow, B., 1999. Effects of dominance and familiarity on behavior and plasma stress hormones in growing pigs during social confrontation. Zentralbl Veterinarmed A 46 (5), 277–292.

Overton, B.W., 1994. Impact of recycling and environmental legislation. Food Additives & Contaminants 11 (2), 285–293.

Owen, E.W., 1975. Trek of the oil finders. American Association of Petroleum Geologists 12.

Pace, N.R., 2001. The universal nature of biochemistry. Proceedings of the National Academy of Sciences of the United States of America 98 (3), 805–808.

Palarea, R.E., Zona, J.C., Langhinrisichsen-Rohling, J., 1999. The dangerous nature of intimate relationship stalking threats, violence and associated risk factors. Behavioral Sciences & the Law 17, 269.

Panter-Brick, C., Layton, R.H., Rowley-Conwy, P. (Eds.), 2001. Hunter-Gatherers: An Interdisciplinary Perspective. Cambridge University Press.

Parent, A., Carpenter, M.B., 1995. Carpenter's Human Neuroanatomy. Williams & Wilkins.

Parrington, J., 2015. Working together Is Part of What Makes Us Human. Socialist Worker. https://socialistworker.co.uk/art/39292/Working+together+is+part+of+what+makes+us+human.

Parrish, M.D., 1984. Factors influencing aggression between foraging yellowjacket wasps, Vespula spp. (Hymenoptera: Vespidae). Annals of the Entomological Society of America 77 (3), 306–311.

Parylak, J., 2014. 8 incredibly strange religions from around the world. http://whatculture.com/offbeat/8-incredibly-strange-religions-around-world.

Patterson, N., Richter, D.J., Gnerre, S., Lander, E.S., Reich, D., 2006. Genetic evidence for complex speciation of humans and chimpanzees. Nature 441 (7097), 1103–1108.

Pauly, D., Palomares, M.L., 2005. Fishing down marine food webs: it is far more pervasive than we thought. Bulletin of Marine Science 76 (2), 197–211.

Pauly, D., Christensen, V., Dalsgaard, J., Froese, R., Torres Jr., F.C., 1998. Fishing down marine food webs. Science 279, 860–863.

Payne, J., Prudente, C., 2008. Orangutans: Behavior, Ecology and Conservation. New Holland Publishers.

Pennisi, E., 2007. Breakthrough of the year: human genetic variation. Science 318 (5858), 1842–1843.

Perelman, M., 2014. The corrosive qualities of inequality: the roots of the current meltdown. Challenge 51 (5), 40–64.

Peta, 2015. Hunting. http://www.peta.org/issues/animals-in-entertainment/cruel-sports/hunting/.

Peterson, D., Wrangham, R., 1997. Demonic Males: Apes and the Origins of Human Violence. Houghton.

Petterle, D., 2009. Are We All Capable of Violence? BBC News Magazine. http://news.bbc.co.uk/2/hi/uk_news/magazine/8043688.stm.

Philips, J., 1988. The Great Ridley Rescue. Mountain Press.

Phillips, C., Axelrod, A., 2004. Encyclopedia of Wars. Facts on File.

Pickett, K.M., Osborne, D.M., Wahl, D., Wenzel, J.W., 2001. An enormous nest of Vespula squamosa from Florida, the largest social wasp nest reported from North America, with notes on colony cycle and reproduction. Journal of the New York Entomological Society 109 (3 & 4), 408–415.

Picton, J.A., 1905. Pantheism: Its Story and Significance. Archibald Constable & Co Ltd.

Piot, P., Russell, M.A.S., Larson, H., 2007. Good politics, bad politics: the experience of AIDS. American Journal of Public Health 97 (11), 1934–1936.

Plumwood, V., 2000. Integrating ethical frameworks for animals, humans, and nature: a critical feminist eco-socialist analysis. Ethics and the Environment 285–322.

Podles, L.J., 2007. Sacrilege: Sexual Abuse in the Catholic Church. Crossland Ptress (Investigative Staff, 2015; Miles, 2012; Podles, 2007).

Polis, G.A., 1981. The evolution and dynamics of intraspecific predation. Annual Review of Ecology and Systematics 12, 225–251.

Posewitz, J., 1994. Chase: The Ethic and Tradition of Hunting. The Globe Pequot Press.

Postacchini, F., Massobrio, M., 1983. Idiopathic coccygodynia. Analysis of 51 operative cases and a radiographic study of the normal coccyx. Journal of Bone and Joint Surgery 65 (8), 1116–1124 (American volume Ed.).

Poston, D., 2010. Population and Society: An Introduction to Demography. Cambridge University Press.

Pounds, J.G., Long, G.J., Rosen, J.F., 1991. Cellular and molecular toxicity of lead in bone. Environmental Health Perspectives 91, 17.

Powers, J., 2007. Introduction to Tibetan Buddhism, rev. ed. Snow Lion Publications, pp. 26–27.

Pozzi, L., Hodgson, J.A., Burrell, A.S., Sterner, K.N., Raaum, R.L., Disotell, T.R., 2014. Primate phylogenetic relationships and divergence dates inferred from complete mitochondrial genomes. Molecular Phylogenetics and Evolution 75, 165–183.

Pratto, F., Stallworth, L.M., Sidanius, J., Siers, B., 1997. The gender gap in occupational role attainment: a social dominance approach. Journal of Personality and Social Psychology 72 (1), 37–53.

Price, T.L., 2005. Understanding Ethical Failures in Leadership. Cambridge Studies in Philosophy and Public Policy, Cambridge.

Price, D.H., 2011. Weaponizing Anthropology, Social Science in Service of the Militarized State. Counterpunch and A. K. Press.

Pringle, H., 2013. Long live the humans. Scientific American 309 (4), 48–55.

Prothero, D.R., Dott Jr., R.H., 2004. Evolution of the Earth, seventh ed. McGraw-Hill.

Purdum, E.D., 2002. Florida Waters. A Water Resources Manual from Florida's Water Management Districts. http://www.sfwmd.gov/portal/page/portal/xrepository/sfwmd_repository_pdf/florida_waters.pdf.

Pusey, A.E., Packer, C., 1997. The ecology of relationships. In: Krebs, J.R., Davies, N.B. (Eds.), Behavioral Ecology: An Evolutionary Approach. Blackwell Science, pp. 254–283.

Queiroz, C., 2014. The human genome project: some social and eugenic implications. Global Bioethics 10 (1–4), 91–100.

Quora, 2015. Why Do Some People Find Hunting an Enjoyable Hobby? https://www.quora.com/Why-do-some-people-find-hunting-an-enjoyable-hobby.

Rabalais, N.N., 2011. Hypoxia: Gulf of Mexico. In: Encyclopedia of Water Science, second ed.

Rabin, S., 2007. Nicolaus Copernicus. Stanford Encyclopedia of Philosophy.

Rachels, J., Rachels, S., 2011. The Elements of Moral Philosophy, seventh ed. McGraw-Hill.

Raichle, M.E., Gusnard, D.A., 2014. Appraising the brain's energy budget. Proceedings of the National Academy of Sciences 99 (16), 10237–10239.

Raine, A., 2002. Biosocial studies of antisocial and violent behavior in children and adults: a review. Journal of Abnormal Child Psychology 30 (4), 311–326.

Ramachandran, V.S., 2011. The Tell-Tale Brain: A Neuroscientist's Quest for What Makes Us Human. W. W. Norton & Company.

Rance, P., 2004. The fulcum, the Late Roman and Byzantine testudo: the germanization of Roman infantry tactics? Greek, Roman and Byzantine Studies 44, 265–326.

Rand, M.R., 2005. The national crime victimization survey: 32 years of measuring crime in the United States. http://www.stat.fi/sienagroup2005/sienarand.pdf.

Rapoport, A., Chammah, A.M., 1966. The game of chicken. American Behavioral Scientist 10 (3), 10–28.

Rasmussen, M.A., Casey, T.A., 2008. Environmental and food safety aspects of Escherichia coli O157:H7 infections in cattle. Critical Reviews in Microbiology 27 (2), 57–73.

Ratey, J.J., 2002. A User's Guide to the Brain, Perception, Attention and the Four Theaters of the Brain. Vintage Books.

Ratnieks, F.L.W., Helantera, H., 2009. The evolution of extreme altruism and inequality in insect societies. Philosophical Transactions of the Royal Society B 364 (1553), 3169–3179.

Rau, P., 1940. Co-operative nest-founding by the wasp, Polistes annularis Linn. Annals of the Entomological Society of America 33 (4), 617–620.

Raup, D., Sepkoski, J., 1982. Mass extinctions in the marine fossil record. Science 215, 1501–1503.

Raviv, M., Medina, S., Krasnovsky, A., Ziadna, H., 2013. Organic matter and nitrogen conservation in manure compost for organic agriculture. Compost Science & Utilization 12 (1), 6–10.

Rayner, C., Keashley, L., 2005. Bullying at work: a perspective from Britain and North America. In: Counterproductive Work Behavior: Investigations of Actors and Targets. American Psychological Association, pp. 271–296.

Reader, S.M., Laland, K.N., 2002. Social intelligence, innovation, and enhanced brain size in primates. Proceedings of the National Academy of Sciences 99 (7), 4436–4441.

Reagans, R., Argote, L., Brooks, D., 2005. Individual experience and experience working together: predicting learning rates from knowing who knows what and knowing how to work together. Management Science 51 (6), 869–881.

Reber, A.S., Reber, E., 2002. The Penguin Dictionary of Psychology. Penguin Books.

Redclift, M., Woodgate, G., 1997. International Handbook of Environmental Sociology. Edgar Elgar.

Reed, G.E., July–August 2004. Toxic Leadership. Military Review.

Rees, M., 2003. Our Final Hour: A Scientist's Warning: How Terror, Error and Environmental Disaster Threaten Humankind's Future in the Century-on Earth and Beyond. Basic Books.

Reeve, H.K., Emlen, S.T., Keller, L., 1998. Reproductive sharing in animal societies: reproductive incentives or incomplete control by dominant breeders? Behavioral Ecology 9 (3), 267–278.

Reinhardt, F.L., 1998. Bringing the environment down to earth. Harvard Business Review 77 (4), 149–157.

Reiss, S., 2009. The Normal Personality: A New Way of Thinking about People. Cambridge University Press.

Rendu, W., Beauval, C., Crevecoeur, I., Bayle, P., Balzeau, A., Bismuch, T., Bourguignon, L., Delfour, G., Faivre, J.P., Lacrampe-Cuyaubere, F., Tavormina, C., et al., 2014. Evidence supporting an intentional Neandertal burial at La Chapelle-aux-Saints. Proceedings of the National Academy of Sciences of the United States of America 111 (1), 81–86.

Reynolds, M.D., et al., 2007. Testosterone levels and sexual maturation predict substance use disorders in adolescent boys: a prospective study. Behavioral Psychology 61 (11), 1223–1227.

Ricardo, A., Szostak, J.W., September 2009. The origin of life on Earth. Scientific American.

Rice Jr., E.F., 1970. The Foundations of Early Modern Europe: 1460–1559. W.W. Norton & Co.

Richards, S.M., 1974. The concept of dominance and methods of assessment. Animal Behaviour 22, 914–930.

Richerson, P.J., Boyd, R., 1998. The evolution of human ultra-sociality. Indoctrinability, Ideology, and Warfare: Evolutionary Perspectives 71–95.

Ridley, M., 1999. Genome: The Autobiography of a Species in 23 Chapters. Perennial.

Rightmire, G.P., 1988. Homo erectus and later middle Pleistocene humans. Annual Review of Anthropology 17 (1988), 239–259.

Rilling, J.K., Insel, T.R., 1999. The primate neocortex in comparative perspective using magnetic resonance imaging. Journal of Human Evolution 37, 191–223.

Ring, K., 2000. Religious wars in the NDE movement: some personal reflections on Michael Sabom's light & death. Journal of Near-Death Studies 18 (4), 215–244.

Ringwald, E., 2015. List of Endangered Animals in Latin America. eHow. http://www.ehow.com/list_6568486_list-endangered-animals-latin-america.html.

Rivera, M.A., Aufderheide, A.C., CartmellC, L.W., Torres, M., Langsjoen, O., 2005. Antiquity of coca-leaf chewing in the south central Andes: a 3,000 year archaeological record of coca-leaf chewing from northern Chile. Journal of Psychoactive Drugs 37 (4), 455–458.

Robertson, G.A., 2013. Avoiding the next deepwater horizon: the need for greater statutory restrictions on offshore drilling off the Arctic coast of Alaska. George Washington Journal of Energy & Environmental Law 4, 107.

Robinson, E.E., 2002. Community frame analysis in Love Canal: Understanding messages in a contaminated community. Sociological Spectrum 22 (2), 139–169.

Robinson, R.A., Stokes, R.H., 2002. Electrolyte Solutions. Courier Corporation.

Rogers, M.D., 2011. Genetically modified plants and the precautionary principle. Journal of Risk Research 7 (7–8), 675–688.

Rohm, R., 2016. Personality Insights. http://www.personality-insights.com/robert-rohm/.

Rolston, H., 1975. Is there an ecological ethic? Ethics 93–109.

Rose, J., 1992. Marie Stopes and the Sexual Revolution. Faber and Faber, pp. 102–103.

Rose, V.J., 2013. Job satisfaction and personal growth. Journal of Knowledge & Human Resource Management 5 (12), 121–129.

Röseler, P.F., Röseler, I., Strambi, A., 1985. Role of ovaries and ecdysteroids in dominance hierarchy establishment among foundresses of the primitively social wasp, *Polistes gallicus*. Behavioral Ecology and Sociobiology 18 (1), 9–13.

Ross, C.N., French, J.A., Ortí, G., 2007. Germ-line chimerism and paternal care in marmosets (*Callithrix kuhlii*). Proceedings of the National Academy of Sciences of the United States of America 104 (15), 6278–6282.

Ross, G.N., 2001. Monarch, what's in a name? Louisiana Wildlife Federtion 29 (4), 13–16.

Ross, G.N., 2010. The monarch's trans – gulf express, a clockwork orange. Southern Lepidopterists' News 32 (1), 11–24.

Roth, G., Dicke, U., 2005. Evolution of the brain and intelligence. Trends in Cognitive Sciences 9 (5), 250–257.

Rousseau, J.J., 1973. The Social Contract and Discourses. Everyman's Library.

Rowell, T.E., 1974. The concept of social dominance. Behavioral Biology 11 (2), 131–154.

Rubenstein, W.B., 2004. The real story of U.S. hate crimes statistics: an empirical analysis. Tulane Law Review 78, 1213–1246.

Rubin, M., Hewstone, M., 2004. Social identity, system justification and social dominance: commentary of Reicher. Jost et al., and Sidanius, et al. Political Psychology 25, 823–844.

Rubin, L.C., 2008. Popular Culture in Counseling, Psychotherapy and Play-Based Interventions. Springer Publishing Company.

Ruhlen, M., 1994. The Origin of Language: Tracing the Evolution. John Wiley & Sons, Inc.

Runciman, W.G., 2012. The Social Animal. Kindle Ebook.

Russell, E., 2001. War and Nature: Fighting Humans and Insects with Chemicals from World War I to Silent Spring. Cambridge University Press.

Rutledge, R.W., Basore, B.L., Mulholland, R.J., 1976. Ecological stability: an information theory viewpoint. Journal of Theoretical Biology 57 (2), 355–371.

Rylands, A.B., Mittermeier, R.A., 2009. The diversity of the new world primates (Platyrrhini). In: Garber, P.A., Estrada, A., Bicca-Marques, J.C., Heymann, E.W., Strier, K.B. (Eds.), South American Primates: Comparative Perspectives in the Study of Behavior, Ecology, and Conservation. Springer.

Ryle, G., 1970. Bertrand Russell 1872–1970. In: Proceedings of the Aristotelian Society, pp. 77–84.

Saguaro, S., 2013. The Republic of Arborea': trees and the perfect society. Green Letters 17 (3), 236–250.

Sahlins, M.D., 1960. The origin of society. Scientific American 203 (3), 76–87.

Sahlins, M., 1972. Stone Age Economics. Routledge.

Sahney, S., Benton, M.J., 2008. Recovery from the most profound mass extinction of all time. Proceedings of the Royal Society: Biological 275 (1636), 759.

Salinger, P., Laurent, E., 1991. Secret Dossier: The Hidden Agenda Behind the Gulf War. Penguin Group.

Samenow, S.E., 1998. Straight Talk about Criminals: Understanding and Treating Antisocial Individuals. Jason Aronson.

Sanchez-Mazas, A., 2008. Genetic Diversity in Africa. John Wiley & Sons, Ltd. Published Online.

Sanderson, S.K., 2001. The Evolution of Human Sociality: A Darwinian Conflict Perspective. Rowman & Littlefield.

Sangodoyin, A.Y., 2008. Considerations on contamination of groundwater by waste disposal systems in Nigeria. Environmental Technology 14 (10), 957–964.

Sarich, V.M., Miele, F., 2004. Race: The Reality of Human Differences. Westview Press.

Sauri-Pujol, D., 2007. Putting the environment back into human geography: a teaching experience. Journal of Geography in Higher Education 17 (1), 3–9.

Schacter, D., Gilbert, D., Wegner, D., 2009. Psychology, second ed. Worth Publishers.

Schauss, A.G., et al.. 1979. A critical analysis of the diets of chronic juvenile offenders, Part 2. Orthomolecular Psychiatry, 8 (4), 222–226.

Scheel, D.E., Graves, H.B., Sherritt, G.W., 1977. Nursing order, social dominance and growth in swine. Journal of Animal Science 45 (2), 219–229.

Schino, G., 2001. Grooming, competition and social rank among female primates: a meta-analysis. Animal Behaviour 62 (2), 265–271.

Schjelderup-Ebbe, T., 1922. Beiträge zur sozialpsychologie des haushuhns. Zeitschrift für Psychologie und Physiologie der Sinnesorgane. Abt. 1. Zeitschrift für Psychologie.

Schnable, P.S., Ware, D., Fulton, D.R.S., et al., 2009. The B73 maize genome: complexity, diversity and dynamics. Science 326 (5956), 1112–1115.

Schneewind, J.B., 1977. Sidgwick's Ethics and Victorian Moral Philosophy. Oxford University Press. ISBN: 978-0198245520, p. 122.

Schneewind, J.B., 2002. Moral Philosophy from Montaigne to Kant. Cambridge University Press.

Schneirla, T.C., 1973. Instinct and aggression. Man and Aggression 59–64.

Schnug, E., Haneklaus, S., Schnier, C., Scholten, L.C., 2008. Issues of natural radioactivity in phosphates. Communications in Soil Science and Plant Analysis 27 (3–4), 829–841.

Schoenemann, T., 2006. Evolution of the size and functional areas of the human brain. Annual Review of Anthropology 35, 379–406.

Schoenthaler, S.J., 1983. Diet and juvenile delinquency. Nutrition Today 18 (6), 34.

Schumm, S.A., Dumont, J.F., Holbrook, J.M., 2002. Active Tectonics and Alluvial Rivers. Cambridge University Press.

Seltzer, L.F., 2008. Evolution of the Self, on the Paradoxes of Personality. Psychology Today.

Senneles, N., 2014. Sexual Abuse Widespread among Muslims. http://10news.dk/?p=553.

Seroczynski, A.D., Bergeman, C.S., Coccaro, E.F., 1999. Etiology of the impulsivity/aggression relationship: genes or environment? Psychiatry Research 86 (1), 41–57.

Serpell, J., 2012. One man's meat. In: Further Thoughts on the Evolution of Animal Food Taboos. On the Human. http://onthehuman.org/2011/11/one-mans-meat/.

Sexton, J.B., Makary, M.A., Tersigni, A.R., et al., 2006. Teamwork in the operating room: frontline perspectives among hospitals and operating room personnel. Anesthesiology 105 (5), 877–884.

Shaffer, H.B., Fellers, G.M., Voss, S.R., Oliver, J., Pauly, G., 2004. Species boundaries, phylogeography and conservation genetics of the red-legged frog (*Rana aurora/draytonii*) complex. Molecular Ecology 13 (9), 2667–2677.

Shaofeng, C., 2013. Carrying capacity: an overview. Chinese Journal of Population Resources and Environment 2 (1), 35–40.

Shea, M.M., Mench, J.A., Thomas, O.P., 1990. The effect of dietary tryptophan on aggressive behavior in developing and mature broiler breeder males. Poultry Science 69 (10), 1664–1669.

Sherif, M., 1936. The Psychology of Social Norms. Harper & Row.

Sherman, P.W., Jarvis, J.U., Alexander, R.D., 1991. Biology of the naked mole-rat. In: Monographs in Behavior and Ecology (USA).

Sherman, P.W., Lacey, E.A., Reeve, H.K., Keller, L., 1995. Forum the eusociality continuum. Behavioral Ecology 6 (1), 102–108.

Shoshani, J., Eisenberg, J.F., 1982. Elephas maximus. Mammalian Species 182 (182), 1–8.

Shoshani, J., 2005. Order Proboscidea. In: Wilson, D.E., Reeder, D.M. (Eds.), Mammal Species of the World: A Taxonomic and Geographic Reference, vol. 1. third ed. Johns Hopkins University Press.

Shultz, S., Dunbar, R.I., 2007. The evolution of the social brain: anthropoid primates contrast with other vertebrates. Proceedings of the Royal Society of London B 274 (1624), 2429–2436.

Shuster, C.N., Barlow, R.B., Brockmann, H.J., 2003. The American Horseshoe Crab. Harvard University Press, pp. 163–164.

Sidanius, J., Pratto, F., 1999. Social Dominance: An Intergroup Theory of Social Hierarchy and Oppression. Cambridge University Press.

Sidanius, J., Pratto, F., 2001. Social Dominance: An Intergroup Theory of Social Hierarchy and Oppression. Cambridge University Press.

Simms, M., 1977. Review: A History of the Malthusian League 1877–1927. New Scientist.

Simon, S., 1999. The Brain. HarperTrophy.

Simon, R.I., 2008. Bad Men Do What Good Men Dream: A Forensic Psychiatrist Illuminates the Darker Side of Human Behavior. American Psychiatric Publishing, Inc.

Singer, P., 2001. A utilitarian defense of animal liberation. In: Wadsworth (Ed.), Environmental Ethics, Louis Pojman, p. 35.

Sinha, M., Milligan, S., 2010. Section 4: Family Violence Against Seniors. Statisstics Canada,. http://www.statcan.gc.ca/pub/85-002-x/2012001/article/11643/11643-4-eng.htm.

Sinha, M., Nussinov, R., 2001. Point mutations and sequence variability in proteins: redistributions of preexisting populations. Proceedings of the National Academy of Sciences 98 (6), 3139–3144.

Sircus, M., 2007. Transdermal Magnesium Therapy: A New Modality for the Maintenance of Health. Phaelos Books.

Skelton, R.R., McHenry, H.M., 1992. Evolutionary relationships among early hominids. Journal of Human Evolution 23 (4), 309–349.

Skulan, J., 2000. Has the importance of the amniote egg been overstated? Zoological Journal of the Linnean Society 130 (2), 235–261.

Skwarzec, B., Strumińska-Parulska, D.I., Borylo, A., Katbat, K., 2012. Polonium, uranium and plutonium radionuclides in aquatic and land ecosystem of Poland. Journal of Environmental Science and Health, Part A 47 (3), 479–496.

Smart, S.M., Thompson, K., Marrs, R.H., et al., 2006. Biotic homogenization and changes in species diversity across human-modified ecosystems. Proceedings of the Royal Society B 273, 2659–2665.

Smelser, N.J., Baltes, P.B., 2001. Evolution of Sociality in International Encyclopedia of the Social & Behavioral Sciences. Elsevier.

Smith, J.M., Parker, G.A., 1976. The logic of asymmetric contests. Animal Behaviour 24 (1), 159–175.

Smith, V.A., 1901. Asoka – the Buddhist Emperor of India. Rulers of India series. Oxford at the Clarendon Press, p. 130.

Smith, W., 1954. William Paley's theological utilitarianism in America. The William and Mary Quarterly Third Series, Omohundro Institute of Early American History and Culture 11 (3), 402–424.

Smith, C.M., 1987. Sediment, phosphorus and nitrogen in channelized surface run-off from a New Zealand pastoral catchment. New Zealand Journal of Marine and Freshwater Research 21 (4), 627–639.

Smith, R.K., 1992. Conscience, coercion and the establishment of religion: the beginning of an end to the wandering of a wayward judiciary. Case Western Reserve Law Review 43, 917.

Smith, D.L., 2007. The Most Dangerous Animal, Human Nature and the Origins of War. St. Martin's Press.

Smuts, G.L., 1982. Lion. Macmillan South Africa.

Sobsey, M.D., Khatib, L.A., Hill, V.R., Alocilja, E., Pillai, S., 2012. Pathogens in Animal Wastes and the Impacts of Waste Management Practices in Their Survival, Transport and Fate. mwpshq.org.

Solomon, S., Qin, D., Manning, M., Chen, Z., Marquis, M., Averyt, K.B., Tignor, M., Miller, H.L., 2007. Climate change 2007: the physical science basis. In: 4th Assessment Report of the Intergovernmental Panel on Climate Change. Cambridge University Press.

Sørli, M.E., Gleditsch, N.P., Strand, H., 2005. Why is there so much conflict in the Middle East? The Journal of Conflict Resolution 49 (1), 141–165.

Soubiran, A., 1969. Diary of a Woman in White, english ed. Avon Books, p. 61.

Spence, J.T., 1985. Achievement American style: the rewards and costs of individualism. American Psychologist 40 (12), 1285.

Spikins, P.A., Rutherford, H.E., Needham, A.P., 2010. From hominity to humanity: compassion from the earliest archaic to modern humans. Time and Mind 3 (3), 303–325.

Springer, A.M., McRoy, C.P., Flint, M.V., 1996. The Bering Sea Green Belt: shelf-edge processes and ecosystem production. Fisheries Oceanography 5 (3–4), 205–223.

Srikanthan, A., Reid, R.L., 2008. Religious and cultural influences on contraception. Journal of Obstetrics and Gynaecology 30 (2), 129–137.

Stacey, N.H., Cantilena, L.R., Klaassen, C.D., 1980. Cadmium toxicity and lipid peroxidation in isolated rat hepatocytes. Toxicology and Applied Pharmacology 53 (3), 470–480.

Stam, J.H., 1976. Inquiries into the Origins of Language. Harper and Row, p. 255.

Stamm, K.R., Clark, F., Eblacas, P.R., 2000. Mass communication and public understanding of environmental problems: the case of global warming. Public Understanding of Science 9, 219–237.

Stamps, J.A., 1988. Conspecific attraction and aggregation in territorial species. American Naturalist 131, 329–347.

Stark, R., Brainbridge, W., 1996. A Theory of Religion. Rutgers University Press, p. 124.

Stark, R., 2005. The Victory of Reason. Random House.

Stein, B.A., Kutner, L.S., Adams, J.S., 2000. Precious Heritage: The Status of Biodiversity in the United States. Oxford Univ. Press.

Steinbruner, J., Dyson, F., 2007. Preventing doomsday. Bulletin of the Atomic Scientists 63 (1), 59–64.

Steinfeld, H., Gerber, P., Wassenaar, T., Castel, V., Rosales, M., de Haan, C., 2006. Livestock's Long Shadow: Environmental Issues and Options. Food and Agriculture Organization of the United Nations, Rome.

Sternberg, R.J., 1988. The Triangle of Love: Intimacy, Passion, Commitment. Basic Books.

Sterling, P., Eyer, J., 1988. Allostasis: A new paradigm to explain arousal pathway. In: Fisher, S., Reason, J.T. (Eds.), Handbook of Life Stress, Cognition, and Health. Wiley.

Stevens, J.M., 1974. Gynaecology from ancient Egypt: the papyrus Kahun: a translation of the oldest treatise on gynaecology that has survived from the ancient world. The Medical Journal of Australia 2 (25–26), 949–952.

Steyer, S., Benateau, A., 2012. Earth Before the Dinosaurs. Indiana University Press.

Stokes, A.W., 1974. Territory. Dowden, Hutchinson & Ross.

Stokstad, E., 2000. Hominid ancestors may have knuckle walked. Science 287 (5461), 2131–2132.

Stone, M.H., 1998. Sadistic personality in murderers. In: Psychopathy, Antisocial, Criminal and Violent Behavior. The Guilford Press.

Stone, M.H., 2009. The Anatomy of Evil. Prometheus Books.

Stotzer, R., 2007. Comparison of Hate Crime Rates Across Protected and Unprotected Groups. Williams Institute.

Strait, D.S., Weber, G.W., Neubauer, S., et al., 2009. The feeding biomechanics and dietary ecology of *Australopithecus africanus*. PNAS 106 (7), 2124–2129.

Strassmann, J.E., 1981. Wasp reproduction and kin selection: reproductive competition and dominance hierarchies among *Polistes annularis* foundresses. Florida Entomologist 74–88.

Strassmann, J.E., 1989. Group colony foundation in *Polistes annularis* (Hymenoptera: Vespidae). Psyche 96 (3), 223–236.

Striedter, G.F., 2004. Principles of Brain Evolution. Sinauer Associates.

Strier, K., 2007. Primate Behavioral Ecology, third ed. Allyn & Bacon.

Stringer, C.B., 1992. Evolution of Early Humans. The Cambridge Encylopedia of Human Evolution.

Stringer, C., 2003a. Human Evolution: Out of Ethiopia. Nature Publishing Group.

Stringer, C., 2003b. Human evolution: out of Ethiopia. Nature 423 (6941), 692–695.

Stringer, C., 2012a. Lone Survivors, How We Came to Be the Only Humans on Earth. Times Books, Henry Holt and Company.

Stringer, C., 2012b. What makes a modern human. Nature 485 (7396), 33–35.

Stutz, A.J., 2012. Culture and politics, behavior and biology: seeking synthesis among fragmentary anthropological perspectives on hunter-gatherers. Reviews in Anthropology 41 (1), 23–69.

Suedfeld, P., 2003. Reactions to societal trauma: distress and/or eustress. Political Psychology 18 (4), 849–861.

Suhre, S., 1999. Misguided morality: the repercussions of the International Whaling Commission's shift from a policy of regulation to one of preservation. Georgetown Environmental Law Review 12, 305.

Susanne, C., 2014. Eugenics and eugenism. Global Bioethics 10 (1–4), 101–110.

Susko, M., 2003. The fragility of evolution: part two. World Futures 59 (7), 495–534.

Sussman, R.W., 2003. Primate Ecology and Social Structure. Pearson Custom Publishing.

Suttner, R., 2006. Party dominance 'theory': of what value? Politikon 33 (3), 277–297.

Svendsen, G.E., 1974. Behavioral and environmental factors in the spatial distribution and population dynamics of a yellow-bellied marmot population. Ecology 55, 760–771.

Swartz, R., Lederman, N., 2008. What scientists say: scientists' views of nature of science and relation to science context. International Journal of Science Education 30 (6), 727–771.

Saudi Trade, 2014. Roaring Trade for Wild and Exotic Animals in Jazan. Jazan Arab News. http://www.arabnews.com/saudi-arabia/news/630921.

Sweeney, M.S., 2014. Brain: The Complete Mind, How It Develops, How It Works and How to Keep It Sharp. National Geographic Society.

Sword, G.A., 2008. Gregarious Behavior in Insects. Springer, Netherlands.

Tal, A., 2014. Overpopulation Is Still the Problem. Urban Progress. http://www.huffingtonpost.com/alon-tal/overpopulation-is-still-t_b_3990646.html.

Tallerman, M., Gibson, K.R., 2012. The Oxford Handbook of Language Evolution. Oxford University Press.

Tan, K.C., 2000. Toleration, Diversity and Global Justice. Cambridge University Press.

Tattersall, I., 1999. Becoming Human: Evolution and Human Uniqueness. Harcourt Inc.

Teaford, M.F., Walker, A., 2005. Quantitative differences in dental microwear between primate species with different diets and a comment on the presumed diet of Sivapithecus. American Journal of Physical Anthropology 64 (2), 191–200.

Tebeau, C.W., 1963. They Lived in the Park: The Story of Man in the Everglades National Park. FL. University of Miami Press, Coral Gables.

Tenaza, R., 1984. Songs of hybrid gibbons (*Hylobates lar* × *H. muelleri*). American Journal of Primatology 8 (3), 249–253.

Thompson, R.F., 2000. The Brain: An Introduction to Neuroscience. Worth Publishers.

Thorington Jr., R.W., Hoffman, R.S., 2005. In: Wilson, D.E., Reeder, D.M. (Eds.), Mammal Species of the World: A Taxonomic and Geographic Reference. Johns Hopkins University Press, pp. 754–818.

Thorpe, N., Shirmohammad, A., 2007. Herbicides and nitrates in groundwater of Maryland and childhood cancers: a geographic information systems approach. Journal of Environmental Science and Health, Part C 23 (2), 261–278.

Tierney, A.J., 1986. The evolution of learned and innate behavior: contributions from genetics and neurobiology to a theory of behavioral evolution. Animal Learning & Behavior 14 (4), 339–348.

Tobias, P.V., 1987. The brain of *Homo habilis*: a new level of organization in cerebral evolution. Journal of Human Evolution 16 (7–8), 741–761.

Tokdemir, M., Polat, S.A., Acik, Y., Gursu, F., Cikim, G., Deniz, O., 2003. Blood zinc and copper concentrations in criminal and noncriminal schizophrenic men. Arch Androl 49 (5), 365–368.

Tomlinson, R., 1975. Demographic Problems: Controversy over Population Control, second ed. Dickenson.

Totton, N., 2015. Living on Earth: embodiment and ecopsychology. Self & Society 35 (3), 15–24.

Traverse, A., 1988. Plant evolution dances to a different beat. Historical Biology 1 (4), 277–301.

Travis, H., 2012. Extremely violent societies: mass violence in the twentieth-century world. Journal of Genocide Research 14 (1), 99–104.

Tremblay, R.E., 2000. The development of aggressive behaviour during childhood: what have we learned in the past century? International Journal of Behavioral Development 24 (2), 129–141.

Trivers, R.L., 1971. The evolution of reciprocal altruism. The Quarterly Review of Biology 46, 35–57.

Trivers, R.L., 2011. The Folly of Fools: The Logic of Deceit and Self-Deception in Human Life. Basic Books.

Trojan, P., 1984. Ecosystem Homeostasis. Dr. Junk Publishers.

Trupin, S.R., 2015. Abortion. Emedicinehealth. http://www.emedicinehealth.com/abortion/page7_em.htm.

Turnbull, C., 1987. The Forest People. Touchstone.

Turney-High, H.H., 1971. Primitive War. University of South Carolina Press.

Tyson, P., 2008. Meet Your Ancestors. NOVA scienceNOW. PBS.

Underhill, J.W., 2012. Ethnolinguistics and Cultural Concepts: Truth, Love, Hate and War. Cambridge University Press, UK.

Unno, T., 2014. Problems of overpopulation and depopulation. Japanese Economic Studies 3 (3), 59–87.

Urbani, C.B., 1979. Territoriality in social insects. In: Hermann, H.R. (Ed.), Social Insects, vol. 1. Academic Press (Chapter 4).

Van Halderen, M.D., Bhatt, M., Berens, G.A., Brown, T.J., Van Riel, C.B., 2014. Managing impressions in the face of rising stakeholder pressures: examining oil companies' shifting stances in the climate change debate. Journal of Business Ethics 1–16.

Valentino, K., Cicchetti, D., Rogosch, F.A., Toth, S.L., 2008. True and false recall and dissociation among maltreated children: the role of self-schema. Development and Psychopathology 20 (1), 213–232.

Valeriano T'ula, E., 2003. Tinku. Patrimonio cultural del altiplano central. In: En: Anales de la Reunion Annual de Etnologia. MUSEF.

Van Andel, T.H., 2002. The climate and landscape of the middle part of the Weichselian Glaciation in Europe: the stage 3 project. Quaternary Research 57 (1), 2–8.

Van der Pluijm, B.A., Marshak, S., 2004. Earth Structure – An introduction to Structural Geology and Tectonics, second ed. W. W. Norton.

Van Os, J., Kapur, S., 2009. Schizophrenia. Lancet 374 (9690), 635–645.

Van Schaic, C., Fox, E., Sitompul, A., 1996. Manufacture and use of tools in wild Sumatran orangutans. Naturwissenschaften 83, 186–188.

VandenBos, G.P., 2006. APA Dictionary of Psychology. American Psychological Association.

Vekua, A., Lordkipanidze, D., et al., 2002. A new skull of early *Homo* from Dmanisi, Georgia. Science 297 (5578), 85–89.

Viets, F.G., Lunin, J., 2009. The environmental impact of fertilizers. CRC Critical Reviews in Environmental Control 5 (4), 423–453.

Vitousek, P.M., Mooney, H.A., Lubchenco, J., Melillo, J.M., 1997. Human domination of Earth's ecosystems. Science 277 (5325), 494–499.

Verda Vivo, 2012. http://verdavivo.wordpress.com/2009/01/19/top-10-worlds-worst-pollution-problems/.

Voet, D., Voet, J.G., 2011. Biochemistry, fourth ed. John Wiley & Sons Inc.

Voigts, E., Schaff, B., Pietrzak-Franger, M., 2014. Reflecting on Darwin. Ashgate.

Von Eckardt, B., 1996. What Is Cognitive Science? MIT Press.

Von Frisch, K., Lindauer, M., 1973. The "language" and orientation of the honey bee. Perception: An Adaptive Process 1, 70.

Votier, S.C., Hatchwell, B.J., Beckerman, A., McCleery, R.H., Hunter, F.M., Pellatt, J., Birkhead, T.R., 2005. Oil pollution and climate have wide-scale impacts on seabird demographics. Ecology Letters 8 (11), 1157–1164.

Wackernagel, M., Rees, W., 1998. Our Ecological Footprint: Reducing Human Impact on the Earth. New Society Publishers. Number 9.

Waid, D.E., 2007. Incineration of organic materials by direct gas flame for air pollution control. American Industrial Hygiene Association Journal 30 (3), 291–297.

Walker, A., Shipman, P., 2005. The Ape in the Tree: An Intellectual & Natural History of Proconsul. The Belknap Press of Harvard University Press.

Wallechinsky, D., 2006. Tyrants: The World's 20 Worst Living Dictators. Harper-Collins Publishers.

Waller, J., 2007. Becoming Evil, How Ordinary People Commit Genocide and Mass Killing. Oxford University Press.

Wallerstein, I., 2011. The Modern World – System I, Capitalist Agriculture and the Origins of the European World-Economy in the Sixteenth Century. University of California Press.

Wals, A.E., 1990. Caretakers of the environment: a global network of teachrs and students to save the Earth. The Journal of Environmental Education 21 (3), 3–7.

Warburton, M., 2014. Wildlife poaching and trafficking of endangered species alarming. Liberty Voice. http://guardianlv.com/2014/03/wildlife-poaching-and-trafficking-of-endangered-species-alarming/.

Ward, S.K., Dziuba-Leatherman, J., Stapleton, J.G., L Yodonis, C., 1994. Acquaintance and Date Rape: An Annotated Bibliography. Greenwood Press.

Ward, A., 1978. Sor Osaac Newton. Science Activities: Classroom Projects and Curriculum Ideas 15 (2), 11–12.

Ward, P.S., 2007. Phylogeny, classification and species-level taxonomy of ants (Hymenoptera: Formicidae). Zootaxa 1668, 549–563.

Washburn, S.L., 2013. Classification and Human Evolution. Routledge.

Watson, J.D., Crick, F.H., 1953a. A structure for deoxyribose nucleic acids. Nature 171 (4356), 737–738.

Watson, J.D., Crick, F.H., 1953b. The structure of DNA. Cold Spring Harbor Symposia on Quantitative Biology 18, 123–131.

van der Weerden, T.J., Luo, J., Dexter, M., Rutherford, A.J., 2014. Nitrous oxide, ammonia and methane emissions from dairy cow manure during storage and after application to pasture. New Zealand Journal of Agricultural Research 57 (4), 354–369.

Wegner, D.M., Erber, R., 1993. Social Foundations of Mental Control. Prentice-Hall, Inc.

Weir, I., Green, P.J., 1994. Modelling data from single-photon emission computerized tomography. Journal of Applied Statistics 21 (1–2), 313–337.

Weiskopf, M., 1990. Cappadocia. Encyclopaedia Iranica 4 (7-8), 780–786.

Wellbery, D.E., 1984. Lessing's Laocoon: Semiotics and Aesthetics in the Age of Reason. Cambridge University Press.

Wells, S., Read, M., 2002. The Journey of Man – A Genetic Odyssey. Random House.

Wells, H., Wells, P.H., Cook, P., 1990. The importance of overwinter aggregation for reproductive success of monarch butterflies (*Danaus plexippus* L.). Journal of Theoretical Biology 147 (1), 115–131.

Wells, M.J., 1962. Brain and Behaviour in Cephalopods. Heinemann.

Wells, K.D., 1977. The social behavior of anuran amphibians. Animal Behaviour 25, 666–693.

Wenban-Smith, F., 2013. The Ebbsfleet Elephant. Oxford Archaeology Monograph.

Wender, E.H., Solanto, M.V., 1991. Effects on sugar on aggressive and inattentive behavior in children with attention deficit disorder with hyper-activity and normal children. Pediatrics 88 (5), 960–966.

Werbach, M., 1992. Nutritional influences on aggressive behavior. Journal of Orthomolecular Medicine 7, 45–51.

Werbach, M.R., 1995. Nutritional influences on aggressive behavior. Journal of Orthomolecular Medicine 7 (1), 46–51.

West, M.J., 1967. Foundress associations in polistine wasps: dominance hierarchies and the evolution of social behavior. Science 157, 1584–1585.

West, R., 1998. Daniel Defoe: The Life and Strange, Surprising Adventures. Carroll & Graf.

Wheeler, P., 1984. The evolution of bipedality and loss of functional body hair in hominids. Journal of Human Evolution 13, 91.

Whewell, W., 1840. The Philosophy of the Inductive Sciences. John W. Parker, West Strand.

Whicker, M.L., 1996. Toxic Leaders: When Organizations Go Bad. Quorum Books.

White, T.D., Berhane, A., Yonas, B., et al., 2009. Ardipithecus Ramidus and the Paleobiology of Early Hominids 326 (5949), 75–86 64.

White, P., 2012. Ground Penetrating Radar Finds Hidden Cities. http://members.tripod.com/~Ravenwoods/index-55Giza.html.

WHO, 2002. World Report on Violence and Health, vol. 6. World Health Organization, Geneva, p. 149.

Widom, C.S., Maxfield, M.G., 2001. An Update on the "Cycle of Violence.". National Institute of Justice. https://www.ncjrs.gov/pdffiles1/nij/184894.pdf.

Wilford, J.N., 2015. Stone Tools from Kenya Are Oldest Yet Discovered. The New York Times.

Willer, D.E., Hermann, H.R., 1989. Multiple foundress associations and nest switching among females of Polistes exclamans (Hymenoptera: Vespidae). Sociobiology 16, 197–216.

Willer, D.E., 1988. Behavioral Ecology and Population Biology of Polistes Exclamans on Sapelo Island, Georgia (Dissertation). Univ. of Georgia, Athens.

William, H., Dyson, F.J., 2012. Iterated Prisoner's Dilemma Contains Strategies that Dominate Any Evolutionary Opponent. PNAS Early Edition.

Williams, K., Haslam, C., Williams, J., 1992. Ford versus Fordism: the beginning of mass production? Work, Employment & Society 6 (4), 517–555.

Williams, R.S., 2000. A modern Earth narrative: what will be the fate of the biosphere? Technology in Society 22 (3), 303–339.

Williamson, R.K., 2005. American History. Parragon Incorporated.

Williamson, K.D., 2013. Political Self-Interest Is No Less Selfish than Economic Self-interest. National Review. http://www.nationalreview.com/article/365100/problem-selfishness-kevin-d-williamson.

Wilson, E.O., Hölldobler, B., 2005. Eusociality: origin and consequences. Proceedings of the National Academy of Sciences of the United States of America 102 (38), 13367–13371.

Wilson, D.E., Reeder, D.M., 2005. Mammal Species of the World: A Taxonomic and Geographic Reference, third ed. Johns Hopkins University Press.

Wilson, E.O., Carpenter, F.M., Brown, W.L., 1967. The first Mesosoic ant, with the description of a new subfamily. Psyche 74 (1), 1–19.

Wilson, E.O., 1975. Sociobiology: The New Synthesis. Harvard Univ. Press.

Wilson, E.O., 1999. Consilience: The Unity of Knowledge. Vintage Books.

Wilson, E.O., 1999. The Diversity of Life. WW Norton & Company.

Wilson, B., 2004. The Hive: The Story of the Honeybee. John Murray.

Wilson, E.O., 2012. The Social Conquest of Earth. W. W. Norton & Company.

Wilson, J.G., 2013. Alfred Russel Wallace and Charles Darwin: perspectives on natural selection. Transactions of the Royal Society of South Australia 137 (2), 90–95.

Winston, R., Wilson, D.E., 2004. Human. Smithsonian Institution, DK Publishing.

Winter, L., September 2014. Destruction of Amazon rainforest visible from space. Environment.

Winterhalder, B., Smith, E.A., 2000. Analysing adaptive strategies: human behavioral ecology at twenty-five. Evolutionary Anthropology: Issues, News, and Reviews 9 (2).

Winterhalder, B., 1981. Optimal foraging strategies and hunter-gatherer research in anthropology: theory and models. In: Hunter-Gatherer Foraging Strategies: Ethnographic and Archaeological Analyses, pp. 13–35.

Wollerman, L., Wiley, R.H., 2002. Background noise from a natural chorus alters female discrimination of male calls in a neotropical frog. Animal Behaviour 63 (1), 15–22.

Wolpoff, M.H., Hawks, J., Senut, B., Pickford, M., Ahern, J., 2006. An ape or the ape: is the Toumai cranium TM 266 a hominid? PaleoAnthropology 36–50.

Wong, M.A., Frank, R., Allsup, R., 2015. An analysis of online discussions by white supremacists. Information & Communications Technology Law 24 (1), 41–73.

Wong, M., 2011. Sex and Sociality in Fishes: The Evolution of Social and Reproductive Behavior. Lambert Academic Publishing.

Woodward, D.G., 1997. Life cycle costing—theory, information acquisition and application. International Journal of Project Management 15 (6), 335–344.

Wrangham, R., Peterson, D., 1996. Demonic Males, Apes and the Origins of Human Violence. Houghton Mifflin Company.

Wrangham, R.W., 1996. Chimpanzee Cultures. Harvard University Press.

Wyndham, D., 2012. Norman Haire and the Study of Sex. Foreword by Michael Kirby. Sydney University Press.

Yang, X., Kalluri, U.C., Jawdy, S., Gunter, L.E., Yin, T., Tschaplinski, T.J., Weston, D.J., Ranjan, P., Tuskan, G.A., 2008. The f-box gene family is expanded in herbaceous annual plants relative to woody perennial plants. American Society of Plant Biologists 148 (3), 1189–1200.

Yang, T.T., Simmons, A.N., Matthews, S.C., et al., 2010. Adolescents with major depression demonstrate increased amygdale activation. Journal of the American Academy of Child and Adolescent Psychiatry 49 (1), 42–51.

Yanowitch, R., Coccaro, E.F., 2011. The neurochemistry of human aggression. Adv Genet 75, 151–169.

Yeates, D., Wiegmann, B.M., 2005. The Evolutionary Biology of Flies. Columbia University Press.

Yilmaz, E., Cancino, B., Parra, W.R., 2007. Statistical analysis of solar radiation data. Energy sources, Part A: recovery. Utilization and Environmental Effects 29 (15), 1371–1383.

Young, S.N., Levton, M., 2002. Serotonin. Pharmacology, Biochemistry and Behavior 71 (4), 857–865.

Zalasiewicz, J., Williams, M., 2012. The Goldilocks Planet. Oxford University Press.

Zatta, P.F., Alfrey, A.C., 1998. Aluminium Toxicity in Infants' Health and Disease. World Scientific Publishing Co.

Zeki, S., Romaya, J.P., 2008. Neural correlates of hate. In Lauwereyns. PLoS One 3 (10), 3556.

Zhang, Y., Love, N., Edwards, M., 2009. Nitrification in drinking water systems. Critical Reviews in Environmental Science and Technology 39 (3), 153–208.

Zhang, W.X., 2003. Nanoscale iron particles for environmental remediation: an overview. Journal of Nanoparticle Research 5 (3–4), 323–332.

Zhao, G., Sun, M., Wilde, S.A., Li, S.Z., 2004. A Paleo-Mesoproterozoic supercontinent: assembly, growth and breakup. Earth-Science Reviews 67, 91–123.

Zimmer, C., 2000. In search of vertebrate origins: beyond brain and bone. Science 287 (3), 1576–1579.

Zimmer, C., July 2013. Genes Are Us. And Them. National Geographic, pp. 103–104.

Zuckerman, M., 1991. Psychobiology of Personality. Cambridge University Press (No. 10).

Zuckerman, M., 1995. Good and bad humors: biochemical bases of personality and its disorders. Psychological Science 6 (6), 325–332.

Zullo, L., Sumbre, G., Agnisola, C., Flash, T., Hochner, B., 2009. Nonsomatotopic organization of the higher motor centers in octopus. Curr. Biol. 19 (19), 1632–1636.

Glossary

Acanthostega A group of tetrapodomorph fish that appeared in the Upper Devonian period, about 365 million years ago. Their limbs represent an intermediate stage between lobe fins and the chiridian limbs of true tetrapods.

Acoelomate An animal that possesses all three primordial tissues (ectoderm, endoderm, and mesoderm) but has no chambers in the mesoderm that can be considered as belonging to a coelom (body cavity). The condition is referred to as acoelomic. Organisms that are typical acoelomates are the flatworms, phylum Platyhelminthes.

Actinopterygians Ray-finned fish.

AD An abbreviation that stands for the Latin phrase *anno domini*, which means "in the year of our Lord." It is used to mean after the death of Christ.

Adventurousness Inclined to undertake new and daring enterprises; hazardous; risky.

Affiliation A relationship between two individuals in which one approaches, interacts, and remains with the other in their group.

Aggression In its broadest definition, aggression is a behavioral expression involving such variables as verbal attack, discriminatory behavior, and economic exploitation. It is a forceful action or procedure demonstrated by one individual or group toward another, generally used in the process of domination. Mazur and others state that a person is said to "act aggressively if his or her apparent intent is to threaten or inflict physical injury and/or psychological pain on another individual." Four approaches to understanding the origins or causes of aggression in humans have been recognized:

- It may stem from differences among humans because of physiological variation resulting from childhood experiences.
- It may stem from economic deprivation or social conflicts.
- It may relate to dominance interactions between groups of humans.
- It may stem from a genetic predisposition.

Based on these considerations, it is obvious that aggression is a complex phenomenon, multidetermined, and almost never due to a single factor.

Agonistic behavior Competitive behavior between two members of the same species, involving offensive or defensive aggression, threats, appeasement/conciliation, or retreat/flight. Agonistic behavior may terminate in a ranking of the competing individuals in a dominance hierarchy.

Agnosticism The belief that one is unsure about the belief in a deity.

Agonistic The range of activities associated with aggressive encounters between members of the same species, including threat, attack, appeasement, or retreat. Agonistic encounters between social animals result in the formation of a dominance hierarchy.

Agreeableness A personality trait that reflects individual differences in cooperation and social harmony.

Agricultural revolution A change in the way that humans obtained their food. The agricultural revolution occurred around 10,000 years ago as humans transitioned from a hunter–gatherer lifestyle to an agricultural way of life.

Alloparent An individual in a social species other than the parent who assists in the care of young, whether they are genetically related or not.

Allopatric populations Two populations that are separate (do not overlap). According to the definition of a species, allopatric populations naturally represent different species.

Allostasis the process by which the body responds to stressors in order to regain homeostasis. According to Sterling and Eyer (1988), adaptation in the face of potentially stressful challenges involves activation of neural, neuroendocrine, and neuroendocrine-immune mechanisms. They refer to this as "allostasis" or "stability through change." Allostasis is an essential component of maintaining homeostasis. "When these adaptive systems are turned on and turned off again efficiently and not too frequently, the body is able to cope effectively with challenges that it might not otherwise survive. However, there are a number of circumstances in which allostatic systems may either be overstimulated or not perform normally, and this condition has been termed 'allostatic load' or the 'price of adaptation' (McEwen and Stellar, 1993)." "Allostatic load can lead to disease over long periods. Types of allostatic load include: (1) frequent activation of allostatic systems; (2) failure to shut off allostatic activity after stress; (3) inadequate response of allostatic systems leading to elevated activity of other, normally counter-regulated allostatic systems after stress."

Alpha individual An individual who becomes the most dominant following an agonistic session. All others that enter agonistic confrontations with the alpha individual occupy a subordinate position.

Alpha radiation "An alpha particle is a nuclear particle composed of two protons and two neutrons. The loss of an alpha particle…reduces the parent material by four mass units and reduces the atomic number by two. As an example, the initial step in the radioactive series U-238-Pb-206, in which the parent, uranium-238, is changed to thorium-234. The complete transition to lead-206 involves 8 alpha and 6 beta emissions."

Altruism A form of behavior performed by an individual who benefits another but at some expense to the altruist. Altruism is most clearly expressed between related individuals. According to Hamilton, natural selection in favor of an altruistic act, involving degree of relatedness, must exceed the cost to the altruist.

Amanorrhea A condition expressing the absence or suppression of menstruation.

Ambition In humans, an ardent desire to seek a dominant position that provides rewards, such as high rank, power, or resources.

Amniotic egg An egg with an amniotic cavity, several extraembryonic membranes (such as those surrounding the amniotic cavity, the yolk sac, and the allantois) and a shell. It was the amniotic egg, which first showed up in the reptiles, that allowed tetrapods to invade and exist in a terrestrial environment without maintaining a connection to the water. The amniotic egg or modifications of it are found in all vertebrates from reptiles up, including humans.

Amorality An unawareness of, indifference toward or disbelief in any set of moral standards or principles.

Amphibians Organisms that arose from tetrapod fish and invaded the terrestrial landscape. While most adult amphibians are able to breathe with lungs, their eggs must be deposited in water or close by, and the larvae breathe through gills. Certain neotenic amphibians are known to retain their gills in their adult form.

Amygdala Part of the limbic system of the brain that controls some of the more basic drives, such as aggression and sexuality.

Aneuploidy A genetic condition that is caused when there are abnormal number of chromosomes in an organism, due to loss of a chromosome (monosomy) or presence of an extra copy of a chromosome (e.g., trisomy and tetrasomy).

Anthropocentric Regarding humans as the central element of the universe, and interpreting reality exclusively in terms of human values and experience. Withgott and Brennen add that it "is a human-centered view of our relationship with the environment; it denies or ignores the rights of any nonhuman entity, and measurers costs and benefits of actions solely according to their human impact."

Anthropoid A primate in the suborder Anthropoidea, including monkeys, apes, and humans.

Antioxidant A substance that inhibits oxidation, especially one used to counteract the deterioration of stored food products; a substance such as vitamin C or E that removes potentially damaging oxidizing agents in a living organism.

Antisocial personality disorder Antisocial personality disorder (ASPD) is described by the American Psychiatric Association's Diagnostic and Statistical Manual, fourth edition (DSM-IV-TR), as an Axis II personality disorder characterized by "… pervasive pattern of disregard for, and violation of, the rights of others that begins in childhood or early adolescence and continues into adulthood." Psychopathy and sociopathy (terms that have been treated as synonyms but used in different contexts by different psychologists) are terms considered as subsets of ASPD. The American Psychiatric Association's Diagnostic and Statistical Manual of Mental Disorders incorporated various concepts of psychopathy, sociopathy, and antisocial personality in early versions but, starting with the DSM-III in 1980, used instead the term Antisocial Personality Disorder. See Chapter 9 for a discussion of these disorders.

Apatheists People who do not care about whether there is a deity.

Arboreal Adapted to living in trees.

Asocial An asocial person is one who lacks a strong motivation to engage in social interaction and/or has a preference for solitary activities. Synonyms used by developmental psychologists are: nonsocial, unsocial, and social disinterest. An asocial condition is considered distinct from an antisocial one, as the latter implies an active dislike or antagonistic view toward other people or society. The condition is often confused with misanthropy, which is defined as a general hatred, mistrust, or dislike of the human species or human nature. Frequently observed in schizophrenia patients, it is characterized by an inability to empathize, feel intimacy, or form close relationships with others. Asociality is not necessarily perceived as a totally negative trait by society, since expressing asociality has been used as a way to express independence of the mind from prevailing ideas (dissent). An asocial attitude is found in many individuals who like to spend time mostly with themselves and away from crowds.

ASPD Antisocial personality disorder, involving broad behavioral patterns, based on clinical observation.

Atheism The belief that there is no deity of any kind.

ATP Adenosine triphosphate, a nucleotide that is important in providing energy to living organisms. ATP is formed through the citric acid cycle from the breakdown of carbohydrates.

Attribute Used as a noun, attribute is a quality or feature regarded as a characteristic or inherent part of someone or something. As a verb, attribute regards something as being caused by (someone or something).

Authoritativeness The quality of possessing authority; the quality of trustworthiness and reliability.

Autosome A chromosome that is not an allosome (not a sex chromosome).

Aunting A form of behavior in which a female other than an infant's mother assists in infant care. Aunting may involve carrying the infant or defending it. Also see *Alloparent*.

Autonomic responses These are automatic responses to situations, involving heart rate, blood pressure, respiration, sweating, papillary dilation, tearing, and hormone secretion.

Autosomal dominant genetic disorders A genetic condition caused by inheriting a defective gene from a single parent (Griffiths et al., 2012). This defective gene belongs to an autosome. An autosome is a chromosome that is not an allosome (i.e., not a sex chromosome). This condition is also known as autosomal dominant pattern of inheritance.

Autosomal recessive genetic disorders Genetic disorders that manifest themselves only when an individual has two defective alleles of the same gene, one from each parent (Barroso et al., 1999). These disorders are inherited via the autosomal recessive pattern of inheritance.

Aunting A form of behavior in which a female other than an infant's mother assists in infant care. Aunting may involve carrying the infant or defending it. Also see Alloparent.

Aventurousness An inclination to undertake new and daring enterprises; hazardous; risky.

Battering Domestic violence among humans. According to Wrangham and Peterson, battering most often involves "a man attacking and beating a woman with whom he has or has had an ongoing relationship."

BC Years numbered back from the supposed year of the birth of Christ.

BCE Before the Common Era.

Behavioral plasticity An organism's ability to alter its behavior in response to environmental changes.

Beta individual An individual in a dominance hierarchy who is subordinate to the alpha individual but dominates other members of a social group.

Beta radiation Beta decay is a high-energy electron released from the atom converting a neutron into a proton. Rubidium-87 in decaying to strontium-87 is an example of losing an electron and increasing the atomic number by 1.

Binocular vision A characteristic of primates in which the eyes are pointed forward, giving the individual a form of vision in which depth perception is enhanced.

Biotic potential The maximum number of individuals that a species can produce. All species have a biotic potential that they never reach because of environmental influences. Reaching a biotic potential is generally impossible because of eventual food shortages, predation, and parasitism. These are the factors that influence a species' environmental carrying capacity.

Bipedalism A condition in which an organism walks upright on two feet.

Blastula An embryological structure in the form of a hollow ball of cells in which all cells remain undifferentiated. It is the stage following the morula, which is a solid ball of cells. In medicine, the blastula is referred to as the blastocyst.

Bottleneck A time in the past in which a species had significantly low numbers. Bottlenecks in human evolution are detected by modern investigators by investigating DNA samples, recording mutations, and measuring genetic diversity (e.g., Alu sequences and mitochondrial DNA being especially important). Environmental catastrophes, for example, extremely divergent temperature swings, the eruption of supervolcanoes, or asteroid impacts, may result in population bottlenecks.

Box genes See *F-box genes; T-box genes.*

Brachiation A form of movement in which an individual swings with their arms, hand-over-hand, between one branch of a tree to another. It is common in arboreal primates, such as gibbons.

Bridging coalitions Coalitions in which one coalitionary male is high-ranking and the other is lower-ranking than the target individual.

Callitrichines New World monkeys of the subfamily Callitrichinae, including marmosets and tamarins.

Callous leadership Leadership in which some followers are uncaring or unkind.

Care of young A condition in which at least one parent provides food and protection for its offspring. While it is characteristic of all social animals, many animals that are considered solitary in their behavior also take care of their young.

Castes Individuals in a social group that have different functions, such as the females of a social insect colony. Queens and workers represent two different castes. Males are a different sex and thus do not represent a caste.

Catarrhines Primates in the infraorder Catarrhini. It has four families, including African monkeys, gibbons, the great apes, and humans.

Cathemeral A form of behavior that relates to periodic and sporadic activity throughout both day and night. Some lemur species demonstrate cathemeral behavior.

Cercopithicines Old World monkeys in the subfamily Cercopithecinae that includes baboons, macaques, langurs, and guenons.

Cheated personality type A personality that is much like the hostile type, being hostile, antisocial, and rejecting the norms and mores of society, but for different reasons.

Chimera A single organism composed of cells from different zygotes. This can result in male and female organs, two blood types, or subtle variations in form. Animal chimæras are produced by the merger of multiple fertilized eggs. In plant chimæras, however, the distinct types of tissue may originate from the same zygote, and the difference is often due to mutation during ordinary cell division. Normally, chimærism is not visible on casual inspection; however, it has been detected in the course of proving parentage (Norton and Zehner, 2008). There are different types of chimera, including tetragametic chimerism, microchimerism, dymbiotic chimerism, graft chimerism, chromosomal chimerism, nuclear gene–differential chimerism, plastid gene–differential chimerism, hematopoietic chimerism, and germline chimerism.

Chiridian limb A limb with well-defined digits at its distal end. This term is often used in distinguishing between tetrapods (which have a chiridian limb) and tetrapodomorphs (which may have a limb-like lobe fin but do not have a chiridian limb). It is believed that the chiridian limb evolved in tetrapodomorph ichthyans (lobe-finned fish) and functioned in maneuvers in their aquatic habitat. Certain tetrapods that had a chiridian limb later ventured onto land, and it was one of these groups that gave rise to the amphibians.

The development of different chiridian limbs follows paths almost identical to those in a mouse: first buds, followed by mitten-like modifications, ping–pong–paddle-like modifications, and finally webs. Special zones (called signal centers) control the formation of specific structures along the proximodistal axis, referred to as the apical ectodermal ridge, the anteroposterior axis (referred to as a zone of polarizing activity), and the dorsoventral axis.

Clade A set of organisms sharing a common ancestor.

Climax vegetation The most stable and dominant vegetation in communities of a particular geographical area (Clements, 1916).

Cline In biology, a cline is a continuum of species, populations, races, varieties, or forms of organisms that exhibit gradual phenotypic and/or genetic differences over a geographic area, typically as a result of environmental heterogeneity.

Coalition Two or more individuals cooperating against a target individual. See *Conservative coalitions*; *Bridging coalitions*; *Revolutionary coalitions*.

Cocaine A tropane alkaloid that is obtained from the leaves of the coca plant.

Codominance A relationship between two animals that share an equal ranking in a dominance hierarchy. Codominants share the rewards of their position, including such benefits in feeding, sexual partners, status, and so on.

Coelomate An organism that has a well-developed body cavity (a coelom). Annelid worms and more advanced organisms are coelomates.

Cofounding females A term used in describing two or more fertilized female wasps that leave their parental nest to establish a new nest away from the parental one. Most cases of cofounding are found in tropical locations.

Cognition In biology, cognition refers to mental processes, including attention, memory, producing and understanding language, solving problems, and making decisions. Cognition, or cognitive processes, can be natural or artificial or conscious or unconscious.

Community A group of populations that interact in specific ways, often due to coevolution. Withgott and Brennen point out that "when an invasive species moves in, these relationships are disrupted. Because nonnative species are sometimes better competitors for resources because of a lack of factors limiting their population growth, they can cause population reductions and even the extinction of native species that occupy a niche similar to that of the invader. The alien species can also have a huge impact on the entire food web of a community if it is a producer or primary consumer because it may decrease food resources for native organisms at higher trophic levels.

Companionate love Love that is often found in marriages in which the passion has died out of the relationship, but a deep affection and commitment remain.

Connected breath The connected breath, hinted at in Don Campbell's definition of music and regarded by Native Americans as a symbol of life, in many ways, links all of us to our past, to our roots, and to the roots of all things. It is an expression of oneness that ties all life and all nonliving constituents of our biosphere together.

Conscientiousness A distinguishing feature of mental life, variously characterized as: (1) a state of awareness as well as the content of the mind, that is, the ever-changing stream of immediate experience, comprising perceptions, feelings, sensations, images, and ideas; (2) a content effect of neural reception; (3) a relation of self to environment; and (4) the totality of an individual's experiences at any given moment.

Consilience Consilience, a term that reflects on the unity of knowledge, refers to the principle that calculating a result by two different methods should lead to the same answer. While the word consilience was coined by William Whewell in *The Philosophy of the Inductive Sciences* in 1840, it has recently become important in the biological sciences because of a book called *Consilience* by E.O. Wilson. Wilson uses the term consilience primarily to describe the synthesis of knowledge from different specialized fields of human endeavor.

Consensual discrimination Discrimination that is driven by the need for accuracy and reflects stable and legitimate intergroup status hierarchies (favoring a high-status in-group).

Conservative coalitions Coalitions in which one coalitionary male is high-ranking and the other is lower-ranking than the targeted individual.

Consummate love A complete form of love, representing the ideal relationship toward which many people strive but which very few actually achieve.

Contact call A form of vocalization that provides information on the whereabouts of an individual.

Cooperative breeding A form of behavior in animal societies in which certain adults breed while others participate in care of the young.

Corrupt leadership Leadership in which the leader and at least some followers lie, cheat, and steal.

Crack The freebase form of cocaine.

Craniate An organism with not only a head but one with a recognizable brain. For instance, lancelets (prochordates) have a dorsal hollow nerve cord, but the anterior section shows no particular swelling that we can call a brain. Jawless fish, however, a group of organisms that arose from a lancelet-like predecessor possesses a recognizable brain and thus are craniates. All organisms in the jawless fish group and higher taxonomic groups can be considered craniates, including humans.

Craniometric analysis Measurement and comparison of skulls of animals from the same or different species.

Crepuscular Active during dusk and dawn.

Cyber-terrorist a programmer who breaks into computer systems in order to steal, change, or destroy information (also called cyberpunk and hacker).

Dead Zones Dead zones are hypoxic (low-oxygen) areas in the world's oceans, the observed incidences of which have been increasing since oceanographers began noting them in the 1970s. These occur near inhabited coastlines, where aquatic life is most concentrated. (The vast middle portions of the oceans, which naturally have little life, are not considered "dead zones.") The term can also be applied to the identical phenomenon in large lakes. Aquatic and marine dead zones can be caused by an increase in chemical nutrients (particularly nitrogen and phosphorus) in the water, known as eutrophication. These chemicals are the fundamental building blocks of single-celled, plant-like organisms that live in the water column, and whose growth is limited in part by the availability of these materials. Eutrophication can lead to rapid increases in the density of certain types of these phytoplankton, a phenomenon known as an algal bloom. Although these algae produce oxygen in the daytime via photosynthesis, during the night hours they continue to undergo cellular respiration and can therefore deplete the water column of available oxygen. In addition, when algal blooms die off, oxygen is used up further during bacterial decomposition of the dead algal cells. Both of these processes can result in a significant depletion of dissolved oxygen in the water, creating hypoxic conditions. Dead zones can be caused by natural and by anthropogenic factors. Use of chemical fertilizers is considered the major human-related cause of dead zones around the world. Natural causes include coastal upwelling and changes in wind and water circulation patterns. Runoff from sewage, urban land use, and fertilizers can also contribute to eutrophication.

Deception Deception (also referred to as beguilement, deceit, bluff, mystification, bad faith, and subterfuge) are behavioral acts that propagate beliefs that are not true or not the whole truth (as in half-truths or omission). It is a major relational transgression from 'normal' behavior which often leads to feelings of betrayal and distrust between relational partners. It violates relational rules and is considered to be a negative violation of expectations. Most people expect friends, relational partners, and even strangers to be truthful most of the time, resulting in their being deceived. While a significant amount of deception occurs between romantic and relational partners, it is found to be well engrained in the behavior of most humans and other animals, as pointed out by Robert Trivers in *The Folly of Fools* and in many reports by other authors. Deception is broadly practiced by people with antisocial behavior disorders.

Decisive dominance A clear settling of agonistic confrontations, in which there is an alpha and a beta individual.

Deference Deference (also referred to as subordinance, submission, or passivity) is the condition of submitting to a more dominant individual or group. It has been studied extensively by political scientists, sociologists, psychologists, and animal behaviorists. In social animal groups, a dominant (alpha) individual enters agonistic interactions with other individuals, resulting in their subordination of deference. Alpha individuals often obtain resources (e.g., mates, status, attention, food, sleeping, and resting places) that subordinates cannot have, but they may sometimes receive such resources through an equalitarian arrangement or deception. I have primarily used the terms domination and subordinance in the text.

Degree of tolerance In certain social insects (polistine wasps), I have found that there are varying degrees of tolerance expressed by dominant individuals toward their nest mates. In some cases, the results of agonistic behavior results in a linear dominance hierarchy in which individuals may be recognized as taking an alpha to omega position. Poorly expressed dominance interactions may result in a poorly defined dominance hierarchy. Some dominant individuals express a despotic dominance in which they tolerate no one on the nest. A similar arrangement can be found in many groups of social organisms, including humans. See Chapter 13, Dominance in Politics.

Deism The belief that reason and observation of the natural world are sufficient to determine the existence of god, a dominant individual who has made the world and its biota and guides the populace in the right direction.

Deletions A genetic condition in which there is a loss of a part of chromosome, as in the case of Jacobsen syndrome.

Despotic dominance A form of dominance in which the alpha individual has complete control over all subordinates. In some situations, despotic dominance may be viewed as an overdominant condition that may lead to a loss of control of subordinates.

Deuterostome Organisms in which the blastopore (an early opening in the embryo) develops into the anus. Echinoderms (stars, sand dollars, and urchins) and chordates are deuterostomes.

Diploblastic An organism or condition in which the body consists of two primordial germ layers, the ectoderm and endoderm. Jellies (both medusa and polyp stages) and other members of the phylum Cnidaria have a diploblastic body.

Diploid cells Cells (generally somatic) with a full complement of chromosomes as opposed to haploid or half the number of chromosomes (as in the germinal cells).

Dipnoi A group of organisms (like lungfish) that are tetrapodomorphs. They have both pulmonary and gilled respiration and lungs with alveoli.

Direct fitness A measure of personal fitness as expressed in an individual's genes. Individuals maximize their fitness in various ways by producing the maximum number of viable offspring.

Disaffiliated personality type A personality in which individuals develop antisocial traits and have an inability to relate emotionally to others.

Disempathetic personality type A personality in which individuals are capable of demonstrating affection and attachments to relatives, friends, or spouses, but they are prone to relate to others as objects.

Dispersal Movement of individuals of a species out of their range (range expansion).

Dispersal of humans out of Africa Early human dispersal began about two million years ago with its first movement out of Africa by *Homo erectus*. This was followed by the dispersal of other premodern humans including *Homo heidelbergensis*, the likely ancestor of both modern humans and Neanderthals. *Homo sapiens* ventured out of Africa around 200,000 years ago, spread across Asia from about 75,000 years ago, and arrived on new continents and islands at subsequent times.

Diurnal Active during daytime.

Division of labor A condition in a social group in which certain individuals have one functional role while others have other functional roles. For instance, a queen in social insects deposits the eggs while workers do other chores, such as constructing and defending the nest and caring for the young.

DNA Deoxyribonucleic acid, the molecule which acts as a template for the process of replication and protein synthesis. DNA consists of a long series of nucleotides. See *Nucleotide*. Chromosomes and DNA are not the same. Chromosomes are composed of both DNA and protein.

Domesticated pets Any animal that has been tamed and made fit for a human environment.

Dominance A condition in which one organism or a group of organisms dominates another individual or group. Mazur states that a person is said to "act dominantly if its apparent intent is to achieve or maintain high status" (as in the acquisition of power, influence, or valued prerogatives) over a member of the same species.

Dominance hierarchy A ranking (peck order) of individuals in a social group, generally through agonistic behavior. Dominance hierarchies may be decisive when dominant or surbordinate rank is clearly evident, for example, in linear dominance, in which there is a clear ranking from the most dominant (alpha) individual to the most subordinate (omega) individual or when there is a despotic dominant who is intolerant toward other individuals who strive for the alpha position. In nondespotic decisive dominance, there is a well-defined alpha who coordinates societal functions, and the alpha tolerates other societal members who rank lower on the hierarchical scale. In nondespotic dominance, dominance can also be indecisive when members of the social group cannot display a clear hierarchy, in which the alpha individual totally dominates the group and a linear ranking beneath the alpha individual is unclear. This latter condition has been termed triangular dominance by some investigators (Ehrlich and Ehrlich, 2008; Hayaki, 1983; Houpt et al., 1978; MacLachlan et al., 2010; Scheel et al., 1977).

Dominant personality The term "dominant personality type" is a descriptive phrase "used for people who generally like to take control of a situation or, in a [more aggressive] way, other people. An individual with a dominant personality may have several characteristics common among naturally dominant people." For example, "he or she may enjoy controlling or bossing others around, and may get angry if expected to obey others, or he or she might be laid back in relationships with other people but be extremely task-oriented and focused on achieving goals."

"Some individuals with a dominant personality will be more likely than others to take it beyond a simple issue of control and into an abusive situation. Conversely, some naturally dominant individuals have no desire to hurt or control others' lives, but simply enjoy being able to take charge and get things done if the need arises." (Definition is from WiseGeek, http://www.wisegeek.com/what-does-it-mean-to-have-a-dominant-personality.htm.)

It is a fairly common belief that men are more likely than women to have a dominant personality. To a certain degree, some people believe it is more acceptable in society for a man to have a dominant personality than if a woman acted in the same way. For example, in some relationships, the man expects to control the finances and the major decisions simply because he is the male in the relationship. If the roles were reversed, some people would find that odd, and assume that the man must be unhappy in that situation, having to 'obey' a woman. In reality, both men and women can have a dominant personality, and it is no healthier or less healthy in one gender than the other.

Some of these misconceptions may stem from a confusion between a 'dominant' person and someone who takes it too far or acts aggressively toward people to whom he or she feels superior, sometimes referred to as a 'domineering' person. Where a dominant personality would be very useful, for example, in a business situation where a charismatic, influential person can successfully lead others, someone who is domineering, or dominant to an unhealthy extent, might try to micromanage others and be unwilling to listen to the opinions of coworkers. Dominant people are extremely important to the way businesses, relationships, and even whole societies work as long as they do not become domineering.

At the opposite end of the spectrum from a dominant personality would be a submissive personality. A submissive person is more likely to enjoy being controlled and taken care of than he or she would be to take charge. For this reason, relationships often form between people with opposing types, where one partner is dominant and the other submissive. This kind of relationship has the potential to be unhealthy if taken to an abusive extent, but it also has the potential to be a happy relationship for both partners because they are each in the role they find most comfortable.

Duplications A genetic condition in which there is a multiplication of a portion of a chromosome that results in extra genetic material.

Ecology The scientific analysis and study of interactions among organisms and their environment, such as the interactions organisms have with each other (biotic components) and their abiotic environment.

Ecological footprint An ecological footprint expresses the environmental impact of an individual or a population in terms of the cumulative amount of land and water required to provide the raw materials consumed and to recycle the waste produced.

Egalitarian A relationship between social individuals in which neither consistently wins in agonistic confrontations. Egalitarian societies are made up of people who share equal status. They are primarily found among mobile bands of hunter–gatherers, like the !King societies of South Africa, the Ainu society of Japan, and the Inuit of northwestern North America. The maintenance of an egalitarian system may be due to:

- Nomadic life that prevents an accumulation of material possessions.
- Foraging for food which requires a large adequate amount of land for community members to use collectively (there is no long ownership).

Egocentricity A concern for one's self over others in a society.

Emotions various body sensations that respond to certain behavioral traits, such as mood, temperament, personality, disposition, and motivation.

Emotional Responses There are three basic components to emotional responses to various situations: (1) behavioral, involving situationally appropriate muscle movements; (2) autonomic, which facilitates the behaviors and energy stores that are required to carry them out; and (3) hormonal, involving chemicals that reinforce autonomic responses.

Empathy The capacity for understanding another's feelings and having compassion for them.

Empty love Love in which a stronger love deteriorates, the commitment remaining but the intimacy and passion have died.

Entropy The degradation of matter and energy in the universe to an ultimate state of inert uniformity.

Environment Withgott and Brennan define environment as "the sum total of our surroundings. It includes both living and nonliving things. The fundamental insight of environmental science is that we are part of the natural world and we are dependent on a healthy, functioning planet."

Environmental carrying capacity The maximum number of individuals that an environment can support. Environmental carrying capacity is influenced by factors such as food abundance, predation, and parasitism.

Epigenetic The study of heritable changes in gene expression or cellular phenotype caused by mechanisms other than changes in the underlying DNA sequence (nucleotides). Examples are DNA modifications and the process of methylation with regards to chromosome-related proteins called histones, both of which serve to regulate gene expression without altering the underlying DNA sequence. In cellular differentiation during morphogenesis in eukaryotic cells, totipotent stem cells (cells that can develop into more differentiated cells) develop into pluripotent cell lines (cells that can develop into all differentiated cells other than extraembryonic ones) of the embryo, which in turn become fully differentiated cells.

Epigenetic variation In genetics, variation in the chemical tags that attach to DNA and affect how genes get read.

Ethics The formal study of moral standards and conduct.

Eugenics The science of improving a human population by controlled breeding to increase the occurrence of desirable heritable characteristics. Developed as a method of improving the human race, it fell into disfavor only after the perversion of its doctrines by the Nazis.

Eukaryotes Organisms which have cells that house a well-defined nucleus. Eukaryotes are more advanced anatomically than prokaryotes, which are organisms that do not have a membrane-enclosed nucleus. All organisms that are not bacteria or archaeans are eukaryotes. See *Prokaryotes*.

Eusocial A condition within a group of organisms that has three prime expressions of colony cohesion, namely care of the young, an overlap of generations, and a division of labor. Organisms that possess eusocial status are ants, honeybees, bumblebees, certain other bees, paper wasps, one other species of wasps, termites, and naked mole rats. There are many animals that possess one or two of these characteristics, and they are generally given some status such as presocial, parasocial, subsocial, and so on. Animals that possess none of these qualities are considered nonsocial.

Eutrophication Eutrophication (more precisely hypertrophication) is an ecosystem's response to the addition of artificial or natural substances (e.g., nitrates and phosphates), generally through fertilizers or sewage, to an aquatic system. One example is the "bloom" (a sudden increase of phytoplankton in a water body as a response to increased levels of nutrients). Negative environmental effects include hypoxia (the depletion of oxygen in the water), which induces reductions in aquatic species. Certain species (such as Nomura's jellyfish in Japanese waters) have been found to experience an increase in population that negatively affects other species.

Evil leadership Leadership in which the leader and at least some followers commit physical, psychological, or both atrocities.

Evolution Few nonbiologists really understand evolution. Simply stated, it is a process whereby organisms with characteristics that are beneficial to a species are selected to be carried along in the gene pool of that species. The process of evolution is dependent upon a number of abiotic and biotic factors: (1) we first must recognize that changes occur in the world, including climatic changes, continental drift (including the separation and fusion of land masses), mountain formation, earthquakes, volcanic activities, global warming, global cooling, and polar shifts. Fauna and flora exist in populations that respond to these changes. We must also recognize that all populations demonstrate variability. Variability may be expressed in various forms: physiological, behavioral, anatomical, and genetic. Such variability results from mutations within the populations. When changes occur, they put stress on populations, causing them to change. It is the populations rather than the individuals that change. Natural selection is a process whereby those individuals within a population that are the fittest to survive under the influence of these changes will survive and be the most likely to contribute their genes to succeeding generations of that species. While the mechanisms within populations and their reactions to environmental changes can be very complex, the principle of evolution is rather simple and a beautiful explanation of what has occurred throughout the billions of years that life has evolved on our planet.

Exons Exons are sections of DNA (genes) which are important in an organism's genetics. They are separated from one another by other nongenetic sections called introns and represented in the mature form of an RNA molecule either after portions of a precursor RNA (introns) have been removed by *cis*-splicing or when two or more precursor RNA molecules have been ligated by *trans*-splicing.

Extant Living at the present, as opposed to extinct in which the organism no longer exists.

Extroversion Also spelled extraversion, this trait is "a tendency to direct interests and energies toward the outer world of people and things rather than the inner world of subjective experience."

Fatuous love Love exemplified by a whirlwind courtship and marriage in which a commitment is motivated largely by passion, without the stabilizing influence of intimacy.

F-box genes According to X. Yang et al., F-box genes influence a variety of biological processes, such as leaf senescence, branching, self-incompatibility, and responses to biotic and abiotic stresses.

Female philopatry A phenomenon in which females remain in their groups or home ranges that they were born in while males leave the groups at sexual maturity. The females of the group are thus related, such as mothers, daughters, sisters, aunts, or cousins, while males entering the group are unrelated.

Fission-fusion groups A pattern of social grouping in which individuals of a group form stable temporary subgroups. Movement between subgroups occurs which frequently changes the size and composition of the main group.

Fitness A term signifying the condition an individual is in which affects its possibility for being successful in contributing genes to subsequent populations. The most highly fit individuals have the greatest potential for passing their genes to successive populations.

Formative years The first few years of postnatal life.

Fragmentation of habitats The breaking of areas of land into smaller fragments, decreasing the available habitat of organisms that need larger territories for survival and reproduction and increasing the amount of "edge" habitat.

Frontal cortex This section of the brain is the center of personality and emotion, and the nervous tissue within it performs tasks that involve logic and reasoning.

Frugivore An animal that feeds primarily on leaves.

Game A form of play or sport, especially a competitive one played according to rules and decided by skill, strength, or luck; a type of activity or business, especially when regarded as a game. In the context of this book, the word "game" pertains to the way humans and other animals compete through agonistic confrontations to establish a dominance hierarchy and how humans have dominated the biosphere and its resources.

Gametes Gametes are cells found in the testes and ovaries. When fully developed, they are represented by the spermatozoa in the testes and ova in the ovaries. Each gamete contains one set of chromosomes and is haploid (it has half the number of chromosomes of a somatic cell). When gametes fuse, they form a zygote that is the original stem cell from which an embryo develops.

Gamma radiation "Energy release which is similar to X-rays."

Ganglion An enlarged structure composed of nervous tissue. Ganglia are often found in invertebrates, numerous ones arranged intersegmentally. The brain can be considered as a tremendously large ganglionic mass.

Gastrula An embryonic stage in which the blastula (blastocyst) has invaginated (caved in) on one side, forming an internal chamber (called an archenteron) and an opening (called a blastopore) in one side. The blastopore later develops either into a mouth (as in protostomes) or an anus (as in deuterostomes). The gastrula stage is diploblastic (composed of two primordial germ cell layers, the ectoderm and endoderm). However, mesoderm (the third primordial germ tissue) develops later in most organisms, and the condition with three germ layers is referred to as triploblastic.

Genito-genital rubbing The rubbing together of genitals, as seen in female bonobos.

Gene flow The movement of genes through a population. While genes (and their characteristics) do flow through a population, the process is rather slow, resulting in variations from one end of the population to the other (a cline).

Genetic diversity Variation within the genetic makeup of a species which provides a means for the population to adjust to changing environmental conditions.

Genetic variability A measure of the tendency of individual genotypes in a population to vary from one another.

Geological time scale Periods of time, sequenced in phylogenetic order (arranged in a sequence from earliest to present and representing the most primitive to most advanced groups). The most recent geologic eras have been: the Paleozoic in which the most dominant terrestrial biota were mosses, ferns, and amphibians; the Mesozoic in which the most dominant terrestrial biota were gymnosperms and reptiles; and the Cenozoic in which the most dominant terrestrial biota are angiosperms, mammals, birds, and arthropods. While the Mesozoic era was dominated by reptiles, mammals were present, and they diversified in the Cenozoic, following a mass extinction of reptilian forms.

Germline chimerism A form of chimerism that occurs when the germ cells (e.g., sperm and egg cells) of an organism are not genetically identical to its own. As an example, it has recently been discovered that marmosets (mammals that almost always give birth to fraternal twins) can carry the reproductive cells of their (fraternal) twin siblings because of placental fusion during development (Hooper, 2007; Ross et al., 2007).

Gestation The period of time in which an embryo is developing (from fertilization to birth) in its female parent.

Gingival smile Broadly exposing teeth and gums.

Grandiosity An exaggerated condition that a person feels they are special and important.

Great apes Members of the family Hominidae, often referred to simply as hominids. Hominidae is a taxonomic family of primates that includes seven extant species in four genera: *Pongo*, the Bornean and Sumatran orangutan; *Gorilla*, the eastern and western gorilla; *Pan*, the chimpanzee and bonobo; and *Homo*, the human.

Group endogamy The practice of marrying within a specific ethnic group, class, or social group.

Hate speech Speech perceived to disparage a person or group of people based on their social or ethnic group, for example, race, sex, age, ethnicity, nationality, religion, sexual orientation, gender, identity, disability, language ability, ideology, social class, occupation, appearance, mental capacity, and any other distinction that might be considered by some as a liability.

Haploidy A process in which organisms reproduce parthenogenetically from haploid cells, such as in male ants, bees, and wasps. In these insects, the process of reproducing offspring is often referred to as a haploid–diploid or haplodiploid process. Unfertilized eggs develop into males, and fertilized eggs develop into females.

Haplogroup In its simplest terms, a haplotype is a specific group of genes or alleles that progeny inherited from one parent. Similar haplotypes (haploid genotypes) share a common ancestor having the same single nucleotide polymorphism (SNP) mutation in all haplotypes. They pertain to deep ancestral origins dating back thousands of years. In human genetics, the haplogroups most commonly studied are Y-chromosome (Y-DNA) haplogroups and mitochondrial DNA (mtDNA) haplogroups, both of which can be used to define genetic populations. Y-DNA is passed solely along the patrilineal line, from father to son, while mtDNA is passed down the matrilineal line, from mother to offspring of both sexes. Neither recombines, and thus Y-DNA and mtDNA change only by chance mutation at each generation with no intermixture between parents' genetic material.

Haplotype The word haplotype (haploid genotype) has been defined in several ways: (1) a specific group of genes or alleles that are inherited by progeny from one of its parents; (2) a set of single-nucleotide polymorphisms (SNPs) on a single chromosome that have a tendency to always occur together; and (3) a collection of specific mutations within a given genetic segment. The term "haplogroup" refers to the single-nucleotide polymorphisms/unique-event polymorphism (UEP) mutations that represent the clade to which a collection of particular human haplotypes belong. See *Haplogroup*.

Hatred A deep, enduring, intense, and emotional dislike toward individuals, groups, entities, objects, or ideas.

Heroin An opiate drug (Meade, 1999).

Hippocampus This section of the brain is at the core of the temporal lobes. It controls the more primitive pathways of pleasure and aversion. Some long-term memory is stored in this section of the brain as well.

Histones Highly alkaline proteins found in eukaryotic cell nuclei which package and order the DNA into structural units called nucleosomes. Histones represent the primary type of protein components of chromatin.

Homeobox genes A homeobox is a DNA sequence found within genes (also referred to as hox genes) that are involved in the regulation of patterns of anatomical development (morphogenesis) in animals, fungi, and plants. They were discovered in *Drosophila*. See *F-box genes*; *Hox genes*; *T-box genes*.

Hominan A member of the Hominidae, Homininae, Homini, and the subtribe Hominina, which includes modern humans and their extinct relatives.

Hominids The great apes, including humans but not including gibbons. Hominids belong to the family Hominidae. It includes the following extant genera: *Homo* (humans), *Pan* (chimpanzees and bonobos), *Gorilla* (gorillas), and *Pongo* (orangutans).

Hominines Also spelled hominins, they are humans and humanlike apes that belong to the family Hominidae and subfamily Homininae, including all hominids that arose after the split from orangutans (subfamily Ponginae). Thus, it includes such hominids as gorillas (tribe Gorillini), chimpanzees, bonobos, and humans (the latter three belonging to the tribe Hominini).

Hominoid Primates that are commonly referred to as apes. They are members of the superfamily Hominoidea. Extant members are the great apes (family Hominidae) and lesser apes (gibbons) (family Hylobidae).

Homologous A condition in which two structures have the same origin. As an example, the forelegs of a dog and forearms of a human are said to be homologous. So are the forelegs of a lizard and the wings of a bird. The forearms of humans are homologous to bird wings, bat wings, and the forelegs of all mammals. However, they are not homologous to the wings of insects, which have had an unrelated origin. Structures that are homologous may or may not be analogous (same function). See *Analogous*.

Honor killing A homicide of a member of a family or social group by other members, commonly as the result of a belief by the perpetrators that the victim has brought dishonor upon the family or community.

Hostile personality type A personality in which individuals are angry, resentful, and aggressive; purposefully reject the social norms and mores of society; and display antisocial and traditional psychopathic traits as a result of their hostile beliefs.

Hox genes Hox genes are a group of related genes that determine the basic structure and orientation of an animal. Hox genes are defined as having the following properties: (1) they contain within them a DNA sequence known as the homeobox; (2) most of them are linked together in a sequential cluster in the chromosome; and (3) they are organized on the sequential cluster in the order of their expression pattern along the anterior–posterior axis of the organism.

Human A primate that belongs to the superfamily Hominoidea, family Hominidae, subfamily Homininae, tribe Hominini, subtribe Hominina, genus *Homo*, and species *sapiens*. Its scientific name is thus *Homo sapiens*, the single species of humans living on Earth today. Humans are chordates, mammals, primates, and simians, related most closely to chimpanzees and bonobos. There have been a long line of hominines that are represented in the fossil records.

Human evolutionary genetics The study of how one human genome differs from another human genome, the evolutionary past that gave rise to it, and its current effects.

Hypothalamus This structure in the brain is the primary center for emotion, sex, anger, temperature control, hormone release, eating, drinking, sleep, and pleasure pathways.

Immorality The active opposite of morality.

Imprinting An innate response by an offspring in which it recognizes its parent. Such recognition generally occurs during a limited critical period.

Inbreeding depression A condition in which individuals within a population demonstrate a reduction in overall health due to inbreeding over multiple generations.

Inclusive fitness A form of fitness which is the result of aiding the survival and reproduction of nondescendant relatives (e.g., siblings).

Incompetent leadership Leadership in which the leader and at least some followers lack the will or skill (or both) to sustain effective action.

Indecisive dominance A poorly defined or unclear settling of agonistic confrontations, in which there is no clear alpha and beta individual.

Industrial revolution A revolution in human behavior, demarking changes in the way that humans processed their natural resources. The industrial revolution began in the mid-1700s, shifting from a rural, agricultural life to an urban society powered by fossil fuels, which are nonrenewable energy sources such as oil, coal, and natural gas.

Infatuated love Love at first sight, without the intimacy and the commitment components of love.

Inherent/erudite aggression Aggression that is expressed by subhuman animals that dominate others, and in some human alternative behaviors is aggression generally from within the aggressor (a genetic propensity) (Buss, 1961), although certain environmental stimuli may provide a base that provokes or influences inner aggression.

Innate A feature of any organism which pertains to qualities that are inborn, as stemming from the organism's genetics, rather than features that are derived from learning.

Inspiration An inspiring or animating action or influence; something inspired, as an idea; a result of inspired activity; or a thing or person that inspires.

Insular leadership Leadership in which the leader and at least some followers minimize or disregard the health and welfare of those outside the group or organization for which they are directly responsible.

Intemperate leadership Leadership in which the leader lacks self-control and is sided and abetted by followers who are unwilling or unable to effectively intervene.

Intimidation To make timid or fill with fear.

Intrauterine device See *IUD*.

Introns Sections of DNA which are not part of an organism's genetic makeup. Introns separate exons that are segments of an organism's genetic makeup. Introns, pseudogenes, and junk DNA are all nongenetic parts of a DNA molecule.

Invasive species Species that are introduced to an area that they did not formerly occupy. Withgott and Brennen mention that "Human intervention may bring in a very large number of invasive species to native communities at a very fast rate, faster than the native species can evolve to adapt to them." When this happens, the alien species can have a huge impact on the entire food web of a community because it may decrease food resources for native organisms. As a consequence, exotic (introduced invasive) species can cause population reduction and even the extinction of native species that occupy a niche similar to that of the invader.

Inversions A genetic condition in which the nucleotide sequence is altered because a portion of a chromosome has broken off, got inverted, and became reattached at the original location of the chromosome.

Islam One of the three major religions, founded in 622 BC.

Isotopes Chemical elements that are unstable, the nucleus decaying spontaneously to form other elements. Certain isotopes are used in radiometric dating of fossils and rocks.

IUD Intrauterine device, a small contraceptive device that is often "T"-shaped and contains either copper or levonorgestrel, which is inserted into the uterus. An IUD represents one form of long-acting reversible contraception which are the most effective types of reversible birth control.

Jawless fish A group of fish that lack jaws, have a cartilaginous endoskeleton, and feed as scavengers on the bottom of their aquatic habitats. Some (like lampreys) have developed an ectoparasitic way of life. It is within this group (superclass Agnatha) that vertebrates arose. Certain extant jawless fish (hagfish), for instance, are considered invertebrates because they do not possess vertebrae, but others (e.g., lampreys) are vertebrates. Jawless fish are considered very primitive chordates, have arisen from a lancelet-like (prochordate) ancestor. They gave rise to the group known as cartilaginous fish.

Joule One joule is defined as the amount of work done by a force of 1 N moving an object through a distance of 1 m.

Judaism The oldest formal religion known, possibly founded around the 1st century BC.

Junk DNA DNA within a cell's nucleus which does not seem to have any connection to an organism's genetic makeup. It has been referred to as selfish DNA by biologists, for example, Richard Dawkins.

Kin selection The selection of individuals that increase the fitness of related individuals rather than themselves.

Kinship Relatedness within a group of organisms.

Knuckle walking Walking on all four limbs, with support of the body resting at times on the knuckles of the forelimbs, as in the locomotion of orangutans.

Lancelet A primitive chordate or prochordate organism that has chordate features in its immature and adult stages. It is more advanced than tunicates, which have chordate features in its immature stage but not in its adult stage. It is at the base of the chordate line, which includes all chordates from jawless fish to humans, but its brain is poorly developed or nonexistent. Thus, it is not a craniate. See *Craniate*.

Lesser ape Gibbons.

Linear dominance hierarchy A dominance hierarchy in which there is a linear ranking of individuals from most dominant (alpha) to most subordinate (omega).

Lobe fins Lobe fins are fins in certain bony fish (like coelacanths, mostly 370–510 million years ago) that sometimes functioned in facilitating movements along the bottom of aquatic habitats. They are found in many primitive types of bony fish and in certain contemporary fish, for example, lungfish. Certain lobe fins developed into more complex limbs that had homologous features in the basal bones (humerus, radius, and ulna) of true tetrapods, and thus they gave rise to the tetrapodomorph fish, some of which subsequently gave rise to the tetrapods.

Lungfish A group of extant fish, generally found in South America, Africa, and Australia. They are considered tetrapodomorph fish because they have limbs that allow them to walk in their aquatic environment and sometimes on land. They are not connected to the tetrapodomorphs that led to tetrapods 370–380 million years ago. They are not considered tetrapods because they have not developed the chiridian limb (limb with digits) which is characteristic of tetrapods. Some unusual features of lungfish are the possession of paired lungs that have alveoli as well as gills. Because of their respiratory features, they belong to a group referred to as the dipnoi. Members of the dipnoi have some social behavior.

Male philopatry A social phenomenon in which males remain in the groups in which they were born while females leave the group when they become sexually mature. The males of such a group are thus related while females enter the group from the outside as nonrelatives.

Mate guarding A behavior in which an individual remains near its mate and prevents it from copulating with someone else.

Matrilineal A phenomenon in which a lineage is derived through the mother rather than the father. Social groups with matrilineal components revolves around female kinship.

Matrilineal society Matrilineal societies are societies in which descent is traced through mothers rather than through fathers. In matrilineal societies, property is often passed from mothers to daughters and the custom of matrilocal residence may be practiced. A matrilocal residence refers to a societal system in which a married couple resides with or near the wife's parents.

Matrilines A line of descent from a female ancestor to a descendant (of either sex) in which the individuals in all intervening generations are mothers.

Meritocratic A ranking of an individual based on their merit or success rather than on the class that they initially belonged to. Social hierarchies within meritocratic populations are fluid, and meritocratic principles result in social aspiration.

Metapopulation A set of discrete populations of the conspecific individuals, in the same general geographic area, that may exchange individuals through migration, dispersal, or human-mediated movement.

Minnesota multiphasic personality inventory A personality inventory, which is one of the most widely used self-report tools for assessing personality. According to the APA Dictionary of Psychology, it has broad applications across a range of mental health, medical, substance abuse, forensic, and personnel screening settings as a measure of psychological maladjustment.

Misanthropy A form of hatred, a general intense dislike, mistrust, or disdain of human nature.

Molecular variance Molecular variation within a single species. Analysis of molecular variance (AMOVA) is a statistical model for the molecular variation in a species, which was developed by Laurent Excoffier, Peter Smouse, and Joseph Quattro at Rutgers University in 1992.

Monogamous A relationship in which one male and one female mate exclusively with one another.

Monotheism The belief in the existence of a single God or in the oneness of God (McGrath, 2003). The various forms of monotheism, all of which have significant parallels in polytheistic religions, are generally classified into three major religions: Judaism, Christianity, and Islam.

Moral absolution A form of morality in which certain acts may be right or wrong, regardless of their context.

Moral code A system of morality, which is established with reference to a particular philosophy, religion, or culture.

Morality R. Audi defines morality as "an informal public system applying to all rational persons, governing behavior that affects others, having the lessening of evil or harm as its goal, and including what are commonly known as the moral rules, moral ideals, and moral virtues." It generally includes such features as empathy, reciprocity, altruism, cooperation, and a sense of fairness. Audi further states that "although morality must include the commonly accepted moral rules such as those prohibiting killing and deceiving, different societies can interpret these rules somewhat differently."

Morula An embryonic stage, following the zygote stage, in which the embryo is represented by a solid ball of cells, all of which are undifferentiated. The morula stage leads to the blastula (blastocyst) stage, in which all internal cells move to the periphery of the ball.

Moral realism A form of morality that is believed to exist outside of human opinion.

Moral relativism A form of morality in which the codes are thought to be a function of human values and social phenomenon.

Moral universalism A form of morality that is a compromise between moral absolutism and moral relativism, favoring the thought that there is a common universal core of morality.

Multifactorial genetic disorders Disorders that are the result of genetic as well as environmental factors (Bodmer and Bonilla, 2008).

Multigenerational group A Social group that consists of individuals with widely varying ages.

Mutations Sudden inheritable changes in the nucleotide sequence of a gene.

Naked mole rats The only mammals considered to be eusocial, naked mole rats (*Heterocephalus glaber*) lives in underground colonies consisting of a queen, several males referred to as kings, and numerous nonreproductive individuals. As in most social groups, members of a naked mole rat colony are closely related.

Narcissistic trait Love of one's own body.

Natal group The group into which an individual is born. Also sometimes referred to as natal community or natal territory.

Natural resources Withgott and Brennan define natural resources as "the various substances and forces we need in order to survive." Renewable natural resources are perpetually available or can be replenished by the environment over short periods of time. Nonrenewable natural resources are in finite supply and are not replenished or are formed much more slowly than we use them. Some renewable resources may turn nonrenewable if we deplete them too drastically.

Natural selection A natural process of nature in which those individuals of a population most suited to environmental changes are selected to contribute to succeeding generations of that species.

Nature The physical world collectively, including plants, animals, the landscape, and other features and products of the Earth, as opposed to humans or human creations; the basic or inherent features of something, especially when seen as characteristic of it.

Neocortex The neocortex is the outer layer of the cerebral hemispheres in mammals, and made up of six layers. Also referred to as the neopallium and isocortex, it is part of the cerebral cortex which belongs to the limbic system. In humans, it is involved in functions such as sensory perception, generation of motor commands, spatial reasoning, conscious thought, and language.

Neotenic A condition in which an organism is reproductively active but anatomically immature. Certain salamanders are reproductively active but have retained their gills, which is an immature feature.

Neurosis According to a Wikipedia article on this subject, neurosis is a type of functional mental disorder involving distress but neither delusions nor hallucinations. Neurotic behavior is typically within socially acceptable limits. Neurosis may also be called psychoneurosis or neurotic disorder.

Neuroticism A pattern of traits and tendencies that increase susceptibility to neurosis of one type or another.

New World monkeys Monkeys naturally living in Central and South America.

Nitrogenous base A chemical that is part of a nucleotide in DNA and RNA. There are four types of nitrogenous bases in DNA and RNA. DNA has adenine, thymine, cytosine, and guanine. RNA has adenine, uracil, cytosine, and guanine. Nitrogenous bases pair up in both nucleic acids in the following fashion: In DNA, adenine and thymine pair up and cytosine and guanine pair up. In RNA, adenine pairs up with uracil, and cytosine pairs up with guanine.

Nonprehensile tail A tail that is not able to grasp objects. Old World monkeys have a nonprehensile tail.

Nucleotide A chemical composed of a phosphate group, pentose sugar, and a nitrogenous base. They represent the repeating units of DNA and RNA. ATP is also a nucleotide.

Nurture Nurture is defined as the sum of environmental factors that influence the behavior and traits expressed by an organism. Many behavioral features have connections to both nature (genetic) and nurture (learning).

Old World monkeys Monkeys native to Asia or Africa.

Olfactory A term relating to smell.

Omega individual The most subordinate (least dominant) in a linear dominance hierarchy.

Omnivore An animal that feeds upon a variety of foods, including both animal and plant material.

Ontological Embryonic.

Oophagy Feeding upon eggs.

Open mindedness A readiness to consider different points of view.

Opportunistic individual An individual that is able to exploit newly available habitats and resources.

Opposable thumb A thumb that is positioned on the hand or foot so that it can make contact with the other digits on that limb, allowing the animal to grasp objects. Most primates also have opposable big toes.

Opposition A term used by primatologists to describe the movement of the thumb (especially well developed in humans), in which the pulp surface of the thumb is placed squarely in contact with–or diametrically opposite to—the terminal pads of one or all of the remaining digits.

Organelles Living structures within cells, for example, the nucleus, Golgi apparatus, mitochondria, plastids, plasma and nuclear membranes, and endoplasmic reticulum. Eukaryotes generally have a number of organelles in their cells. See *Eukaryotes*.

Orthologous Homologous sequences are said to be orthologous if the individuals that possess them have undergone speciation. When a species diverges into two separate species, the copies of a single gene in the two resulting species are said to be orthologous.

Ovarian cycle The cycle involving the development and rupture of oocytes within the ovary and accompanying uterine development and breakdown.

Overlap of generations A condition within a social group in which at least one parent lives long enough to live with the next generation.

Ovulation The formation and discharge of mature ova from the ovary in mammalian reproductive systems.

Ovum The female gamete, produced in the female germinal tissues of eukaryotic organisms. Spermatozoa, which are produced in male germinal tissues, unite with ova to produce zygotes. A zygote then divides mitotically to form the embryo of whatever species is developing.

Pantheism The belief that everything composes an all-encompassing, immanent god and that nature and the universe are identical with divinity (Picton, 1905).

Parable A short allegorical story, designed to illustrate or teach some truth, religious principle, or moral lesson.

Parturition The process of giving birth.

Patriline A social grouping revolving around male kinship or derivation of lineage through the father instead of the mother.

Pecking order A linear dominance hierarchy, with individuals ranked from alpha to omega.

Perfectionism In psychology, perfectionism is a personality disposition that is characterized by an individual who is striving for flawlessness and setting excessively high performance standards, accompanied by overly critical self-evaluations and concerns regarding evaluations by others. Perfectionism is best conceptualized as a multidimensional characteristic, consisting of both negative and positive aspects. In a maladaptive form, it may drive individuals to achieve an unobtainable goal, resulting in psychological distress if they fail to meet their expectations. Adaptive perfectionism is expressed in people who reach their goals, and thus they derive pleasure from doing so.

Personality A description of a person, involving both innate and learned expressions. Personality is affected by temperament (innate), as well as learned behavior throughout life, but particularly in the formative years, during adolescence and all traumatic experiences.

Perspective The state of one's ideas, the facts known to one, in establishing a meaningful interrelationship.

Pessary A small soluble block that is inserted into the vagina to treat infection or as a contraceptive.

Phylogenetic Basically, this term refers to evolution, but instead of being confined to the process of evolution, it is more inclusive in involving the history of evolution.

Physiology A study of the organs, tissues, and cells of an organism.

Piloerection The erection of hairs on a mammal when reacting to a stimulus.

Pituitary This brain structure secretes somatotrophic (growth) hormones and controls the action of certain other glands, for example, the thyroid gland, adrenal glands, ovaries, and testes. It is generally under the influence of the hypothalamus.

Pleistocene megafauna Numerous species of large animals that lived on Earth during the geologic period referred to as the Pleistocene epoch. These animals subsequently became extinct during the Quaternary extinction event. They appear to have died off as humans dispersed out of Africa and southern Asia, indicating that humans may have had a hand in their extinction. Africa and southern Asia remain as the only two continents that still support diverse populations of megafauna comparable to what was lost throughout the rest of the world. The Americas, northern Eurasia, Australia, and many larger islands lost the vast majority of their larger and all of their largest mammals.

Plesiomorphic This is a cladistic term, referring to a character state that is present in both out-groups and in the ancestors. It is used to refer to primitive features in more modern groups.

Pluripotency The ability of a stem cell to differentiate into any of the three germ layers: endoderm (interior stomach lining, gastrointestinal tract, and the lungs), mesoderm (muscle, bone, blood, and urogenital), or ectoderm (epidermal tissues and nervous system) but not to extraembryonic tissue, such as the placenta.

Polyandry A reproductive term that centers on a female that copulates with multiple males.

Polydactyly The condition in which a chiridian limb (limb with digits) has more than five digits. Polydactyly appeared in the Devonian, Triassic, and Jurassic periods. *Acanthostega* (an early amphibian), for instance, had an eight-digit hand and a seven-digit foot. Some tetrapodomorph fish possessed polydactylous limbs, and the condition sometimes shows up as a plesiomorphic feature in modern tetrapods. *Tulerpeton* was a polydactylous tetrapod with six digits.

Polygamous group A social group in which males and females may have more than one mate.

Polygynous group A group of animals in which one male mates with multiple females.

Polytheism The worship or belief in multiple deities (portrayed as gods and goddesses), along with their religions and rituals.

Population A group of organisms belonging to a single species.

Population bomb This term was coined by Paul Ehrlich (1968) who predicted that a rapidly increasing human population would bring widespread famine and conflict. He claimed that population control was the only way to avoid starvation and war. Many who support his ideas predict a global food crisis in the near future.

Population growth Most scientists who study the growth of the human population feel that it is out of control, and many feel we are reproducing beyond our environmental carrying capacity. Thomas Malthus (1798) claimed that unless population growth was controlled, the number of people would outgrow the food supply. He argued that a growing population would eventually be checked by starvation, disease, or war. All three possibilities complement one another and are features of all organismic existence.

Precraniate chordates At the beginning of chordate evolution, cephalochordates possessed a dorsal hollow nerve cord, but there was no enlargement that could be referred to as a brain. As these early organisms evolved to higher forms, the brain developed at the anterior end of the nerve cord, and the individuals with a noticeable brain, plus all organisms that came about later, are referred to as craniates. All organisms beyond the cephalochordates are craniates, including humans.

Prehensile tail A tail that is able to grasp objects. Among the primates, they are found in the New World monkeys.

Premonition A feeling of anticipation of or anxiety over a future event; presentiment; a forewarning.

Precocial A condition, in which a young bird or other animal hatches or is born in an advanced state and able to move around and feed itself almost immediately. Also called nidifugous. Often contrasted with the word "altricial"; the term is sometimes applied to a particular species that has precocial young.

Primate A mammal in the order Primates. The group includes prosimians and simians. Prosimians are ancient members of the group, (including such forms as lemurs, lorisoids, adapiformes, tarsiers, and their extinct relatives). Extant simians include Old World monkeys, New World monkeys, and apes. Differences between these two primate groups reveal an evolutionary trend toward a reduced snout. The group referred to as apes include lesser apes (represented by gibbons) and great apes (represented by orangutans, gorillas, chimpanzees, bonobos, and humans). All primates have binocular vision, in which the eyes are on the front of the skull, allowing accurate distance perception. They generally have five digits on each limb (pentadactyly), with opposable thumbs and keratin nails at the end of each finger and toe (although there are exceptions, e.g., a claw found in certain prosimians and the loss of thumbs in gibbon apes). The primate collar bone is a prominent element of the pectoral girdle, allowing the shoulder joint extensive mobility. Bones in the arm and forearm have also been modified for greater movement that is found in most mammals. They range in size from Madame Berthe's mouse lemur, weighing only 30 g (1 oz), to the eastern lowland gorilla, weighing over 200 kg (440 lb). The group arose from ancestors that lived in the trees of tropical forests. While fossil evidence indicates that the earliest known true primates date to 55.8 million years ago, other studies suggest that the primate branch may have originated near the Cretaceous–Paleogene boundary (around 63–74 mya).

Projective personality technique A technique in which a personality test is designed to let a person respond to ambiguous stimuli, presumably revealing hidden emotions and internal conflicts projected by the person into the test. This is sometimes contrasted with a so-called "objective test" or "self-report test" in which responses are analyzed according to a presumed universal standard (for example, a multiple choice exam), and are limited to the content of the test. The responses to projective tests are content-analyzed for meaning rather than being based on presuppositions about meaning, as is the case with objective tests. Projective tests have their origins in psychoanalytic psychology, which argues that humans have conscious and unconscious attitudes and motivations that are beyond or hidden from conscious awareness.

Prokaryotes Organisms, for example, bacteria and archaeans that do not have a membrane-bound nucleus. They also lack most of the organelles found in eukaryotes. See *Eukaryotes*.

Prosimian A mammal (pre-monkey) which shares the characteristics which distinguish primates with other mammals (opposable thumbs and binocular vision), including the bush babies, lemurs, lorises, and tarsiers.

Protein synthesis A process in which peptides are formed through the use of DNA and RNA, including other processes, for example, transcription and translation in the cytoplasm (to form peptides) and a final step to form proteins in the Golgi apparatus.

Protostome Organisms in which the blastopore (an embryological opening that develops in the gastrula stage) develops into a mouth. Most invertebrates are protostomes, except echinoderms. See *Deuterostome*.

Pseudocoelomic A condition in an organism in which there are pseudocoelomic sacs but no true and complete coelom (body cavity). Roundworms typically have a pseudocoelom in adults, and all organisms that develop a coelom go through metamorphic changes in which pseudocoelomic sacs arise but later fuse with one another to form a true coelom.

Pseudogenes DNA sequences include both introns (segments that do not appear to be related to a cell's genetics) and exons (segments that make up a cell's genetic material). Pseudogenes are segments of DNA which are considered dysfunctional relatives of genes that have lost their protein-coding ability or are otherwise no longer expressed in the cell. Although some do not have introns or promoters (pseudogenes are copied from mRNA and incorporated into the chromosome and are called processed pseudogenes), most have some gene-like features. Nevertheless, they are considered nonfunctional because of their lack of protein-coding ability resulting from various genetic disablements (premature stop codons, frameshifts, or a lack of transcription) or their inability to encode RNA (such as with rRNA pseudogenes). Because pseudogenes are generally thought of as the last stop for genomic material that is to be removed from the genome, they are often labeled as junk DNA. In spite of what may seem to be their insignificance, pseudogenes contain fascinating biological and evolutionary histories within their sequences, due to a their shared ancestry with functional genes and may have an important function in the concept of Ontogeny Recapitulates Phylogeny (developmental stages pass through various phases that occurred in the process of evolution). Pseudogenes may have some relationship with genetic entities referred to as selfish genes, as brought out by Richard Dawkins in 1976.

Psychopathic behavior Psychopathic behavior is behavior expressed by a person referred to as a psychopath, an individual who has a personality disorder characterized by a pervasive pattern of disregard for the feelings of others and often the rules of society. Psychopaths have a lack of empathy and remorse and have very shallow emotions. Chapter 9 discusses this condition. While there are behavioral similarities, the terms antisocial personality disorder (ASPD) and psychopathy are not considered synonymous. A diagnosis of antisocial personality disorder is said to be based on behavioral patterns, whereas psychopathy measurements also include personality characteristics. The diagnosis of antisocial personality disorder covers two to three times as many prisoners as are rated as psychopaths in a system of behavioral measurement referred to as Hare's scale.

Psychopathy A mental (antisocial) disorder in which an individual manifests amoral and antisocial behavior, shows a lack of ability to love or establish meaningful personal relationships, expresses extreme egocentricity, and demonstrates a failure to learn from experience and other behaviors associated with the condition.

Quadrupedal A form of terrestrial locomotion in animals (also referred to as a pronograde posture), using four limbs or legs. An animal or machine that usually moves in a quadrupedal manner is known as a quadruped, meaning "four feet" (from the Latin quad for "four" and ped for "foot").

Quasi Having some resemblance to something, usually by possession of certain attributes.

Quasi-coexistence Quasi is defined as having some resemblance usually by possession of certain attributes. Therefore, quasi-coexistence refers to organisms that appear to possess some ability to coexist in their environments, while others may not have this ability. Or they may be more able to do so. The term implies that populations of organisms appear to have some sort of balance in their relationships with one another, although the relationships may not be perfect.

Radiometric dating A form of dating fossils and rocks on a scale of absolute time. Such dating is based on the half-life of certain radioactive isotopes.

As pointed out by Morris Petersen and Kieth Rigby in *Interpreting Earth History, A Manual in Historical Geology*, "the basis for radiometric dating is radioactivity, the spontaneous decay of certain isotopes, called parent isotopes, to produce unique end products called daughter products, which accumulate as the parent material disintegrates. Most radioactive isotopes decay by alpha, beta, or gamma radiation.

Radiocarbon dating is useful for measuring the time interval 500–50,000 years. It is based on the acquisition of Carbon-14 by organisms that received it in the food chain and its subsequent breakdown to return to Carbon-12.

Rank acquisition The acquisition of establishing a position in a dominance hierarchy.

Rationality An exercise of reason.

Ratios of relatedness Ratios of relatedness in biology involve the closeness in which individuals are related, especially in haplodiploidy or arrhenotoky. In social insects, for instance, females are diploid and males are haploid. The relatedness of females to the queen and her offspring (as well as predecessors and descendants) is often used to discuss certain social phenomena, for example, altruism.

Ray fins The type of fins found in most cartilaginous and bony fish. Certain of the bony fish over millions of years developed lobe fins from their ray fins, which led to the rise of walking appendages.

Realistic competition Competition that is driven by self-interest and is aimed at obtaining material resources for the in-group (favoring an in-group in order to obtain more resources for its members, including one's self).

Reciprocal altruism A form of altruistic behavior in social groups in which an altruist benefits at a later time when the current beneficiary reciprocates.

Relatedness The state of being related, especially by kinship. In biology, kinship typically refers to the degree of genetic relatedness or coefficient of relationship between individual members of a species.

Religion An organized collection of belief and cultural systems that relate humanity to spirituality and sometimes to moral issues.

Replication The duplication of DNA, a process found in cells that are approaching division to form two cells.

Respect As a noun, a feeling of deep admiration for someone or something elicited by their abilities, qualities, or achievements; a particular aspect, point, or detail. As a verb, respect means to admire (someone or something) deeply, as a result of their abilities, qualities, or achievements.

Revolutionary coalitions Coalitions in which the two coalitionary males are both lower-ranking than the target individual.

Rigid leadership Leadership in which the leader and at least some followers are stiff and unyielding.

Romantic love A love bonded emotionally and physically through passionate arousal.

Rorschach inkblot test A psychological test in which subjects' perceptions of inkblots are recorded and then analyzed using psychological interpretation, complex algorithms, or both. Some psychologists use this test to examine a person's personality characteristics and emotional functioning. It has been employed to detect underlying thought disorder, especially in cases where patients are reluctant to describe their thinking processes openly. The test is named after its creator, Swiss psychologist Hermann Rorschach (Chapman and Chapman, 1982; Gacono et al., 2007).

Sadism As stated by R.J. Corsini in The Dictionary of Psychology, sadism (1) denotes "cruelty in general. (2) Preference during erotic activity to punish, humiliate or hurt the sexual partner; sometimes extended to include animals. Although technically a deviation, some degree of sadism is an accepted part of sexual foreplay in some cultures. Also known as active algolagnia." Sadomasochism involves "being sexually aroused by combined sadism and masochism, usually in role-playing."

Sadomasochistic This term describes a combination of sadistic and masochistic behavior. Masochism is a term that has different meanings. In the context of this book, sexual masochism is described as "a psychological disorder in which people apparently prefer to be mistreated before, during and after sex. Purportedly, more men prefer this sort of sex than do women. The man is usually humiliated by a woman, known as a mistress or dominatrix, and may be blindfolded, bound, spanked, whipped [or] punished, leading to the desired sexual climax."

Sagittal crest A prominent point along the midline at the top of the skull.

Sarcopterigians Tetrapodomorph fish with paddle-like lobe fins but no chiridian limbs.

Science A systematic process for learning about the world and testing our understanding of it.

Scientific method A traditional approach to scientific research, involving a technique for testing kedas with observations and involves several assumptions and a series of interrelated steps. Withgott and Brennen point out that the assumptions are (1) the universe functions in accordance with fixed natural laws; (2) all events arise from some cause and cause other events; and (3) we can use our senses and reasoning abilities to detect and describe natural laws.

Scientology A body of beliefs and related practices created by science fiction writer L. Ron Hubbard (1911–86), beginning in 1952 as a successor to his earlier self-help system, Dianetics. Hubbard characterized Scientology as a religion, and in 1953 incorporated the Church of Scientology in Camden, New Jersey.

Secondary sexual characteristics Physical characteristics of a male or female, often appearing at their peak following sexual maturity. Examples are found in certain color expressions, hair growth, body size, and muscle development.

Self-assertion in humans An insistence on or an expression of one's own importance, wishes, needs, opinions, or the like (indicating a dominant and narcissistic personality); it, like dominance, does not have to be aggressive.

Self-schema A self-image. Self-schemas store information and influence the way we think and remember.

Semantic communication The usage of distinct signals with reference to objects in the environment.

Serial killers People who kill three or more people over a period of more than a month, usually with a cooling-off period between killings.

Sex-linked disorders Genetic disorders related to sex chromosomes (X and Y) or genes in them.

Sexual dichromism Color differences in males and females.

Sexual dimorphism Anatomical differences between males and females.

Sign stimulus An external sensory stimulus that triggers a behavioral series of events referred to as a fixed action pattern.

Simians Primates that are more advanced than prosimians. It includes such groups as New World and Old World monkeys, gibbons, orangutans, gorillas, chimpanzees, bonobos, and humans.

Sneak mating Copulation in which a female is bred by a male from a different social group.

Social In a biological context, the term refers to organisms that live and function in a group. In such groups, individuals cooperate in colony functions, for example, constructing a nest (if one is made), caring for the young and their reproductive female and/or dominant male, and in colony defense. The highest level of sociality among animals is considered to include: (1) care of the young, (2) reproductive division of labor, and (3) an overlap of generations. See Chapters 1 and 2 in which sociality is discussed.

Social competition Competition that is driven by the need for self-esteem and is aimed at achieving a positive social status for the in-group relative to comparable out-groups (favoring an in-group in order to make it better than an out-group).

Social intelligence hypothesis Social intelligence describes the capacity of humans to effectively navigate and negotiate complex social relationships and environments. It involves a person's competence to comprehend his or her environment optimally and react appropriately for socially appropriate conduct. The social intelligence hypothesis asserts that complex socialization (social intelligence, as used in such activities as politics, romance, family relationships, quarrels, retribution, collaboration, reciprocity, and altruism) was the driving force in developing the size of human brains, which contemporarily enables humans to use their large brains in complex social circumstances.

Sociality The state of being social. Being social generally is characterized by grouping behavior, care of the young, and an overlap of generations (parents and offspring living together). Certain behaviors are expressed in societies, including the establishment of dominance hierarchies and demonstration of overt territorial and defensive displays. Animals that demonstrate all of these qualities, in addition to a well-defined division of labor, are considered to be eusocial.

Social spiders Certain spiders that group and care for their young. No spider species are considered eusocial.

Social traits Traits that bring and keep individuals together, resulting in their interaction.

Sociobiology A study of social behavior of all animals from a biological view point, including evolutionary theory.

Socio-cognition The encoding, storage, retrieval, and processing of information in the brain. Contemporary use of the term involves psychology (e.g., the development of social cognition abilities) and cognitive neuroscience (e.g., the biological basis of social cognition).

Sociopathic behavior A sociopath is a person who completely disregards and violates the rights of others while refusing to conform to societal norms. Features that characterize sociopathic behavior are:

1. Exhibiting a lack of conscience. Signs of sociopathic behavior are generally present in childhood, indicators being torturing or killing animals, showing no emotions when something bad happens to someone else, or showing no guilt or remorse for any of his or her own actions.
2. A pattern of irresponsible or poor behavior, including doing poorly in school or on the job. Other behaviors to look for are recklessness, impulsivity, and participation in illegal activities.
3. A person's personal relationships. Many sociopathic people have an inability to love or have lasting personal relationships. This is because they can be very manipulative.
4. Patterns of pathological lying. Sociopaths will continue to lie about things even if they are caught doing them. They can also be very charming and get others who are blind to their behaviors to side with them.
5. An inflated sense of self-importance or narcissism. A sociopath behaves as if she is the only person who matters and she will have complete disregard for everyone else. Although she has the ability to charm people, she will take advantage of them at the same time.
6. A need for stimulation by engaging in risky or dangerous behaviors. These behaviors can be sexual or simply thrill-seeking. Sociopaths tend to get bored easily, which is why they seldom complete tasks and seek out forms of excitement.

Species A group of organisms that breed together under natural conditions and are isolated from other such groups. All species generally show some forms of variation within their range, their variation being expressed in clines. Geographical and other forms of isolation that separate segments of a population of a species are responsible for speciation (the formation of new species) over long spans of time.

Stegocephalia An old taxonomic term for early (generally large) amphibians, comprising a group of extinct large amphibians with a more-or-less salamander-like build.

Stochastic events An event or system which is unpredictable because of random variables.

Subordination The process of dominating another individual. In agonistic interactions, the individual that loses in a confrontation is subordinated and becomes the subordinate.

Sub-population A group of organisms which show differences from other conspecific groups within a population.

Sustainability A sustainable system is one that can continue in its current mode of operation indefinitely. Withgott and Brennen bring up the question: "Does the present generation have an obligation to conserve resources for future generations? This ethical question is at the core of the notion of sustainability, living within our planet's means. This means that the Earth and its resources can sustain us and the rest of Earth's living things for the foreseeable future."

Sympatric A condition in which there are two populations of organisms that overlap. Whether they breed within the zone of overlap determines if they belong to the same or different species.

T-box genes According to L.A. Naiche et al., T-box genes are involved in early embryonic cell fate decisions, regulation of the development of extraembryonic structures, embryonic patterning, and many aspects of organogenesis. In humans, mutations in T-box genes are responsible for developmental dysmorphic syndromes, and several T-box genes have been implicated in neoplastic processes. T-box transcription factors function in many different signaling pathways, notably bone morphogenetic protein and fibroblast growth factor pathways.

Teleost fish Ray-finned fishes in the class Actinopterygii, which make up about 98% of all fish. The group arose in the Triassic period.

Temperament The usual vigor of a person's response to things that are going on around them. Temperament is a genetic component of a person's personality as opposed to the learned component.

Teratogenic substance A substance or agent that can interfere with normal embryonic development which, in connection with pregnancy, causes a fetus to be born with a condition termed fetal alcohol syndrome (FAS).

Teratology Study of disruptions in normal development which leads to malformation.

Terrorism The use of violent acts to frighten and/or hurt people in an area as a way of trying to achieve a political goal.

Terrorist A radical individual who employs terror as a political weapon.

Territory A territory is an area defended by an individual to exclude other members of its own species. It differs from home range in that territories are generally a portion of a home range. Typically, a territory is used for feeding, mating, and rearing the young. They are most overtly expressed in social species. Territories are usually defended through agonistic behavior, and some form of behavior is expressed in proclaiming a territory. Proclaiming a territory may involve scent marks, oral signals, or a variety of behavioral expressions.

Tetrapodomorphs Certain types of bony fish that have developed a limb from lobe fins but do not possess digits at the limb's distal tip. The best known of early tetrapodomorphs are *Acanthostega* and *Ichthyostega* which existed 370–380 million years ago. Tetrapodomorph limbs represent a stage between the lobe fins of certain bony fish and the limbs of chiridian species (species with well-defined digits). It was in the tetrapodomorphs that a pelvic girdle and an isolation of the head from the trunk occurred.

Tetrapods Chordates that have developed four legs. The first tetrapods were fish that developed their locomotor appendages from lobe fins, often using them to move through their aquatic environments. The limbs of later tetrapods developed cartilage and bones that were more useful in walking, one such group leading to the amphibians. Tetrapods emerged from aquatic habitats and moved to land about 330 million years ago.

Theism The belief that at least one deity exists. Theism and the concepts that define it may be divided into subcategories, for example, deism, pantheism, polytheism, and monotheism.

Theist A person who believes in some sort of deity.

Thematic apperception test A projective psychological test in which proponents of the technique assert that subjects' responses, in the narratives they make up about ambiguous pictures of people, reveal their underlying motives, concerns, and the way they see the social world. Historically, the test has been among the most widely researched, taught, and used of such techniques.

Thermoregulation An ability of an endothermic organism (such as birds and mammals) in which a body temperature is maintained regardless of environmental changes.

Tiktaalik A Late Devonian amphibious tetrapodomorph.

Tolerance Tolerance or toleration is the practice of permitting a thing of which one disapproves, such as social, ethnic, sexual, or religious practices. There are many forms of tolerance. It is used in this book primarily with a relationship to social behavior expressed by an alpha individual to subordinates in the group. Dominant individuals may be despotic (intolerant) toward their subordinates, which affects the group's social organization, whereas a tolerant dominant will provide an environment in which all members of the group may get along. The world is full of intolerant despots who cause societal disharmony.

Totipotency Totipotency is defined as the ability of an undifferentiated single cell to produce all the differentiated cells in an organism, including those in extraembryonic tissues. Totipotent cells formed as a result of sexual and asexual reproduction include spores and zygotes.

Toxic manager A manager who bullies, threatens, and yells at his subordinates.

Trait dominance Characteristics related to generalized feelings of dominance versus submissiveness.

Translation The process of forming transfer RNA (tRNA) in a cell's nucleus. Once formed, tRNA moves into the cell's cytoplasm to function in the process referred to as protein synthesis. Its main objective in this process is to collect amino acids.

Translocations A genetic condition in which there is an interchange between chromosome segments. In some cases, a portion of a chromosome may simply get attached to another chromosome.

Transcription The process of forming messenger RNA (mRNA) in a cell's nucleus. Once formed, mRNA moves into the cell's cytoplasm to function in the process referred to as protein synthesis. Its main objective is to represent a code for the lining up of transfer RNAs.

Trophallaxis The social regurgitation of chemicals between members of a social group. The transference of chemicals represents a form of communication between societal members and often influences the development and behavior of society members.

Totipotency The ability of a single cell to divide and produce all of the differentiated cells in an organism, including the extraembryonic tissues.

Trophic level A feeding or energy level, the term often used in discussing food chains. Each type of organism (plant, herbivore, and carnivore) in a food chain represents a trophic level. Energy consumed when one organism feeds upon another organism is lost to the environment at each trophic level.

Tunicate A marine organism that has chordate features in its immature stage, but they are lost in the adult stage. Because of the placement of the notochord, tunicates are often referred to as urochordates.

Twinning The production of two simultaneous offspring during the process of birthing.

Variability The amount of variation seen in a particular population. The variability of a trait is how much that trait tends to vary in response to environmental and genetic influences.

Vertebrate An organism that has vertebrae. The first vertebrates were lamprey-like organisms in the group referred to as jawless fish. All vertebrates are chordates. See *Chordate*.

Vestigial A structure which had more extensive form and function in predecessors but which is currently small and functionless (in remnant form).

Vestigiality Those traits (e.g., organs and behaviors) occurring in organisms that have lost all or most of their original function through evolution. Examples of human vestigiality are numerous, including anatomical (e.g., appendix, tailbone, wisdom teeth, and inside corner of the eye), behavioral (e.g., goose bumps and palmer grasp reflex), sensory (e.g., decreased olfaction), and molecular (noncoding DNA).

The coccyx (tailbone), a remnant of a lost tail, is an interesting example of vestigiality in humans. All mammals (including humans) have a tail at some point in their embryonic development. In humans, it is present for a period of 4 weeks, being most prominent in embryos that are 31–35 days old. Located at the end of the spine, it has lost its original function in assisting balance and mobility, though it still serves some secondary functions, for example, being an attachment point for muscles, which explains why it has not become degraded further.

In rare cases, a congenital defect results in a short tail-like structure being present at birth. Twenty-three cases of human babies born with such a structure have been reported in the medical literature since the late 1880s. In these rare cases, the spine and skull were determined to be entirely normal, and the only abnormality was the presence of a tail approximately 12 cm long. Surgical removal allowed the individuals to resume normal lives.

Vienna School A group of philosophers, scientists, and mathematicians who met during the 1920s and 1930s to discuss the foundations of science and philosophy.

Violence According to R. Audi, violence is (1) the use of force to cause physical harm, death, or destruction (physical violence); (2) the causing of severe mental or emotional harm, as through humiliation, deprivation, or brainwashing, whether using force or not (psychological violence); (3) more broadly, profaning, desecrating, defiling, or showing disrespect for (i.e., doing violence to) something valued, sacred, or cherished; and (4) extreme physical force in the natural world, as in tornados, hurricanes, and earthquakes.

Vitamins Organic or similar compounds that are required as vital nutrients in small amounts by an organism.

Vomeronasal A system found in some monkey species in which they process chemical signals by employing a specialized organ located in the nasal cavity. Such chemical signals may reveal information about another animal's social status, sex, or maturity.

Wean The process of changing a young animal's dependency from mostly parental care to obtaining food in a more independent fashion.

Zygote A cell that results from the fusion of a spermatozoan and an ovum. It is the first cell of the embryo which subsequently undergoes mitotic divisions to eventually form a morula. The zygote is the ultimate form of a stem cell, and all cells in the body of an organism arise from the zygote or its descendants.

Index